Cognitive Wireless
Communication Networks

Ekram Hossain • Vijay Bhargava
Editors

Cognitive Wireless Communication Networks

 Springer

Edited by:

Ekram Hossain
Department of Electrical &
Computer Engineering
University of Manitoba
75A Chancellor's Circle
Winnipeg MB R3T 5V6
CANADA

Vijay Bhargava
Department of Electrical &
Computer Engineering
2332 Main Mall
University of British Columbia
Vancouver V6T 1Z4
CANADA

ISBN 978-0-387-68830-5   e-ISBN 978-0-387-68832-9

Library of Congress Control Number: 2007930304

Printed on acid-free paper.

9 8 7 6 5 4 3 2 1

springer.com

To our families

Preface

A Brief Journey through "Cognitive Wireless Communication Networks"

Ekram Hossain, University of Manitoba, Winnipeg, Canada
Vijay Bhargava, University of British Columbia, Vancouver, Canada

Introduction

Cognitive radio has emerged as a promising technology for maximizing the utilization of the limited radio bandwidth while accommodating the increasing amount of services and applications in wireless networks. A cognitive radio (CR) transceiver is able to adapt to the dynamic radio environment and the network parameters to maximize the utilization of the limited radio resources while providing flexibility in wireless access. The key features of a CR transceiver are awareness of the radio environment (in terms of spectrum usage, power spectral density of transmitted/received signals, wireless protocol signaling) and intelligence. This intelligence is achieved through learning for adaptive tuning of system parameters such as transmit power, carrier frequency, and modulation strategy (at the physical layer), and higher-layer protocol parameters.

Development of cognitive radio technology has to deal with technical and practical considerations (which are highly multidisciplinary) as well as regulatory requirements. There is an increasing interest on this technology among the researchers in both academia and industry and the spectrum policy makers. The key enabling techniques for cognitive radio networks (also referred to as dynamic spectrum access networks) are wideband signal processing techniques for digital radio, advanced wireless communications methods, artificial intelligence and machine learning techniques, and cognitive radio-aware adaptive wireless/mobile networking protocols.

This book on Cognitive Wireless Communication Networks includes a rich set of research articles featuring recent advances in theory, design, and analysis of cognitive wireless networks. The aim of this book is to provide a unified view on the state-of-the-art of cognitive wireless communication networking technology. It consists of 15 invited articles from distinguished contributors in this field. The contributed articles cover a wide range of topics from the fundamental challenges and issues in designing cognitive radio systems to the information-theoretic analysis of such systems, spectrum sensing and co-existence issues, adaptive physical layer protocols and link adaptation techniques for cognitive radio, orthogonal frequency division multiple access (OFDM) and ultra wide band (UWB)-based cognitive radio, different techniques for spectrum access by distributed cognitive radio, cognitive medium access control (MAC) protocols, decentralized learning-based dynamic spectrum access methods as well as microeconomic models for spectrum management in cognitive radio.

Cognitive Radio: Signal Processing and Communication – Theoretic Perspective

Chapter 1, authored by Simon Haykin, provides a comprehensive survey on the fundamentals of cognitive radio and the major research challenges mostly from a signal-processing and communication-theoretic perspective. The author presents several ideas and/or algorithms for some of the fundamental tasks of cognitive radio including spectral estimation of a radio frequency (RF) band/spectrum hole identification, extraction of channel state estimation, and transmitter power control. Spectral information of the radio environment (or RF stimuli) and spectrum hole identification in the neighborhood of a receiver as well as information on the evolution of the spectrum holes are required for the cognitive radio transmitter for efficient utilization the radio spectrum. These information can be used by the cognitive radio transmitter, for example, to select the appropriate modulation and coding format, and transmission power level. The transmit power control problem is basically to determine the transmit power levels for the cognitive transmitters to maximize their data transmission rates under constrained interference temperature limits in the frequency bands occupied by the spectrum holes. The channel state information is required by the cognitive radio receiver for efficient reception/detection performance. For this, the receiver may use a semi-blind training procedure. In this procedure, the receiver has two modes of operations: *supervised training mode* to estimate the channel state, and *tracking mode* which is performed in an unsupervised manner during the course of data transmission.

For a multiuser distributed cognitive radio network, *cooperation* and *competition* are two basic mechanisms which enable it to achieve self-organization. By cooperation (either through a distributed or a centralized way) the cognitive nodes can share network information among each other to achieve a coordinated and efficient spectrum management. However, it may require synchronization among the nodes which results in a more complex network design. In contrast, a competitive (or non-cooperative) approach may simplify the network design, however, at the expense of network performance. The author formulates the distributed transmit power control problem in a non-cooperative environment using a game-theoretic framework where

each user follows the criterion of competitive optimality to maximize its own transmission rate.

Cognitive Radio: Information Theory Perspective

Chapter 2, authored by Natasha Devroye et al., addresses the problem of determining
the capacity of a cognitive radio network from an information theoretic point of view.
This chapter summarizes some of the most important results in this topic considering three types of cognitive radio behavior: interference mitigating cognitive behavior, interference-free (or collaborative) cognitive behavior, and interference avoiding
cognitive behavior. In the interference mitigating behavior, a cognitive radio transmitter can transmit simultaneously with other cognitive radio transmitter or a primary
transmitter. However, the interference among the users can be mitigated by utilizing
the spectrum sensing information, for example, through asymmetric transmitter cooperation (i.e., asymmetric cognitive behavior) or symmetric transmitter cooperation
(i.e., cooperative behavior). Again, there could be no cooperation at all among the
transmitters (i.e., competitive behavior). In the collaborative behavior, cognitive radio devices collaborate with the primary users by acting as relays (and therefore,
there is no interference in the system). In interference avoiding behavior, a cognitive
radio user transmits only when there is no other transmission, and therefore, it avoids
interference completely.

The authors show that with asymmetric cognitive behavior and single antenna
at each node, a cognitive radio user is able to communicate at a non-zero rate in
a 2 sender, 2 receiver Gaussian noise channel without affecting the primary user,
and therefore, higher spectral efficiency can be achieved. However, the multiplexing
gain (which intuitively represents the number of information streams that can be
pushed through the channel) of this cognitive radio channel is observed to be 1.
It is shown, however, that, there are other cognitive channels (2 sender, 2 receiver
Gaussian noise channel) in which partial asymmetric knowledge about the messages
at the transmitters can provide a multiplexing gain greater than 1 which is achievable
through appropriate coding and multiple antenna at the transmitters. The authors
refer to such a channel as the cognitive X-channel.

With collaborative cognitive behavior, the network capacity depends on the channel quality between the source and the relay nodes. In case of interference avoidance
cognitive behavior, the communication between the cognitive transmitter and the receiver would be successful if they both choose the same spectrum hole. Coordination
of this selection of spectrum holes, however, may result in a loss of network capacity.

Coexistence Scenarios in Cognitive Radio Networks

Chapter 3, authored by Sofie Pollin, presents a taxonomy and classification of coexistence (or dynamic spectrum sharing) scenarios in cognitive radio networks. Dynamic

spectrum sharing can be of two types: horizontal spectrum sharing and vertical spectrum sharing. In the former case, all of the users have equal regulatory status to access the radio spectrum where the nodes may use similar wireless access technology (i.e., homogeneous networks) or different wireless access technique (i.e., heterogeneous networks). In case of vertical spectrum sharing, the radio spectrum is licensed to the primary user only, while the secondary user(s) can access the spectrum opportunistically without affecting the primary users' performance.

Irrespective of the type of coexistence model, three tasks, namely, spectrum sensing (or detection of spectrum holes and/or higher layer information in the transmission protocol stack), spectrum analysis (i.e., building a model of the RF scene), and spectrum decision (i.e., whether and how to access the spectrum), are fundamental to any spectrum sharing process. The author outlines the major challenges related to spectrum sensing, spectrum analysis, and spectrum decision and some of the possible solutions from a system's perspective. A centralized or a distributed network architecture can be used for spectrum analysis and spectrum decision. Again, spectrum decision can be taken in a cooperative way (i.e., based on global optimization) or a non-cooperative way (i.e., based on local optimization). To coordinate the spectrum decision, a common control channel and/or a synchronization mechanism between the transmitter and the receiver can be used.

The author also presents an overview of the IEEE 802.22 standard for Wireless Regional Area Networks (WRAN) which is the first cognitive radio standard. IEEE 802.22 networks will coexist with the legacy TV transmissions. The spectrum sensing, spectrum analysis, and spectrum decision for IEEE 802.22 standard are summarized. To this end, as a case study, the author also addresses the dynamic spectrum sharing problem between IEEE 802.15.4 sensor networks and IEEE 802.11g WLANs both operating in the 2.4 GHz ISM band. Specifically, distributed channel selection algorithms to optimize the performance of 802.15.4 networks in presence of interference from 802.11g networks are discussed. These algorithms take into consideration the energy cost of sensing operation and operate without the use of a common control channel. Also, learning-based spectrum analysis (i.e., quality of channels) and spectrum decision methods are discussed.

Spectrum Sensing in Cognitive Radio Systems

Chapter 4, authored by Khaled Ben Letaief and Wei Zhang, presents a survey of spectrum sensing techniques for cognitive radios. Spectrum sensing is basically a binary hypothesis testing problem and the key performance measures for a spectrum sensing method are the probability of correct detection, probability of false alarm, and probability of miss. The pros and cons of the several signal detection techniques, namely, energy detection, matched filter detection, cyclostationary detection, and wavelet detection methods are discussed. Due to the wireless fading and shadowing effects, spectrum sensing performance can be degraded significantly. To overcome this problem, cooperative spectrum sensing can be used where multiple cognitive radios cooperate on spectrum sensing to achieve sensing diversity gain.

In a cooperative spectrum sensing scheme, distributed cognitive radios perform local spectrum sensing independently and then fuse their local decisions together. However, impairments in the reporting channels (i.e., channels used to convey the sensing information from cognitive radios to the common receiver) may degrade the performance of cooperative spectrum sensing. By exploiting cooperative diversity, where the neighboring cognitive radios exchange their spectrum sensing decisions and then send all the decisions to a common receiver, cooperative spectrum sensing performance can be improved. Another technique to improve the performance of cooperative spectrum sensing is to exploit multiuser diversity where distributed cognitive radios are organized into different clusters and only the cluster head with the highest channel quality in the reporting channel participates in the spectrum sensing decision fusion process.

One important issue in cooperative spectrum sensing is the communication overhead for transmitting the local sensing decisions to the common receiver. By using a censored decision approach, where unreliable decisions are censored and excluded from the final decision, the communication overhead can be significantly reduced.

Radio Link Adaptation Techniques for Cognitive Radio

Chapter 5, authored by Michael B. Pursley and Thomas C. Royster IV, presents a framework for adaptive modulation and coding, and power adjustment at the cognitive radio transmitters (i.e., spectrum decision) in order to achieve reliable communication without causing interference to other radios. In the proposed framework, the cognitive radios do not employ any channel estimation and/or measurement technique for dynamic link adaptation. Instead, *demodulator and decoder statistics* for a packet derived in the receiver (such as the average Euclidean distance for the modulation symbols in a packet, number of errors in the binary symbols at the demodulator output, number of iterations performed by the decode for a packet), are exploited.

For dynamic link adaptation, the authors propose two spectrum etiquette measures, namely, the *time-bandwidth product* (i.e., the product of the bandwidth and the session's data transmission time) and the *resource consumption*. These etiquette measures are functions of bandwidth, transmission time, and power. The modulation and coding scheme can be selected such that the resource consumption is minimized subject to the QoS constraints. Alternatively, for a given modulation scheme and a fixed bandwidth, the code rate can be selected such that the transmission time is minimized. In general, there is a tradeoff between these two etiquette measures. That is, a modulation and coding scheme with a small time-bandwidth product has a large value of resource consumption and vice versa.

A power adjustment protocol is necessary at the beginning of a new session to adjust the transmission power at a satisfactory level within the first few packets. The power adjustment protocol presented in this chapter uses some adaptation statistics derived in the destination's demodulator and decoder. Power adjustment is also necessary when the channel attenuation becomes too high to be offset by adjusting the modulation and the coding rate. The authors demonstrate the performances of the proposed link adaptation protocols for AWGN channels with fixed but unknown propagation loss and for dynamic channels with time-varying propagation loss.

Agile Physical Layer Transmission Techniques for Cognitive Radio

OFDM-Based Cognitive Radio

Chapter 6, authored by Rakesh Rajbanshi et al., focuses on the design and implementation of multicarrier modulation (MCM)-based transmission technique for cognitive radio. Specifically, one variant of orthogonal frequency division multiplexing (OFDM) called non-contiguous OFDM (NC-OFDM) is considered in which subcarriers can be selectively deactivated based on the spectrum sensing measurements. Efficient implementation of NC-OFDM transceivers is the focus of this chapter.

OFDM-based transceivers employ fast Fourier transform (FFT) and inverse fast Fourier transform (IFFT) algorithms for efficient modulation and demodulation of subcarriers. For NC-OFDM transceivers, methods are required to efficiently implement the FFT blockes when several subcarriers are deactivated. In a wideband cognitive radio system with a highly sparse spectrum occupancy, the relative number of zero valued inputs in the FFT can be quite large, and therefore, computation time and/or hardware complexity can be significantly reduced by *pruning* the FFT algorithm (i.e., by removing operations on input values which are zeros and on output values which are not required). This is very desurable for hardware-constrained cognitive radios operating in a dynamic environment. Again, with NC-OFDM-based transmission, since the power of the deactivated subcarriers can be redistributed to active subcarriers, the bit error rate (BER) performance can be improved as compared to conventional OFDM.

Similar to OFDM signals, NC-OFDM signals also suffer from the peak-to-average power ratio (PAPR) problem. PAPR characterizes the envelope variations of the signal in time domain. With high PAPR, the digital-to-analog (D/A) converters and power amplifiers at the NC-OFDM transmitter require a large dynamic range in order to avoid amplitude clipping. This increases both power consumption and component cost of the NC-OFDM transceiver. Reduction in PAPR can be achieved only at the expense of increased system complexity, reduced data rate, or degraded BER performance. Also, due to the different statistical properties of the PAPR for NC-OFDM signals compared to those for OFDM signals, the PAPR reduction techniques proposed for conventional OFDM systems may need to be modified. In a cognitive radio environment, the PAPR reduction techniques should be able to adapt to the dynamic variations in the total number of active subcarriers and their locations.

System performance in an NC-OFDM-based cognitive radio system can be enhanced by non-uniform bit allocation across the subcarriers under given objective function and constraints. The complexity of the non-uniform bit allocation can be reduced by exploiting the information on incumbent spectral occupancy.

Chapter 7, authored by Gaurav Bansal et al., presents some of the research challenges along with their solutions related to the design of link adaptation algorithms (e.g., adaptive power and bit loading) for OFDM-based cognitive radio systems. Different power, bit or both power and bit loading schemes can be designed for cognitive radios (i.e., secondary users) which exploit the time varying nature of fading gains across the OFDM subcarriers. However, one of the challenges here is to ensure that the interference caused to the primary users remains below the target interference threshold. Therefore, not only the fading gains, but also the spectral distance of the subcarriers from the primary user's band need to be considered.

The authors formulate the power and bit loading problem in an OFDM-based cognitive radio system as a constrained optimization problem the objective of which is to maximize the transmission capacity of the cognitive radio users while keeping the interference caused to the primary users below a specified threshold. The amount of transmit power and the transmission capacity depend on the interference threshold. It is observed that, for a given interference threshold, assuming that the transmission rate can be varied continuously, the optimal scheme increases the transmission capacity of cognitive radio users compared to the uniform bit loading scheme as well as the water-filling scheme. For a given interference threshold, compared to the two other schemes, the optimal scheme allows more power to be transmitted. Note that, with the uniform power loading policy, the total transmit power is allocated equally among all the subcarriers while with the water-filling policy, more power is allocated to the subcarriers having better channel quality.

The authors also investigate the case of discrete (or integer) rate adaptation. A sub-optimal scheme for integer bit loading is presented which approximates the optimal continuous rate value to the nearest integer. The authors also present two modifications of the two well-known algorithms for integer bit loading in order to minimize interference to the primary users.

In an OFDM-based cognitive radio system, interference to the primary users can be reduced by nulling the adjacent subcarriers. However, if the time-varying nature of the fading gains is not considered while nulling the adjacent subcarriers, it may affect the achievable transmission capacity of cognitive radio users. The authors demonstrate that, for a given interference threshold, nulling degrades the transmission capacity compared with the optimal scheme.

UWB Cognitive Radio Transmission

Chapter 8, authored by Hüseyin Arslan and Mustafa E. Şahin, investigates the potential of the UWB technology to implement cognitive radio networks. Ultra wideband (UWB) wireless technology is considered as one of the enabling technologies for cognitive radio networks due to its potential to fulfill some of the key cognitive radio requirements. UWB is defined as a wireless technology using a bandwidth greater than 500 MHz or a fractional bandwidth (i.e., $\frac{2(f_h - f_l)}{f_h + f_l}$, where f_h and f_l denote, respectively, the upper and lower edge frequencies) greater than 0.2. According to the current FCC regulations in USA, UWB systems are allowed to operate in the 3.1–10.6 GHz band without any licensing requirement. To ensure that the UWB systems do not affect the licensed operators in this band, the maximum transmission power should be limited to -42 dBm/MHz (FCC Part 15 limit).

UWB is commonly implemented as an impulse radio-based UWB (IR-UWB) or OFDM-based UWB (UWB-OFDM). In IR-UWB, extremely low power pulses which are on the order of nanoseconds are transmitted. IR-UWB provides flexibility in modulation methods (e.g., on–off keying (OOK), pulse amplitude modulation (PAM), pulse position modulation (PPM), phase shift keying (PSK) etc.), transmit power, receiver types (e.g., coherent receivers, non-coherent receivers), and pulse shaping. It also possesses an excellent multipath resolving capability. In UWB-OFDM, the entire UWB bandwidth is divided into multiple subbands with each subband containing a number of orthogonal subcarriers. To avoid interference to licensed users, in a UWB-OFDM transmitter, the subcarriers can be selectively turned off.

UWB can be implemented in both underlay and overlay modes. Adaptability in spectrum occupancy and data rate can be achieved by varying the pulse duration and/or shape (in IR-UWB systems) or by turning off subcarriers (in UWB-OFDM systems). Adaptability in multiple access can be achieved, for example, by changing the number of chips in a frame (in IR-UWB systems) or by changing the number of subcarriers assigned to each user (in UWB-OFDM systems).

Besides the above adaptive communication capabilities, UWB technology may offer other supplementary services for cognitive radio networks. Among these services include estimating the locations and speeds of the nodes and providing awareness of the physical environment (e.g., moving objects).

Dynamic Spectrum Access Techniques in Cognitive Radio Networks

One of the major challenges in developing cognitive radio networks is the design of dynamic spectrum access techniques among the distributed cognitive radio users. A range of dynamic spectrum access techniques can be developed based on the degree of cooperation among the cognitive radios. These techniques may vary from completely distributed non-cooperative scheme on one extreme to a completely centralized scheme at the other extreme.

Degrees of Cooperation in Dynamic Spectrum Access

Chapter 9, authored by Zhu Han, describes different dynamic spectrum access techniques exploiting different degrees of cooperation, possible network scenarios they are suitable for, and the underlying tradeoffs.

In a non-cooperative spectrum access environment, the cognitive radios act selfishly in a distributed fashion without cooperating with each other. Dynamic spectrum access in this case can be modeled as a game where each player of the game (i.e., a cognitive radio) tries to optimize its own utility function. Nash equilibrium is a well-known solution concept for a game which denotes a set of strategies, one for each player, such that no player has any incentive to unilaterally deviate from its strategy. In the equilibrium, each player selects a utility-maximizing strategy given the strategies of other players. There might be more than one Nash equilibrium in which case the optimal one needs to be selected based on an optimality criterion.

Correlated equilibrium is another solution concept (more general than Nash equilibrium). The correlated equilibrium defines a probability distribution for the strategy profile. The Nash equilibrium corresponds to the special case where the probability distribution is a product of each individual player's probability for different actions. In a dynamic opportunistic spectrum access model for cognitive radios, there could be solutions corresponding to correlated equilibria which achieve strictly better performance compared to the Nash equilibrium in terms of spectrum utilization and fairness. The optimal correlated equilibrium can be found, for example, based on criterion such as the maximization of the sum of utilities of the cognitive radios or ensuring max–min fairness.

In a distributed cognitive environment, the game solutions can be achieved through an adaptive learning process. For a class of algorithms called regret-matching (or no-regret) algorithms, the probability for a player i to take action $\mathbf{r}_i$ is a linear function of the *regret* which can be interpreted as the average payoff that user i would have obtained if it had played $\mathbf{r}_i'$ instead of $\mathbf{r}_i$. The stationary solution of such a algorithm exhibits no regret. Learning algorithms for distributed opportunistic spectrum access can be also developed based on finite-state Markov decision process (MDP), reinforcement learning algorithms such as Q-learning, etc.

When the distributed learning-based approaches lead to undesired equilibria, the outcome of non-cooperative competition can be improved by using a referee-based approach where the referee collects the information necessary to improve the equilibrium and instruct the cognitive radios to change the game rules. The author in Chapter 8 illustrates the application of this approach to perform channel assignment, adaptive modulation, and power control for a multi-cluster cognitive network.

In general, non-cooperative dynamic spectrum access (e.g., based on Nash equilibrium) results in inefficient system performance. By enforcing cooperation among the distributed cognitive radios, the system performance can be significantly improved. The repeated game framework provides a mechanism to enforce cooperation. In a repeated game, the players are able to obtain information on the other players' strategies in the past and exploit those information to obtain better equilibrium. To enforce cooperation in such a repeated game, a method would be required to introduce punishment to the defecting users. For this, the cognitive radios need to determine if any user is deviating from cooperation (e.g., through information such as successful transmission rate, network throughput etc.). Upon detection of any deviating behavior, a cognitive radio switches to a punishment phase (which lasts for an optimal punishment time) during which the players play a static Nash equilibrium the outcome of which is much worse than that generated by a cooperative strategy. Therefore, the deviating cognitive radios will have much lower payoff in the punishment phase. If the punishment time is designed such that the gains by the deviating cognitive radios during the non-cooperation phase is outweighed by the punishment, they will not have any motivation to deviate from cooperation. The author illustrates the application of the above framework for optimal rate control in a distributed cognitive radio network.

Auction theory can be also used for radio resource allocation and management in cognitive wireless networks. For this, some nodes (e.g., cluster heads) can serve as auctioneers to handle the bidding among the cognitive nodes for spectrum allocation. Methods for spectrum allocation among the bidders (i.e., cognitive nodes) based on some design criteria (e.g., maximize total utility, minimize variance in utility) and payments (i.e., how the bidders pay the auctioneer) need to be designed. The cognitive nodes follow the instructions from the auctioneer for spectrum usage.

For cooperative resource allocation among the cognitive radios, the idea of mutual benefits using bargaining can be used where the cognitive radios exchange information among themselves locally for the bargaining. Fair solutions for dynamic spectrum access can be achieved, for example, through Nash bargaining solution, in a self-organized way.

In a cooperative spectrum access scenario, the mutual benefits can be also obtained through coalition. For this, all of the cognitive radios "put their cards on the table" and determine the best strategies for coalition. That is, a contract is signed among the cognitive radio users which ensures that the benefit of coalition is higher than that without coalition. The benefits of cooperation are shared among the cognitive users in a fair way.

Dynamic spectrum access in a cognitive radio network can be performed in a centralized manner where the cognitive radios communicate with a common controller node (e.g., a base station/access point), and thereby, they fully cooperate and follow the instructions from the centralized node. The resource allocations problems among the cognitive radios are then formulated as constrained optimization problems and solved by the central controller. Methods such as Lagrangian method, mathematical programming (e.g., linear programming, convex programming, nonlinear programming, dynamic programming) can be used to obtain the solution of the optimization formulation. Such a centralized scheme can offer the best network performance, but at the expense of considerable signaling and computation overhead.

Decentralized Medium Access Control (MAC) Protocols for Dynamic Spectrum Access: Decision Theoretic Framework

Chapter 10, authored by Qing Zhao, Yunxia Chen, and Ananthram Swami, presents a cross-layer design framework for decentralized MAC protocols for cognitive radios in a non-cooperative dynamic spectrum access scenario. Specifically, the joint optimization of the physical layer (PHY) sensing policy and the MAC layer access policy is sought for when the issues such as channel fading, activity of the primary users, sensing errors, and energy constraints and hardware limitations in the cognitive radio need to be also taken into consideration. While the sensing policy specifies which channel to sense, the access policy determines whether to transmit or not based on the sensing outcome. The access policy can be an aggressive one or a conservative one depending on the sensor operating characteristics. With an aggressive spectrum access policy, miss detections by the spectrum sensor (i.e., a busy channel is sensed to be idle) may cause collisions with primary users. With a conservative spectrum access policy, false alarms may cause wasted spectrum opportunities.

The authors consider a system model involving a spectrum sensor at the physical layer and a sensing policy and an access policy at the MAC layer. The joint PHY-MAC design problem is formulated as a constrained Partially Observable Markov Decision (POMDP) process for which the knowledge of the current spectrum occupancy state based on the past sensing and access actions can be summarized by a *belief state*. The sensing policy is given by a sequence of functions each of which maps this belief state at the beginning of a time interval (e.g., time slot) to a channel to be sensed. An access policy is given by a sequence of functions where each function maps the belief state and the sensing outcome of the chosen channel to an access action. The objective of the POMDP formulation is to obtain the optimal sensing and access policies along with the optimal sensor operating point such that the total expected throughput of a cognitive radio is maximized over a finite time interval.

The optimal transmission probabilities depend only on the sensor operating characteristics (e.g., probability of miss detection) and the maximum allowed probability of collision (and hence independent of belief state). When the probability of miss detection is high, the access policy should be conservative to ensure that the probability of collision remains below the threshold. In this case, even when the channel is sensed to be idle, the cognitive radio should transmit with a probability less than 1. In contrast, when the probability of false alarm is high (i.e., an idle channel is sensed to be busy), the access policy should be aggressive. That is, it should always transmit when the channel is sensed to be idle and transmit with a non-zero probability even when the channel is sensed to be busy.

The authors show that the joint design problem can be separated into two problems, namely, the design of optimal spectrum sensor operating point and spectrum access policy, and design of the sensing policy. The optimal sensor operating point is the point where the maximum allowed probability of collision is equal to the probability of miss detection and the optimal access policy is to always trust the sensing outcome (i.e., deterministic). The optimal sensing policy is one of sequential decision making which exploits the entire history of sensing outcome (i.e., it depends on the belief state).

The authors also study the impact of channel fading on the optimal sensing and access policies when the energy constraint in a cognitive radio is also taken into account. Depending on the fading condition, a cognitive radio may have to use different power levels for spectrum access. The problem of opportunistic spectrum access under energy constraint is formulated as an unconstrained POMDP. A policy of this POMDP is a sequence of functions each of which maps the belief state (i.e., past decisions and observations as well as the residual energy level) to a sensing decision and a set of access decisions. The design objective here is to obtain the optimal policy, i.e., the policy for which, the total expected number of information bits that can be transmitted successfully by a cognitive radio during its battery lifetime (i.e., total expected reward), can be maximized. The computational complexity of the optimal policy increases exponentially with the number of channels and battery lifetime. A suboptimal solution to the problem is then proposed which reduces the computational complexity significantly with only moderate loss in performance.

The essence of the optimal sensing and access policy is that it achieves a tradeoff among gaining instantaneous reward, gaining information for future use, and conservation of energy. Specifically, the optimal sensing decision (i.e., to sense or not to sense) strikes a balance between gaining reward and conserving energy. The optimal access policy depends on the energy consumption due to sensing and the channel fading conditions. If the energy consumption due to sensing is negligible, under poor channel conditions, a cognitive radio should wait for the best channel condition. On the other hand, if the sending energy consumption is large, it should access the channel regardless of the fading condition.

Decentralized Dynamic Spectrum Access: Game Theoretic Learning Model

As has been stated before, game theory provides useful models for developing dynamic spectrum access techniques which involve multiple competing cognitive radios. Chapter 11, authored by Michael Maskery, Vikram Krishnamurthy, and Qing Zhao, presents a game theoretic learning model for dynamic spectrum access in a distributed cognitive radio environment. Specifically, a decentralized game theoretic reinforcement learning scheme is proposed for slotted carrier sense multiple access (CSMA)-based channel access for cognitive radios. Each radio competes for opportunistic access to a number of channels used by the primary users. The objective of each radio is to select a subset of the unoccupied channels to satisfy its current demand level. Each cognitive radio aims at maximizing its utility which is formulated as a function of its demand level, channel usage pattern of primary users, number of competing users for a given channel, pricing for the available radio channels, and the throughput achieved. The utility function is used to guide a modified regret-based reinforcement learning procedure. This decentralized procedure allows each cognitive radio to perform channel access satisfactorily through repeated channel selections and corresponding performance measurements. Regret-based learning procedures are simple to implement and their convergence properties are well understood.

The proposed learning algorithm is game theoretic in nature since each cognitive radio adopts a channel allocation that maximizes its own utility in response to the actions by other cognitive radios (i.e., competitive optimality). In a regret-based learning algorithm, a regret matrix is used by each user, which tracks for every pair of actions a_i, a_j the difference in utility if the user had taken action a_k in the past everywhere he took action a_j. Given that the action taken by the user at time instant n is a_j, the probability of choosing action a_k at time instant $n + 1$ is proportional to the regret value from a_j to a_k. The learning process proceeds through exploration by switching among different actions. Since the algorithm requires that a use knows the utility he would have received for each action even if that action was not taken, which might not be feasible in practice, a stochastic approximation is used to compute the regret values. The modified regret-based learning algorithm converges to a set of correlated equilibria. As we described before, correlated equilibrium is a more general equilibrium concept than Nash equilibrium.

The performance of a decentralized spectrum access scheme may be improved by occasionally adjusting the behavior of the cognitive radios, for example, by periodically broadcasting parameter (e.g., spectrum price) updates from a central controller. These updates can be used by the cognitive radios to update their utility function. This adjustment may improve the equilibrium behavior from a global perspective, for example, improve the spectral efficiency of the cognitive radio system (i.e., average proportion of radio channels used by the cognitive radios). The pricing parameter should be determined such that it maximizes the global utility which is a function of the equilibrium behavior of the cognitive radios under the utility determined as a function of this pricing parameter. Since it is difficult (if not impossible) to determine the equilibrium behavior a priori, a stochastic approximation approach is used to find the optimal value of the pricing parameter.

Decentralized Dynamic Spectrum Access in Cognitive Networks: A Local Coordination/Bargaining-Based Approach

Chapter 12, authored by Haitao Zheng and Lili Cao, presents a decentralized approach for spectrum sharing where the cognitive users perform local coordinations to modify their spectrum usage (to adapt to the variations in network topology) to achieve a conflict-free spectrum assignment which maximizes the system utility. In the considered system model, cognitive radios select communication channels and adjust transmission power accordingly so that they do not cause interference to the primary users. The system utility is measured in terms of proportional fairness (i.e., total logarithmic user throughput) among the cognitive radios. For local coordination, the cognitive radios consider both the needs of neighboring devices and spectrum availability to determine their spectrum usage. Upon detection of suboptimal spectrum usage (e.g. due to user mobility), the coordination/bargaining protocol is triggered to apply local adjustments among the neighboring radios. During local coordination, sets of neighboring cognitive radios self-organize themselves into coordination/bargaining groups and each such group modifies spectrum assignment within the group to improve system utility. The major challenge in designing a local coordination protocol is to ensure that the local improvements approach the global optimal solution and the convergence speed is fast.

Two different approaches can be used for local coordination, namely, the *explicit bargaining-based approach* and the *implicit rule-based approach*. In the former approach, the cognitive radios negotiate spectrum usage through message exchange, while in the latter approach the radios observe the behaviors of neighboring nodes and independently adjust spectrum usage following predefined rules. The explicit approach will incur overheads in terms of communication signaling and transmission energy. The implicit approach may be simpler to implement with significant reduction in signalling traffic among the radios and power consumption.

To facilitate bargaining in the explicit bargaining-based approach, the authors propose two constraints, namely, the *limited neighbor bargaining constraint* and the *self-contained group bargaining constraint*. With the former constraint, bargaining within a group is coordinated by a central leader, and it is limited among the leader and its k-hop neighbors. With the latter constraint, bargaining inside each group should not disturb the spectrum assignments in other groups. Therefore, a local improvement should lead to a system improvement. Under this constraint, the number of channels which are exchanged among nodes inside a bargaining group is limited and also the members of any two bargaining groups are not directly connected.

The authors propose a local bargaining protocol, namely, *Fairness Bargaining with Feed Poverty* based on the explicit negotiation approach. In this protocol, a cognitive radio willing to improve its spectrum usage starts bargaining with its neighbors on a one-to-one basis to improve system utility. However, if there is no negotiable channel found between it and any of its neighbors, the node initiates a *Feed Poverty Bargaining* so that the neighboring nodes can collaborate together to feed it with some channels. For this protocol, the lower bound on the throughput performance (i.e., *poverty line*) for each of the nodes is obtained. Simulation results show that compared to a centralized approach (e.g., based on a graph multi-coloring approach), the proposed local bargaining approach incurs much lower communication overhead while achieving similar performance (in terms of fairness utility). Note that, each iteration of spectrum assignment/bargaining involves a four-way handshake (i.e., request, acknowledgment, action, acknowledgment) among neighbors. As expected, the complexity of the bargaining approach increases with increase in the rate of change of network topology (i.e., user mobility). System utility scales inversely with increase in user density. With a fixed user density, the system overhead scales linearly with the number of users. Therefore, the local bargaining-based approach would be suitable for large scale networks.

MAC Protocols for Hardware-Constrained Cognitive Wireless Networks

Chapter 13, authored by Qian Zhang, Juncheng Jia, and Xuemin Shen, presents a single radio multi-channel MAC protocol for hardware-constrained cognitive wireless networks. The cognitive radios need to sense spectrum before transmission. The MAC layer in a cognitive radio determines when and which channel it should sense and then physical layer techniques (e.g., energy detection, matched filter detection, etc.) are used to detect the primary users' signal. The hardware constraints in a cognitive radio are due to the *sensing constraint* and *transmission constraint*. Since at a given time a practical cognitive radio may be able to sense only a small portion of the radio spectrum, this gives rise to the sensing constraint. The transmission constraint arises due to the spectrum fragmentation (e.g., a cognitive radio may be able to spread the transmission signal within a limited number of spectrum fragments). Again, there is a constraint on the maximum amount of time a primary user can tolerate interference from the secondary user (as in IEEE 802.22) – this is referred to as the transmission parameter limitation. This parameter dictates how quickly a cognitive radio must be able to detect incumbents. The problem is then to optimize the sensing decision during each sensing and transmission interval. Note that, the more spectrum is sensed, the more spectrum opportunity can be explored. The proposed hardware-constrained MAC protocol takes the sensing constraint and sensing overhead as well as the transmission parameter limitation into account.

The authors model the sensing process as an optimal stopping problem with a finite horizon which can be described as the problem of choosing a time to take an action based on sequentially observed random variables in order to maximize/minimize an expected payoff/cost. Such a problem can be solved by using the backward induction principle. However, it incurs exponential computational complexity. To reduce the complexity for a practical MAC protocol, the authors use a k-stage look-ahead rule at each stage to decide whether to stop or continue sensing.

In the considered system model, a common control channel is used for contention-based random access (e.g., based on IEEE 802.11 DCF) by the cognitive radios to reserve the time interval for spectrum sensing and subsequent data transmission between a cognitive radio transmitter and receiver. The time frame in HC-MAC is divided into three parts: contention, sensing, and transmission. The contention part is used to reserve the sensing period. During the sensing part, the cognitive radio transmitter and the receiver sense the spectrum channels and determine whether a channel is available at both the sides. The optimal stopping rule is used to decide when to stop sensing. In the transmission part, actual data transmission takes place. Simulation results show that, under different system configurations, the proposed HC-MAC protocol achieves better system throughput compared to a static sensing scheme where the cognitive radios always sense a fixed number of channels.

Microeconomic Models for Dynamic Spectrum Management

Chapter 14, authored by Dusit Niyato and Ekram Hossain, investigates the problem of dynamic spectrum sharing and pricing using the *oligopoly market models* from microeconomic theory. In an oligopoly market model, a number of producers (i.e., oligopolists) compete with each other independently to maximize their utility by controlling the quantity and/or the price of the supplied commodity. The classical oligopoly models include *Cournot*, *Stackelberg*, and *Bertrand* models which differ from one another in terms of the nature of the competition. In the Cournot model, producers compete in terms of amount of commodity to be supplied to the market. In the Stackelberg model, some producers referred to as leaders are able to make decisions on the amount of supplied quantity before other producers referred to as followers. The followers make decision on the amount of supplied quantity by taking the leader's decision into account. In the Bertrand model, all producers make decision simultaneously in terms of price. All of these models can be analyzed by using game theory.

Oligopoly market models can be used for analyzing dynamic spectrum management in cognitive radio networks. One of the objectives of spectrum management here is to maximize the profits of the cognitive radios (e.g., in terms of spectrum share) and/or the primary service providers/users (e.g., in terms of spectrum price). This is the same as the objective of an oligopoly competition. The oligopoly market models were well studied in economics. Also, they are computationally simple, and therefore, suitable for implementation in resource-limited software defined radio transceiver.

The authors demonstrate the applications of Cournot, Stackelberg, and Bertrand models of competition for spectrum/bandwidth sharing and pricing in cognitive wireless networks. Specifically, these three different models for oligopoly are applied to obtain the optimal size of spectrum/bandwidth sharing and the charging price. The Cournot game model is used for the case where multiple cognitive radios share the spectrum/bandwidth with a primary user and the objective is to maximize the profit of the cognitive radios. Here, all the cognitive radios can completely observe the strategies and the payoffs of other secondary users. The profit of a cognitive radio is a decreasing function of the amount of spectrum requested (and subsequently allocated) to other users (i.e., their strategies). Again, the profit is an increasing function of the amount of spectrum allocation and the quality of the corresponding spectrum opportunity. The Nash equilibrium is considered as the solution of the spectrum competition which ensures that all of the cognitive radios are satisfied with the solution. The Nash equilibrium is obtained by using the best response functions.

In the Bertrand model, several service providers (or primary users) compete with each other in terms of price to gain the highest profit under QoS constraints for the primary users. Here, the bandwidth demand of the cognitive radios is established based on a utility function which depends on the quality of transmission (i.e., channel quality) in the available spectrum as well as the spectrum price charged by the primary user/service provider. The payoffs for the primary users are determined based on the spectrum price and the spectrum demand from the secondary users and the cost of sharing the spectrum with the secondary users. In addition, the authors consider *spectrum substitutability* in the formulation which represents the ability of a cognitive radio to switch among the frequency spectra offered by different primary users. Again, the Nash equilibrium is obtained as the solution of this price competition.

The Stackelberg leader-follower competition is used to model the problem of optimal sharing and pricing under elastic bandwidth demand from the cognitive radios. The authors illustrate the application of this competition model in the context of spectrum sharing in an integrated WiMAX/WiFi network where the WiFi nodes share the licensed WiMAX spectrum for broadband Internet access. The WiMAX base stations (BSs) and the WiFi access points (APs) are operated by different service providers. The WiMAX BS and the WiFi APs are the leader and the followers, respectively. The Stackelberg equilibrium is considered as the solution of this spectrum sharing game.

Numerical performance evaluation results are presented for all of these oligopoly competition models to show their efficacy in allocating radio resource in cognitive radio environments.

Analysis of Cognitive Radio Dynamics

Chapter 15, authored by Maria-Gabriella Di Benedetto et al., presents a mathematical framework for analyzing the behavior of a self-organizing cognitive radio network where continuous dynamics and discrete processes tightly interact. This framework is based on the hybrid system modeling approach. In a hybrid system, continuous and discrete variables interact and determine the system evolution. The system state is made of a discrete state (from a finite set) and a continuous state. The evolution of the discrete state is governed by using a discrete finite-state automaton, where for each state, state-specific rules of operation govern the evolution of the network. There are both discrete and continuous control inputs and disturbance variables as well as both discrete and continuous output variables. The authors illustrate the application of the hybrid system formalism in a cognitive radio network where the cognitive nodes adapt the transmission parameters such as pulse shape, transmission power in response to RF stimuli from the environment.

Conclusion

We have provided a summary of the contributed articles in this book. We hope this summary would be helpful to follow the rest of the book easily. We believe that the rich set of references in each of the articles will be invaluable to the researchers. We would like to express our sincere appreciation to all of the authors for their excellent contibutions and their patience during the publication process of the book. We hope this book will be useful to both researchers and practitioners in this emerging area.

Contents

1 Fundamental Issues in Cognitive Radio
Simon Haykin 1

2 Information Theoretic Analysis of Cognitive Radio Systems
Natasha Devroye, Patrick Mitran, Masoud Sharif, Saeed Ghassemzadeh,
Vahid Tarokh 45

3 Coexistence and Dynamic Sharing in Cognitive Radio Networks
Sofie Pollin 79

4 Cooperative Spectrum Sensing
Khaled Ben Letaief, Wei Zhang 115

**5 A Protocol Suite for Cognitive Radios in Dynamic Spectrum Access
Networks**
Michael B. Pursley, Thomas C. Royster IV 139

**6 OFDM-Based Cognitive Radios for Dynamic Spectrum Access
Networks**
Rakesh Rajbanshi, Alexander M. Wyglinski, Gary J. Minden 165

7 Link Adaptation in OFDM-Based Cognitive Radio Systems
Gaurav Bansal, Md. Jahangir Hossain, Vijay K. Bhargava 189

8 UWB-Based Cognitive Radio Networks
Hüseyin Arslan, Mustafa E. Şahin 213

**9 Degrees of Cooperation in Dynamic Spectrum Access for Distributed
Cognitive Radios**
Zhu Han 231

10 Cognitive MAC Protocols for Dynamic Spectrum Access
Qing Zhao, Yunxia Chen, Ananthram Swami 271

11 Game Theoretic Learning and Pricing for Dynamic Spectrum Access in Cognitive Radio
Michael Maskery, Vikram Krishnamurthy, Qing Zhao 303

12 Decentralized Spectrum Management Through User Coordination
Haitao Zheng, Lili Cao 327

13 Optimal Spectrum Sensing Decision for Hardware-Constrained Cognitive Networks
Qian Zhang, Juncheng Jia,, Xuemin (Sherman) Shen 365

14 Microeconomic Models for Dynamic Spectrum Management in Cognitive Radio Networks
Dusit Niyato, Ekram Hossain 391

15 Analysis of Cognitive Radio Dynamics
Maria-Gabriella Di Benedetto, Maria Domenica Di Benedetto, Guerino Giancola, Elena De Santis 425

Index 439

List of Contributors

Simon Haykin
McMaster University, Canada
Haykin@mcmaster.ca

Natasha Devroye, Patrick Mitran, Masoud Sharif, Saeed Ghassemzadeh, and Vahid Tarokh
Harvard University, USA; Boston University, USA; AT&T Labs-Research, USA
ndevroye@deas.harvard.edu
mitran@deas.harvard.edu
vahid@deas.harvard.edu sharif@bu.edu
saeedg@research.att.com

Sofie Pollin
Inter-university Micro-Electronics Center (IMEC), Belgium, and University of California, Berkeley, USA
pollins@eecs.berkeley.edu
pollins@imec.be

Khaled Ben Letaief and Wei Zhang
Hong Kong University of Science and Technology, Hong Kong
eekhaled@ece.ust.hk
eewzhang@ece.ust.hk

Michael B. Pursley and Thomas C. Royster IV
Clemson University, USA
pursley@ces.clemson.edu
troyste@ces.clemson.edu

Rakesh Rajbanshi, Alexander M. Wyglinski, and Gary J. Minden
The University of Kansas, USA
rajbansh@ittc.ku.edu
alexw@ittc.ku.edu gminden@ittc.ku.edu

Gaurav Bansal, Md. Jahangir Hossain, and Vijay K. Bhargava
The University of British Columbia, Canada
gauravbs@ece.ubc.ca jahangir@ece.ubc.ca vijayb@ece.ubc.ca

Hüseyin Arslan and Mustafa E. Şahin
University of South Florida, USA
arslan@eng.usf.edu
mesahin@eng.usf.edu

Zhu Han
Boise State University, USA
zhuhan@boisestate.edu

Qing Zhao, Yunxia Chen, and Ananthram Swami
University of California at Davis, USA; Army Research Laboratory, USA
qzhao@ece.ucdavis.edu
yxchen@ece.ucdavis.edu
aswami@arl.army.mil

Michael Maskery, Vikram Krishna-murthy, and Qing Zhao
The University of British Columbia,
Canada; University of California at
Davis, USA
mikem@ece.ubc.ca
vikramk@ece.ubc.ca
qzhao@ece.ucdavis.edu

Haitao Zheng and Lili Cao
University of California at Santa
Baarbara, USA
htzheng@cs.ucsb.edu lili-cao@cs.ucsb.edu

Qian Zhang, Juncheng Jia, and Xuemin (Sherman) Shen
Hong Kong University of Science and
Technology, Hong Kong; University of
Waterloo, Canada

qianzh@cse.ust.hk jiajc@cse.ust.hk
xshen@bbcr.uwaterloo.ca

Dusit Niyato and Ekram Hossain
University of Manitoba, Canada
tao@ee.umanitoba.ca
ekram@ee.umanitoba.ca

Maria-Gabriella Di Benedetto, Maria Domenia Di Benedetto, Guerino Giancola, and Elena De Santis
University of Rome La Sapienza, Italy;
University of L'Aquila, Italy
dibenedetto@newyork.ing.
   uniroma1.it
dibenede@ing.univaq.it
giancola@newyork.ing.
   uniroma1.it
desantis@ing.univaq.it

1

Fundamental Issues in Cognitive Radio

Simon Haykin

McMaster University, Canada
Haykin@mcmaster.ca

1.1 Introduction

The electromagnetic radio spectrum is a natural resource, the use of which by transmitters and receivers (transceivers) is licensed by government agencies. However, this resource is presently underutilized. In particular, if we were to scan the radio spectrum, including the revenue-rich urban areas, we would find that some frequency bands in the spectrum are unoccupied some of the time, some other frequency bands are only partially occupied, and the remaining frequency bands are heavily used. It is therefore not surprising to find that underutilization of the radio spectrum is being challenged on many fronts, including the Federal Communications Commission (FCC) in the United States of America.

Cognitive radio[1] offers a novel way of solving spectrum underutilization problems. It does so by sensing the radio environment with a twofold objective: identifying those subbands of the radio spectrum that are underutilized by the primary (i.e., legacy) users and providing the means for making those bands available for employment by unserviced secondary users. To achieve these goals in an autonomous manner, multiuser cognitive radio networks would have to be self-organized. Moreover, there would have to be a paradigm shift from transmitter-centric wireless communications to a new mode of operation that is receiver-centric, so as to maintain a limit on the interference produced by secondary user.

The underutilized frequency bands of the radio spectrum, owned by legally licensed (primary) users, are referred to as *spectrum holes*, which are formally defined as follows [1]:

---

[1] Cognitive radio is a constituent of the emerging discipline: *Cognitive Dynamic Systems*; see the point-of-view article in [2]. This discipline, motivated by the human brain, includes other constituents: *cognitive radar* and *cognitive immunity*. Unlike traditional radar, cognitive radar includes feedback from the receiver to the transmitter, resulting in immense benefits to radar performance. The purpose of cognitive immunity is to resist cyber attack in dynamic software systems.

A spectrum hole is a band of frequencies assigned to a primary user, but at a particular time and specific geographic location, the band is not being utilized by that user.

The operation of cognitive radio hinges on the availability of spectrum holes. The identification and exploitation of spectrum holes presents technical challenges grouped under two categories, one rooted in computer software and the other rooted in signal-processing and communication technology. These technical challenges are further compounded by the fact that the spectrum holes come and go in a stochastic manner.

Much of the material presented in this article focuses on signal-processing and communication-theoretic aspects of cognitive radio. Specifically, the material is organized as follows. The notion of cognition is discussed in Sect. 1.2. Section 1.3 describes two complementary visions of cognitive radio, one addressing software architectural aspects of cognitive radio and the other addressing signal-processing and communication-theoretic aspects of the subject. Section 1.4 deals with radio-scene analysis, which encompasses the sensing of the radio environment and identifying the specific locations of spectrum holes in the radio spectrum. Section 1.5 deals with two related issues: channel-state estimation and predictive modeling, both of which are fundamental to efficient utilization of the radio spectrum and coherent detection of the information-bearing signal at a user's receiver. Information gathered by the receiver on its local environment is sent to the transmitter via a low bit-rate feedback channel, which is discussed in Sect. 1.6.

Up to this point in this chapter, the discussion is focused on issues relating largely to a single user (i.e., transmitter linked to its receiver). The rest of the chapter, beginning with Sect. 1.7, is devoted to self-organized multiuser cognitive radio networks, with emphasis on the complementary use of cooperation and competition. Section 1.8 discusses the function of dynamic spectrum management, where the use of orthogonal frequency-division multiplexing (OFDM) based on cooperative communication is advocated. Based on this encoding strategy, Sect. 1.9 describes a statistical model of cognitive radio networks, which sets the stage for formulation of the transmit-power control problem in Sect. 1.10. Section 1.11 views the multiuser cognitive radio network, operating in a non-cooperative manner, as a game-theoretic problem. Section 1.12 describes an iterative waterfilling algorithm for resolving the issue of transmit-power control, followed by Sect. 1.13 on the emergent behavior of cognitive radio networks. Section 1.14 briefly discusses a plan for distributed traffic coordination of cognitive radio users in an ad hoc network environment. Then the chapter concludes with some final remarks.

1.2 Cognition

In a way, it can be argued that cognitive radio draws its inspiration from cognitive science. The roots of *cognitive science* are intimately linked to two scientific meetings that were held in 1956 [3]:

- The Symposium on Information Theory, which was held at the Massachusetts Institute of Technology (MIT). That meeting was attended by leading authorities in the information and human sciences, including Allen Newell (computer scientist), the Nobel Laureate, Herbert Simon (political scientist and economist), and Noam Chomskey (linguist). As a result of that symposium, linguists began to theoretize about language, which was to be found subsequently in the theory of computers: the language of information processing.
- The Dartmouth Conference, which was held at Dartmouth College, New Hampshire. The conference was attended by the founding fathers of artificial intelligence, namely, John McCarthy, Marvin Minsky and Allen Newel. The goal of this second meeting was to think about intelligent machines. The Dartmouth Conference was also attended by Frank Rosenblatt (psychologist), the founder of (artificial) neural networks. At the conference, Rosenblatt described a novel method for supervised learning, which he called the perceptron.[2] However, interest in neural networks was short lived: in a monograph published in 1969, Minsky and Papert used mathematics to demonstrate that there are fundamental limits on what Rosenblatt's perceptron could compute. The Minsky–Papert monograph, coupled with a few other factors, contributed to the dampening of interest in neural networks in the 1970s. We had to wait for the pioneering contributions of John Hopfield on neurodynamic systems and Rumlehart, Hinton and Williams on supervised learning in the 1980s for the revival of research interest in neural networks.[3]

In a book entitled "The Computer and the Mind," Johnson-Laird [4] postulated the following tasks of a human mind:

- To perceive the world
- To learn, to remember and to control actions
- To think and create new ideas
- To control communication with others
- To create the experience of feelings, intentions and self-awareness

Johnson-Laird, a prominent psychologist and linguist, went on to argue that theories of the mind should be modeled in computational terms.

Much of what has been identified by Johnson-Laird as the mind's main tasks and their modeling in computation terms apply equally well to cognitive radio. Indeed, we can go on to offer the following definition for cognitive radio involving multiple users.

The cognitive radio network is an intelligent multiuser wireless communication system that embodies the following list of primary tasks:

- To perceive the radio environment (i.e., outside world) by empowering each user's receiver to sense the environment on continuous time

---

[2] The perceptron provided the inspiration for Widrow and Hoff to develop the least-mean-square (LMS) algorithm, which has established itself as the workhorse for adaptive filtering for close to 50 years.

[3] For a historical account of neural networks, see Haykin [5].

- To learn from the environment and adapt the performance of each transceiver to statistical variations in the incoming RF stimuli
- To facilitate communication between multiple users through cooperation in a self-organized manner
- To control the communication processes among competing users through the proper allocation of available resources
- To create the experience of intentions and self-awareness

The primary objective of all these tasks, performed in real time, is twofold:

- To provide highly reliable communication for all users
- To facilitate efficient utilization of the radio spectrum in a fair-minded way

1.3 Two Complementary Visions of Cognitive Radio

In the first doctoral dissertation on cognitive radio published in 2000, Joseph Mitola described how a cognitive radio could enhance the flexibility of personal wireless services through a new language called the radio knowledge representation language [6]. Mitola followed this dissertation with the publication of a book on cognitive radio architecture [7]. A distinctive feature of both publications is a *cognitive computer cycle*, which encapsulates the various actions expected from a cognitive radio, as depicted in Fig. 1.1. Through deployment of the right software control, it is envisioned that a cognitive radio could orient itself by establishing priorities, then create plans decide and finally take the appropriate action in response to sensing of the RF environment. As envisioned in Fig. 1.1, provisions are also made for the cognitive radio to do two things:

- Bypass the planning phase and go directly to the decision phase in the event of an urgent situation
- Bypass the two phases of planning and decision-making by proceeding immediately to the action phase in the event of an emergency.

In the first journal paper published in 2005, Simon Haykin presented detailed expositions of the signal-processing, adaptive and learning procedures that lie at the heart of cognitive radio [2]. In particular, the paper identifies three specific tasks:

1. *Radio-scene analysis* (RSA), which encompasses
   - Estimation of interference temperature of the radio environment localized around a user's receiver
   - Detection of spectrum holes
   - Predictive modeling of the environment.
2. *Channel identification*, which is needed for improved spectrum utilization and coherent detection of original information-bearing signal at the user's receiver.
3. *Dynamic spectrum management* (DSM) and *transmit-power control* (TPC), which culminates in decision-making and action taken by the user's transmitter in response to the analysis of RF stimuli picked up by the receiver.

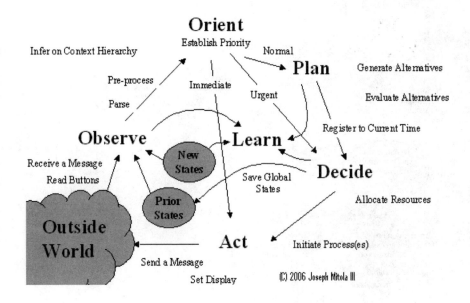

Fig. 1.1. Cognitive computer cycle. (©*2006 Joseph Mitola III. Reprinted, with permission from [7, p. 135]*).

Tasks (1) and (2) are performed in the receiver, and task (3) is performed in the transmitter, as depicted in the *cognitive signal-processing cycle* in Fig. 1.2; the depiction is presented in the context of a multiuser network.

For the transmitter to work harmoniously with the receiver,[4] there is an obvious need for a *feedback* channel connecting the receiver to the transmitter as shown in Fig. 1.2. Through the feedback channel, the receiver is enabled to convey to the transmitter two essential forms of information:

- Information on the performance of the forward link for adaptive modulation
- Information on the spectral state of the RF environment in the local neighborhood of the receiver

The cognitive radio is therefore, by necessity, an example of a *global closed-loop feedback control system*.

---

[4] Every node of the network is equipped with a transceiver (i.e., transmitter/receiver combination). Accordingly, the transmitting part of the node can analyze the radio scene in its local neighborhood, and thereby identify the spectrum holes available for use by the transmitter for communication with the receiver of some other node.

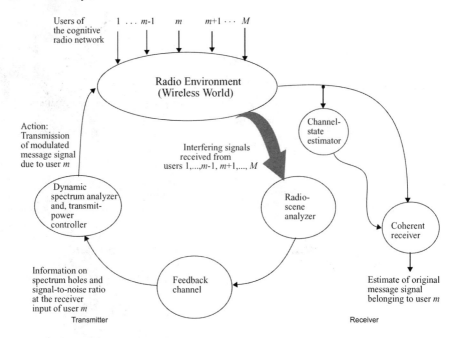

Fig. 1.2. Basic signal-processing cycle for user m in a cognitive radio network; the diagram also includes elements of the receiver of user m.

The pioneering contributions made by Mitola and Haykin are in fact complementary, with Mitola's contribution focusing on software computer aspects of cognitive radio, and Haykin's contribution focusing on signal-processing and communication-theoretic aspects of this exciting multidisciplinary subject.

One other relevant comment is in order. A broadly defined cognitive radio technology accommodates a *scale of differing degrees of implementation*. At one end of the scale, the user may simply find a spectrum hole and build its cognitive cycle around that hole. At the other end of the scale, the user may employ multiple implementation technologies to build its cognitive cycle around a wideband spectrum hole or a set of narrowband spectrum holes to provide the best expected performance in terms of spectrum management and transmit-power control, data rate, and reliable communication, and do all this in the most secure manner feasible.

1.4 Radio-Scene Analysis

With the background material covered in the previous three sections at hand, we are now ready to address the issues involved in radio-scene analysis (RSA). This section is organized as follows:

- We first describe the notion of interference temperature, followed by the issue of the non-stationary character of RF stimuli.

- Next, we describe the multitaper method as the preferred method for non-parametric estimation of the power spectrum of incoming RF stimuli.
- We then describe a spatio-temporal procedure for estimating the interference temperature across the prescribed frequency band.
- We next describe how the occupancy of the radio spectrum in its contiguous subbands is classified.
- Finally, the need for a predictive model describing the evolution of spectrum holes is addressed.

1.4.1 Interference Temperature

Currently, the wireless communication environment is *transmitter-centric*, in the sense that the transmitted power is designed to approach a prescribed noise floor at a certain distance from the transmitter. However, it is possible for the RF noise floor to rise due to the unpredictable appearance of new sources of interference, thereby causing a progressive degradation of the signal coverage. To guard against such a possibility, the FCC Spectrum Policy Task Force [8] has recommended a paradigm shift in interference assessment, that is, a shift away from largely fixed operations in the transmitter and toward *real-time interactions between the transmitter and receiver in an adaptive manner*. The recommendation is based on a new metric called the *interference temperature*,[5] which is intended to quantify and manage the sources of interference in a radio environment. Moreover, the specification of an *interference-temperature limit* provides a "worst-case" characterization of the RF environment in a particular frequency band and at a particular geographic location, where the receiver is expected to operate satisfactorily.

The FCC's recommendation is made with two key benefits in mind:

1. The interference temperature at a receiving antenna provides an accurate measure for the acceptable level of RF interference in the frequency band of interest; any transmission in that band is considered to be "harmful" if it would increase the noise floor above the interference-temperature limit.
2. Given a particular frequency band in which the interference temperature is not exceeded, that band could be made available to unserviced users; the interference-temperature limit would then serve as a "cap" placed on potential RF energy that could be introduced into that band. Logically, the licensed legacy users (i.e., primary owners of the radio spectrum) would be responsible for setting the interference-temperature limit.

---

[5] We may also introduce the concept of interference temperature density, which is defined as the interference temperature per capture area of the receiving antenna [9]. The interference temperature density could be made independent of the receiving antenna characteristics through the use of a reference antenna.

In a historical context, the notion of radio noise temperature is discussed in the literature in the context of microwave background, and also used in the study of solar radio bursts [10,11].

What about the unit for interference temperature? Following the well-known definition of equivalent noise temperature of a receiver [12,13], we may state that the interference temperature is measured in *degrees Kelvin*. Moreover, the interference-temperature limit, $T_{\max}$, multiplied by *Boltzmann's constant*, $\kappa = 1.3807 \times 10^{-23}$ Joules per degree Kelvin, yields the corresponding upper limit on permissible power spectral density in a frequency band of interest, and that density is measured in watts per hertz.

Summarizing, we may therefore say:

> Given an estimate of the power spectral density in a specific subband of the radio spectrum, we may determine the corresponding value of the interference temperature in that subband by dividing the estimate by Boltzmann's constant κ.

This statement emphasizes the need for reliable estimation of the power spectral density of the received RF signal.

1.4.2 Stochastic Approach for Dealing with Non-stationarity

The stimuli generated by radio emitters are *non-stationary spatio-temporal signals* in that their statistics depend on both time and space. Correspondingly, the passive task of radio-scene analysis involves space–time processing, which encompasses two adaptive, spectrally related functions, namely, estimation of the interference temperature and detection of spectrum holes, both of which are performed at a user's receiver.

Unfortunately, the statistical analysis of non-stationary signals, exemplified by RF stimuli, has had a rather mixed history. Although the general second-order theory of non-stationary signals was published during the 1940s by Loève [14,15], it has not been applied nearly as extensively as the theory of stationary processes published only slightly previously and independently by Wiener and Kolmogorov.

To account for the non-stationary behavior of a signal, we have to include time (implicitly or explicitly) in a statistical description of the signal. Given the desirability of working in the frequency domain for well-established reasons, we may include the effect of time by adopting a *time-frequency distribution* of the signal. During the last three decades, many papers have been published on various estimates of time–frequency distributions; see, for example [16] and the references cited therein. In most of this work, however, the signal is assumed to be deterministic. In addition, many of the proposed estimators of time–frequency distributions are constrained to match time and frequency marginal density conditions. However, the frequency marginal distribution is, except for a scaling factor, just the periodogram of the signal. At least since the early work of Lord Rayleigh [17], it has been known that *the periodogram is a badly biased and inconsistent estimator of the power spectrum*. We therefore do not consider matching marginal distributions to be important. Rather, we advocate a stochastic approach to time–frequency distributions which is rooted in the works of Loève [14,15] and Thomson [18].

For the stochastic approach, we may proceed in one of two ways:

1. The incoming RF stimuli are divided into a continuous sequence of successive sections (blocks), with each section being short enough to justify pseudo-stationarity and yet long enough to produce an accurate spectral estimate.
2. Time and frequency are considered jointly under the Loève transform.

Approach (1) is well suited for wireless communications by virtue of the fact that the transmitted signal is typically transmitted on a packet-by-packet basis; we may thus form each section from several adjacent packets, depending on the desired accuracy. In any event, we need a *non-parametric* method for spectral estimation that is both accurate and principled. For reasons that will become apparent in what follows, multitaper spectral estimation is considered to be the method of choice.

1.4.3 Multitaper Spectral Estimation

In the spectral estimation literature, it is well known that the estimation problem is made difficult by the *bias-variance dilemma*, which encompasses the interplay between two points:

- Bias of the power-spectrum estimate of a time series, due to the sidelobe leakage phenomenon, is reduced by tapering (i.e., *windowing*) the time series.
- The cost incurred by this improvement is an increase in variance of the estimate, which is due to the loss of information resulting from a reduction in the effective sample size.

How can we resolve this dilemma by mitigating the loss of information due to tapering? The answer to this fundamental question lies in the principled use of *multiple orthonormal tapers (windows)*,[6] an idea that was first applied to spectral estimation by Thomson in 1982 [18]. The idea is embodied in the *multitaper spectral estimation procedure*.[7] Specifically, the procedure linearly expands the part of the time series in a fixed bandwidth $f - W$ to $f + W$ (centered on some frequency f) in a special family of sequences known as the *Slepian sequences*.[8] The remarkable property of Slepian sequences is that their Fourier transforms have the *maximal energy concentration* in the bandwidth $f - W$ to $f + W$ under a finite sample-size constraint. This property, in turn, allows us to trade spectral resolution for improved spectral characteristics,

---

[6] Another method for addressing the bias-variance dilemma involves dividing the time series into a set of possible overlapping sections, computing a periodogram for each tapered (windowed) section, and then averaging the resulting set of power spectral estimates, which is what is done in *Welch's method* [19]. However, unlike the principled use of multiple orthogonal tapers, Welch's method is rather ad hoc in its formulation.

[7] In the original paper by Thomson [18], the multitaper spectral estimation procedure is referred to as the *method of multiple windows*. For detailed descriptions of this procedure, see [18] and Chap. 7 of the book by Percival and Walden [20].

[8] The Slepian sequences are also known as *discrete prolate spheroidal sequences*. For detailed treatment of these sequences, see the original paper by Slepian [21], the appendix to Thomson's paper [18] and Chap. 8 of the book by Percival and Walden [22].

namely, reduced variance of the spectral estimate without compromising the bias of the estimate.

Given a time series $\{x_t\}_{t=1}^N$, representing the *baseband* version of the received RF signal with respect to the center frequency of the RF band under scrutiny, the multitaper spectral estimation procedure determines two things:

1. An orthonormal sequence of K *Slepian tapers* denoted by $\{w_t^{(k)}\}_{t=1}^N$
2. The associated *eigenspectra* defined by the Fourier transform

$$Y_k(f) = \sum_{t=1}^N w_t^{(k)} x(t) e^{-j2\pi ft}, \quad k = 0, 1, \ldots K - 1. \tag{1.1}$$

The energy distributions of the eigenspectra are concentrated inside a *resolution bandwidth*, denoted by $2W$. The *time–bandwidth product*

$$p = 2NW \tag{1.2}$$

defines the *degrees of freedom* available for controlling the variance of the spectral estimator. The choice of parameters K and p provides a tradeoff between spectral resolution and variance.[9] A natural spectral estimate, based on the first few eigenspectra that exhibit the least sidelobe leakage, is given by [18]

$$\hat{S}(f) = \frac{\sum_{k=0}^{K-1} \lambda_k(f) |Y_k(f)|^2}{\sum_{k=0}^{K-1} \lambda_k(f)} \tag{1.3}$$

where λ_k is the eigenvalue associated with the kth eigenspectrum. The denominator in (1.3) makes the estimate $\hat{S}(f)$ unbiased.

The multitaper spectral estimator of (1.3) is intuitively appealing in the way it works: as the number of tapers, K, increases, the eigenvalues decrease, causing the eigenspectra to be more contaminated by leakage. But, the eigenvalues themselves counteract by reducing the weighting applied to higher leakage eigenspectra.

It is also noteworthy that in [24], Stoica and Sundin show that the multitaper spectral estimation procedure can be interpreted as an "approximation" of the *maximum-likelihood* power spectrum estimator. Moreover, they show that for wideband signals, the multitaper spectral estimation procedure is "nearly optimal" in the sense that it

---

[9] For an estimate of the variance of a multitaper spectral estimator, we may use a resampling technique called *jackknifing* [23]. The technique bypasses the need for finding an exact analytic expression for the probability distribution of the spectral estimator, which is impractical because time-series data (e.g., stimuli produced by the radio environment) are typically non-stationary, non-Gaussian, and frequently contain outliers. Moreover, it may be argued that the multitaper spectral estimation procedure results in nearly uncorrelated coefficients, which provides further justification for the use of jackknifing.

almost achieves the Cramèr–Rao bound for a non-parametric spectral estimator.[10] Most important, unlike the maximum-likelihood spectral estimator, the multitaper spectral estimator is computationally feasible.

1.4.4 Adaptive Modification of Multitaper Spectral Estimation

While the lower-order eigenspectra have excellent bias properties, there is some degradation as the order K increases toward the time–bandwidth product $2NW$. In [18], Thomson introduces a set of adaptive weights, denoted by $\{d_k(f)\}$, which downweight the higher order eigenspectra. Using a mean-squared error optimization procedure, the following formula for the weights is derived:

$$d_k(f) = \frac{\sqrt{\lambda_k}S(f)}{\lambda_k S(f) + \mathbf{E}[B_k(f)]}, \quad k = 0, 1, \ldots, K-1 \tag{1.4}$$

where $S(f)$ is the true power spectrum, $B_k(f)$ is the broadband bias of the kth eigenspectrum, and $\mathbf{E}$ is the expectation operator. Moreover,

$$\mathbf{E}[B_k(f)] \leq (1 - \lambda_k)\sigma^2, \quad k = 0, 1, \ldots, K-1 \tag{1.5}$$

where σ^2 is the *process variance* defined by

$$\sigma^2 = \frac{1}{N} \sum_{t=0}^{N-1} |x(t)|^2. \tag{1.6}$$

In order to compute the adaptive weights $d_k(f)$ using (1.4), we need to know the true spectrum $S(f)$. But if we did, then there would be no need to perform any spectrum estimation at all. Nevertheless, the formula of (1.4) is useful in setting up an *iterative procedure for computing the adaptive spectral estimator*

$$\hat{S}(f) = \frac{\displaystyle\sum_{k=0}^{K-1} |d_k(f)|^2 \hat{S}_k(f)}{\displaystyle\sum_{k=0}^{K-1} |d_k(f)|^2} \tag{1.7}$$

where

$$\hat{S}_k(f) = |Y_k(f)|^2, \quad k = 0, 1, \ldots, K-1. \tag{1.8}$$

Note that if we set $\{d_k(f)\}^2 = \lambda_k$ for all k, then the estimator of (1.7) reduces to that of (1.3).

---

[10] In [22], a comparative evaluation of the multitaper method (MTM) and maximum-likelihood (ML) method is presented for angle-of-arrival estimation in the presence of multipath. The results reported therein give consistent results for low grazing angles. The MTM is found to be slightly superior to ML, but the difference between them is not overwhelming.

Next, setting $S(f)$ equal to the spectrum estimator $\hat{S}_k(f)$ in (1.4), then substituting the new equation into (1.7) and collecting terms, we get (after simplifications)

$$\sum_{k=0}^{K-1} \frac{\lambda_k(\hat{S}(f) - \hat{S}_k(f))}{(\lambda_k \hat{S}(f) + \hat{B}_k(f))^2} = 0 \qquad (1.9)$$

where $\hat{B}_k(f)$ is an estimate of the expectation $\mathbf{E}[B_k(f)]$. Using the upper bound of (1.5), we have

$$\hat{B}_k(f) = (1 - \lambda_k)\sigma^2, \quad k = 0, 1, \ldots, K - 1. \qquad (1.10)$$

We now have all that we need to solve for the null condition of (1.9) via the *recursion*

$$\hat{S}^{(j+1)}(f) = \left[\sum_{k=0}^{K-1} \frac{\lambda_k \hat{S}_k(f)}{(\lambda_k \hat{S}^{(j)}(f) + \hat{B}_k(f))^2} \right] \left[\sum_{k=0}^{K-1} \frac{\lambda_k}{(\lambda_k \hat{S}^{(j)}(f) + \hat{B}_k(f))^2} \right]^{-1}$$
$$(1.11)$$

where j denotes an iteration step. To initialize this recursion, we may set $S^{(j)}(0)$ equal to the average of the two lowest order eigenspectra. Convergence of the recursion is usually rapid, with successive spectral estimates differing by less than 5% in 5–20 iterations. For a more accurate (also more complex) estimate of $B_k(f)$, see [18,22]. In any event, the result obtained from (1.11) is substituted into (1.4) to obtain the desired weights, $d_k(f)$.

A useful by-product of this adaptive spectral estimation procedure is a *stability measure of the estimates*, given by

$$v(f) = 2 \sum_{k=0}^{K-1} |d_k(f)|^2 \qquad (1.12)$$

which is the approximate number of degrees of freedom for the estimator $\hat{S}_k(f)$ expressed as a function of frequency f. If $\bar{v}$, denoting the average of $v(f)$ over frequency f, is significantly less than $2K$, then the result is an indication that either the window W is too small, or additional prewhitening of the time series $x(n)$ should be used.

The importance of *prewhitening* cannot be stressed enough for RF data. In essence, prewhitening reduces the dynamic range of the spectrum by filtering the data, prior to processing. The resulting residual spectrum is nearly flat or "white." In particular, leakage from strong components is reduced, so that the fine structure of weaker components is more likely to be resolved. In actual fact, most of the theory behind spectral estimation is smooth, almost white-like spectra to begin with, hence the need for "prewhitening" [22].

1.4.5 Summarizing Remarks I

1. Estimation of the power spectral density based on the multitaper method of (1.3) is said to be *incoherent*, because the kth magnitude spectrum $|Y_k(f)|^2$ ignores phase information for all k.

2. For the parameters needed to compute the multitaper spectral estimator (1.3), recommended values are:
   - Time-bandwidth product: $NW = 6$, possibly extending up to 10.
   - Number of Slepian tapers: $K = 10$, possibly extending up to 16.

   These values are needed, especially when the dynamic range of the RF data is large.

   As an illustrative example, in [25] describing the application of the multitaper method to radar sea-clutter classification, the number of available samples in each section of the radar data was relatively small, namely, 256. Reasonably good results were obtained using $NW = 6$ and $K = 10$ within each section.
3. If and when the number of tapers is increased toward the time–bandwidth product $2NW$, then the adaptive multitaper spectral estimator should be used.
4. Whenever possible, prewhitening of the data, prior to processing, should be applied.

1.4.6 Space–Time Processing

With cognitive radio being receiver-centric, it is necessary that the receiver be provided with a reliable spectral estimate of the interference temperature. We may satisfy this requirement by doing two things:

1. *Use the multitaper method to estimate the power spectrum of the interference temperature due to the cumulative distribution of both internal sources of noise and external sources of RF energy.* In light of the findings reported in [24], this estimate is near-optimal.
2. *Employ a large number of sensors to properly "sniff" the RF environment, wherever it is feasible.* The large number of sensors is needed to account for the spatial variation of the RF stimuli from one location to another.

The issue of multiple-sensor feasibility is raised under point (2) because of the diverse ways in which wireless communications could be deployed. For example, in an indoor building environment and communication between one building and another, it is feasible to employ a large number of sensors (i.e., antennas) placed at strategic locations in order to improve the reliability of interference-temperature estimation. On the other hand, in the case of an ordinary mobile unit with limited real estate, the interference-temperature estimation may have to be confined to a few sensors beamed at different directions.

Let M denote the total number of sensors deployed in the RF environment. Let $Y_k^{(m)}(f)$ denote the kth eigenspectrum computed by the mth sensor. We may then construct the M-by-K spatio-temporal complex-valued matrix [26]

$$
\mathbf{A}(f) = \begin{bmatrix}
a_1 Y_0^{(1)}(f) & a_1 Y_1^{(1)}(f) & \cdots & a_1 Y_{K-1}^{(1)}(f) \\
a_2 Y_0^{(2)}(f) & a_2 Y_1^{(2)}(f) & \cdots & a_2 Y_{K-1}^{(2)}(f) \\
\vdots & \vdots & & \vdots \\
a_M Y_0^{(M)}(f) & a_M Y_1^{(M)}(f) & \cdots & a_M Y_{K-1}^{(M)}(f)
\end{bmatrix} \tag{1.13}
$$

where each row is produced using stimuli sensed at a different gridpoint, each column is computed using a different Slepian taper, and the $\{a_m\}_{m=1}^M$ represent variable coefficients accounting for relative areas of the gridpoints.

Each entry in the matrix $\mathbf{A}(f)$ is produced by two contributions, one due to additive ambient noise in the sensor and the other due to the interfering RF stimuli. Insofar as radio-scene analysis is concerned, however, the primary contribution of interest is that due to RF stimuli. An effective tool for denoising is the *singular value decomposition* (SVD), the application of which to the matrix $\mathbf{A}(f)$ yields the decomposition [27]

$$\mathbf{A}(f) = \sum_{k=0}^{K-1} \sigma_k(f)\mathbf{u}_k(f)\mathbf{v}_k^\dagger(f) \tag{1.14}$$

where $\sigma_k(f)$ is the kth *singular value* of matrix $\mathbf{A}(f)$, $\mathbf{u}_k(f)$ is the associated *left singular vector*, and $\mathbf{v}_k(f)$ is the associated *right singular vector*; the superscript $\dagger$ denotes Hermitian transposition. In analogy with principal components analysis, the decomposition of (1.14) may be viewed as one of *principal modulations* produced by the external RF stimuli. According to (1.14), the singular value $\sigma_k(f)$ scales the kth principal modulation of matrix $\mathbf{A}(f)$.

Forming the K-by-K matrix product $\mathbf{A}^\dagger(f)\mathbf{A}(f)$, we find that the entries on the main diagonal of this product, except for a scaling factor, represent the eigenspectrum due to each of the Slepian tapers, spatially averaged over the M sensors. Let the singular values of matrix $\mathbf{A}(f)$ be ordered $|\sigma_0(f)| \geq |\sigma_1(f)| \geq \ldots \geq |\sigma_{K-1}(f)| > 0$. The kth eigenvalue of $\mathbf{A}^\dagger(f)\mathbf{A}(f)$ is $|\sigma_k(f)|^2$. We may then make the following statements:

1. The eigenvalues are proportional to average power, expressed as a function of frequency f. In particular, the largest eigenvalue $|\sigma_0(f)|^2$, measured across the frequency band of interest, provides an estimate of the interference temperature in that band, except for a constant. This estimate would be improved by using a linear combination of the largest two or three eigenvalues: $|\sigma_k(f)|^2$, $k = 0, 1, 2$.
2. The left singular vectors $\mathbf{u}_k(f)$ for $k = 0, 1, \ldots, K - 1$, provide information on the spatial distribution of the interferers. Most importantly, this information could be used for *wavenumber spectrum estimation* or *adaptive beamforming*; here, it is assumed that the number of sensors (i.e., spatial degrees of freedom) is large enough.
3. The right singular vectors $\mathbf{v}_k(f)$ for $k = 0, 1, \ldots, K - 1$, provide the multitaper coefficients for the interferers' waveforms.

1.4.7 Summarizing Remarks II

In space–time processing, the spatial and temporal dimensions are distinct. The RF data therefore represent a multivariate time series, whose spectral structure is summed up in the matrix $\mathbf{A}(f)$ of (1.13). Accordingly, we can make the following statements:

1. The two-dimensional tapers of the time–space processor are the *tensor products* of the standard one-dimensional Slepian tapers.

2. The time–space processor is *coherent* and therefore richer in the extent of information it extracts from the RF environment. Specifically, it is capable of providing *joint* estimates of the interference temperature across a frequency band of interest and the angles-of-arrival of the interfering RF signals emitted by other users.
3. However, this rich source of information on the RF environment is obtained at the expense of a significant increase in computational complexity.

1.4.8 Spectral Classification

In passively sensing the radio scene and thereby estimating the power spectra of incoming RF stimuli, we have a basis for classifying the spectra into three broadly defined types, as summarized here:

1. *Black spaces*, which are occupied by high-power "local" interferers some of the time.
2. *Gray spaces*, which are partially occupied by low-power interferers.
3. *White spaces*, which are free of RF interferers except for *ambient noise*, made up of natural and artificial forms of noise:
   - Broadband thermal noise produced by external physical phenomena such as solar radiation
   - Transient reflections from lightening, plasma (fluorescent) lights and aircraft
   - impulsive noise produced by ignitions, commutators and microwave appliances
   - thermal noise due to internal spontaneous fluctuations of electrons at the front end of individual receivers

White spaces (for sure) and gray spaces (to a lesser extent) are potential candidates for use by unserviced operators. Of course, black spaces are to be avoided whenever and wherever the RF emitters residing in them are switched ON. However, when at a particular geographic location those emitters are switched OFF and the black spaces assume the new role of "spectrum holes," cognitive radio provides the opportunity for creating significant "white spaces" by invoking its dynamic-coordination capability for spectrum sharing.

From the picture of the radio scene presented in this section, it is apparent that a *reliable strategy for the detection of spectrum holes* is of paramount importance to the design and practical implementation of cognitive radio systems. Moreover, the multitaper method combined with singular-value decomposition, hereafter referred to as the *MTM-SVD method*,[11] provides the method of choice for solving this detection problem by virtue of its accuracy and near-optimality.

---

[11] Mann and Park [26] discuss the application of the MTM-SVD method to the detection of oscillatory spatial-temporal signals in climate studies. They show that this new methodology avoids the weaknesses of traditional signal detection techniques. In particular, the methodology permits a faithful reconstruction of spatio-temporal patterns of narrowband signals in the presence of additive spatially correlated noise.

By repeated application of the MTM-SVD method to the RF stimuli at a particular geographic location and from one section of data to the next, a time–frequency distribution of that location is computed. The dimension of time is quantized into discrete intervals separated by the section duration. The dimension of frequency is also quantized into discrete intervals separated by resolution bandwidth of the multitaper spectral estimation procedure.

Let L denote the number of largest eigenvalues considered to play important roles in estimating the interference temperature, with $|\sigma_l(f,t)|^2$ denoting the lth largest eigenvalue produced by the section (block) of RF stimuli received at time t. Let N denote the number of frequency resolutions of width $\Delta f = 2W$, which occupy the frequency subband (space) under scrutiny. Then, setting the discrete frequency

$$f = f_{\text{low}} + v \cdot \Delta f, \quad v = 0, 1, \ldots, N-1$$

where f_{low} denotes the lowest end of a black, gray or white space, we may define the *decision statistic* for classifying the subbands as

$$D(t) = \sum_{l=0}^{L-1} \sum_{v=0}^{N-1} |\sigma_l(f_{\text{low}} + v \cdot \Delta f, t)|^2 \Delta f. \tag{1.15}$$

Let $D_{\min}$ denote the *minimum possible value* that could be assumed by the decision statistic $D(t)$ due to the ambient noise floor, and let $D_{\max}$ denote its *maximum permissible value* corresponding to the prescribed temperature limit. Let D_{av} denote the average value of $D(t)$, computed over a number of successive sections of the incoming RF signal. We may then classify the frequency subband (space) under scrutiny as follows:

- If $D_{\max} - \delta_1 \leq D_{\text{av}} \leq D_{\max}$, then the subband is said to be a black space.
- If $D_{\min} \leq D_{\text{av}} \leq D_{\min} + \delta_2$, then the subband is said to be a white space.
- Otherwise, the subband is declared to be a gray space.

The parameters δ_1 and δ_2 are chosen by the system designer, depending on how fine a spectral classification is described. Moreover, the specifications of $D_{\max}$ and $D_{\min}$ are *location-specific*. For example, if the spectral classification is performed in the basement of a building, then the spacing between $D_{\max}$ and $D_{\min}$ is expected to be significantly smaller than in an open environment.

1.4.9 Spatio-temporal Evolution of Spectrum Holes

From a cognitive radio user's viewpoint, the following pieces of information are needed:

1. The location of spectrum holes
2. The variance of the interference plus noise in each spectrum hole
3. The duration for which the spectrum hole is likely to be available for use

The MTM-SVD method addresses points (1) and (2). To address point (3), we need a predictive model of the evolution of spectrum holes over time, as discussed next.

The existence of spectrum holes is directly related to the primary user's traffic patterns, which can be of a deterministic or stochastic kind:

- *Deterministic traffic patterns* are attributed to television and AM/FM radio stations and/or air-traffic control radar and weather radar installations. The traffic patterns produced by these primary users are known on a daily basis, which makes their predictability a straightforward matter.
- *Stochastic traffic patterns* arise from wireless communication devices.

The availability of traffic patterns at different times of the day and different geographic locations, desirably provided by legacy users and/or government agencies, could form the basis of a *spatio-temporal prediction model* of traffic behavior. Such a model could make it possible to predict the duration of time for which spectrum holes are likely to be employable, thereby enhancing coexistence between legacy users and secondary users.

1.5 Extraction of Channel-State Information (CSI)

Section 1.4 on radio-scene analysis dealt with issues pertaining to spectral information on the radio environment, which is needed by the transmitter for efficient utilization of the radio spectrum. In this section, we deal with another function of the receiver, namely, the *extraction of channel-state information* (CSI), which is needed by a user's receiver for coherent detection of the transmitted information-bearing signal. This section is organized as follows:

- First, we set the stage for semi-blind training, which offers a compromise between two extreme approaches: differential detection for unsupervised transmission and pilot-assisted transmission for supervised training.
- Next, we describe a channel-tracking procedure that is basic to the semi-supervised training procedure.

1.5.1 Semi-supervised Training

To deal with the channel-state estimation problem, traditionally we have proceeded in one of two ways:

- *Differential detection*, which lends itself to implementation in a straightforward fashion, using M-ary phase modulation.
- *Pilot-assisted transmission*, which involves the periodic transmission of a pilot (training sequence) known to the receiver.

The use of differential detection offers robustness and simplicity of implementation, but at the expense of a significant degradation in the frame-error rate (FER) versus signal-to-noise ratio (SNR) performance of the receiver. On the other hand, pilot-assisted transmission (PAT) offers improved receiver performance, but the use of a pilot is wasteful in both transmit power and channel bandwidth, the very thing we should strive to avoid. What then do we do, if the receiver requires knowledge of CSI for efficient receiver performance? The answer to this fundamental question lies in the use of *semi-blind training* of the receiver, which distinguishes itself from the differential detection and PAT procedures in that the receiver has two modes of operations:

1. *Supervised training mode.* During this mode, the receiver acquires an estimate of the channel state, which is performed under the supervision of a short training sequence, consisting of fewer symbols than that required with PAT. As with PAT, the training sequence is known to the receiver. It is sent over the channel for a limited duration by the transmitter prior to the actual data-transmission session; the pilot transmission is repeated periodically.
2. *Tracking mode.* Once a reliable estimate of the channel state has been achieved, the training sequence is switched off, actual data transmission is initiated and the receiver is switched to the tracking mode; this mode of operation is performed in an unsupervised manner on a continuous basis during the course of data transmission.

1.5.2 Channel Tracking

The evolution of CSI with time is governed by a *state-space model* comprised of two equations:

1. *Process equation.* The state of a wireless link is defined as the *minimal set of data on the past behavior of the link that is needed to predict the future behavior of the link.* For the sake of generality, we consider a *multiple-input, multiple-output (MIMO) wireless link*[12] of a narrowband category. Let $h_{jk,t}$ denote the

---

[12] The use of a MIMO link offers several important advantages:
  1. *Spatial degree of freedom,* defined by $N = \min\{N, L\}$, where N and L denote the numbers of transmit and receive antennas, respectively [28].
  2. *Increased spectral efficiency,* which is asymptotically defined by

$$\lim_{N \to \infty} \frac{C(N)}{N} = \text{constant}$$

  where $C(N)$ is the ergodic capacity of the link, expressed as a function of $L = N$. This asymptotic property provides the basis for a spectacular increase in spectral efficiency by increasing the number of transmit and receive antennas.
  3. *Diversity,* which is asymptotically defined by

$$\lim_{\rho \to \infty} \frac{\log \text{FER}(\rho)}{\log \rho} = -d_o$$

  where d_o is the diversity order and $\text{FER}(\rho)$ is the frame-error rate expressed as a function of the signal-to-noise ratio ρ.
  These benefits (gained at the expense of increased complexity) commend the use of MIMO links for cognitive radio, all the more so considering the fact that the primary motivation

channel coefficient from the kth transmit antenna to the jth receive antenna at time t, with $k = 1, 2, \ldots, N$ and $j = 1, 2, \ldots, L$. We may then describe the scalar form of the state equation as

$$h_{jk,t+1} = \sum_{l=0}^{M} (\beta_{l,t}) (h_{jk,t-1}) + v_{jk,t} \tag{1.16}$$

where the $\beta_{l,t}$ are *time-varying autoregressive (AR) coefficients* and $v_{jk,t}$ is the corresponding *dynamic noise*, both at time t. The AR coefficients account for the *memory* of the channel due to the multipath phenomenon. The upper limit of summation in (1.16) namely, M, is the *model order*.

2. *Measurement equation.* The measurement equation for the MIMO wireless link, also in scalar form, is described by

$$y_{j,t} = \sum_{k=1}^{N} (s_{k,t}) (h_{jk,t}) + n_{j,t} \quad \text{for} \quad j = 1, 2, \ldots, L \tag{1.17}$$

where $s_{k,t}$ is the *encoded symbol* transmitted by the kth antenna at time t, and $n_{j,t}$ is the corresponding *measurement noise* at the input of jth receive antenna at time t. The $y_{j,t}$ is the *signal observed* at the output of the jth antenna at time t.

The state-space model comprised of (1.16) and (1.17) is *linear*. The property of linearity is justified in light of the fact that the propagation of electromagnetic waves across a wireless link is governed by Maxwell's equations that are inherently linear.

What can we say about the AR coefficients, the dynamic noise, and measurement noise, which collectively characterize the state-space model of (1.16) and (1.17)? The answers to these questions determine the choice of an appropriate tracking strategy. In particular, we say the following:

1. *AR model.* A *Markov model* of order one offers simplicity and sufficient accuracy to model a Rayleigh-distributed time-varying channel.
2. *Noise processes.* The dynamic noise in the process equation is Gaussian, but the noise in the measurement equation is likely to be non-Gaussian due to the presence of impulsive noise generated in the radio environment. (The impulsive noise is attributed to different sources such as automobile engine noise in an outdoor environment and microwave devices in an indoor environment.)

Point (1) directly affects the design of the *predictive model*, which is an essential component of the channel tracker. Point (2) prompts the search for a tracker outside of the classical Kalman filters, whose theory is rooted in Gaussian statistics.

Two different channel-tracking procedures are described in [29] and [30]; herein, we briefly highlight the procedure described in [30].

Rewriting (1.16) in matrix form, under the assumption of an AR model of order one, we have

---

for cognitive radio is the attainment of improved spectral efficiency. Simply put, a MIMO wireless link is not a necessary ingredient for cognitive radio but a highly desirable one.

$$\mathbf{h}_{t+1} = \beta_{0,t}\mathbf{h}_t + \mathbf{v}_t \tag{1.18}$$

where $\mathbf{h}_t$ denotes the vector representation of the channel matrix $\mathbf{H}_t$ by stretching the columns of $\mathbf{H}_t$ one over another, and $\mathbf{v}_t$ denotes the corresponding vector representation of the dynamic noise. The key objective of the channel tracker is to estimate the update equation for posterior probability density function of the sequence

$$\mathbf{h}_{1:t} = \{\mathbf{h}_i\}_{i=1}^t$$

when we are given the entire set of measurements

$$\mathbf{y}_{1:t} = \{\mathbf{y}_i\}_{i=1}^t.$$

That is, the posterior density of the channel state is updated in accordance with the equation

$$p\left(\mathbf{h}_{1:t}|\mathbf{y}_{1:t}\right) = \frac{p(\mathbf{y}_t|\mathbf{h}_t)p(\mathbf{h}_t|\mathbf{h}_{t-1})}{p(\mathbf{y}_t|\mathbf{y}_{1:t-1})}p(\mathbf{h}_{1:t-1}|\mathbf{y}_{1:t-1}). \tag{1.19}$$

Reference [30] describes a novel procedure for computing this update equation, using a particle filter. The main idea of the procedure is to introduce a correction factor in the predicted estimate of the channel state, with the correction being based on an approximate *maximum-likelihood* (ML) estimate of the channel state. Specifically, the corrected channel estimate is defined by the convex combination of the old value of the channel state and the current maximum-likelihood estimate, as shown by

$$\mathbf{h}_{t|t-1}^{\mathcal{C}} = (1 - \alpha)\mathbf{h}_{t|t-1} + \alpha\mathbf{h}_t^{\mathrm{ML}} \tag{1.20}$$

where α, lying in the range between zero and one, is a weight given to the confidence in the ML estimate; the value assigned to this weight depends on the signal-to-noise ratio (SNR) and the fading rate of the wireless environment. For example, at low SNR, we are less confident in the current estimation of the channel state and therefore a small value is assigned to α. Likewise, for a highly time-selective channel, we have less confidence in the ML estimate, in which case we also assign less weight to α. Choosing the "optimal" α is problem-specific and may therefore require the inclusion of an adaptive loop in the estimation procedure for online operation.

The motivation behind the convex-combined predictive channel estimate is to "guide" the particles in the tracking filter toward a high probability region of the density; as such, it may be viewed as a more refined approach than that taken in [29]. As the combined step in the state update incorporates recent measurements, the state space is efficiently exploited so as to improve the sampling efficiency. Indeed, in [30], Monte Carlo simulation results are presented for a radio environment assuming:

- The use of a frequency-flat time-selective channel based on the *Jakes model*
- The use of a *Middleton Class-A model* for an impulsive measurement noise

The simulations compare the performance of a wireless system using two channel trackers, one incorporating the approximate ML channel estimate in the particle filter to select *informative particles* as described herein and the other incorporating *gradient information* in the selection of the particles as described in [29]. The results of the simulation presented in [30] reveal that unlike the channel tracker based on gradient particle filtering, the asymptotic *performance gap* between the genie scenario (assuming that the channel state is known) and the corresponding scenario involving the use of the new channel tracker is essentially uniform for increasing SNR, which is desirably how it should be.

1.6 Feedback Channel

As pointed out previously, the primary motivation of cognitive radio is improved utilization of the radio spectrum, hence the requirement for identification of spectrum holes in the local neighborhood of a user's receiver. Having performed this function by the radio-scene analyzer in the receiver, we need a feedback channel to send relevant information of the receiver to the user's transmitter for appropriate action by that user. This information consists of two constituents:

- The center frequencies and bandwidths of the spectrum holes
- The combined variance of interference and thermal noise in each spectrum hole

Later on, we will also find that there is an additional role for the feedback channel:

- A measure of the signal-to-noise ratio at the output of the transmitter–receiver wireless link, which is needed by the adaptive modulator in the transmitter.

Rather than send the actual values of the various parameters identified here, the practical approach is to feed their respective *quantized* values back to the transmitter. To do this, a predetermined list of quantized values pertaining to the following parameters is kept in the receiver:

- Center frequencies and bandwidths of all possible spectrum holes
- Variance of interference plus noise in each possible spectrum hole
- Output signal-to-noise of the pertinent wireless link

Given such a list, the receiver picks the closest entries in the list that are less than the actual values of the parameters. In so doing, the bit rate of the feedback channel is minimized.

Putting it altogether, the feedback channel plays a fundamental role in the design and operation of cognitive radio. Indeed, we may go on to say that feedback is the *facilitator of intelligence*, without which the radio loses its cognitive capability.

1.7 Multiuser Cognitive Radio Networks

As it is with every other communication network, the deployment of a *cognitive radio network* can be justified in financial terms if, and only if, the network is utilized by a multiple users.

Mobile wireless communication networks are *centralized*, in that an infrastructure of base stations is deployed to route calls from one user to another. In contrast, for both civilian and military applications, it is desirable for cognitive radio networks to be *decentralized.* In other words, the network is configured in a *self-organized* manner [31,32], which makes it possible to dispense with the need for a costly pre-established infrastructure. With this objective in mind, the adoption of *ad hoc networks* [33,34] is the logical basis for cognitive radio networks.

From what we know about brain theory [35] and neural networks [5], self-organization builds on two basic mechanisms: *cooperation* and *competition*; these two mechanisms operate in a complementary manner so as to "bring order in the network out of disorder." In a similar sort of way, we may envision a self-organized cognitive radio network, in which cooperation and competition are purposely configured to perform complementary functions. Specifically:

- Cooperation is used to facilitate *communication* across the nodes of the network without any fixed infrastructure.
- Competition is used to provide *control* over the power transmitted from each individual node of the network to maintain the interference temperature at a receiving node below a prescribed limit.

In a cognitive radio network built on ad hoc wireless principles, the network is basically an association of nodes that cooperate. Insofar as network coordination is concerned, for example, we may simply require *each pair of neighboring nodes* be in direct communication. Thus, in the communication scenario, each node creates a *transmit–receive schedule*; the schedule is communicated to a nearest neighbor only when a source node's schedule and that of a neighboring node permit the source node to transmit the schedule and the neighboring node is able to receive it. In [36], it is shown that under reasonable assumptions, such a completely decentralized network can scale to an almost arbitrary number of nodes. It is therefore feasible to develop a dynamic frequency selection policy that supports utilization of the network by more users through a built-in cooperative mechanism. The capacity of wireless networks is discussed in [37,38].

Turning next to the benefit that could be gained from competition, it will be shown in Sect. 1.12 that by adopting a non-cooperative (i.e., competitive) game-theoretic approach, it is possible to design an efficient and effective transmit-power control policy. Most important, this policy does not require synchronization among the multiple users, thereby simplifying the design of the network.

1.8 Dynamic Spectrum Management

The primary motivation of cognitive radio is to improve utilization of the radio spectrum, subject to two requirements:

1. Secondary users of the spectrum's unoccupied subbands must coexist with the primary users.
2. Interference temperature at the receiver input of each user in the network does not exceed a prescribed limit.

Requirement (2) is considered later in Sect. 1.12. In this section, we address requirement (1).

We first note that by having the network operate in a decentralized cooperative manner, information-bearing signals could hop from one node of the network to a neighboring node, thereby facilitating communication across the entire network. Moreover, the spectrum holes come and go. Accordingly, we may formulate the dynamic spectrum management problem as follows:

> Given a set of spectrum holes detected by the radio-scene analyzer and whose composition is likely to change from one time instant to another, devise a decentralized dynamic spectrum management policy that enables secondary users to employ these spectrum holes without disruption to the primary users.

For the policy to be decentralized, we need a *random (probabilistic) multiple-access technique*. Here we have the choice between two protocols: *Aloha* and *carrier-sense multiple-access* (CSMA). For terrestrial networks, CSMA is the preferred choice [39].

In its simplest form, CSMA operates as follows:

1. If the wireless channel is sensed to be *idle* (i.e., a spectrum hole is available), the user transmits its packets.
2. If the channel is sensed to be *busy* (i.e., the spectrum hole has become occupied), the transmission of packets is scheduled for a later time according to a specified random distribution.
3. At the new point in time, the user senses the channel and repeats the algorithm.

If the transmissions were instantaneous, then collisions would occur in the CSMA protocol only if two users transmitted at exactly the same time; this should be a rare occurrence but nevertheless, it could happen.

In a modified form of CSMA called *carrier-sense multiple-access with collision avoidance* (CSMA/CA) each node of the network must inform other nodes in the network of the intent to transmit packets, and it is only then that transmission can take place. In so doing, packet collisions are prevented, because all nodes in the network have been made aware of packet transmission before it occurs. Such a protocol is indeed feasible by virtue of the cooperative communication built across the network.

1.8.1 Modulation Format

The next issue to be considered is the choice of a modulation format for the actual transmission of packets over the selected spectrum hole. For this purpose, we consider *orthogonal frequency-division multiplexing* as a method of choice [40,41]. We say so for the following reasons:

- OFDM is a bandwidth-efficient signaling scheme, which converts a difficult frequency-selective channel into a parallel collection of frequency-flat subchannels, whose subcarrier frequencies form an orthogonal set.

- Unlike ordinary frequency-division multiplexing (FDM), the spectra of the individually modulated subcarriers in OFDM overlap mutually and thereby optimally occupy the frequency response of the channel.
- The choice of OFDM fits perfectly into the design of the transmit-power controller.

Orthogonality of the subcarriers over the duration of a symbol is achieved by having the frequency spacing between them equal to the reciprocal of the symbol duration.

There are some other practical requirements that need to be satisfied:

1. The modulation format must be *impervious* to that of the primary user, so as not to violate the coexistence requirement.
2. The modulation format must be *adaptive*, so as to account for the time-varying conditions of the radio environment.
3. Due to channel noise and interference from other users, *reliable* communication must be maintained between the wireless link that connects the transmitting node to the receiving node.

To satisfy these requirements, we may use the *concatenated coding scheme* depicted in Fig. 1.3, where the concatenation of channel encoder and space–time encoder is performed on a symbol-by-symbol basis [40].

To explain, the data bits produced by the OFDM are first channel-coded by a *turbo convolutional encoder* [42], which is followed by a pseudo-random block interleaver. Next, the adaptive *quadrature amplitude modulator* (QAM) selects a mode of modulation from the set (for example):

- Binary phase-shift keying (BPSK)
- Quadrature phase-shift keying (QPSK)
- 16-QAM
- 64-QAM

This selection is made by the adaptive modulator in response to the quality of signal reception measured at the receiving node; in effect, *feedback* is needed between each pair of neighboring nodes in the network for adaptive modulation to be feasible.

Finally, the modulated symbol is space–time block encoded [43], with the encoding being performed in the frequency domain. Here, it is assumed that a set of

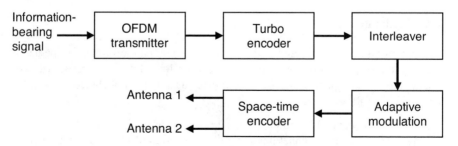

Fig. 1.3. Block diagram of adaptive OFDM transmitter.

adjacent subcarriers in the OFDM signal, belonging to the same space–time encoding block, have approximately the same signal-to-noise ratio. The space–time code, involving the use of multiple transmit as well as receive antennas, provides *diversity* to combat multipath; reliability of communication across the network is thereby further enhanced.

1.9 Statistical Modeling of Cognitive Radio Networks

To set the stage for formulating the transmit-power control problem considered in the next section, we need a *statistical model* for cognitive radio networks. In what follows, we assume that the wireless channel linking one node in the network to a neighboring node is *frequency-selective*. As pointed out in the previous section, OFDM is well suited for dealing with such a channel, the use of which converts a frequency-selective channel into a set of frequency-flat subchannels whose individual subcarriers are ideally orthogonal to each other. In practice, however, we find that OFDM is sensitive to frequency offset in the channel, which arises because the subcarriers are inherently closely spaced in frequency, compared to channel bandwidth; consequently, the tolerable frequency offset is a small fraction of the channel bandwidth.[13]

Let f_0 denote the frequency spacing between adjacent subcarriers, and Δf denote the frequency offset. The baseband version of the OFDM signal radiated by the transmitter of a user labeled m is thus defined by

$$s(m, n) = c_n \exp[j2\pi n(f_0 - \Delta f)] \tag{1.21}$$

where n denotes one of the N subcarriers in the OFDM signal, and c_n is the modulated amplitude of the nth subcarrier. To simplify the notation, we have omitted the dependence on time t in $s(m, n)$. Correspondingly, the signal picked up by the intended receiver of user m is given by[14]

$$y(m, n) = \sum_{k=1}^{M} g(m, k, n)s(k, n) + w(m, n) \tag{1.22}$$

where M is the total number of users. The term $g(m, k, n)$ is the combined effect of two factors:

- Propagation path loss from the transmitter of user k to the receiver of user m at subcarrier n; this loss also includes the effects of lognormal and Rayleigh fading phenomena.
- Subcarrier amplitude reduction due to the frequency offset Δf.

---

[13] Moose [44] describes an algorithm, based on maximum-likelihood estimation, for frequency offset correction.

[14] In the notation used herein, user m refers to the combination of a transmitter at one end of a wireless link and its intended receiver at the other end of the link.

The $w(m, n)$ denotes the zero-mean Gaussian thermal noise at the receiver input of user m on the nth subcarrier.

Next, we isolate the contribution due to $k = m$ in the summation term in (1.22) and rewrite that equation in the desired form:

$$y(m, n) = g(m, m, n)s(m, n) + \sum_{\substack{k=1 \\ k \neq m}}^{M} g(m, k, n)s(k, n) + w(m, n). \quad (1.23)$$

The first term on the right-hand side of (1.23) is due to user m acting alone. The second term is the *total interference* produced at the receiver of user m due to the signals transmitted by all the other users: $1, 2, \ldots, m - 1, m + 1, \ldots, M$; this interference is attributed to the frequency offset Δf as well as other imperfections in the network.

Let $P(m, n)$ denote the average power transmitted by user m on the nth subcarrier n, and $\sigma_w^2(m, n)$ denote the variance of zero-mean thermal noise $w(m, n)$. We may then express the *signal-to-interference plus noise ratio* (SINR) at the receiver input of user m on the nth subcarrier as

$$\text{SINR}(m, n) = \frac{|g(m, m, n)|^2 P(m, n)}{\displaystyle\sum_{\substack{k=1 \\ k \neq m}}^{M} |g(m, k, n)|^2 |s(k, n)|^2 + \sigma_w^2(m, n)}$$

$$= \frac{P(m, n)}{\displaystyle\sum_{\substack{k=1 \\ k \neq m}}^{M} \alpha(m, k, n)P(k, n) + v(m, n)} \quad (1.24)$$

where, in the last line, the denominator is *normalized* with respect to the factor $|g(m, m, n)|^2$ that pertains completely to user m. Specifically, we have:

$$P(k, n) = |s(k, n)|^2 \quad (1.25)$$

$$\alpha(m, k, n) = \frac{|g(m, k, n)|^2}{|g(m, m, n)|^2} \quad (1.26)$$

and

$$v(m, n) = \frac{\sigma_w^2(m, n)}{|g(m, m, n)|^2}. \quad (1.27)$$

The numerator of (1.24) represents power transmission and reception by user m over a *direct lossless path*. The denominator of this equation represents the normalized value of the total interference plus noise measured at the receiver input of user m.

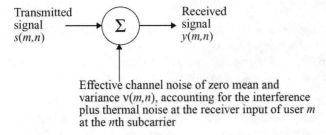

Transmitted signal $s(m,n)$ → Σ → Received signal $y(m,n)$

Effective channel noise of zero mean and variance $v(m,n)$, accounting for the interference plus thermal noise at the receiver input of user m at the nth subcarrier

Fig. 1.4. Depiction of the equivalent additive noise model for user m operating on subcarrier n in the OFDM format.

Examination of this equation also leads us to make another important observation. Insofar as user m of the cognitive radio network is concerned, we may view the wireless channel connecting its receiver to the transmitter as the *equivalent of a single-user additive-noise channel*, as depicted in Fig. 1.4, where the noise variance refers to the variance of total interference plus thermal noise (i.e., the denominator of (1.24)). For analytic purposes, it is assumed that the channel noise in this figure is zero-mean Gaussian. It would be tempting to justify this assumption by recognizing the large number of users responsible for the overall interference, and therefore invoking the central limit theorem. Typically, however, we find that a few of the interferers are dominant and a large number of them are weak. Hence, in reality, the additive noise in the model of Fig. 1.4 may not be strictly Gaussian.

1.10 Formulation of the Transmit-Power Control Problem

Under the assumptions made in Sect. 1.9, we may now invoke *Shannon's* celebrated *information capacity theorem* for an additive Gaussian noise channel [45] to express the maximum achievable rate of data transmission over the wireless channel connecting the transmitter of user m to its receiver as follows:

$$R(m, n) = \log_2[1 + \text{SINR}(m, n)] \quad \text{bits per use of subchannel } n \qquad (1.28)$$

where $\text{SINR}(m, n)$ is the signal-to-interference plus noise ratio defined in (1.24). The multiuser coding scheme needed to achieve the data-transmission rate $R(m, n)$ is implementable, since the only item that needs to be measured is the variance of the interference plus noise at the receiver input of user m for each n. In other words, from a practical perspective, no user in the cognitive radio network would need to identify the sources of interference or noise affecting its operation; rather, it is sufficient for the user to merely measure the variance of the overall interference plus thermal noise at its receiver input for each subcarrier frequency n. This measurement is the function of the radio-scene analyzer to undertake.

Consider then a non-cooperative multiuser cognitive radio network using OFDM for data transmission among its M users. The *transmit-power control problem* for this network may now be stated as follows:

Given:

1. a set of spectrum holes known to be adequate to support the data-transmission needs of M secondary users, and
2. measurements of the variance of interference plus noise at the receiver input at each of the N subcarriers of the OFDM for every user,

determine the transmit-power levels of the M secondary users so as to jointly maximize their data-transmission rates, subject to the constraint that the interference-temperature limits in the subfrequency bands occupied by the spectrum holes are not violated.

It may be tempting to suggest that the solution of this problem is attained by simply increasing the transmit-power level of each secondary user. However, increasing the transmit-power level of any one user has the undesirable effect of also increasing the levels of interference to which the receivers of all the other users are subjected. The conclusion to be drawn from this reality is that it does not make practical sense to represent the overall performance of the cognitive radio network by means of a single index of performance. Rather, we have to adopt a *tradeoff* among the data rates of all secondary users in some computationally tractable fashion.

Ideally, we would like to find a global solution to the constrained optimization of the joint set of data-transmission rates under study. Unfortunately, finding this *global* solution requires an exhaustive search through the space of possible power allocations for all M users, which is impractical for two reasons:

- The computational complexity needed to attain the global solution may assume a prohibitively high level.
- The time needed to find the solution could become unacceptably long.

To mitigate these practical difficulties, we relax the statement for global optimality by adopting *competitive optimality* as the criterion for solving the transmit-power control problem. Specifically, we now state:

> Given a multiuser non-cooperative cognitive radio network using OFDM as described above, optimize the performance of secondary user m, regardless of what all the other secondary users do, subject to the constraint that the interference-temperature limit at the receiver input of user m is not violated.

This formulation of the distributed transmit-power control problem leads to a solution that is of a local nature. Although, of course, the solution is suboptimum, it is not only insightful but also practically feasible. Most important, the *local* optimization envisioned here is a basic ingredient of self-organization.

To set the stage for presenting an iterative procedure (based on competitive optimality) for solving the transmit-power control problem, we find it informative to digress briefly to first think in terms of game theory.

1.11 The Multiuser Non-cooperative Cognitive Radio Network Viewed as a Game-Theoretic Problem

Game theory[15] is a well-established discipline; it deals with the mathematical modeling of practical situations, which involve the following ingredients:

- *Multiple players* who, by virtue of their responsibilities as decision-makers, are required to take specific *actions*.
- The actions may lead to consequences, which could be of mutual conflict to the players themselves.

The formulation of a mathematical framework for a non-cooperative game rests on three basic realities:

- *State space* that is the product of the individual players' states
- *State transitions* that are functions of *joint actions* taken by the players
- *Payoffs* to individual players that depend on joint actions as well

This framework is found in stochastic games [46], which, also occasionally appear under the name "Markov games" in the computer science literature.

A stochastic game is described by the five tuple $\{\mathcal{N}, \mathcal{S}, \overrightarrow{\mathcal{A}}, \mathcal{P}, \overrightarrow{\mathcal{R}}\}$, where

- $\mathcal{N}$ is a set of players, indexed $1, 2, \ldots, M$.
- $\mathcal{S}$ is a set of possible states.
- $\overrightarrow{\mathcal{A}}$ is the *joint-action space* defined by the product set $\overrightarrow{\mathcal{A}}_1 \times \overrightarrow{\mathcal{A}}_2 \times \ldots \times \overrightarrow{\mathcal{A}}_M$, where $\overrightarrow{\mathcal{A}}_m$ is the set of actions available to the mth player.
- $\mathcal{P}$ is a probabilistic transition function, an element of which for joint action a satisfies the condition

$$\sum_{s \in \mathcal{S}} \mathcal{P}^a_{ss'} = 1 \quad \text{for all } s' \in \mathcal{S} \text{ and } a \in \overrightarrow{\mathcal{A}}. \tag{1.29}$$

- $\overrightarrow{\mathcal{R}} = r_1 \times r_2 \times \ldots \times r_M$, where r_m is the payoff for the mth player and which is a function of the joint actions of all M players.

One other notational issue: the action of player $m \in \mathcal{M}$ is denoted by a_m, while the joint actions of the other $M - 1$ players in the set $\mathcal{M}$ are denoted by a_{-m}.

Stochastic games are the *supersets* of two kinds of decision processes, namely, *Markov decision process* and matrix games, as illustrated in Fig. 1.5. A *Markov decision process* is a special case of a stochastic game with a single player, that is, $M = 1$. On the other hand, a *matrix game* is a special case of a stochastic game with a single state, that is, $|\mathcal{S}| = 1$.

---

[15] In a historical context, the formulation of game theory may be traced back to the pioneering work of John von Neumann in the 1930s, which culminated in the publication of the co-authored book entitled "Theory of Games and Economic Behavior" [47]. For modern treatments of game theory, see the books under [46,48]. Game theory is widely used in the study of economics [49]; it has also been applied in other areas such as machine learning [50] and neuroscience [51]. For the use of game theory in cognitive radio, see [52].

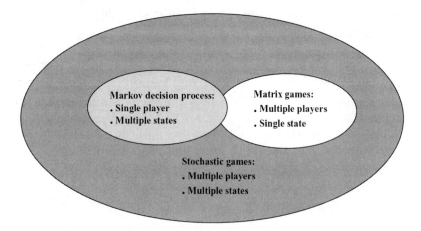

Fig. 1.5. Highlighting the differences between Markov decision processes, matrix games and stochastic games.

1.11.1 Nash Equilibrium

In [53,54], John Nash focused his study of game theory on a class of games described as *non-cooperative, simultaneous-move, one-shot, finite games with complete information*, where

- "Simultaneous-move" means that each player picks a strategy without knowledge of the other players' strategies
- "One-shot" implies that the game is played once and once only
- "Finite game" refers to the fact the game involves a finite number of players, with each player taking only a finite number of possible actions

In the context of this background, Nash introduced the concept of an equilibrium of a game, which is defined as follows:

> A *Nash equilibrium* is defined as an action profile (i.e., vector of players' actions) in which each action is a *best response* to the actions of all the other players [53].

Consider, for example, a *multiple-access game* [55] involving two transmitters (i.e., players) p_1 and p_2 who respectively want to send data packets to their receivers r_1 and r_2 over a shared channel. In each time slot, each player can decide to transmit a packet or to remain quiet (i.e., not to transmit); these two actions are denoted by T and Q, respectively. Let c denote the *cost* incurred in the transmission of a packet, where $0 < c < 1$. With the channel being shared, transmissions by both players result in a collision, in which case, packets are lost. Thus, in strategic terms, the multiple-access is represented by Fig. 1.6. From this figure, it is apparent that the optimal solution to the multiple-access game is as follows:

p_1/p_2	Q	T
Q	$(0,0)$	$(0, 1\text{-}c)$
0	$(1\text{-}c, 0)$	(c, c)

Fig. 1.6. Tabular representation of the multiple-access game.

- If player p_1 decides to transmit, then the best response for player p_2 is to remain quiet.
- Conversely, if player p_2 decides to transmit, the best response for player p_1 is to remain quiet.

From this example, we see that Nash equilibrium is a *stable operating* (i.e., *equilibrium) point* in the sense that there is no incentive for any player involved in a finite game to change strategy, given that all the other players continue to follow the equilibrium policy. The important point to note here is that the Nash-equilibrium approach provides a powerful tool for modeling non-stationary processes. Simply put, it has had an enormous influence on the evolution of game theory by shifting its emphasis toward the study of equilibria as a *predictive concept*.

The Nash equilibrium features prominently in the study of game theory; indeed, it earned John Nash the Nobel Prize in Economics in 1994. This concept works perfectly well provided two assumptions are satisfied:

1. The players engaged in a game are all *rational.*
2. The underlying structure of the game is *common knowledge* to all the players.

Under these two assumptions, the Nash equilibrium offers an intuitively satisfying approach that predicts the equilibrium outcome of the game as follows. Any player, being "rational," will play a "best-response" strategy. Moreover, under the "common knowledge" assumption, this strategy is known to all the other players and, being rational, they will therefore play their own "best-response" strategies, which therefore leads the game to a Nash equilibrium.

1.12 Iterative Waterfilling Algorithm

Now that we understand the importance of the Nash equilibrium in the study of game theory, we can proceed with the solution to the transmit-power control problem in a non-cooperative multiuser cognitive radio network using OFDM. We begin with the statement:

> When users of such a network operate under the common knowledge that each user will follow the criterion of competitive optimality for maximizing its own data-transmission rate, subject to an interference-temperature constraint, the strategy so adopted will lead to a Nash equilibrium.

In information-theoretic terms, maximization of the data-transmission rate of each user in accordance with (1.28) over each subcarrier frequency of the user is similar to the idea of *waterfilling*. In the classical description of waterfilling [45], water is poured over the inverse of noise variance at each subcarrier frequency. In our situation, on the other hand, water is poured over the inverse of the combined interference plus noise. We may therefore modify the above statement as follows:

> If the users of a non-cooperative radio network perform "waterfilling" with respect to the combined variance of interference plus noise at each subcarrier frequency of the OFDM, subject to an interference-temperature constraint, then the network will reach a Nash equilibrium.

Although the multiuser solution produced by this strategy is *suboptimal*, it offers the practical virtue of eliminating the need for synchronization among users of the network insofar as transmit-power control is concerned. The fundamental question is: How do we perform the waterfilling procedure in an efficient manner?

This very question is addressed in the *iterative waterfilling algorithm* for multiuser data transmission systems. The algorithm was originally described in [56,57] in the context of discrete multitones (DMT); it is expanded on in [58]. Much of that theory is also applicable to frequency-selective channels using OFDM, since DMT and OFDM belong to the same family of multichannel transmission systems [13,59].

To simplify the presentation of the iterative waterfilling algorithm for a multiuser cognitive radio network using OFDM, we assume that each iteration of the algorithm starts with user 1 and ends with user M. Each iteration consists of an inner loop followed by an outer loop. In the inner loop of iteration j, say, each user maximizes its data transmission rate, subject to an interference-temperature constraint. In the outer loop of iteration j, the power allocation among the M users is adjusted up or down. The iterative waterfilling computation is terminated after a total of J iterations when a prescribed tolerance ϵ is attained.

Note also that at iteration j, the interference plus noise (IN) at the receiver input of receiver m at subcarrier frequency n has the fixed value [see the denominator of (1.24)]

$$\mathrm{IN}^{(j)}(m, n) = \sum_{k=1}^{m-1} \alpha(m, k, n) P^{(j)}(k, n)$$

$$+ \sum_{k=m+1}^{M} \alpha(m, k, n) P^{(j-1)}(k, n) + v^{(j)}(m, n). \qquad (1.30)$$

The first summation term of (1.30) represents the normalized contributions made by users 1 to $m - 1$ processed during the current iteration j, and the second summation term represents the normalized contributions made by users $m + 1$ to M processed during the previous iteration $j - 1$. With $\mathrm{IN}^{(j)}(m, n)$ fixed, it follows that placing a limit on the total interference temperature at the receiver input of user m is actually equivalent to the imposition of a corresponding limit on the total transmit power of user m.

We are now ready to describe the iterative waterfilling algorithm as follows:

1. *Initialization $j = 0$*
   Unless prior knowledge is available, the power distribution across the users, $m = 1, 2, \ldots, M$, is set equal to zero.
2. *Inner loop: iteration $j = 1, 2, \ldots$*
   In this iteration, user m maximizes its total data transmission rate through waterfilling, subject to a total power constraint. In mathematical terms, for iteration j we write:

$$\text{Maximize} \quad R^{(j)}(m) = \sum_{n=1}^{N} \log_2 \left(1 + \frac{P^{(j)}(m, n)}{\text{IN}^{(j)}(m, n)} \right)$$

$$\text{subject to the constraint} \sum_{n=1}^{N} P^{(j)}(m, n) \le \bar{P}(m) \tag{1.31}$$

where the permissible transmitter power $\bar{P}(m)$ is determined as follows: the total power measured at the receiver input of user m, summing the contributions due to its own transmission and ambient noise plus those due to the remaining $M - 1$ interferers, is defined by:

$$P_{\text{total}}(m) = \sum_{n=1}^{N} \left(\sum_{k=1}^{M} |g(m, m, n)|^2 P(k, n) + \sigma_w^2(m, n) \right). \tag{1.32}$$

Given that the interference-temperature limit $T_{\max}$ must not be exceeded by user m, we require:

$$P_{\text{total}}(m) \le \kappa T_{\max} B_m \tag{1.33}$$

where κ is Boltzmann's constant and B_m is the bandwidth of the spectrum hole being occupied by user m. Using the definition of (1.27) and (1.28) and recognizing that $\alpha(m, m, n) = 1$ for all n, we write:

$$\sum_{n=1}^{N} P(m, n) \le \bar{P}(m) \tag{1.34}$$

where $\bar{P}(m)$ is defined by:

$$\bar{P}(m) = \frac{\kappa T_{\max} B_m}{|g(m, m, n)|^2} - \sum_{n=1}^{N} \left(\sum_{\substack{k=1 \\ k \ne m}}^{M} \alpha(m, k, n) P(k, n) + v(m, n) \right). \tag{1.35}$$

Here it is presumed that the spectrum hole being occupied by user m is , at least, partially filled to permit $\bar{P}(m)$ to assume a positive value. Bearing in mind that cognitive radio is receiver-centric, the determination of $\bar{P}(m)$ requires knowledge of two quantities for user m:

(i) *Total interference plus noise* measured at its own receiver input.

(ii) The *path loss* $|g(m, m, n)|^2$ from its transmitter to the receiver.

The measurement of item (i) is performed at the receiver and supplied to the respective transmitter via the feedback channel. The calculation of item (ii) is performed by the transmitter itself, knowing how far away its own receiver is from it.

The above-stated constrained maximization problem is a *convex optimization problem*, coupled across the set of N subcarrier frequencies. It may therefore be solved using dual decomposition [60]. Specifically, we first set up the Lagrangian

$$L^{(j)}(m) = \sum_{n=1}^{N} \log_2 \left(1 + \frac{P^{(j)}(m, n)}{\text{IN}^{(j)}(m, n)} \right) - \lambda^{(j)}(m) \left(\sum_{n=1}^{N} P^{(j)}(m, n) - \bar{P}(m) \right) \quad (1.36)$$

where $\lambda^{(j)}(m)$ is the Lagrangian multiplier for user m at iteration j. Next, invoking the orthogonality property of the OFDM subcarriers, the convex optimization problem is decomposed into N suboptimization problems, as shown by

$$\text{Maximize} \quad \log_2 \left(1 + \frac{P^{(j)}(m, n)}{\text{IN}^{(j)}(m, n)} \right) - \lambda^{(j)}(m) P^{(j)}(m, n) \quad (1.37)$$

$$\text{for} \quad n = 1, 2, \ldots, N.$$

Solutions of this optimization are obtained by waterfilling [45]. A subgradient search is used to find the optimal value of the Lagrange multiplier $\lambda^{(j)}(m)$ for each user m; this optimal value is denoted by $\lambda^{*(j)}(m)$.

3. *Outer loop: iteration* $j = 1, 2, \ldots$

After the inner loop of iteration j is completed, the power allocation among the M users is adjusted. Specifically, for user m the optimal power

$$P^{*(j)}(m, n) = \left(\frac{1}{\lambda^{*(j)}(m)} - \text{IN}^{(j)}(m, n) \right)^{\dagger} \quad (1.38)$$

is computed, such that the total power constraint

$$\sum_{n=1}^{N} P^{*(j)}(m, n) = \bar{P}(m)$$

is satisfied.

4. *Confirmation step*

After the power adjustments for the M users have been made, the condition

$$\sum_{m=1}^{M} \sum_{n=1}^{N} |P^{(j)}(m, n) - P^{(j-1)}(m, n)| < \epsilon \quad (1.39)$$

is checked for the prescribed tolerance ϵ at iteration j. If this *tolerable condition* is satisfied, the computation is terminated at $j = J$. Otherwise, the iterative process (encompassing both the inner and outer loops) is repeated.

1.12.1 Robustification of the Algorithm

In describing the iterative waterfilling algorithm, we have made a fundamental assumption:

- The normalized parameter $\alpha(m, k, n)$, denoting the combined effect of (1) propagation-path loss from the transmitter of user k to the receiver of user m at subcarrier n, and (2) frequency offset in the OFDM, is maintained constant throughout the entire sequence of iterations $j = 1, 2, ..., J$ of the algorithm.

This assumption is highly likely to be violated in practice, particularly when dealing with a *rapidly changing wireless channel*. It could also be aggravated by variations in the frequency offset Δf with time. The implication of these realities is that the multiuser cognitive radio problem should be modeled as a *partially observable Markov decision process*. For a possible cure, we could mitigate the effect of these sources of uncertainty by including a *signal-to-noise ratio gap* in formulating the data-transmission rate of each user $m = 1, 2, \ldots, M$. In effect, this gap is chosen large enough to provide reliable communication under practical operating conditions of the multiuser cognitive radio environment. Let the signal-to-noise ratio gap be denoted by Γ. We then rewrite the information capacity formula of (1.28) in the expanded form

$$R(m, n) = \log_2 \left[1 + \frac{\text{SINR}(m, n)}{\Gamma} \right] \quad \text{bits per use of subchannel } n \qquad (1.40)$$

which applies to all users $m = 1, 2, \ldots, M$ and subcarrier frequencies $n = 1, 2, \ldots, N$. Accordingly, the iterative waterfilling procedure is modified in a corresponding way.

1.12.2 Summarizing Remarks

Based on the criterion of competitive optimality, the iterative waterfilling algorithm is *user-centric* and therefore a *selfish*, *greedy*, and *sub-optimal algorithm* for solving the transmit-power control problem in a multiuser cognitive radio network using OFDM. Nevertheless, practical virtues of the algorithm include:

- The algorithm functions in a self-organized manner, thereby making it possible for the network to assume an ad hoc structure.
- It avoids the need for communication links (i.e., synchronization) among the multiple users, thereby significantly simplifying the design of the network.
- By using convex optimization [60], the algorithm tends to converge relatively rapidly to a Nash equilibrium; however, once this stable condition is reached, no user is permitted to change its transmit-power control policy unilaterally.
- Computational complexity of the algorithm is relatively low, being on the order of two numbers: the number of secondary users and the number of spectrum holes available for utilization.

1.13 Emergent Behavior of Cognitive Radio Networks

In light of the material presented in the preceding sections, we may characterize the multiuser cognitive radio network as a *complex, stochastic* and *time-varying feedback control system* that exhibits the following unique combination of attributes (among others): partial observability, adaptivity, learning, self-organization, cooperation, competition and exploitation. Given this characterization, we may wonder about the emergent behavior of a cognitive radio environment by virtue of what we know on two relevant fields: *self-organizing systems* and *evolutionary games.*

First, we note that the emergent behavior of a cognitive radio environment, viewed as a self-organized network is influenced by the *degree of coupling* that may exist between the actions of different users (i.e., transmitter–receiver linkages) operating in the network. The coupling may have the effect of *amplifying* local perturbations in a manner analogous with *Hebb's postulate of learning*, which accounts for self-amplification in self-organizing systems [5]. Clearly, if they are left unchecked, the amplifications of local perturbations would ultimately lead to *instability*. From the study of self-organizing systems, we also know that competition among the constituents of such a system can act as a stabilizing force [5]. By the same token, we expect that competition among the cognitive radio users for limited resources (e.g., transmitted power) may have the influence of a *stabilizer*, provided, of course, that the competition is carried out on the basis of the common application of the competitive optimality criterion by all the users. However, the tendency of one or more users to exploit the limited resources for selfish interests may drive the network into instability and possibly a chaotic state.[16]

For additional insight, we next look to evolutionary games. The idea of evolutionary games, developed for the study of ecological biology, was first introduced by Maynard Smith in 1974. In his landmark work [62,63], Maynard Smith wondered whether the theory of games could serve as a tool for modeling conflicts in a population of animals. In specific terms, two critical insights into the emergence of so-called *evolutionary stable strategies* were presented by Maynard Smith, as succinctly summarized in [51,63]:

- The animals' behavior is stochastic and unpredictable, when it is viewed at the microscopic level of actions taken by individual animals.
- The theory of games provides a plausible basis for explaining the complex and unpredictable patterns of the animals' behavior.

---

[16] The traditional method of studying the stability of a time-varying feedback control system is to apply the Lyapunov stability theory [61]. To apply this theory, we need to formulate a Lyapunov function for a multiuser cognitive radio network, which can be a hard task to do. The problem is complicated further by the stochastic nature of the network. For these reasons, we advocate the approach described in this section on evolutionary games.

Two key issues are raised here:

1. *Complexity.*[17] The emergent behavior of an evolutionary game may be *complex*, in the sense that a change in one or more of the parameters in the underlying dynamics of the game can produce a dramatic change in behavior. Note that the dynamics must be nonlinear for complex behavior to be possible.
2. *Unpredictability.* Game theory does not require that animals be fundamentally unpredictable. Rather, it merely requires that the individual behavior of each animal be *unpredictable with respect to its opponents.*

From this brief discussion on evolutionary games, we may conjecture that the emergent behavior of a multiuser cognitive radio network is explained by the unpredictable action of each user, as seen individually by the other users (i.e., opponents).

1.13.1 State of the World

In light of the conflicting influences of cooperation, competition and exploitation on the emergent behavior of a cognitive radio environment, we may identify two possible end-results for the state of the (wireless) world [64]:

1. *Positive emergent behavior*, which is characterized by *order*, and therefore a harmonious and efficient utilization of the radio spectrum by all primary and secondary users of the cognitive radio. (The positive emergent behavior may be likened to Maynard Smith's evolutionary stable strategy).
2. *Negative emergent behavior*, which is characterized by *disorder*, and therefore a culmination of traffic jams, chaos,[18] and therefore unused radio spectrum.

From a practical perspective, what we therefore need are, first, a reliable criterion for the early detection of negative emergent behavior (i.e., disorder) and, second, corrective measures for dealing with this undesirable behavior. With regards to the first issue, we recognize that cognition, in a sense, is an exercise in assigning probabilities to possible behavioral responses, in light of which we may say the following. In the case of positive emergent behavior, predictions are possible with nearly complete confidence. On the other hand, in the case of negative emergent behavior, predictions

---

[17] The new sciences of complexity (whose birth was assisted by the Santa Fe Institute, New Mexico) may well occupy much of the intellectual activities in the twenty-first century [64–67]. In the context of complexity, it is perhaps less ambiguous to speak of complex behavior rather than complex systems [68]. A non-linear dynamic system may be complex in computational terms, but it is incapable of exhibiting complex behavior. By the same token, a non-linear system can be simple in computational terms, but its underlying dynamics are rich enough to produce complex behavior.

[18] The possibility of characterizing negative emergent behavior as a chaotic phenomenon needs some explanation. Idealized chaos theory is based on the premise that dynamic noise in the state-space model (describing the phenomenon of interest) is zero. However, it is unlikely that this highly restrictive condition is satisfied by real-life physical phenomena. So, the proper thing to say is that it is feasible for a negative emergent behavior to be *stochastic chaotic* [69].

are made with less confidence. There is therefore the need to formulate a likelihood function based on predictability as a criterion for the onset of negative emergent behavior. The key question is how to do it effectively and efficiently?

Given a multiuser non-cooperative cognitive radio network based on OFDM and designed along the lines described in Sect. 1.13 on iterative waterfilling, we know the following: when all the users of the network use competitive optimality as their common criterion to satisfy their individual transmit-power control requirements, the network will reach a Nash equilibrium, that is, an orderly behavior throughout the network. On the other hand, when any of the users exploit the limited resources (i.e., transmitted power and spectrum holes) for selfish interest, there is the likelihood that the network will assume a disorderly behavior. It would therefore seem logical to look to the Nash equilibrium as the basis for designing a maximum-likelihood processor capable of detecting the emergence of disorderly behavior in the network; recall that the Nash equilibrium is a prediction concept.

To summarize, what we are advocating here is an expansion of the game-theoretic viewpoint of multiuser cognitive radio networks to embrace evolutionary games as described originally by Maynard-Smith. By so doing, we may be able to quantify the predictability of individual users' behavior. In particular, the expansion could facilitate the design and development of a maximum-likelihood processor for detecting the onset of the disorderly utilization of limited resources in the network due to the misbehavior of one or more users.

1.14 Distributed Traffic Coordination in Cognitive Radio Networks

The material presented up to this point in this chapter has focused on signal-processing and communication-theoretic issues relating to the identification of spectrum holes, the extraction of channel-state information, dynamic spectrum management, and transmit-power control. With the emphasis on a self-organized ad hoc network as the structure for building a cognitive radio network, we need a protocol for the distributed traffic coordination of secondary users of the network in such an environment. Needless to say, the development of this protocol is a challenging task.

Basically, the issue to be addressed is summed up as follows:

In a self-organized and decentralized cognitive radio network, how can we establish the dissemination of control traffic signals between neighboring secondary users of the network, which is rapid, robust and efficient?

The requirement that the dissemination of control traffic signals be *rapid* is essential, because the secondary user could be faced with a limited duration of time for which spectrum holes are likely to be available. The dissemination has to be robust with respect to external attack not only for reasons of security but also to prevent disruptions in network use due to traffic congestion. Lastly, it has to be *efficient* so as to minimize the use of energy and computing resources.

For self-organized coordination among neighboring secondary users to be feasible, we expect two provisions:

1. Each user has knowledge of all the spectrum holes that are locally available.
2. Neighboring users have reasonably similar views of their respective spectral scenes, so as to guarantee the availability of common wireless channels.

Given these two provisions, which are provided by the radio-scene analyzer, we may then envision a *self-organized traffic-coordination* protocol that proceeds as follows [70]:

1. By exploiting the availability of similar spectrum holes, each local group of neighboring users forms a *mini-multihop network* with a common coordination channel. This could be achieved through broadcasting a beacon and recursive voting procedure, whereby the channel with the highest level of connectivity is selected by all the users in the mini-multihop network as the common coordination channel.
2. Through the availability of one or more spectrum holes common to adjoining mini-multihop networks, communication across the cognitive radio network is established.
3. Through eavesdropping on coordination messages from "bridge" users, a new user may join an existing mini-multihop network and thereby quickly subscribe to appropriate channels.

In a loose sense, point 1 of the procedure described herein is similar to what goes on in the formation of a self-organizing map in neural networks [5].

For an alternative solution to the traffic-coordination problem, one may look to the use of an out-of-band licensed channel as the dedicated common channel. In [70], Zhao, Zheng and Yang present simulation results that compare the performance of the self-organized coordination approach against an approach based on the dedicated common channel; the results presented therein appear to show that the self-organized approach outperforms the dedicated common channel approach, in terms of both throughput and processing delay.

Conclusion

Cognitive radio holds the promise of a new and exciting frontier in wireless communications. Most importantly, the development of an orderly dynamic spectrum-sharing process will make it possible to improve the utilization of radio spectrum under constantly changing user conditions. For the spectrum-sharing process to become a reality, two basic issues have to be in place:

1. There has to be a paradigm shift in wireless communications from transmitter-centricity to receiver-centricity, which, in turn, means that interference power at the receiver rather than transmitted power at the transmitter is regulated.

2. A new generation of wireless communication systems is developed, in which awareness of the radio environment and the ability to adapt to the environment and learn from it feature prominently.

Specifically, from a signal-processing and communication-theoretic perspective, we need to develop new algorithms that operate satisfactorily and in a robust manner in a wireless communications environment to perform the following functions:

- Identification of spectrum holes for employment by secondary users
- Channel-state estimation for improved utilization of the radio spectrum
- Adaptive modulation format that is impervious not only to the modulation format of the primary user but also to varying received signal-to-noise conditions
- Transmit-power control to support the transmission needs of multiple users
- Development of a decentralized radio network that is efficient in the use of resources and effective in performance
- Detection of the onset of instability whenever the network is misused
- Coordination of distributed traffic in the network.

The ideas and algorithms described in this chapter (building on [1]) should be viewed as starting points for a long road ahead, which will occupy the ingenuity and extensive research and development efforts of numerous researchers.

This immense effort is justified, given the potential of cognitive radio to make a significant difference to wireless communications; hence the reference to it as a "disruptive, but unobtrusive technology." In the final analysis, however, the key issue that will shape the evolution of cognitive radio in the course of time, be that for civilian or military applications, is trust. By this we mean, trust by users of cognitive radio, and trust by all other users who could be interfered with. For this trust to be a reality, cognitive radio will not only have to improve spectrum utilization but also do so in a robust, reliable and affordable manner.

References

1. S. Haykin, "Cognitive dynamic systems," *Proc. IEEE*, vol. 94, pp. 1910–1911, Nov. 2006.
2. S. Haykin, "Cognitive radio: Brain-empowered wireless communications," *IEEE J. Select. Areas Commun.*, vol. 23, no. 2, pp. 201–220, Feb. 2005.
3. R. Pfeifer and C. Scheier, *Understanding intelligence*, pp. 5–6. MIT Press, 1999.
4. S. Haykin, *Neural networks: A comprehensive foundation*, 2nd ed. Prentice-Hall, 1999.
5. P. N. Johnson-Laird, *The Computer and the mind: An introduction to cognitive science.* Harvard University Press, 1988.
6. J. Mitola, "Cognitive radio: An integrated agent architecture for software defined radio," Dissertation, Doctor of Technology, Royal Institute of Technology (KTH), Sweden, May 8, 2000.
7. J. Mitola, *Cognitive radio architecture: The engineering foundations of radio XML.* Wiley, 2006.
8. Federal Communications Commission, "Spectrum Policy Task Force," Report ET Docket No. 02,135, Nov. 2002.
9. "Wolfram Research." http://scienceworld.Wolfram.com/physics/antennatemperature.html.

10. B. Bale et al., "Noise in wireless systems produced by solar radio bursts," *Radio Sci.*, vol. 37, 2002.

11. L. J. Lanzerotti et al., "Engineering issues in space weather," in M. A. Stucthly, editor, *Modern Radio Science*, pp. 25–50, Oxford University Press, 1999.

12. S. Haykin and M. Moher, *Introduction to analog and digital communications*. Wiley, 2001.

13. S. Haykin, *Communication systems*, 4th ed., p. 61. Wiley, 2001.

14. M. Loève, "Fonctions alatoires de second ordre," *Rev. Sci., Paris,* vol. 84, pp. 195–206, 1946.

15. M. Loève, *Probability theory.* Van Nostrand, 1963.

16. L. Cohen, *Time–frequency analysis.* Prentice-Hall, 1995.

17. Lord Rayleigh, "On the spectrum of an irregular disturbance," *Philos. Mag.*, vol. 41, pp. 238–243, 1903. (Note: This paper is reproduced in the Scientific Papers by Lord Rayleigh, Volume V, Article 285, pp. 98–102, Dover Publications, 1964.)

18. D. J. Thomson,"Spectrum estimation and harmonic analysis," *Proc. IEEE*, vol. 20, pp. 1055–1096, Sept. 1982.

19. P. D. Welch, "The use of fast Fourier transform for the estimation of power spectra: A method based on time-averaging over short, modified periodograms," *IEEE Trans. Audio Electroacoust.*, vol. AU-15, pp. 70–73, 1967.

20. D. B. Percival and A. T. Walden, *Spectral analysis for physical applications.* Cambridge University Press, 1993.

21. D. Slepian, "Prolate spheroidal wave functions, Fourier analysis and uncertainty", *Bell Syst. Tech. J.*, vol. 57, pp. 1371–1430, 1978.

22. A. Drosopoulos and S. Haykin,"Angle-of-arrival estimation in the presence of multipath," in S. Haykin, editor, *Adaptive Radar Signal Processing*, pp. 11–89, Wiley, 2007.

23. D. J. Thomson and A. D. Chave, "Jackknifed error estimates for spectra, coherences, and transfer functions," in S. Haykin, editor, *Advances in Spectrum Analysis and Array Processing*, vol. 1, pp. 58–113, Prentice-Hall, 1991.

24. P. Stoica and T. Sundin, "On nonparametric spectral estimation," *Circuits Syst. Signal Process.*, vol. 16, pp. 169–181, 1999.

25. D. J. Thomson and S. Haykin, "Time-frequency analysis of sea clutter," in S. Haykin, editor, *Adaptive Radar Signal Processing*, pp. 91–115, Wiley, 2007.

26. M. E. Mann and J. Park, "Oscillatory spatiotemporal signal detection in climate studies: A multiple-taper spectral domain approach," in R. Dnowska and B. Saltzman, editors, *Advances in Geophysics*, vol. 41, pp. 1–131, Academic Press, 1999.

27. G. H. Golub and C. F. VanLoan, *Matrix computations*, 3rd ed. Johns Hopkins University Press, 1996.

28. G. J. Foschini and M. J. Gans, "On limits of wireless communications in a fading environment when using multiple antennas," *Wireless Pers. Commun.*, vol. 6, pp. 311–335, 1998.

29. S. Haykin, K. Huber, and Z. Chen, "Bayesian sequential state estimation for MIMO wireless communications," *Proc IEEE*, Special Issue on Sequential State Estimation, vol. 92, pp. 439–454, 2004.

30. I. Arasarathnam and S. Haykin, "Improved channel tracking for wireless PAT," submitted to *IEEE Trans. Commun.*

31. A. Ephirenides and T. Truong, "Schedule broadcasts in multihop radio networks," *IEEE Trans. Commun.*, vol. 38, pp. 456–460, 1990.

32. K. Scott and N. Bambos, "Formation and maintenance of self-organizing wireless networks," Conference Record 3rd Asilomar Conference on Signals, Systems, and Computers, vol. 1, pp. 31–35, Nov. 1997.

33. C. E. Perkins, *Ad hoc networking*. Addison-Wesley, 2001.
34. O. K. Tonguz and G. Ferrari, *Ad hoc wireless networks*. Wiley, 2006.
35. M. A. Arbib, *The handbook of brain theory and neural networks*, 2nd ed. MIT Press, 2003.
36. T. J. Shepard, "Decentralized channel management in scalable multihop spread-spectrum packet radio networks," PhD Thesis, MIT, July 1995.
37. P. Gupta and P. R. Kumar, "The capacity of wireless networks," *IEEE Trans. Inf. Theory*, vol. 46, no. 2, pp. 388–404, 2000.
38. P. Gupta and P. R. Kumer, "Internets in the sky: The capacity of three-dimensional wireless networks," *Commun. Inf. Syst.*, vol. 1, pp. 39–49, 2001.
39. S. Haykin and M. Moher, *Modern wireless communications*. Prentice-Hall, 2003.
40. L. Hanzo and T. Keller, *OFDM and MC-CDMA*. Wiley, 2006.
41. J. A. C. Bingham, *ADSL, VDSL, and multicarrier modulation*. Wiley, 2000.
42. C. Berrou, "The ten-year old turbo codes are entering into service," *IEEE Commun. Mag.*, vol. 42, pp. 110–116, Aug. 2003.
43. V. Tarokh, H. Jafarkhni, and A. R. Calderbank, "Space–time block coding for wireless communication: Performance results," *IEEE J. Select. Areas Commun.*, vol. 17, pp. 451–460, May 1999.
44. P. Moose, "A technique for orthogonal frequency division multiplexing frequency offset correction," *IEEE Trans. Commun.*, vol. 42, pp. 2908–2914, 1994.
45. T. M. Cover and J. A. Thomas, *Elements of information theory*. Wiley, 1991.
46. J. von Neumann and O. Morgenstein, *Theory of games and economic behavior*. Princeton University Press, 1947.
47. D. Fudenberg and D. K. Levine, *The theory of learning in games*. MIT Press, 1999.
48. T. Basar and G. J. Olsder, *Dynamic noncooperative game theory*, 2nd ed. SIAM, 1999.
49. M. J. Osborne and A. Rubinstein, *A course in game theory*. MIT Press, 1994.
50. G. Gordon, "No-regret algorithms for structured prediction problems," Technical report 112, Carnegie-Mellon University, Center for Automated Learning and Discovery, 2005.
51. P. W. Glimcher, *Decisions, uncertainty, and the brain: The science of neuroeconomics*. MIT Press, 2003.
52. A. B. McKenzie, L. Dasilva, and W. Tranter, *Game theory for wireless engineers*. Morgan and Claypool Publishers, 2006.
53. J. F. Nash, "Non-cooperative games," *Ann. Math.*, vol. 54, pp. 286–295, 1951.
54. J. F. Nash, "Equilibrium points in n-person games," in *Proc. Natl Acad. Sci.*, vol. 36, pp. 48–49, 1950.
55. M. Felegyhazi and J. P. Hubaux, "Game theory in wireless networks: A tutorial," EPFL Technical report, LCA-REPORT-2006-002, EPFL, Switzerland.
56. W. Yu, "Competition and cooperation in multi-user communication environments," Doctoral Dissertation, Stanford University, 2002.
57. W. Yu, G. Ginis, and J. M. Cioffi, "Distributed multiuser power control for digital subscriber lines," *IEEE J. Select. Areas Commun.*, vol. 20, pp. 1105–1115, June 2002.
58. S. T. Chung, "Transmission schemes for frequency selective Gaussian interference channels," Dissertation, Doctor of Philosophy, Stanford University, CA, Nov. 2003.
59. T. Starr, J. M. Cioffi, and P. J. Silverman, *Understanding digital subcarrier line technology*. Prentice-Hall, 1999.
60. A. Boyd and L. Vandenbarghe, *Convex optimization*. Cambridge University Press, 2004.
61. H. K. Khalil, *Nonlinear systems*. Prentice-Hall, 1992.
62. J. Maynard Smith, "The theory or games and the evolution of animal conflicts," *J. Theor. Biol.*, vol. 47, pp. 209–221, 1974.

63. J. Maynard Smith, *Evolution and the theory of games.* Cambridge University Press, 1982.
64. H. G. Schuster, *Complex adaptive systems: An introduction.* Springer-Verlag, 2001.
65. D. L. Stein, editor, *Lectures in the Sciences of Complexity.* Addison-Wesley, 1989.
66. E. Jen, editor, 1989 *Lectures in Complex Systems.* Addison-Wesley, 1990.
67. G. G. Weisbunch, *Complex system dynamics.* Addison-Wesley, 1991.
68. G. Nicolis and I. Progogine, *Exploring complexity: An introduction.* W. H. Freeman and Company, 1989.
69. S. Haykin, editor, *Adaptive Radar Signal Processing*, pp. 153–155. Wiley, 2007.
70. J. Zhao, H. Zheng, and G. H. Yang, "Distributed coordination in dynamic spectrum allocation networks," in *IEEE Workshop* (Baltimore, MA), pp. 259–268, 2005.

2

Information Theoretic Analysis of Cognitive Radio Systems

Natasha Devroye[1], Patrick Mitran[1], Masoud Sharif[2], Saeed Ghassemzadeh[3], and Vahid Tarokh[1]

[1] Division of Engineering and Applied Sciences, Harvard University, USA
{ndevroye, mitran, vahid}@deas.harvard.edu
[2] Department of Electrical and Computer Engineering, Boston University, USA
sharif@bu.edu
[3] Communications Technology Research Department, AT&T Labs-Research, USA
saeedg@research.att.com

2.1 Introduction

Cognitive radios have recently emerged as a prime candidates for exploiting the increasingly flexible licensing of wireless spectrum. Regulatory bodies have come to realize that most of the time, large portions of certain licensed frequency bands remain empty [1]. To remedy this, legislators are easing the way frequency bands are licensed and used. In particular, new regulations would allow for devices which are able to sense and adapt to their spectral environment, such as cognitive radios, to become *secondary users*.[1] Such users are wireless devices that opportunistically employ the spectrum already licensed to *primary users*. Primary users generally associate with the primary spectral license holder, and thus have a higher priority right to the spectrum.

The intuitive goal behind secondary spectrum licensing is to increase the spectral efficiency of the network, while, depending on the type of licensing, not affecting higher priority users. The exact regulations governing secondary spectrum licensing are still being formulated [2], but it is clear that networks consisting of heterogeneous devices, both in terms of physical capabilities and in the right to the spectrum, will emerge.

Among the many questions that remain to be answered about cognitive networks, is that of the *fundamental limits* of possible communication. Although this may be defined in various ways, information theory is an ideal tool and approach from which to explore the underlying, implementation-independent limits of such heterogeneous networks. In this chapter, we will outline the current state of the art in information theoretic analysis of cognitive systems.

---

[1] In this chapter, we will use the terms secondary user and cognitive user interchangeably. Cognitive radio will be clearly defined in Sect. 2.1.2, and can be thought of as "smart" radios which are able to adapt to their environment for now.

2.1.1 Secondary Spectrum Licensing

The emergence of the FCC's Secondary Markets Initiative (SMI, [2]) was brought on by both the obvious desire for spectral efficiency, as well as empirical measurements showing that most of the time certain licensed frequency bands remain unused. The goal of the SMI is to remove unnecessary regulatory barriers to new secondary market oriented policies such as

- Spectrum leasing, which allows non-licensed users to lease any part, or all of the spectrum from the licensed user.
- Dynamic spectrum leasing, which is a temporary and opportunistic usage of spectrum rather than a longer-term sub-lease.
- Private commons, whereby a licensee could allow non-licensed users access to his/her spectrum without a contract, optionally with an access fee.
- Interruptible spectrum leasing, which would be suitable for a lessor that wants a high level of assurance that any spectrum temporarily in use, or leased, to an incumbent cognitive radio could be efficiently reclaimed if needed. A prime example would be the leasing of the generally unoccupied spectrum allotted to the US government or local enforcement agencies, which in times of emergency could be quickly reclaimed.

Of interest in this chapter is the *dynamic spectrum leasing*, in which some wireless devices opportunistically employ the spectrum rather than opt for a longer term sub-lease. In order to exploit the spectrum, we require a device which is able to sense the communication opportunities, and then take actions based on the sensed information. In this chapter, such actions will include transmitting (or refraining from transmitting) and adapting their modulation and/or coding strategies so as to "better" employ the sensed spectral environment. Cognitive radios are prime candidates for such actions.

2.1.2 Cognitive Radios and Behavior

Over the past few years, the incorporation of software into radio systems has become increasingly common. This has allowed for faster upgrades, and has given these wireless communication devices the ability to transmit and receive using a variety of protocols and modulation schemes (enabled by reconfigurable software rather than hardware). Furthermore, as their name suggests, such radios can even become "cognitive", and, as dictated by the software, adapt their behavior to their wireless surroundings without user intervention. According to the FCC, software defined radios (SDR) encompasses any "radio that includes a transmitter in which operating parameters such as frequency range, modulation type or maximum output power can be altered by software without making any changes to hardware components that affect the radio frequency emissions." Mitola [3] took the definition of an SDR one step further, and envisioned a radio which could make decisions as to the network, modulation and/or coding parameters *based on its surroundings,* and called such a "smart" radio a *cognitive radio.* Such radios could even make decisions based on the availability of nearby collaborative nodes, or on the regulations dictated by their current location and spectral conditions.

The spectral conditions sensed by the cognitive radio may be utilized in many ways. In this chapter, we consider and survey the information theoretic results on three main categories of cognitive behavior:

1. Interference mitigating cognitive behavior: This behavior allows *two users to simultaneously transmit* over the same time or frequency band(s). Under this scheme, a cognitive radio will listen to the channel and, if sensed idle, could transmit during the void, not worrying about interference to the primary user (who is not transmitting). On the other hand, if another sender is sensed, the radio may decide to proceed with simultaneous transmission. The cognitive radio need not wait for an idle channel to start transmission. There will be interference between the primary and secondary users, but as we will show, this could potentially be mitigated. Here, the sensed information is fully utilized as side information, which will be the main aid in interference mitigation.

2. Collaborative behavior (interference-free cognitive behavior): When cognitive devices exist in a network but have no information of their own to transmit, they could potentially act as relays, and *collaborate* with the primary users. Rather than cause interference to the primary link, they boost it. Neglecting any other possibly active cognitive clusters [4], this system is interference-free. Incentives for cognitive radios to collaborate with primary users is beyond the scope of this chapter, but must also be considered. Here the sensing capability of the cognitive radio is used to obtain the message of the primary user, in order to relay it.

3. Interference avoiding cognitive behavior: In current FCC proposals on opportunistic channel usage, the cognitive radio listens to the wireless channel and determines, either in time or frequency, which part of the spectrum is unused [1]. It then adapts its signal to fill this void in the spectral domain, by either transmitting at a different time, or in a different band. A device transmits over a certain time and/or frequency band *only when no other user does*, thus avoiding interference, rather than mitigating it. Such behavior employs the sensing capability to determine a suitable moment, protocol, and band to transmit in.

2.1.3 Chapter Outline

The chapter is structured as follows. In Sect. 2.2, we look at interference-mitigating cognitive behavior, where a prime example is the cognitive radio channel. We outline strategies and their resulting achievable rate regions (for general discrete memoryless cognitive radio channels [5]) and capacity regions (for Gaussian cognitive radio channels [6]). We also demonstrate applicable and related results on interference channels with degraded message sets [7] and interference channels with unidirectional cooperation [8]. In Sect. 2.3 we demonstrate that the multiplexing gain of the cognitive radio channel is 1. This somewhat pessimistic result motivates the definition of the cognitive X-channel in Sect. 2.4. We study an achievable rate region for this channel before demonstrating that it achieves a multiplexing gain of 2

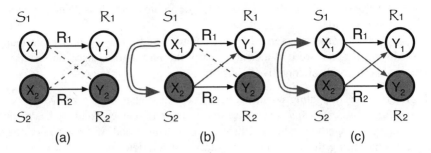

Fig. 2.1. (a) Competitive behavior, the interference channel. The transmitters may not cooperate. **(b)** Cognitive behavior, the cognitive radio channel. Asymmetric transmitter cooperation. **(c)** Cooperative behavior, the two antenna broadcast channel. The transmitters, but not the receivers, may fully and symmetrically cooperate.

in Sect. 2.5. In Sect. 2.6, the limits of collaborative communications [9] are examined. There, the cognitive radio serves as a relay, and many of the previous idealistic assumptions often encountered in the relay channel literature are removed in establishing achievable rate regions. In Sect. 2.7, we take a look at the capacity limits of interference-avoiding cognitive behavior. The problem of tracking and matching the cognitive transmitter and receiver channels in a distributed and dynamic spectral environment is posed, and capacity inner and outer bounds are examined.

2.2 Interference-Mitigating Cognitive Behavior: The Cognitive Radio Channel

We start our discussion by looking at the simplest possible scenario in which a cognitive radio could be employed. We assume there exists a primary transmitter and receiver pair ($S_1 \rightarrow R_1$), as well as the cognitive secondary transmitter and receiver pair ($S_2 \rightarrow R_2$). As shown in Fig. 2.1, there are three possibilities for transmitter cooperation in these two point-to-point channels. We have chosen to focus on transmitter cooperation because such cooperation is often more insightful and general than receiver-side cooperation [10, 11]. We thus assume that each receiver decodes independently. Transmitter cooperation in this figure is denoted by a directed double line. These three channels are simple examples of the cognitive decomposition of wireless networks seen in [4]. The three possible types of transmitter cooperation in this simplified scenario are

1. Competitive behavior: The two transmitters transmit independent messages. There is no cooperation in sending the messages, and thus the two users *compete* for the channel. This is the same channel as the two sender, two receiver interference channel [12].

2. Cognitive behavior: Asymmetric cooperation is possible between the transmitters. This asymmetric cooperation is a result of S_2 knowing S_1's message, but not vice versa. As a first step, we idealize the concept of message knowledge: whenever the cognitive node S_2 is able to hear and decode the message of the primary node S_1, we assume it has full *a priori* knowledge. We call this the *genie assumption*, as these messages could have been given to the appropriate transmitters by a genie. The one-way double arrow indicates that S_2 knows S_1's message but not vice versa. This is the simplest form of asymmetric non-causal cooperation at the transmitters. We use the term "cognitive behavior" to emphasize the need for S_2 to be a "smart" device capable of altering its transmission strategy according to the message of the primary user. We can motivate considering asymmetric side information in practice in three ways:

- Depending on the device capabilities, as well as the geometry and channel gains between the various nodes, certain cognitive nodes may be able to hear and/or obtain the messages to be transmitted by other nodes. These messages would need to be obtained in real time, and could exploit the geometric gains between cooperating transmitters relative to receivers in, for example, a two-phase protocol [5].
- In an Automatic Repeat reQuest (ARQ) system, a cognitive transmitter, under suitable channel conditions (if it has a better channel to the primary transmitting node than the primary receiver), could decode the primary user's transmitted message during an initial transmission attempt. In the event that the primary receiver was not able to correctly decode the message, and it must be re-transmitted, the cognitive user would already have the to-be-transmitted message, or asymmetric side information, at no extra cost (in terms of overhead in obtaining the message).
- The authors in [7] consider a network of wireless sensors in which a sensor S_2 has a better sensing capability than another sensor S_1 and thus is able to sense two events, while S_1 is only able to sense one. Thus, when they wish to transmit, they must do so under an asymmetric side-information assumption: sensor S_2 has two messages, and the other has just one.

3. Cooperative behavior: The two transmitters know each others' messages (two-way double arrows) and can thus fully and symmetrically cooperate in their transmission. The channel pictured in Fig. 2.1(c) may be thought of as a two antenna sender, two single antenna receivers broadcast channel [13].

Many of the classical, well-known information theoretic channels fall into the categories of competitive and cooperative behavior. For more details, we refer the interested reader to the cognitive network decomposition theorem of [4]. We now turn to the much less studied behavior which spans and in a sense interpolates between the symmetric cooperative and competitive behaviors. We call this behavior asymmetric cognitive behavior. In this section, we will consider one example of cognitive behavior: a two sender, two receiver (with two independent messages) interference channel with asymmetric and a priori message knowledge at one of the transmitters, as shown in Fig. 2.1(b). Certain asymmetric (in transmitter cooperation) channels have been considered in the literature: for example in [14], the capacity region of a multiple access channel with asymmetric cooperation between the two transmitters is computed. The authors in [8] consider a channel which could involve asymmetric transmitter cooperation, and explore the conditions under which the capacity of this channel coincides with the capacity of the channel in which *both messages are decoded at both receivers*. In [15] the authors introduced the *cognitive radio channel*, which captures the most basic form of asymmetric transmitter cooperation for the interference channel. We now study the information theoretic limits of interference channels with asymmetric transmitter cooperation, or *cognitive radio channels*.

Our survey on the work on the two sender, two receiver channel with asymmetric cooperation at the transmitters will proceed as follows. First, we will define and demonstrate an achievable rate region for the case of two independent messages for the discrete memoryless cognitive radio channel. This will be followed by the results of [6], who, under certain channel conditions, find the capacity region of the Gaussian interference channel with degraded message sets, a formulation equivalent to the Gaussian cognitive radio channel. We then consider the work of [7] on the general discrete memoryless interference channel with degraded message sets. In particular, they look for conditions under which the derived achievable rate regions are tight. In the Gaussian noise case, their result explicitly equals that of [6]. We then look at the work of [8] on the interference channel with unidirectional cooperation, where the capacity region of the cognitive radio channel when both messages are to be decoded at both receivers, under certain *strong interference* conditions, is derived. We proceed to explore the multiplexing gain of the Gaussian cognitive radio channel, which turns out to be 1. Motivated by this result, we define and derive an achievable rate region for the Gaussian X-channel with partial asymmetric (or cognitive) side information at the transmitter. In this case, the multiplexing gain turns out to be 2.

2.2.1 Cognitive Radio Channel: An Achievable Rate Region

We define a 2×2 *genie-aided cognitive radio channel* C_{COG}, as in Fig. 2.2, to be two point-to-point channels $S_1 \rightarrow R_1$ and $S_2 \rightarrow R_2$ in which the sender S_2 is given, in a non-causal manner (that is, by a genie), the message X_1 which the sender S_1 will transmit. Let X_1 and X_2 be the random variable inputs to the channel, and let Y_1 and Y_2 be the random variable outputs of the channel. The conditional probabilities of the discrete memoryless C_{COG} are fully described by $P(y_1|x_1, x_2)$ and $P(y_2|x_1, x_2)$.

In [16], an achievable region for the interference channel is found by first considering a modified problem and then establishing a correspondence between the achievable rates of the modified and the original channel models. We proceed in the same fashion.

The channel C_{COG}^m, defined as in Fig. 2.2 introduces many new auxiliary random variables, whose purposes can be made intuitively clear by relating them to auxiliary random variables in previously studied channels. They are defined and described in Table 2.1. Standard definitions of achievable rates and regions are employed [15,17] and omitted for brevity. Then an achievable region for the 2×2 cognitive radio channel is given by

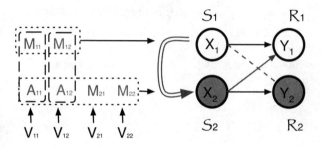

Fig. 2.2. The modified cognitive radio channel with auxiliary random variables M_{11}, M_{12} and M_{21}, M_{22}, inputs X_1 and X_2, and outputs Y_1 and Y_2. The auxiliary random variable A_{11}, A_{12} associated with S_2, aids in the transmission of M_{11} and M_{12}, respectively. The vectors V_{11}, V_{12}, V_{21} and V_{22} denote the effective random variables encoding the transmission of the private and public messages.

Table 2.1. Description of random variables and rates in Theorem 2.1.

(Random) variable names	(Random) variable descriptions
M_{11}, M_{22}	Private information from $S_1 \to R_1$ and $S_2 \to R_2$ resp.
M_{12}, M_{21}	Public information from $S_1 \to (R_1, R_2)$ and $S_2 \to (R_1, R_2)$ resp.
R_{11}, R_{22}	Rate between $S_1 \to R_1$ and $S_2 \to R_2$ resp.
R_{12}, R_{21}	Rate between $S_1 \to (R_1, R_2)$ and $S_2 \to (R_1, R_2)$ resp.
A_{11}, A_{12}	Variables at S_2 that aid in transmitting M_{11}, M_{12} resp.
$V_{11} = (M_{11}, A_{11}), V_{12} = (M_{12}, A_{12})$	Vector helping transmit the private/public (resp.) information of S_1
$V_{21} = M_{21}, V_{22} = M_{22}$	Public and private message of S_2. Also the auxiliary random variables for Gel'fand–Pinsker coding
W	Time-sharing random variable, independent of messages

Theorem 2.1. *Let* $Z \triangleq (Y_1, Y_2, X_1, X_2, V_{11}, V_{12}, V_{21}, V_{22}, W)$, *be as shown in Fig. 2.2. Let* $\mathcal{P}$ *be the set of distributions on* Z *that can be decomposed into the form*

$$
\begin{aligned}
&P(w) \times [P(m_{11}|w)P(m_{12}|w)P(x_1|m_{11}, m_{12}, w)] \\
&\times [P(a_{11}|m_{11}, w)P(a_{12}|m_{12}, w)] \\
&\times [P(m_{21}|v_{11}, v_{12}, w)P(m_{22}|v_{11}, v_{12}, w)] \\
&\times [P(x_2|m_{21}, m_{22}, a_{11}, a_{12}, w)] P(y_1|x_1, x_2)P(y_2|x_1, x_2), \qquad (2.1)
\end{aligned}
$$

where $P(y_1|x_1, x_2)$ and $P(y_2|x_1, x_2)$ are fixed by the channel. Let $T_1 \triangleq \{11, 12, 21\}$
and $T_2 \triangleq \{12, 21, 22\}$. For any $Z \in \mathcal{P}$, let $S(Z)$ be the set of all rate tuples
$(R_{11}, R_{12}, R_{21}, R_{22})$ (as defined in Table 2.1) of non-negative real numbers such
that there exist non-negative reals $L_{11}, L_{12}, L_{21}, L_{22}$ satisfying

$$\bigcap_{T \subset \{11,12\}} \left(\sum_{t \in T} R_t \right) \leq I(X_1; \mathbf{M}_T | \mathbf{M}_{\overline{T}}) \tag{2.2}$$

$$R_{11} = L_{11} \tag{2.3}$$

$$R_{12} = L_{12} \tag{2.4}$$

$$R_{21} \leq L_{21} - I(V_{21}; V_{11}, V_{12}) \tag{2.5}$$

$$R_{22} \leq L_{22} - I(V_{22}; V_{11}, V_{12}) \tag{2.6}$$

$$\bigcap_{T \subset T_1} \left(\sum_{t_1 \in T} L_{t_1} \right) \leq I(Y_1, \mathbf{V}_{\overline{T}}; \mathbf{V}_T | W) + f(\mathbf{V}_T | W) \tag{2.7}$$

$$\bigcap_{T \subset T_2} \left(\sum_{t_2 \in T} L_{t_2} \right) \leq I(Y_2, \mathbf{V}_{\overline{T}}; \mathbf{V}_T | W) + f(\mathbf{V}_T | W) \tag{2.8}$$

where $f(\mathbf{v}_T)$ denotes the divergence between the joint distribution of the random
variables $\mathbf{V}_T$ in (2.1) and their product distribution (where all components are in-
dependent). $\overline{T}$ denotes the complement of the subset T with respect to T_1 in (2.7),
with respect to T_2 in (2.8), and $\mathbf{V}_T$ denotes the vector of V_i such that $i \in T$. Let
S be the closure of $\cup_{Z \in \mathcal{P}} S(Z)$. Then any pair $(R_{11} + R_{12}, R_{21} + R_{22})$ for which
$(R_{11}, R_{12}, R_{21}, R_{22}) \in S$ is achievable for C_{COG}.

Proof outline: The main intuition is as follows: the equations in (2.2) ensure that
when $\mathcal{S}_2$ is presented with X_1 by the genie, the auxiliary variables M_{11} and M_{12} can
be recovered. Equations (2.7) and (2.8) correspond to the equations for two overlapping MAC channels seen between the effective random variables $\mathbf{V}_{T_1} \to \mathcal{R}_1$, and
$\mathbf{V}_{T_2} \to \mathcal{R}_2$. Equations (2.5) and (2.6) are necessary for the Gel'fand–Pinsker [18]
coding scheme to work ($I(V_{21}; V_{11}, V_{12})$ and $I(V_{22}; V_{11}, V_{12})$ are the penalties for
using non-causal side information). The $f(\mathbf{V}_T)$ terms correspond to the highly unlikely events of certain variables being correctly decoded despite others being in
error. Intuitively, the sender $\mathcal{S}_2$ could aid in transmitting the message of $\mathcal{S}_1$ (the A_{11},
A_{12} random variables) or it could dirty paper code against the interference it will
see (the M_{21}, M_{22} variables). The theorem smoothly interpolates between these two
options. Details may be found in [5]).

2.2.2 Achievable Rates for the Gaussian Cognitive Radio Channel

The previous section proposed inner and outer bounds on the capacity of the cognitive radio channel for discrete memoryless channels. Although the regions can be
succinctly expressed, as done in Theorem 2.1, because this expression involves evaluation of the mutual information terms over all distributions of the specified form,

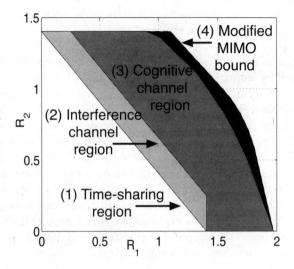

Fig. 2.3. Rate regions (R_1, R_2) for 2×2 wireless channels.

it is unclear what these regions look like in general (and numerically intractable to try all possible input distributions). When the channel is affected by additive white Gaussian noise, as is often done in the literature, one can assume the input distributions to be of a certain form, and thus obtain a possible achievable rate region (not necessarily the largest one). In this section, we use this approach to arrive at the inner and outer bound regions shown in Fig. 2.3.

We consider the 2×2 Gaussian cognitive radio channel described by the input, noise and output relations:

$$Y_1 = X_1 + a_{21}X_2 + Z_1 \tag{2.9}$$
$$Y_2 = a_{12}X_1 + X_2 + Z_2 \tag{2.10}$$

where a_{12}, a_{21} are the crossover (channel) coefficients, $Z_1 \sim \mathcal{N}(0, Q_1)$ and $Z_2 \sim \mathcal{N}(0, Q_2)$ are independent AWGN terms, X_1 and X_2 are channel inputs constrained to average powers P_1 and P_2, respectively, and S_2 is given X_1 non-causally. Thus the Gaussian cognitive radio channel is simply the cognitive radio channel, where we have specified the conditional distributions which describe the channel, $p(y_1, y_2|x_1, x_2)$ to be of the above (2.9), (2.10) form. In order to determine an achievable region for the modified Gaussian cognitive radio channel, specific forms of the random variables described in Theorem 2.2 are assumed, and are analogous to the assumptions.

The resulting achievable region, in the presence of additive white Gaussian noise for the case of identical transmitter powers ($P_1 = P_2$) and identical receiver noise powers ($Q_1 = Q_2$), is presented in Fig.2.3. The ratio of transmit power to receiver noise power is 7.78 dB. The cross-over coefficients in the interference channel are $a_{12} = a_{21} = 0.55$, while the direct coefficients are 1.

In the figure, we see four regions. The time-sharing region (1) displays the result of pure time sharing of the wireless channel between users X_1 and X_2. Points in this region are obtained by letting X_1 transmit for a fraction of the time, during which X_2 refrains, and vice versa. The interference channel region (2) corresponds to the best known achievable region [16] of the classical information theoretic interference channel. In this region, both senders encode independently, and there is no a priori message knowledge by either transmitter of the other's message. The cognitive channel region (3) is the achievable region described here. In this case, X_2 received the message of X_1 non-causally from a genie, and X_2 uses a coding scheme which combines interference mitigation with relaying the message of X_1. We see that both users – not only the incumbent X_2 which has the extra message knowledge – benefit from using this scheme. This is as expected: the selfish strategy boosts R_2 rates, while the selfless one boosts R_1 rates, and so gracefully combining the two will yield benefits to both users. Thus, the presence of the incumbent cognitive radio X_2 can be beneficial to X_1, a point which is of practical significance. This could provide yet another incentive for the introduction of such schemes.

The modified MIMO bound region (4) is an outer bound on the capacity of this channel: the two antenna Gaussian broadcast channel capacity region [13], where we have restricted the form of the transmit covariance matrix to be of the form $\begin{pmatrix} P_1 & c \\ c & P_2 \end{pmatrix}$, to more closely resemble our constraints, intersected with the capacity bound on $R_2 \leq I(Y_2; X_2|X_1)$ for the channel for $X_2 \rightarrow Y_2$ in the absence of interference from X_1. Let $H_1 = [1 \ a_{21}]$ and $H_2 = [a_{12} \ 1]$. Then modified MIMO bound region is explicitly given by the set:

$$\text{Convex hull } \{(R_1, R_2):$$

$$
\begin{aligned}
R_1 &\leq \tfrac{1}{2}\log_2\left(\frac{H_1(B_1+B_2)H_1^T+Q_1}{H_1B_2H_1^T+Q_1}\right) \\
R_2 &\leq \tfrac{1}{2}\log_2\left(\frac{H_2B_2H_2^T+Q_2}{Q_2}\right) \\
R_2 &\leq \tfrac{1}{2}\log_2\left(1+\frac{P_2}{Q_2}\right)
\end{aligned}
\quad \bigcup \quad
\begin{aligned}
R_1 &\leq \tfrac{1}{2}\log_2\left(\frac{H_1B_1H_1^t+Q_1}{Q_1}\right) \\
R_2 &\leq \tfrac{1}{2}\log_2\left(\frac{H_2(B_1+B_2)H_2^T+Q_2}{H_2B_1H_2^T+Q_2}\right) \\
R_2 &\leq \tfrac{1}{2}\log_2\left(1+\frac{P_2}{Q_2}\right)
\end{aligned}
$$

for any 2×2 matrices B_1, B_2 such that

$$B_1 \succeq 0, \quad B_2 \succeq 0$$
$$B_1 + B_2 \preceq \begin{pmatrix} P_1 & c \\ c & P_2 \end{pmatrix}$$
$$c^2 \leq P_1 P_2\}.$$

Here $X \succeq 0$ denotes that the matrix X is positive semi-definite.

2.2.3 Further Results on the Cognitive Radio Channel

Following the introduction of the *cognitive radio channel* Jovicic and Viswanath [7], Wu et al. [8], and Devroye et al. [16] considered the Gaussian cognitive radio channel, albeit under different names, and subsequently obtained its capacity in weak interference. The authors in [8] consider a channel which could involve asymmetric transmitter cooperation, and explore the conditions under which the capacity of this channel coincides with the capacity of the channel in which both messages are decoded at both receivers. We briefly review the results of these three works.

The authors of [6] consider a two-sender two-receiver channel which consists of a primary user and a secondary, or cognitive user. Like in the cognitive radio channel, each has its own independent message to send, and the cognitive user is assumed to know, a priori, the message of the primary user. They term their channel the *interference channel with degraded message sets (IC-DMS)*. This work is particularly interested in determining the maximal rate at which the secondary cognitive user may transmit such that the primary user's rate remains unchanged (that is, the primary user's rate continues to be the same as if there were no interference), in the Gaussian noise channel. This would correspond to a single point in the capacity region of the channel in general. They furthermore require the primary receiver to employ a single-user decoder, which would be the case if no cognitive user were present. In essence, these two conditions, which they term *co-existence conditions,* require the cognitive user to remain transparent to the primary user. In fact, the only difference between the IC-DMS and the cognitive radio channel is that the IC-DMS, and all the results pertaining to it, are only valid in the Gaussian noise case. In addition, the *co-existence conditions* are not explicitly required in cognitive radio channel. In [6], these co-existence conditions are also relaxed (allowing for joint codebook design between primary and secondary users), and the authors show that the capacity achieving coding/decoding scheme in fact satisfy these *co-existence conditions,* that is, that the primary user decoder behaves as a single user decoder.

Let R_1 and R_2 denote the rates achieved by the primary and cognitive users, respectively. The main results of [6] stated in their Theorems 3.1 and 4.1 are summarized in the following single theorem. Here the primary user is expected power limited to P_1, the secondary user is expected power limited to P_2, and the noises at the two receivers are Gaussian of zero mean and variance N_1 and N_2 respectively. The conditions, and notation, which are the same as in the Gaussian cognitive radio channel of Sect. 2.2, save the *co-existence conditions.*

Theorem 2.2. *The capacity region of the IC-DMS defined in (2.9), (2.10) is given by the union, over all $\alpha \in [0, 1]$, of the rate regions*

$$0 \leq R_1 \leq \tfrac{1}{2} \log_2 \left(1 + \frac{(\sqrt{P_1} + a_{21}\sqrt{\alpha P_2})^2}{1 + a_{21}^2(1-\alpha)P_2} \right)$$
$$0 \leq R_2 \leq \tfrac{1}{2} \log_2 \left(1 + (1-\alpha)P_2 \right).$$

In particular, the maximal rate R_2 (or capacity) at which a cognitive user may transmit such that the primary user's rate R_1 remains as in the interference-free regime $(R_1 = \tfrac{1}{2} \log_2 (1 + P_1/N))$ is given by

$$R_1 = \tfrac{1}{2} \log_2 \left(1 + \tfrac{P_1}{N}\right)$$
$$R_2 = \tfrac{1}{2} \log_2 \left(1 + (1 - a^*)P_2\right).$$

as long as $a_{21} < 1$, and a^* is

$$a^* = \left(\frac{\sqrt{P_1}\left(\sqrt{1 + a_{21}^2 P_2(1 + P_1)} - 1\right)}{a_{21}\sqrt{P_2}(1 + P_1)}\right)^{\frac{1}{2}}.$$

Both these results are obtained using a Gaussian encoder at both the primary and cognitive transmitters. For more precise definitions of achievability in this channel, we refer to [6]. We paraphrase their achievability results here. The primary user generates its 2^{nR_1} codewords, X_1^n (block length n), by drawing the coordinates i.i.d. according to $\mathcal{N}(0, P_1)$, where we recall P_1 is the expected noise power constraint. Then, since the cognitive radio knows the message the primary user, it can form the primary user's encoding X_1^n, and performs superposition coding as

$$X_2^n = \hat{X}_2^n + \sqrt{\frac{\alpha P_2}{P_1}} X_1^n$$

where $\alpha \in [0, 1]$. The codeword $\hat{X}_2^n$ encodes one of the 2^{nR_2} messages, and is generated by performing Costa precoding [19] (dirty-paper coding). Costa showed that to optimize the rate achieved by this dirty-paper coding, one selects $\hat{X}_2^n$ statistically independently from X_1^n, and thus i.i.d. Gaussian. Encoding is done using a standard information theoretic binning technique, which treats the message X_1^n as non-causally known interference. In order to satisfy the average power constraint of P_2 on the components of X_2^n, $\hat{X}_2^n$ must be $\mathcal{N}(0, (1-\alpha)P_2)$. A converse, resulting in the capacity region of the cognitive radio channel under weak interference, is given in [6] and is based on the conditional entropy power inequality, and results from [13].

Whereas the paper [6] considers only the Gaussian IC-DMS with specific *co-existence conditions*, the work [7] considers the discrete memoryless IC-DMS (not necessarily Gaussian), and looks at the Gaussian IC-DMS as a special case. The authors in this work are motivated by a sensor network in which one sensor has better sensing capabilities than another. The one with the better channel is thus able to detect two sensed events, while another is only able to detect one. This problem then reduces to the interference channel with degraded message sets (where the message of one user is a subset of the other user's message). The authors define three types of *weak interference* (as opposed to the *very strong* and *strong* interference typically seen in the interference channel literature [12]), an achievable rate region, outer bounds, and conditions under which these outer bounds are tight. They then look at a Gaussian noise example in which their region is tight, and for which the result is as described in the capacity region of [6]. We summarize some of their main results in the single following theorem. It provides an inner and an outer bound on the IC-DMS, which turns out to be the capacity region for the types of interference specified.

Theorem 2.3. *Inner bound Let $\mathcal{R}_{in}$ be the set of all rate pairs (R_1, R_2) (same as in the cognitive radio channel) such that*

$$R_1 \leq I(V, X_1; Y_1)$$
$$R_2 \leq I(U; Y_2) - I(U; V, X_1)$$

for the probability distribution $p(x_1, x_2, u, v, y_1, y_2)$ that factors as

$$p(v, x_1)p(u|v, x_1)p(x_1|u)p(y_1, y_2|x_1, x_2).$$

Then $\mathcal{R}_{in}$ is an achievable rate region for the IC-DMS where transmitter S_2 knows both messages and transmitter S_1 only knows one.
Outer bound: Define $\mathcal{R}_o$ to be the set of all rate pairs (R_1, R_2) such that

$$R_1 \leq I(V, X_1; Y_1)$$
$$R_2 \leq I(X_1; Y_2|X_1)$$
$$R_1 + R_2 \leq I(V, X_1; Y_1) + I(X_2; Y_2|V, X_1)$$

for the probability distribution $p(x_1, x_2, v, y_1, y_2)$ that factors as

$$p(v, x_1)p(x_2|v)p(y_1, y_2|x_1, x_2).$$

Then $\mathcal{R}_o$ is an outer bound for the capacity of the IC-DMS.
Capacity conditions: If there exists a probability transition matrix $q_1(y_2|x_2, y_1)$ such that

$$p(y_2|x_1, x_2) = \sum_{y_1} p(y_1|x_1, x_2)q_1(y_2|x_2, y_1)$$

or if there exists a probability transition matrix $q_2(y_1|x_1, y_2)$ such that

$$p(y_1|x_1, x_2) = \sum_{y_2} p(y_2|x_1, x_2)q_2(y_1|x_1, y_2)$$

then the set of all rate pairs (R_1, R_2) such that

$$R_1 \leq I(V, X_1; Y_1) \tag{2.11}$$
$$R_2 \leq I(X_2; Y_2|V, X_1) \tag{2.12}$$

for the probability distribution $p(x_1, x_2, y_1, y_2)$ that factors as

$$p(v, x_1)p(x_2|v)p(y_1, y_2|x_1, x_2)$$

is the capacity region of the IC-DMS.

Since the channel of [7] is the same as the cognitive radio channel [5], direct comparisons between their respective bounds may be made. Whereas the outer bounds are equivalent, due to the fact that the inner bounds for the discrete memoryless channel involve non-trivial unions over all distributions of a certain form, it is unclear a priori which region will be larger. However, the authors demonstrate that all *Gaussian* weak interference channels satisfy the *capacity conditions* of the theorem, and thus the region of (2.11) and (2.12) is the capacity region. This capacity region in the Gaussian noise case is shown to be explicitly equal to that of [6], and, numerically specialized to the Gaussian noise case.

Finally, the work [8] considers again the cognitive radio channel, referred to as the *interference channel with unidirectional cooperation*. There, one set of conditions for which the capacity region of the channel coincides with that of the channel in which both messages are required at both receivers is derived. Notice that in the cognitive radio channel this added condition, of being able to decode both messages at both receivers, is not assumed. This is related to the work [20] on the compound multiple access channel with common information, in which the capacity region for another set of *strong interference*-type conditions is computed. Notice that whereas [7] considers *weak interference* conditions, [8] considers *strong interference conditions*. Their results on the cognitive radio channel capacity read as follows:

Theorem 2.4. *For an interference channel with unidirectional cooperation satisfying*

$$I(X_2; Y_2 | X_1) \leq I(X_2; Y_1 | X_1)$$
$$I(X_1, X_2; Y_1) \leq I(X_1, X_2; Y_2)$$

for all joint distributions on X_1 and X_2, the capacity region $\mathcal{C}$ is given by

$$\mathcal{C} = \bigcup \{(R_1, R_2):$$
$$R_2 \leq I(X_2; Y_2 | X_1)$$
$$R_1 + R_2 \leq I(X_1, X_2; Y_1)\}$$

where the union is over joint distributions $p(x_1, x_2, y_1, y_2)$.

2.2.4 Cognitive Radio Channel Conclusions

As we have seen, various authors have studied the fundamental information theoretic limits of *cognitive behavior*, albeit sometimes under different names, with the common idea of partial asymmetric side information at one transmitter. In addition, in Gaussian noise, it can be seen that cognitive behavior allows for a secondary user to transmit at a non-zero rate while the primary user remains unaffected. Alternatively, tradeoffs between the primary and secondary users' rates can also be analyzed. The capacity regions are known under certain conditions, but as is the case for the interference channel, the capacity region of the most general discrete memoryless cognitive radio channel remains an open problem.

2.3 The Multiplexing Gain of Cognitive Radio Channels

The previous section showed that when two interfering point-to-point links act in a *cognitive* fashion, or employ asymmetric non-causal side information, interference may be at least partially mitigated, allowing for higher spectral efficiency. That is, it is possible for the cognitive user to communicate at a non-zero rate while the primary user suffers no loss in rate. Thus, at medium SNR levels, there is an advantage to cognitive transmission. One immediate question that arises is how cognitive transmission performs in the high SNR regime. The *multiplexing gain* is defined as the limit of the ratio of the maximal achieved rate to the log(SNR) as the SNR tends to infinity. That is,

$$\text{multiplexing gain} = \lim_{\max \text{ SNR} \to \infty} \frac{R(\text{SNR})}{\log(\text{SNR})}.$$

The multiplexing gain of various multiple input multiple output (MIMO) systems has been extensively studied in the literature [21]. For the single-user point-to-point MIMO channel with M_T transmit and N_R receive antennas, the maximum multiplexing gain is known to be $\min(M_T, N_R)$ [22, 23]. For the two user MIMO multiple access channel with N_R receive antennas and M_{T_1}, M_{T_2} transmit antennas at the two transmitters, the maximal multiplexing gain is $\min(M_{T_1} + M_{T_2}, N_R)$. For its counterpart, the two user MIMO broadcast channel with M_T transmit antennas and N_{R_1}, N_{R_2} receive antennas at the two transmitters, respectively, the maximum multiplexing gain is $\min(M_T, N_{R_1} + N_{R_2})$. These results, as outlined in [21] demonstrate that when joint signal processing is available at either the transmit or receive sides (as is the case in the MAC and BC channels), then the multiplexing gain is significant. However, when joint processing is neither possible at the transmit nor receive side, as is the case for the interference channel, then the multiplexing gain is severely limited. Results for the maximal multiplexing gain when cooperation is permitted at the transmitter or receiver side through noisy communication channels can be found in [24, 25]. In the cognitive radio channel, a form of partial joint processing is possible at the transmitter. It is thus unclear whether this channel will behave more like the cooperative MAC and BC channels, or whether it will suffer from interference at high SNR as in the interference channel. We thus outline results on the multiplexing gain in this scenario, under additive white Gaussian noise [26].

We expect the multiplexing gain (which intuitively corresponds to the number of information streams one can push through a channel) to lie somewhere between 1 and 2, as we have two independent messages, and single antennas at all nodes. One can show that the sum-rate of the Gaussian cognitive radio channel, with two independent messages $\mathcal{S}_1 \to \mathcal{R}_1$ and $\mathcal{S}_2 \to \mathcal{R}_2$, as shown in Fig. 2.4(a) scales at best like $\log P$ (not $2 \log P$). In other words, although partial side information may help the interference channel in a medium SNR-regime [5, 6], at high SNR, one cannot improve the scaling law of the sum-rate.

Theorem 2.5. *Consider a Gaussian interference channel defined in* (2.9), (2.10), *and where additionally* $\mathcal{S}_2$ *has non-causal knowledge of the message of* $\mathcal{S}_1$. *Then the sum-rate capacity of this channel satisfies*

$$\lim_{P \to \infty} \frac{\max_{(R_1, R_2) \in \mathcal{C}} R_1 + R_2}{\log P} = 1 \tag{2.13}$$

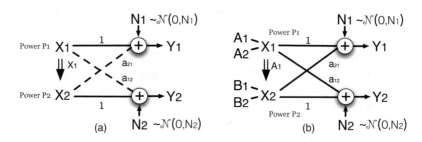

Fig. 2.4. Both channels are additive Gaussian noise interference channels with cross-over parameters α_{12}, α_{21}, transmitted encodings X_1, X_2 with expected transmit power limitations P_1 and P_2, and received signals Y_1 and Y_2. (a) Cognition in the interference channel: there are two information streams $(X_1 \to Y_1)$ and $(X_2 \to Y_2)$, and X_1 is the asymmetric side information known at X_2. (b) Cognition in the X-channel: there are four message streams $(A_1 \to Y_1)$, $(A_2 \to Y_2)$, $(B_1 \to Y_1)$ and $(B_2 \to Y_2)$. A_1 is the partial and asymmetric message knowledge at X_2.

where R_i corresponds to the rates from the i-th source to the i-th receiver, P is the expected transmit power constraint at each transmitter and C is the capacity region of the channel.

Proof. : The $a_{21} \leq 1$ condition ensures that we are operating in the *weak interference* regime. Consider the capacity region denoted by C claimed in (24) and (25) of [6]. Notice that $a_{21} \leq 1$ corresponds to $a \leq 1$, $P_1 = P_p$, $P_2 = P_c$, and $R_1 = R_p, R_2 = R_c$ in the notation of [6]. If $P_p = P_c = P$, then it follows that

$$\lim_{P \to \infty} \frac{\max_{(R_1, R_2) \in C} R_1 + R_2}{\log P} \tag{2.14}$$

$$= \lim_{P \to \infty} \frac{\max_\alpha \frac{1}{2} \log \left(\left(1 + \frac{(\sqrt{P} + a\sqrt{\alpha P})^2}{1 + a^2(1 - \alpha)P}\right) \cdot (1 + (1 - \alpha)P)\right)}{\log P} \tag{2.15}$$

$$= 1. \tag{2.16}$$

For the case when $a_{21} > 1$ the sum-rate again scales like $\log(P)$, which can be seen by using Theorem 2.2.

2.4 The X-Channel with Asymmetric Side Information

Section 2.2 showed that when two non-overlapping single sender, single receiver channels act in a *cognitive* fashion, or employ asymmetric non-causal side information, interference may be at least partially mitigated, allowing for higher spectral efficiency. In this scenario, the two senders and the two receivers were independent. However, at high SNR, the multiplexing gain was limited to 1. This is in fact equal to that of a channel with no cognition. We ask ourselves if there are other *cognitive* channels in which partial asymmetric message knowledge does provide a multiplexing gain greater than 1. The answer, as we will see in the next section, is yes. The channel for which the multiplexing gain using partial asymmetric side information is the cognitive *X-channel*, which we define next. This channel is equivalent to the cognitive version of the X-channel, defined in [21,27], where the degrees of freedom, or multiplexing gain, is considered in the multiple antenna, non-cognitive case. We will ultimately be interested in the multiplexing gain for Gaussian noise channels, and so introduce the Gaussian cognitive radio channel, and the Gaussian cognitive X-channel.

Repeating for clarity, in the cognitive radio channel, defined in Sect. 2.2.1 and shown in Fig. 2.4(a), there are two messages, one from $(S_1 \rightarrow R_1)$, and the other from $(S_2 \rightarrow R_2)$. There is no cross-over information from $(S_1 \rightarrow R_2)$ or $(S_2 \rightarrow R_1)$. Here S_2 knows the message X_1, as seen by the directed double arrow in Fig. 2.4(a). The multiplexing gain of this channel is 1. Consider now the same two sender, two receiver Gaussian noise channel as Fig. 2.4(a) except that here we *do* have cross-over information. That is, each sender has an independent message destined to each receiver, for a total of four messages, as shown in Fig. 2.4(b). S_1 wishes to send message $s_{11} \in \{1, 2, \cdots 2^{nR_{11}}\}$, encoded as $A_1 \in \mathcal{A}_1$ to R_1 (at rate R_{11}) and $s_{12} \in \{1, 2, \cdots, 2^{nR_{12}}\}$, encoded as $A_2 \in \mathcal{A}_2$ to R_2 (at rate R_{12}) in n channel uses. Similarly, S_2 wishes to send message $s_{21} \in \{1, 2, \cdots 2^{nR_{21}}\}$, encoded as $B_1 \in \mathcal{B}_1$ to R_1 (at rate R_{21}) and $s_{22} \in \{1, 2, \cdots, 2^{nR_{22}}\}$ encoded as $B_2 \in \mathcal{B}_2$ to R_2 (at rate R_{22}) in n channel uses. The double arrow from X_1 to X_2 denotes *partial side information*, specifically, that the encoding A_1 is known fully, non-causally (or a priori) to the second transmitter. Notice also that only one of S_1's messages is known to S_2, that is, only *partial* knowledge is used in the following. We could alternatively have allowed A_2 to be known at the second transmitter. This would lead to analogous results when indices are permuted. The channel is still an additive Gaussian noise channel with independent noise at the receivers, so the received signals are

$$Y_1 = A_1 + A_2 + a_{21}(B_1 + B_2) + N_1 \tag{2.17}$$
$$Y_2 = a_{12}(A_1 + A_2) + (B_1 + B_2) + N_2. \tag{2.18}$$

Standard definitions of achievable rates and regions are employed in [15, 17] or chapters 8 and 14 of [17]. Although our achievable rate region will be defined for finite alphabet sets, in order to determine an achievable region for the Gaussian noise channel, specific forms of the random variables described in Theorem 2.6 are assumed. As in [17,20,29], Theorem 2.6 can readily be extended to memoryless channels with discrete time and continuous alphabets by finely quantizing the input, output, and interference variables (Gaussian in this case).

We now outline an achievable region for this Gaussian noise channel. The capacity region of the Gaussian MIMO broadcast channel [13] is achieved by Costa's dirty-paper coding techniques [19]. In the X-channel, at S_1, the encodings A_1 and A_2 may be jointly generated, for example using a dirty-paper like coding scheme. That is, one message may treat the other as non-causally known interference and code so as to mitigate it. At S_2, not only may the encodings B_1 and B_2 be jointly designed, but they may additionally use A_1 as a priori known interference. Thus, transmitter 2 could encode B_2 so as to potentially mitigate the interference Y_2 will experience from A_1 as well as B_1.

We demonstrate an achievable region for the discrete, finite alphabet case in Theorem 2.6 and look at the achieved rate scalings in the Gaussian noise case, assuming specific forms for all involved variables in Theorem 2.6. Let R_{11} be the rate from $A_1 \rightarrow Y_1$, R_{12} from $A_2 \rightarrow Y_2$, R_{21} from $B_1 \rightarrow Y_1$ and R_{22} from $B_2 \rightarrow Y_2$.

Theorem 2.6. *Let $Z \triangleq (Y_1, Y_2, X_1, X_2, A_1, A_2, B_1, B_2)$, and let $\mathcal{P}$ be the set of distributions on Z that can be decomposed into the form*

$$
\begin{aligned}
&p(a_1|a_2)p(a_2)p(b_1)p(b_2|a_1,b_1) \\
&p(x_1|a_1,a_2)p(x_2|a_1,b_1,b_2) \\
&p(y_1|x_1,x_2)p(y_2|x_1,x_2)
\end{aligned}
\tag{2.19}
$$

where we additionally require $p(a_2,b_2) = p(a_2)p(b_2)$. For any $Z \in \mathcal{P}$, let $S(Z)$ be the set of all tuples $(R_{11}, R_{12}, R_{21}, R_{22})$ of non-negative real numbers such that

$$
\left.
\begin{aligned}
R_{11} &\leq I(A_1; Y_1|B_1) - I(A_1; A_2) \\
R_{21} &\leq I(B_1; Y_1|A_1) \\
R_{11} + R_{21} &\leq I(A_1, B_1; Y_1) - I(A_1; A_2)
\end{aligned}
\right\}
\begin{aligned}
&\text{MAC} \\
&(A_1, B_1) \\
&\searrow\!\!\swarrow \\
&Y_1
\end{aligned}
$$

$$
\left.
\begin{aligned}
R_{12} &\leq I(A_2; Y_2|B_2) \\
R_{22} &\leq I(B_2; Y_2|A_2) - I(B_2; A_1, B_1) \\
R_{12} + R_{22} &\leq I(A_2, B_2; Y_2) - I(B_2; A_1, B_1)
\end{aligned}
\right\}
\begin{aligned}
&\text{MAC} \\
&(A_2, B_2) \\
&\searrow\!\!\swarrow \\
&Y_2
\end{aligned}
$$

Let S be the closure of $\cup_{Z \in \mathcal{P}} S(Z)$. Then any element of S is achievable.

Proof. The codebook generation, encoding, decoding schemes and formal probability of error analysis are deferred to the manuscript in preparation [26]. Heuristically, notice that the channel from $(A_1, B_1) \rightarrow Y_1$ is a multiple access channel with encoders that are possibly correlated [29, 30] and employ dirty paper coding [18, 19]. However, by (2.19) we see that A_1 and B_1 are in fact independent, and thus the regular MAC equations hold. A_1 does use a binning scheme with respect to A_2, but this does not alter the $(A_1, B_1) \rightarrow Y_1$ MAC equations other than reduce the rate R_{11} by $I(A_1; A_2)$ (like in Gel'fand–Pinsker [18] coding). Similarly, for the MAC $(A_2, B_2) \rightarrow Y_2$ the encodings A_2 and B_2 are independent (this is true in particular in the Gaussian case of interest in the next subsection, and so we simplify our theorem by ensuring the condition $p(a_2, b_2) = p(a_2)p(b_2)$) so that the regular MAC equations also hold here. Again, there is a penalty of $I(B_2; A_1, B_1)$ for the rate R_{22} incurred in order to guarantee finding an n-sequence b_2 in the desired bin that is jointly typical with any *given* a_1, b_1 pair.

2.5 Multiplexing Gains in Overlapping Cognitive Broadcast Channels

The multiplexing gain of the Gaussian cognitive radio channel was shown to be 1. We now proceed to examine the multiplexing gain of the cognitive Gaussian X-channel. We wish to see how the achievable rate tuple varies as a function of the transmit powers, or equivalently, of the SNRs when the white Gaussian noise variance is held fixed. To do so, the achievable rate region is evaluated in the proof of the following corollary, which emphasizes that the sum-rate of two the X-channel with *partial non-causal side information* has a multiplexing gain of 2.

Corollary 2.1. *Consider the Gaussian X-channel with asymmetric side information described in Theorem 2.6. Then*

$$\lim_{P \to \infty} \frac{\max_{(R_{11}, R_{12}, R_{21}, R_{22}) \in \mathcal{C}_{\text{OBC}}} R_{11} + R_{12} + R_{21} + R_{22}}{\log P} = 2 \qquad (2.20)$$

where $\mathcal{C}_{\text{OBC}}$ is the capacity region of the cognitive X-channel.

Proof: First, note that the multiplexing gain of a single sender, 2 receiver broadcast channel is 2, and as this channel's capacity region provides an upper bound to our channel's region, we cannot have a multiplexing gain larger than 2. We will in fact prove that 2 is achievable using the scheme of Theorem 2.6. To prove this result, we specify forms for the variables, and then optimize the dirty paper coding parameters, similar to Costa's technique [19]. The Gaussian distributions we assume on all variables are of the form

$$
\begin{aligned}
A_1 &= U_1 + \gamma_1 U_2 & U_1 &\sim \mathcal{N}(0, P_{11}) \\
A_2 &= U_2, & U_2 &\sim \mathcal{N}(0, P_{12}) \\
B_1 &= V_1, & V_1 &\sim \mathcal{N}(0, P_{21}) \\
B_2 &= V_2 + \gamma_2(V_1 + a_{12}U_1) & V_2 &\sim \mathcal{N}(0, P_{22})
\end{aligned}
$$

$$
\begin{aligned}
X_1 &= U_1 + U_2 & \sim \mathcal{N}(0, P_1) & \quad P_1 = P_{11} + P_{12} \\
X_2 &= V_1 + V_2 & \sim \mathcal{N}(0, P_2) & \quad P_2 = P_{21} + P_{22} \\
Y_1 &= U_1 + U_2 + a_{21}(V_1 + V_2) + N_1 & & \quad N_1 \sim \mathcal{N}(0, N_1) \\
Y_2 &= a_{12}(U_1 + U_2) + (V_1 + V_2) + N_2 & & \quad N_2 \sim \mathcal{N}(0, N_2).
\end{aligned}
$$

Here the variables U_1, U_2, V_1, V_2 are all independent, encoding the four messages to be transmitted. Notice that here $p(a_1, b_1) = p(a_1)p(b_1)$ and $p(a_2, b_2) = p(a_2)p(b_2)$ as needed in Theorem 2.6. The sum rates $R_1 = R_{11} + R_{21}$ and $R_2 = R_{12} + R_{22}$ to each receiver can be calculated separately. Each can be maximized with respect to the relevant dirty-paper coding parameter (γ_1 for $\mathcal{S}_1$, and γ_2 for $\mathcal{S}_2$). The bounds of Theorem 2.6 may be evaluated by combining the appropriate determinants of sub-matrices of the overall covariance matrix $E[\Theta\Theta^{\text{T}}]$ where $\Theta \stackrel{\triangle}{=} (A_1, B_1, A_2, B_2, Y_1, Y_2)$. The details may be found in [26]. The main idea is that when the dirty paper coding parameters are properly chosen, and when we let the powers $P_{11} = P_{12} = P_{21} = P$ scale like $P \to \infty$ while keeping P_{22} fixed, then the multiplexing gain of 2 is achieved. Keeping P_{22} fixed is crucial for achieving the

$\log P$ scaling in R_1. Intuitively, this is because of asymmetric message knowledge; the interference the second cognitive transmitter causes the first is not mitigated. Keeping P_{22} constant still allows the second transmitter to dirty paper code, or mitigate the interference caused by A_1 and B_1 to the second receiver's signal Y_2, while causing asymptotically (as $P_{11}, P_{12}, P_{21} \to \infty$) negligible interference to Y_1. This is a remarkable fact: only partial side information is needed to attain the full multiplexing gain of a broadcast channel with a two antenna transmitter.

2.6 Collaborative Communications

We now consider another example of cognitive behavior where rather than having two independent messages to be transmitted, there is only one message to be sent from a given source to a given destination, possibly with the help of a relay. This relay help can be considered as asymmetric transmitter cooperation, or cognitive behavior. We first survey some relay channel results before moving onto the case considered in [9], which has removed many of the classical, and somewhat unrealistic constraints,

2.6.1 The Relay Channel

The relay channel, which in its simplest and most classical form is a three-terminal channel with one source, one relay (without its own information to transmit) and one destination, is another example of *cognitive behavior*. Relay channels were introduced by van der Meulen [14], and various variations of the problem were later studied by others [31, 32]. The current state of the art is well summarized in [33].

The classical relay channel is shown in Fig. 2.5. It consists of a source, with information, a relay, with no independent information of its own, and destination. Here, as in the cognitive radio channel, full channel-state information is assumed at all terminals. The paper [31] introduced two fundamental coding schemes for the relay channel often called Decode-and-Forward (DF, Theorem 2.1) and Compress-and-Forward (CF, Theorem 2.6). This formulation may be extended to multiple relays, as done in [32, 34] and improved in [35, 36]. We defer to the very informative and insightful [33] for further information on relay channels.

Three major issues are ignored in the classical relay channel framework: the half-duplex constraint of most practical wireless systems, the compound nature, and the non-degraded nature of most wireless channels. To elaborate,

1. The first constraint often ignored in the classical relay-channel framework is the *duplex* constraint. Most of the results on relay channel assume full-duplex relays, that is, relays which may receive and transmit simultaneously. In realistic wireless channel, this assumption begins to break down, since the intensity of the near-field of the transmitted signal is much higher than that of the far field of the received signal. In essence, a full-duplex relay would, in practice, interfere with itself. Cognitive relay schemes which operate under a *half-duplex* constraint, that is, where a node cannot simultaneously transmit and receive data, must be considered.

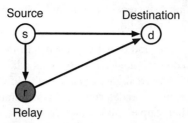

Fig. 2.5. The classical relay channel has a source, with information, a relay, with no information, and a destination. The relay aids the source in transmitting its message to the destination.

Although the capacity of a half duplex relay channel is yet to be found, there
has been a large body of work to understand optimal schemes in the asymptotic
regimes of low and high signal to noise ratio (SNR) in slowly fading wireless
channels [38–41]. In large SNR, the outage capacity of such a channel has been
analyzed in [39, 41]. Interestingly, it is proved that for small multiplexing gains,
the diversity gain achieved by the relay channel matches the maximum diversity
gain achieved by max-flow min-cut bound in Rayleigh fading channels [39]. In
other words, for small multiplexing gains r, i.e., $r \leq \frac{1}{2}$, the relay channel can
provide the same diversity gain as that of a system with two transmit antennas
and a receiver with a single antenna. This result is achieved by a variation of decode and forward (DF) scheme in which the relay starts forwarding the message
as soon as it can decode the message.

As for the low SNR regime, it has been recently shown that the decode and forward scheme is strictly suboptimal in terms of outage capacity [40]. It is further
proved that a bursty variant of the Amplify and Forward cooperation scheme
in which the source broadcasts with a larger power $\frac{P}{\alpha}$ for a short fraction α of
the transmission time and then remains silent for the rest of the time [38, 40], is
outage optimal for Rayleigh fading channels. Intuitively, sending bursty signals
with high power significantly improves the quality of the received signal at the
relay. This scheme turns out to be optimal not only for Rayleigh fading channels,
but also for a wide class of channel distributions, namely the distributions that
are analytic in the neighborhood of zero [37].

2. The second assumption often made in the context of wireless communications
   is the quasi-static fading model. That is, traditionally many authors assume that
   the fading coefficients remain fixed for the entire duration of the transmission
   frame. In an information theoretic framework, where block lengths tend to infinity, all realizations of a channel are thus *not* experienced in a frame, and ergodic
   capacity results seem limited in their applicability. This, in addition to the fact
   that the channel state is often not known to the transmitters but only to receivers
   motivates the study of more realistic *compound* channels [42, 43].

3. Finally, while the degraded relay channel has been completely solved [31, 44],
   in wireless systems most noise is due to thermal noise in the receiver frontend.
   While it may be reasonable to assume that the relay has a better signal to noise

ratio (SNR) than the ultimate receiver, it is unrealistic to assume that the receiver is a degraded version of the relay.

These three drawbacks of traditional approaches to the relay channel motivate the study of non-degraded compound relay channels which satisfy the half-duplex constraint. In [9] the authors investigate a bandwidth efficient decode and forward approach that does not employ predetermined phase durations or orthogonal subchannels to resolve the half-duplex constraint: each relay determines based on its own receive channel when to listen and when to transmit. Furthermore, the transmitters are not aware of the channel and no assumption of degradedness are made: the noise at the relays is independent of that at the destination. Also, as opposed to previous relay and collaborative literature, the results still hold under a bounded asynchronous model. Finally, in the case of multiple relays assisting the source, their approach permits one relay to assist another in receiving the message, a feature not present in much of the early work on communications over compound channels. However, more recent work along this line may be found in [39, 45].

2.6.2 Collaborative Communications

We now present a brief summary of this important and alternate view of the compound relay channel [9], which is a prime example of cognitive behavior in a network where the cognitive nodes do not have information of their own to send. The authors of [9] use the term *collaborative communications* to describe their category of work. This falls into the category of cognitive behavior in the setting considered here.

Spatial diversity is the term often used to capture the potential gain (reliability in this case) of independent paths between sources and destinations which result from spatial separation of nodes or antennas. Of primary interest then is to determine if one can achieve the genie bound on diversity: the diversity gain that would be achieved if all the transmit antennas of the source and relay nodes were in fact connected to a single node (in [41, 46] this is referred to as the transmit diversity bound). For example, consider the three transmit collaborators and one receiver node scenario (each equipped with a single antenna) as illustrated in Fig. 2.6. If all the collaborators were aware of the message a priori, one could in principle achieve the ideal performance of a 3×1 space-time system between the transmit cluster and the receiver node. However, only the source node in the transmit cluster is aware of the message a priori. The other two nodes in the cluster must serve as relays and are not aware of the message a priori. There will be a loss in performance (as measured by the probability of outage) compared to the idealized 3×1 space-time system. In particular, the authors in [9] are interested in determining sufficient conditions on the geometry and signal path loss of the transmitting cluster for which performance close to the genie bound can be guaranteed.

To determine an upperbound on this loss, the authors [9] derive a novel approach to the compound relay channel. This approach is best summarized as follows. In a traditional compound channel, a set of possible channel realizations are given and

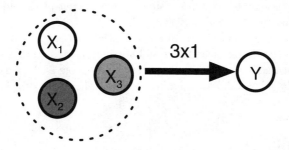

Fig. 2.6. Is an ideal 3×1 space-time gain achievable with three separate transmit nodes and one receive node?

one seeks to prove the existence of a code (with maximal rate) which is simultaneously good on all channel realizations. In [9], the problem is framed in the opposite direction. They fix a rate and ask how large the set of compound channels can be made while guaranteeing that the code is still good.

Consider three nodes denoted as source (s), relay (r) and destination (d) as illustrated in Fig. 2.7 and each equipped with N_s, N_r and N_d antennas, respectively (the results readily generalize to multiple relay nodes).

It is assumed that while listening to the channel, the relay may not transmit, satisfying the half-duplex constraint. Hence, the communications protocol proposed is as follows. The source node wishes to transmit one of 2^{nR} messages to the destination employing n channel uses. While not transmitting, the relay node listens. Due to the relay node's proximity to the source, after n_1 samples from the channel (a number which the relay determines on its own and for which the source has no knowledge), it may correctly decode the message. After decoding the message, it then proceeds to transmit for the remaining $n - n_1$ transmissions in an effort to improve the reception of the message at the destination. The destination is assumed to be made aware of n_1 before attempting to decode the message. This may be achieved by an explicit low-rate transmission from the relay to the destination. Alternatively, if the value of n_1 is constrained to some integer multiple of a fundamental period n_0 (say $n_0 \sim \sqrt{n}$), then the destination may estimate n_1 accurately using power detection methods. De-

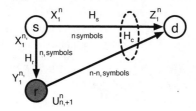

Fig. 2.7. The collaborative communications problem for two transmit collaborators and one receiver.

note the first phase of the n_1 transmissions as the *listening* phase and the last $n - n_1$ transmissions as the *collaboration* phase.

All channels are modeled as additive white Gaussian noise (AWGN) with quasi-static fading. In particular, X and U are column vectors representing the transmission from the source and relay nodes respectively and denote by Y and Z the received messages at the relay and destination respectively. Then during the listening phase,

$$Z = H_s X + N_Z \tag{2.21}$$
$$Y = H_r X + N_Y \tag{2.22}$$

where the N_Z and N_Y are column vectors of statistically independent complex AWGN with variance 1/2 per row per dimension, H_s is the fading matrix between the source and destination nodes and likewise, H_r is the fading matrix between the source and relay nodes. During the collaboration phase,

$$Z = H_c[X^T, U^T]^T + N_Z \tag{2.23}$$

where H_c is a channel matrix that contains H_s as a submatrix (see Fig. 2.7).

It is further assumed that the source has no knowledge of the H_r and H_c matrices (and hence the H_s matrix too). Similarly, the relay has no knowledge of H_c but is assumed to know H_r. Finally, the destination knows H_c.

Without loss of generality, we will assume that all transmit antennas have unit average power during their respective transmission phases. Likewise, the receive antennas have unit power Gaussian noise. If this is not the case, the respective H matrices may be appropriately scaled row-wise and column-wise.

Under the above unit transmit power per transmit antenna and unit noise power per receive antenna constraint, it is well known that a multiple input multiple output (MIMO) system with Gaussian codebook and with rate R bits/channel use can reliably communicate over any channel with transfer matrix H such that $R < \log_2 \det(I + HH^\dagger) \triangleq C(H)$ [2] [22, 24], where I denotes the identity matrix and $H^\dagger$ is the conjugate transpose of H.

Intuition for the above problem then suggests the following. During the listening phase, the relay knowing H_r listens for an amount of time n_1 such that $nR < n_1 C(H_r)$. During this time, the relay receives at least nR bits of information and may reliably decode the message. The destination, on the other hand, receives information at the rate of $C(H_s)$ bits/channel use during the listening phase and at the rate of $C(H_c)$ bits/channel use during the collaborative phase. It may reliably decode the message provided that $nR < n_1 C(H_s) + (n - n_1)C(H_c)$. In the limit as $n \to \infty$, the ratio n_1/n approaches a fraction f and one may conjecture that there exists a "good" code of rate R for the set of channels (H_r, H_c) which satisfy

$$R \leq f C(H_s) + (1 - f)C(H_c) \tag{2.24}$$
$$R \leq f C(H_r) \tag{2.25}$$

---

[2] Here, $C(H)$ does not, in general, designate the capacity of each link as is witnessed by the fact that only for a special subset of matrices is capacity achieved by placing an equal transmit power on each antenna.

for some $f \in [0,1]$. Note that if the channel between the source and the relay is particularly poor, one may fall back on the traditional point-to-point communications paradigm and add the following region to that given in (2.24) and (2.25)

$$R \leq C(H_\mathrm{s}). \tag{2.26}$$

The above intuition is not a proof of achievability but it does provide an upper bound on the performance of the protocol. The essential difficulty in proving that there exists a code which is "good" for any such pair of channels $(H_\mathrm{r}, H_\mathrm{c})$ is two-fold. The problem considered is a *relay channel* which is also a *compound channel*: the authors seek to prove the existence of a code which performs well over an entire set of channels (unknown to the transmitters). The key will be to show the existence of a code that may essentially be refined. Regardless of the actual value of n_1, there exists a codebook for the source which, starting at time $n_1 + 1$, may be layered with the transmission of the relay and perform just as well as if the value of n_1 had been known to the source. For a formal statement and proof of these results, we defer to [9].

The authors simulated the outage probability of their scheme under a quasi-static Rayleigh fading assumption. These numerical and simulation results showed that if the intra-cluster communication has a 10 dB path loss advantage over the receiver at the destination node, in most cases there is essentially no penalty for the intra-cluster communication. Physically, in a two collaborator scenario, this corresponds to a transmit cluster whose radius is 1/3 the distance between the source and destination nodes. By comparison, for a time-division scheme (first the source sends to the relay for a *half* of the time rather than the adjustable fraction f allowed by the authors, then the relay and source send to the destination for the remaining half) with a 5 dB geometric penalty, the allowable cluster size is at most 0.178 times the distance between the source and the destination. This work demonstrates the power of this flexible technique with more realistic assumptions on the wireless channel.

2.7 Interference Avoiding Cognitive Behavior

Up to now the schemes for channels employing cognitive radios have either involved simultaneous transmission, over the same time and frequency, of the primary and secondary users' data (using an interference-mitigating technique), or have not caused any interference at all (collaborative communications). The primary user's message was used as side-information at the secondary transmitter in order to mitigate interference effects. Another way cognitive radios may improve spectral efficiency is by sensing and filling in *spectral gaps*. This can be seen as interference-avoiding cognitive behavior. Suppose the wireless spectrum is populated by some primary users, transmitting on any number of bands. At any point in time, a number of frequency bands will be occupied by primary users, leaving the remainder unoccupied. If a cognitive radio can sense these spectral nulls, it can opportunistically transmit during these times at these frequencies. The work in [47] and [48] addresses issues involved in the opportunistic sensing of and communication over spectral holes. We outline some of these results next.

The authors in [47,48] are interested in deriving capacity inner and outer bounds for a cognitive transmitter–receiver pair acting as secondary users in a network of primary users. The capacity is limited by the *distributed* and *dynamic* nature [47] of the spectral activity which these cognitive radios wish to exploit. To illustrate these points, consider a cognitive transmitter (T) and receiver (R) pair denoted by the grey circles in Fig. 2.8. Each of these is able to sense transmissions within a certain circular radius around themselves, denoted by the dotted circles. Thus, each transmitter and each receiver has a different *local* view of the spectrum utilization. The white circles indicate the primary users (PU), which may or may not be transmitting at a particular point in time. The authors use the term *distributed* to denote the different views of local spectral activity at the cognitive transmitter T and receiver R. In addition to the spectrum availability being location-dependent, it will also vary with time, depending on the data that must be sent at different moments. The authors use the term *dynamic* to indicate the temporal variation of the spectral activity of the primary users.

Communication by the cognitive transmitter–receiver pair takes place as follows. The transmitter senses the channel and detects the presence of primary users. If primary users are detected, the secondary user refrains from transmission. If not, the cognitive user may opportunistically transmit to the receiver. The cognitive receiver may similarly sense the presence of primary users. If none are present, it may opportunistically receive from the secondary transmitter. If primary users are present, in a simplified model, these will cause interference at the receiver, thus making the reception of a cognitive transmission impossible. Cognitive transmission may thus take place when both the cognitive transmitter and the cognitive receiver sense a spectral

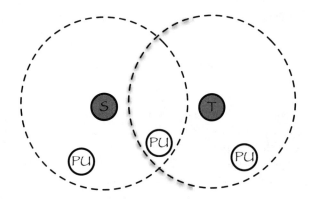

Fig. 2.8. The grey cognitive transmitter (T) receiver (R) pair each have a radius in which they can sense the transmissions of primary users (PU). This leads to different views of local spectral activity, or a *distributed* view on the spectral activity. The PU may change their transmissions over time, leading to *dynamic* spectral activity.

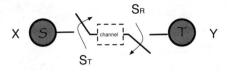

Fig. 2.9. The two switch channel model representing the distributed and dynamic nature of the cognitive channel spectral activity. For successful transmission of the encoded message X to the received message Y in the secondary link $\mathcal{S} \rightarrow \mathcal{T}$, both switches S_T and S_R must be closed, and have a value of 1 in (2.27).

hole. The communication opportunities detected at the transmitter T and the receiver R are in general correlated but not identical. The authors in [47] wish to quantify the effect of this *distributed* nature of the spectral environment. To do so they model the channel as a *switched* channel, shown in Fig. 2.9. The input X is related to the output Y (all of the cognitive link) as

$$Y = (X S_T + N) S_R \qquad (2.27)$$

where N is the additive white Gaussian noise, and $S_T, S_R \in \{0, 1\}$ are binary random variables modeled as switches that represent communication opportunities sensed at the transmitter and the receiver respectively. An S_T or S_R value of 0 indicates that communication is not possible at that end of the cognitive link. The authors proceed to model and analyze this switched model using causal and non-causal side information tools [11]. The capacity of the channel depends on whether the transmitter, the receiver, or both, know the states of the switches S_T and S_R. Knowing whether the switch is open at the transmitter allows it to transmit or remain idle. This *side information* allows the secondary link to transmit more efficiently. Intuitively, if the transmitter lacks this side information (on whether the channel is unoccupied or not), power will be lost in failed transmissions, which are caused by collisions with primary user messages. Similarly, power will also be more efficiently used if the transmitter is aware of the receiver's switch state S_R, as it will refrain from transmission if $S_R = 0$. However, the *distributed* nature of the channel will cause a loss in the capacity of such systems, as analyzed in [47]. The effect of the *dynamic*, or temporal variation in the spectral activity is also considered.

In [47], the capacity limits of a secondary cognitive radio link is explored in terms of how well the spectral holes at the transmitter and the receiver are *matched*, that is, as a function of the state switches S_T and S_R and how well they are known to the cognitive transmitter and receiver. In the work [48], a similar switching framework is used to analyze the effect of spectral hole *tracking*. That is, once the detection of spectral holes is complete, the secondary cognitive user selects one of the locally free spectral segments for opportunistic transmission. The cognitive receiver must also select one of the locally free spectral segments to monitor in order to detect and decode this cognitive message. For communication to be successful, the transmitter and receiver must select the same spectral hole, which must also be empty (of primary users) at both ends. To coordinate the selection of opportunistic spectral holes,

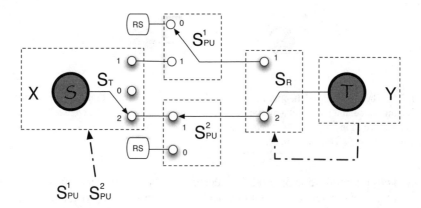

Fig. 2.10. The tracking model of [48]. Secondary transmitter S wishes to communicate with the secondary receiver T on one of two channels. The primary user occupancy on the two channels are modeled as binary random processes S_{PU}^1, $S_{PU}^2 \in \{0, 1\}$. The cognitive user may be in one of three states indicated by S_T, and the cognitive receiver may listen to one of the two channel, as indicated by S_R. For successful communication, S_T must equal S_R (they must be matched).

protocols resulting in transmission overhead could be used. The purpose of [48] is to determine the cost and benefits, in terms of capacity, of these overheads to the cognitive user.

Their model is depicted in Fig. 2.10 for the case of two spectral channels. Here, the primary user occupancy on the two channels are modeled as binary random processes S_{PU}^1, $S_{PU}^2 \in \{0, 1\}$. A value of 0 indicates that a primary user is transmitting on the channel indicated by the superscript, while a 1 indicates that channel is free for the secondary user. These processes are modeled as independent identical Markov chains. The cognitive user may be in one of three states, as indicated by $S_T \in \{0, 1, 2\}$. If $S_T = 0$ then the cognitive transmitter is idle, if is it 1 or 2, it means the cognitive user is transmitting on channel 1 or 2 respectively. The cognitive receiver monitors the channel indicated by $S_R \in \{1, 2\}$. When the cognitive transmitter and receiver states are matched, that is, $S_T = S_R$, then the input and output are related through the channel model (in [48] this is a Q-ary symmetric channel), and when they are not matched the cognitive receiver sees random signals. Thus, it is of interest to calculate the channel capacity assuming that the transmitter knows only S_T and the receiver knows only S_R. They can of course exchange this information, but this would cause a loss in capacity. The goal of [48] is to evaluate this loss.

Capacity inner and outer bounds of this *cognitive tracking channel* are determined and simulated. The inner bounds consist of suggesting particular spectral hole selection strategies at the transmitter and receiver, and seeing what fraction of the time these match up (or track each other). Outer bounds are constructed using a *genie* that gives the transmitter and receiver various amounts of side information, which can only improve what can be achieved in reality. For details, we refer to [48, 49].

Conclusion

Due to their ability to adapt to their spectral environment, cognitive radios allow for much more flexible and potentially more spectrally efficient wireless networks. Heterogeneous networks consisting of both cognitive and non-cognitive devices will soon be a reality. In order to exploit the full capabilities of cognitive radios, many questions must be addressed. One of the foremost, from a physical layer, communications perspective, is that of the fundamental limits of the communication possible over a network when using cognitive devices. In order to effectively study this question, researchers have looked at simplified versions of the problem which capture the essence of the communication characteristics particular to such devices. For example, cognitive devices allow for asymmetric side information between transmitting nodes. Information theoretic limits of cognitive channels have been studied in, among others, [5–8, 15, 47, 48]. In this chapters, we summarized some of the most important results in these works. They all had the property that the primary and secondary users had independent information to transmit, and did so by either mitigating the interference using non-causal side information at the cognitive transmitter, or by filling in spectral gaps. Alternatively, when cognitive radios do not have any information of their own to transmit, they can act as relays, a form of asymmetric behavior. As an example, we outlined the work of [9], where some of the idealistic assumptions of relay channels are removed. The benefits and feasibility of cognitive behavior are intimately linked to the topology of the network: poor primary to secondary user wireless links will make the partial asymmetric side information inherent in cognitive behavior to become very costly to obtain. The value of side-information in wireless networks, in terms of diversity, multiplexing, or delays gains, is another fundamental, and not yet fully understood research problem. In summary, research has thus far looked at simplified scenarios in which cognitive radios may be used. Even there many open problems remain. However, the true question that must be answered in order to understand the limits of communication using cognitive radios, is how their capabilities may be harnessed in order to optimize some network communication utility function. We hope that the research outlined in this chapter serves as a first step to this ultimate goal.

References

1. FCC.
2. FCC, "Secondary markets initiative."
3. J. Mitola, "Cognitive radio," PhD Thesis, Royal Institute of Technology (KTH), 2000.5. N. Devroye, P. Mitran, and V. Tarokh, "Achievable rates in cognitive networks," in *2005 IEEE International Symposium on Information Theory*, Sept. 2005.
4. N. Devroye, P. Mitran, and V. Tarokh, "Cognitive decomposition of wireless networks," in *Proc. of CROWNCOM*, Mar. 2006.
5. N. Devroye, P. Mitran, and V. Tarokh, "Achievable rates in cognitive radio channels," *IEEE Trans. Inf. Theory*, vol. 52, pp. 1813–1827, May 2006.

6. A. Jovicic and P. Viswanath, "Cognitive radio: An information-theoretic perspective," submitted to *IEEE Trans. Inf. Theory*, 2006.

7. W. Wu, S. Vishwanath, and A. Arapostathis, "On the capacity of the interference channel with degraded message sets," submitted to *IEEE Trans. Inf. Theory*, June 2006.

8. I. Maric, R. Yates, and G. Kramer, "The strong interference channel with unidirectional cooperation," in *Information Theory and Applications ITA Inaugural Workshop*, Feb. 2006.

9. P. Mitran, H. Ochiai, and V. Tarokh, "Space-time diversity enhancements using collaborative communication," *IEEE Trans. Inf. Theory*, vol. 51, pp. 2041–2057, June 2005.

10. C. T. K. Ng and A. Goldsmith, "Capacity gain from transmitter and receiver cooperation," in *Proc. IEEE International Symposium on Information Theory*, Sept. 2005.

11. S. Jafar, "Capacity with causal and non-causal side information – a unified view," submitted to *IEEE Trans. Inf. Theory*, Oct. 2005.

12. A. Carleial, "Interference channels," *IEEE Trans. Inf. Theory*, vol. IT-24, pp. 60–70, Jan. 1978.

13. H.Weingarten, Y. Steinberg, and S. Shamai, "The capacity region of the Gaussian MIMO broadcast channel," *IEEE Trans. Inf. Theory*, vol. 52, pp. 3936–3964, Sept. 2006.

14. E. C. van der Meulen, "Three-terminal communication channels," *Adv. Appl. Prob.*, vol. 3, pp. 120–154, 1971.

15. N. Devroye, P. Mitran, and V. Tarokh, "Achievable rates in cognitive radio channels," in *39th Ann. Conf. Inf. Sci. Syst. (CISS)*, Mar. 2005.

16. T. Han and K. Kobayashi, "A new achievable rate region for the interference channel," *IEEE Trans. Inf. Theory*, vol. IT-27, no. 1, pp. 49–60, 1981.

17. T. Cover and J. Thomas, *Elements of Information Theory*. New York: Wiley, 1991.

18. S. Gel'fand and M. Pinsker, "Coding for channels with random parameters," *Probl. Contr. Inf. Theory*, vol. 9, no. 1, pp. 19–31, 1980.

19. M. Costa, "Writing on dirty paper," *IEEE Trans. Inf. Theory*, vol. IT-29, pp. 439–441, May 1983.

20. I. Maric, R. Yates, and G. Kramer, "The strong interference channel with common information," in *Proc. of Allerton Conference on Communications, Control and Computing*, Sept. 2005.

21. S. Jafar, "Degrees of freedom on the MIMO X channel – optimality of zero forcing and the MMK scheme," submitted to *IEEE Trans. Inf. Theory*, Sept. 2006.

22. I. E. Telatar, "Capacity of multi-antenna Gaussian channels," *Eur. Trans. Telecommun.*, vol. 10, no. 6, pp. 585–595, 1999.

23. G. J. Foschini and M. J. Gans, "On limits of wireless communications in a fading environment when using multiple antennas," *Wireless Personal Commun.*, vol. 6, pp. 311–335, 1998.

24. A. Host-Madsen, "Capacity bounds for cooperative diversity," *IEEE Trans. Inf. Theory*, vol. 52, pp. 1522–1544, Apr. 2006.

25. A. Host-Madsen, "The multiplexing gain of wireless networks," in *Proc. of ISIT*, Sept. 2005.

26. N. Devroye and M. Sharif, "The value of partial side information in interfering channels," in preparation.

27. M. Maddah-Ali, A. Motahari, and A. Khandani, "Combination of multi-access and broadcast schemes," in *Proc. IEEE International Symposium on Information Theory* (Seattle, WA), pp. 2104–2108, July 2006.

28. R. G. Gallagher, *Information Theory and Reliable Communication*, ch. 7. New York: Wiley, 1968.

29. T. Cover, A. E. Gamal, and M. Salehi, "Multiple access channels with arbitrarily corre-lated sources," *IEEE Trans. Inf. Theory*, vol. IT-26, pp. 648–657, Nov. 1980.
30. F. Willems and E. van der Meulen, "The discrete memoryless multiple-access channel with cribbing encoders," *IEEE Trans. Inf. Theory*, vol. IT-31, pp. 313–327, Nov. 1985.
31. T. M. Cover and A. E. Gamal, "Capacity theorems for the relay channel," *IEEE Trans. Inf. Theory*, vol. 25, pp. 572–584, Sept. 1979.
32. M. Aref, "Information flow in relay networks," Technical report, Stanford University, 1980.
33. G. Kramer, M. Gastpar, and P. Gupta, "Cooperative strategies and capacity theorems for relay networks," *IEEE Trans. Inf. Theory*, vol. 51, Sept. 2005.
34. P. Gupta and P. R. Kumar, "Towards an information theory of large networks: An achiev-able rate region," *IEEE Trans. Inf. Theory*, vol. 49, pp. 1877–1894, Aug. 2003.
35. L.-L. Xie and P. R. Kumar, "A network information theory for wireless communication: Scaling laws and optimal operation," *IEEE Trans. Inf. Theory*, vol. 50, pp. 748–767, May 2004.
36. L.-L. Xie and P. R. Kumar, "An achievable rate for the multiple level relay channel," submitted to *IEEE Trans. Inf. Theory*, vol. 51, no. 4, April 2005.
37. G. Atia, M. Sharif, and V. Saligrama, "On optimal outage in relay channels with general fading distributions," in *Proc. of Allerton Conference on Communications, Control and Computing*, Oct. 2006.
38. A. E. Gamal, M. Mohseni, and S. Zahedi, "On reliable communication over additive white gaussian noise relay channels," *IEEE Trans. Inf. Theory*, 2006.
39. K. Azarian, H. El Gamal, and P. Schniter, "On the achievable diversity-multiplexing trade-off in half-duplex cooperative channels," *IEEE Trans. Inf. Theory*, Dec. 2005.
40. A. S. Avestimehr and D. N. Tse, "Outage-optimal relaying in the low SNR regime," in *Proc. IEEE International Symposium on Information Theory*, Sept. 2005.
41. J. N. Laneman, D. N. C. Tse, and G. W. Wornell, "Cooperative diversity in wireless net-works: Efficient protocols and outage behavior," *IEEE Trans. Inf. Theory*, 2004.
42. J. Wolfowitz, *Coding Theorems of Information Theory*. New York: Springer-Verlag, 1978.
43. I. Csisz'ar and J. K¨orner, *Information Theory: Coding Theorems for Discrete Memory-less Systems*. New York: Academic Press, 1981.
44. A. Reznik, S. Kulkarni, and S. Verdú, "Capacity and optimal resource allocation in the degraded Gaussian relay channel with multiple relays," in *Proc. of Allerton Conference on Communications, Control and Computing* (Monticello, IL), Oct. 2002.
45. M. Katz and S. Shamai, "Communicating to co-located ad-hoc receiving nodes in a fading environment," in *Proc. IEEE International Symposium on Information Theory* (Chicago, IL), p. 115, July 2004.
46. J. N. Laneman and G. W. Wornell, "Distributed space-time-coded protocols for exploiting cooperative diversity in wireless networks," *IEEE Trans. Inf. Theory*, vol. 49, pp. 2415–2425, Oct. 2003.
47. S. Jafar and S. Srinivasa, "Capacity limits of cognitive radio with distributed dynamic spectral activity," in *Proc. of ICC*, June 2006.
48. S. Srinivasa, S. Jafar, and N. Jindal, "On the capacity of the cognitive tracking channel," in *Proc. of ISIT*, July 2006.
49. S. Srinivasa and S. Jafar, "On the capacity of the cognitive tracking channel," in prepara-tion.

Additional Reading

1. A. E. Gamal, "A capacity of a class of broadcast channels," *IEEE Trans. Inf. Theory*, vol. 25, pp. 166–169, Mar. 1979.
2. K. Marton, "A coding theorem for the discrete memoryless broadcast channel," *IEEE Trans. Inf. Theory*, vol. 25, pp. 306–311, May 1979.
3. C. Shannon, "A mathematical theory of communication," *Bell Syst. Tech. J.*, vol. 27, Jul., Oct. 1948.
4. E. C. van der Meulen, "A survey of multi-way channels in information theory: 1961–1976," *IEEE Trans. Inf. Theory*, vol. 23, pp. 1–37, 1977.
5. W. D. Horne, "Adaptive spectrum access: Using the full spectrum space, in *Proc. Telecommunications Policy Research Conference (TPRC)*, September 2003."
6. F. M. J. Willems, E. C. van der Meulen, and J. P. M. Schalkwijk, "An achievable rate region for the multiple access channel with generalized feedback," in *Proc. Allerton Conference on Communications, Control and Computing*, pp. 284–293, Oct. 1983.
7. J. Peha, "Approaches to spectrum sharing," *IEEE Commun. Mag.*, vol. 43, no. 2, pp. 10–12, 2005.
8. T. Cover, "Broadcast channels," *IEEE Trans. Inf. Theory*, vol. IT-18, pp. 2–14, Jan. 1972.
9. G. Kramer and M. Gastpar, "Capacity theorems for wireless relay channels," in *Proc. Allerton Conference on Communications, Control and Computing*, pp. 1074–1083, 2003.
10. T. Hunter and A. Nostratinia, "Coded cooperation under slow fading, fast fading, and power control," in *Asilomar Conference on Signals, Systems, and Computers*, Nov. 2002.
11. T. Hunter and A. Nosratinia, "Coded cooperation under slow fading, fast fading, and power control," in *Asilomar Conference on Signals, Systems, and Computers*, Nov. 2002.
12. T. Hunter, A. Hedayat, M. Janani, and A. Nostratinia, "Coded cooperation with spacetime transmission and iterative decoding," in *WNCG Wireless Networking Symposium*, Oct. 2003.
13. J. Mitola, "Cognitive radio for flexible mobile multimedia communications," in *Proc IEEE Mobile Multimedia Conference*, 1999.
14. T. Cover, "Comments on broadcast channels," *IEEE Trans. Inf. Theory*, vol. 44, pp. 2524–2530, Sept. 1998.
15. A. Stefanov and E. Erkip, "Cooperative space-time coding for wireless networks," *IEEE Trans. Commun.*, vol. 53, pp. 1804–1809, Nov. 2005.
16. N. Jindal and A. Goldsmith, "Dirty-paper coding versus tdma for mimo broadcast channels," *IEEE Trans. Inf. Theory*, vol. 51, pp. 1783–1794, May 2005.
17. T. Cover and M. Chiang, "Duality between channel capacity and rate distortion," *IEEE Trans. Inf. Theory*, vol. 48, no. 6, 2002.
18. S. Viswanath, N. Jindal, and A. Goldsmith, "Duality, achievable rates and sum rate capacity of the gaussian MIMO broadcast channel," *IEEE Trans. Inf. Theory*, vol. 49, pp. 2658–2668, Oct. 2003.
19. FCC, "FCC ET docket no. 03-108: Facilitating opportunities for flexible, efficient, and reliable spectrum use employing cognitive radio technologies," Technical rep., FCC, 2003.
20. F. C. C. S. P. T. Force, "FCC report of the spectrum efficiency working group," Technical report, FCC, 2002.
21. J. Mitola, "Future of signal processing – cognitive radio," in *Proc IEEE ICASSP*, May 1999. Keynote address.
22. J. Korner and K. Marton, "General broadcast channels with degraded message sets," *IEEE Trans. Inf. Theory*, vol. 23, pp. 60–64, Jan. 1979.
23. J. Korner and K. Marton, "General broadcast channels with degraded message sets," *IEEE Trans. Inf. Theory*, vol. 23, pp. 60–64, Jan. 1979.

24. F. M. J. Willems, "Information theoretic results for the discrete memoryless multiple access channel," PhD Thesis, Katholieke Universiteit Leuven, Oct. 1982.
25. D. Bertsimas and J. Tsitsiklis, *Introduction to Linear Optimization*. Belmont: Athena Scientific, 1997.
26. N. Devroye, P. Mitran, and V. Tarokh, "Limits on communications in a cognitive radio channel," *IEEE Commun. Mag.*, June 2006.
27. R. Ahlswede, "Multi-way communcation channels," in *Proc. Int. Symp. Inf. Theory*, Sept. 1973.
28. A. B. Carleial, "Multiple-access channels with different generalized feedback signals," *IEEE Trans. Inf. Theory*, vol. 28, pp. 841–850, Nov. 1982.
29. T. Berger, "Multiterminal source coding," in G. Longo, editor, *The Information Theory Approach to Communications*. New York: Springer-Verlag, 1977.
30. I. Sason, "On achievable rate regions for the Gaussian interference channel," *IEEE Trans. Inf. Theory*, June 2004.
31. M. Khojastepour, A. Sabharwal, and B. Aazhang, "On capacity of Gaussian 'cheap' relay channel," in *Proc. IEEE Global Telecommun. Conf.*, pp. 1776–1780, Apr. 2003.
32. G. Caire and S. Shamai, "On the achievable throughput of a multi-antenna Gaussian broadcast channel," *IEEE Trans. Inf. Theory*, vol. 49, pp. 1691–1705, July 2003.
33. M. Gastpar and M. Vetterli, "On the asymptotic capacity of Gaussian relay networks," in *Proc. IEEE International Symposium on Information Theory* (Lausanne, Switzerland), p. 195, July 2002.
34. M. Khojastepour, A. Sabharwal, and B. Aazhang, "On the capacity of 'cheap' relay networks," in *Conference on Information Sciences and Systems*, Apr. 2003.
35. M. Gastpar and M. Vetterli, "On the capacity of wireless networks: The relay case," in *Proc. IEEE INFOCOM* (New York, NY), pp. 1577–1586, June 2002.
36. M. Costa, "On the gaussian interference channel," *IEEE Trans. Inf. Theory*, vol. 31, pp. 607–615, Sept. 1985.
37. G. Kramer, "Outer bounds on the capacity of Gaussian interference channels," *IEEE Trans. Inf. Theory*, vol. 50, Mar. 2004.
38. V. Tarokh, N. Seshadri, and A. Calderbank, "Space-time codes for high data rate wireless communication: Performance criterion and code construction," *IEEE Trans. Inf. Theory*, vol. 44, pp. 744–765, Mar. 1998.
39. T. Weiss and F. Jondral, "Spectrum pooling: An innovative strategy for the enhancement of spectrum efficiency," *IEEE Commun. Mag.*, pp. S8–S14, Mar. 2004.
40. O. Simeone, Y. Bar-Ness, and U. Spagnolini, "Stable throughput of cognitive radios with relaying capability," in *Proc. Fourty-Fourth Annual Allerton Conference on Communication, Control, and Computing*, Sept. 2006.
41. W. Yu and J. Cioffi, "Sum capacity of gaussian vector broadcast channels," *IEEE Trans. Inf. Theory*, vol. 50, pp. 1875–1892, Sept. 2004.
42. P. Viswanath and D. Tse, "Sum capacity of the vector gaussian broadcast channel and downlink-uplink duality," *IEEE Trans. Inf. Theory*, vol. 49, pp. 1912–1921, Aug. 2003.
43. S. Vishwanath, N. Jindal, and A. Goldsmith, "The "Z" channel," in *Proc. IEEE Global Telecommun. Conf.*, Dec. 2003.
44. H. Sato, "The capacity of Gaussian interference channel under strong interference," *IEEE Trans. Inf. Theory*, vol. IT-27, Nov. 1981.
45. T. Han, "The capacity of general multiple-access channels with certain correlated sources," *Information and Control*, vol. 40, pp. 37–60, 1979.
46. I. Maric, R. Yates, and G. Kramer, "The discrete memoryless compound multiple access channel with conferencing encoders," in *Proc. IEEE International Symposium on Information Theory*, Sept. 2005.

47. B. Schein and R. G. Gallager, "The Gaussian parallel relay network," in *Proc. IEEE International Symposium on Information Theory* (Sorrento, Italy), p. 22, June 2000.

48. M. Gastpar, G. Kramer, and P. Gupta, "The multiple relay channel: Coding and antennaclustering capacity," in *Proc. IEEE International Symposium on Information Theory* (Lausanne, Switzerland), p. 136, July 2002.

49. J. Mitola, "The software radio architecture," *IEEE Commun. Mag.*, vol. 33, pp. 26–38, May 1995.

50. I. Kang, W. Sheen, R. Chen, and S. L. C. Hsiao, "Throughput improvement with relayaugmented cellular architecture," Sept. 2005.

51. E. van der Meulen, "Transmission of information in a T-terminal discrete memoryless channel," Technical report, University of California, Berkeley, 1968.

52. H. Sato, "Two user communication channels," *IEEE Trans. Inf. Theory*, vol. IT-23, Nov. 1985.

53. A. Sendonaris, E. Erkip, and B. Aazhang, "User cooperation diversity–part I: System description," *IEEE Trans. Commun.*, vol. 51, pp. 1927–1938, Nov. 2003.

54. A. Sendonaris, E. Erkip, and B. Aazhang, "User cooperation diversity–part II: Implementation aspects and performance analysis," *IEEE Trans. Commun.*, vol. 51, pp. 1939–1948, Nov. 2003.

3

Coexistence and Dynamic Sharing in Cognitive Radio Networks *

Sofie Pollin[1,2]

[1] Inter-university Micro-Electronics Center (IMEC), Belgium
[2] University of California, Berkeley, USA
    pollins@{eecs.berkeley.edu,imec.be}

3.1 Introduction

Wireless technology has enabled the development of increasingly diverse applications resulting in an exponential growth in usage and services. To cope with the demand, network design has been focusing on increasing the spectral efficiency by designing more and more complex algorithms to be used by powerful portable devices. In parallel, complex protocols have been developed to adaptively deliver the required quality of service (QoS) to the heterogeneous applications. This has resulted in the introduction of different standards specifying the physical (PHY) or medium access control (MAC) layers for a range of wireless communication technologies.

Clearly, the increasing diversification of possible application, devices and access technologies has resulted in a complex wireless communication scene with various types of networks coexisting. In response to this, physical layer design started focusing on offering flexible solutions, to be used by reconfigurable devices. Combined with adaptive protocol solutions, this potentially allows users to seamlessly roam across wireless access technologies and search for the best-fit solution, given the environment and application requirements. Intelligent dynamic sharing rules are needed to use the spectrum most efficiently.

In this chapter, such intelligent dynamic sharing is discussed, focusing on coexistence or sharing between networks of varying regulatory status and equipped with heterogeneous intelligent capabilities. In Sect. 3.2 the technological trends that motivate the evolution toward dynamic sharing are discussed. In Sect. 3.3 an overview of the various types of coexistence and dynamic sharing in wireless networks is given. In Sect. 3.4, the different tasks involved in dynamic sharing are discussed.

To illustrate the main concepts, two case studies are treated in more detail in Sects. 3.5 and 3.6. The first example focuses on the coexistence of a primary licensed network with secondary or license-exempt users. In this context, it is discussed how

---

 * Portions reprinted with permission from Pollin et al. [1] *[©2006 IEEE]*. Figs. 3.12 and 3.13 reprinted with permission from [2] *[©2005]*. The IEEE disclaims any responsibility or liability resulting from the placement and use in the described manner.

the dynamic or opportunistic channel access is organized in the IEEE 802.22 standard. Secondly, a case study is discussed where networks of equal regulatory status coexist. The focus is on spectrum sharing between 802.15.4 and 802.11 networks.

3.2 Cognitive Radio:
Innovative Concept Building on Technological Trends

Interest in wireless technology has been increasing exponentially over the last decade. New standards are being released at a fast rate to improve the achieved performance, diversify the possible applications and open up new frequency bands [3]. Recently, advances in reconfigurable hardware have paved the way to flexible radios (or software-defined radios (SDR) [4]) that can adapt their air interface and communication protocol to operate using a range of existing standards or access technologies (Fig. 3.1). The advantages to the user are twofold. First, it is now possible to use a range of applications, relying on different wireless communication techniques, in a single portable device. The second added value lies in the fact that users can now seamlessly and opportunistically roam across various wireless access networks in the search for more throughput or cheaper bandwidth. This second advantage from a user point of view, maybe the most attractive.

However, the realization of true seamless handover requires a tight coupling of the hardware flexibility with the higher layer protocol layers. Intelligent schemes for environment awareness, hand-off and distributed QoS control are needed. The com-

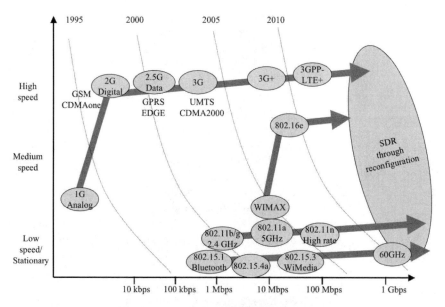

Fig. 3.1. The wireless communication scene.

bination of flexibility and increased protocol intelligence has recently led to the novel concept of a cognitive radio (CR) that adapts to the current environment and spectrum use [5]. In its most generic form, this is an innovative technology to exploit the available dynamics leveraging reconfigurability, increased awareness and intelligent control. In a more restricted definition, cognitive devices use their increased flexibility and awareness to control channel access dynamically, i.e., to achieve a dynamic spectrum access (DSA) [6]. Such flexible behavior can result in a very efficient use of the spectrum. The spectrum efficiency gains increase with the dynamics present in the wireless communication and with the number of spectrum bands opened for such opportunistic access. The latter is taken care of by spectrum regulation authorities, e.g., the Federal Communications Commission (FCC) is opening up the TV bands for opportunistic access as a first initiative [7].

In this section, we first give an overview of the many dynamics present in wireless communication that motivate the need of dynamic sharing. Next, we define in more detail the two important technical requirements to achieve DSA as a first step toward CRs. It will be shown that the innovative concept of a CR for DSA is technically possible today.

3.2.1 There are Dynamics to Exploit

Since wireless technology has become so cheap and reliable, more and more applications use the technology. First, wireless technology was mainly used to replace wires in existing applications. More importantly, a broad range of new mobile applications became possible thanks to the introduction of wireless communication. These range from mobile multimedia applications for mobile terminals, ad hoc file transfer, ad hoc mesh networking applications (such as distributed gaming and disaster management) or even large networks of tiny sensors for environment monitoring. Obviously, the QoS requirements for each of those applications are very different. As a result the spectrum access pattern varies significantly across applications.

Next to the range of possible applications, the demand and QoS requirements vary significantly over the lifetime of some applications. This is the case for, e.g., variable bit rate video applications where the frame size and hence throughput demand is very different from frame to frame. On top of this uncontrolled dynamic behavior, scalable applications exist that actively adapt, e.g., the bit rate, to the current environment [8]. QoS requirements are hence not only application dependent, but also time varying.

On top of that, there clearly exist patterns in the use of wireless applications. Depending on the time of the day, or even day in the week, the use of voice, data or video streaming applications varies drastically. While wireless communication is useful in a surprisingly broad range of average users' occupations, the type of task and involved technology varies significantly, e.g., from cellular over Internet access to DVB. The use pattern of these technologies is typically very predictable.

Wireless communication in itself is very dynamic in nature. Due to mobility of the users or movements in the neighborhood, the channel is varying over time, frequency and space. These variations are enlarged when we add to this the varying impact of other interfering wireless transmissions. Even from the network point of view wireless networks vary since wireless communication allows node mobility. Nodes join and leave the network, resulting in very dynamic traffic scenarios.

Clearly, given all the dynamics present in wireless communication, is it required to design flexible and adaptive solutions for spectrum allocation? Indeed, a static design that is based on the worst case, i.e., largest spectrum needs of all users, would not be feasible since such an amount of spectrum might not exist. Alternatively, a static allocation based on average spectrum requirements might not achieve the required performance at all time. Alternatively, adapting to the dynamics present in wireless communication is clearly a nice opportunity to overcome spectrum scarcity and improve application QoS. Next to spectrum regulation challenges, this opportunity relies on the availability of flexible systems that can be controlled intelligently. In the next section, we introduce two important technology trends that will allow the design of such wireless systems.

3.2.2 Cognitive Radio for Dynamic Access

Next to regulatory issues related to spectrum licensing, this innovative concept also involves some technological challenges. It is required to build flexible hardware solutions that can easily be tuned to the current band or access technique that is the most appropriate. Next, in order to know the most appropriate band or technique, the systems needs to be aware of the environment, and intelligent control solutions need to be added to current protocols. In this section it will be shown that both the trend toward flexible hardware and the evolution toward more adaptive protocols are present in the current wireless technology design advances. As a result the innovative cognitive radio concept can be evolved from those advances.

As illustrated in Fig. 3.2, improvements in wireless processor technology have enabled the development of fast physical layer schemes to improve spectral efficiency and hence throughput [9]. However, in order to achieve the required QoS for a range of applications and users, increasing the throughput is typically not sufficient. Simultaneously, developments in distributed control of channel access and resource control enable distributed QoS control and optimization [10]. Protocols achieve such improved QoS through adapting to the specific QoS requirements of the application.

In a next phase, advances in hardware design have enabled the design of flexible systems, allowing adaptation to the dynamic application demands or wireless communication environment. In [11] for example, a flexible power amplifier (PA) is proposed that allows to effectively trade-off the performance versus effective power consumption of the PA which is a crucial hardware component in each transmitter. With that PA, it becomes possible to implement transmit power control which is an important aspect of spectrum sharing for ad hoc networks [10]. Next, the physical layer standardized in IEEE 802.11a/g [12] enables adaptation of the code rate or modulation order in state-of-the art WLAN transceivers. Considering the IEEE 802.11 medium access control (MAC) layer, QoS extensions are standardized in the IEEE 802.11e and a range of tunable MAC parameters are enabled to tune the QoS achieved in wireless networks [13].

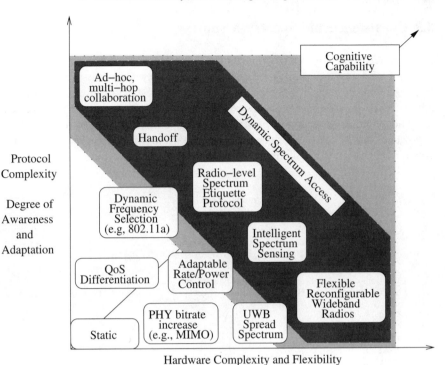

Fig. 3.2. Cognitive radio design space.

The trend toward more flexibility in radio implementations is expected to result in software-defined radios which are, ideally, transceivers that can be tuned to a range of possible networking standards (both physical and medium access control layers) depending on the instantaneous environment scenario. With the help of those radios, it is possible to adaptively tune the wireless communication to that technology or spectrum access method that is currently the most efficient in terms of throughput or cost. However, to be effective, this SDR technology needs to be complemented with intelligent spectrum scanning, seamless hand-off, a spectrum etiquette protocol and distributed ad hoc and multi-hop networking functionalities. A combination of these important tasks enables dynamic spectrum access (DSA). Eventually, this process toward more reconfigurability, awareness and intelligent control will lead to the concept of true cognitive radio. In this chapter, the focus is, however, on dynamic spectrum sharing as an important milestone in this evolution toward more and more cognitive capability. Concepts such as dynamic spectrum access and dynamic spectrum sharing are introduced in the next section.

3.3 Coexistence and Spectrum Sharing

As introduced in the previous section, dynamic spectrum access technology is needed to achieve a better use of the spectrum, given the many variations present in wireless communication. DSA is the opposite of the current static spectrum management policy. However, various approaches are possible to make the spectrum management more adaptive, as presented in Fig. 3.3. In this chapter, the focus will be on those approaches that involve coexistence, or dynamic spectrum sharing. The different flavors for DSA are first briefly defined.

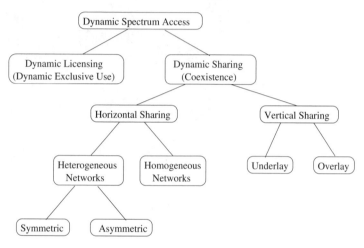

Fig. 3.3. Dynamic spectrum access, classification along regulatory status.

3.3.1 Dynamic Licensing

Dynamic licensing results in a dynamic spectrum allocation that gives exclusive use to the technology or network that currently has the most profit of spectrum use. It is similar to the current spectrum regulation in that it licenses spectrum bands for exclusive use. This dynamic licensing is, however, much more flexible, to be able to adapt to the wireless communication dynamics. Two approaches have been introduced: spectrum property rights [14, 15] and dynamic spectrum allocation [16]. The first approach allows licensed users to sell spectrum freely. The main idea behind this strategy is that economy will automatically drive the users toward more effective and hence profitable use of the spectrum. So, although spectrum sharing is not mandatory in this model, licensed users will be motivated to share following the rules of supply and demand. The second approach is followed by the European DRiVE project [16]. It dynamically adapts spectrum allocation as a function of the variations present in the wireless communication scene. This adaptation can be in time or in space, at a much finer granularity than the current spectrum regulation variations over time or space. The efficiency will depend on the ability to predict traffic load (or spectrum occupancy).

3.3.2 Coexistence or Dynamic Sharing

The above-presented models are still based on an exclusive-use model. As a result, they are expected to be limited regarding the adaptation speed. Ideally, spectrum sharing should adapt very fast to all dynamics present in wireless communication, which can be caused by the channel variations or because of the bursty application demands. Coexistence or dynamic sharing allows such sharing, in theory, on a packet per packet basis since it licenses spectrum to networks simultaneously, while relying on in-network spectrum sharing techniques to avoid conflicts. In this section, the various coexistence scenarios that can be encountered are classified.

3.3.2.1 Horizontal Spectrum Sharing

This model for spectrum sharing assumes that all networking nodes have equal regulatory status. As a result, this model is also referred to as open sharing model [17] or as spectrum commons [18, 19]. Medium access protocols for wireless networks are working according to this model, and considerable literature can be found on both centrally controlled or distributed access techniques for spectrum sharing between nodes of a single network. Next to that, techniques exist for spectrum planning between different networks using the same access technology. Also, techniques are being developed for spectrum sharing across heterogeneous networks. This is especially useful in the unlicensed ISM and U-NII bands for which a broad range of technologies exist. We briefly give an overview of the main techniques for spectrum sharing across homogeneous networks and heterogeneous networks. The focus is on inter-network sharing only.

Homogeneous Networks

Spectrum sharing between networks has typically been solved by careful spectrum planning. With the introduction of the 802.11a standard for WLAN communication in the U-NII 5 GHz frequency band, dynamic frequency selection (DFS) was introduced. 802.11a base stations can autonomously and dynamically select the best channel for their WLAN access network while avoiding interference to existing communication. Next to the current channel, transmit power control was added for interference mitigation between neighboring 802.11a networks.

A distributed spectrum sharing scheme for wireless Internet service providers that share the same spectrum is proposed in [20]. A distributed QoS based dynamic channel reservation (D-QDCR) is proposed. Depending on the QoS requirement of its users, a base station of a certain service provider will compete with interfering base stations for spectrum. Control and data channels are split, which means that spectrum competition is done using a dedicated common control channel (CCC).

Various competition policies are proposed as a function of the traffic type of the users.

Game-theoretic concepts are used in [21] to determine the transmit power settings of users in a distributed way. Cooperative and non-cooperative (i.e., when the users are considered to be selfish) algorithms exist for distributed sharing solutions. Next to distributed solutions, centrally controlled techniques are often proposed.

For dynamic spectrum sharing between different cognitive radio networks, a spectrum policy server (SPS) for central spectrum coordination is proposed in [21].[3] Each operator bids for the spectrum indicating the cost it will pay for the duration of the usage. The SPS then allocates the spectrum by maximizing its profit from these bids. The operators also determine the price for the users, upon which users can freely select which operator to use for a given traffic type. When compared to the case where each operator is assigned an equal share of the spectrum, a higher throughput, which means lower price or higher revenue, is achieved.

Heterogeneous Networks

Due to the success of the unlicensed ISM and U-NII bands, spectrum sharing between heterogeneous networks of equal regulatory status is possible. The problem was first noticed in the coexistence of 802.11b and 802.15.1 (Bluetooth) networks. To address this, the IEEE 802.15.2 working group has been established to solve the coexistence problems. The 802.15.1 PHY is based on FHSS (frequency hopping spread spectrum), which means that every physical layer symbol is formed by a frequency hopping sequence. To avoid the harmful interference of 802.11b networks, adaptive frequency hopping has been proposed for Bluetooth. Alternatively, approaches that rely on a cooperation between 802.11 and 802.15.1 have been proposed, such as deterministic frequency nulling or a time division multiple access scheduling of both technologies.

Alternatively, the common spectrum coordination channel (CSCC) [22] etiquette protocol is proposed for coexistence of IEEE 802.11b and 802.16a networks. It requires that users of both technologies have cognitive capabilities, i.e., each node has to be equipped with a cognitive radio and a low bit rate control radio. The coexistence is achieved by sending information to the different networks through broadcast messages over the control channel. Using this information, each user locally decides the channel it can use and the appropriate power level.

In the above-discussed case for coexistence, both considered networks adapt their transmission schemes as function of the environment. This is because both networks can benefit from avoiding the mutual interference, and both networks have (limited) adaptive or cognitive capabilities. In this case, spectrum sharing can be classified as symmetric in nature. Alternatively, it is possible that only one of the involved

---

[3] It should be noted that a central spectrum policy server acting as a superbasestation for spectrum access is also used in dynamic licensing. However, the SPS considered here coordinates access between homogeneous networks, and can hence be considered as an in-network spectrum sharing solution.

networks dynamically adapts its spectrum access. This can be because the other co-existing network has no incentives to adapt, or because the other network has no adaptive capabilities. The former case will be discussed in more detail in Sect. 3.6, where the coexistence of powerful 802.11 networks with low-power 802.15.4 net-works will be discussed. The latter case is possible when the coexistence of legacy technology with adaptive cognitive radio technology is considered.

This asymmetric spectrum sharing, in which only one of the technologies is adaptive, is somewhat similar to vertical spectrum sharing as will be introduced next.

3.3.2.2 Vertical Spectrum Sharing

The initial definition of spectrum sharing assumes the existence of a primary and a secondary user. While the spectrum has been licensed to the primary user only, the secondary user can use it opportunistically provided this does not affect the primary users' performance. Two approaches exist for spectrum access to minimize the interference caused to the primary users by the secondary users' communication: spectrum overlay and spectrum underlay (Fig. 3.4). The underlay approach severely constrains the transmission power of the secondary users, so that the interference for the primary users is below a certain level. This is possible by spreading the communication signals over a very wide band (i.e., ultra-wideband communication). This approach is, however, not yet adaptive to the communication statistics of the primary users? Indeed, the transmission levels are still based on the worst case assumption that primary users communicate constantly. In dynamic sharing, the goal is, however, to adapt to the communication dynamics.

Spectrum overlay was first envisioned by Mitola [23], using the terminology spectrum pooling. In this access method, radios seek spectrum holes for their communication. A spectrum hole is defined in space, time and frequency. Within such a

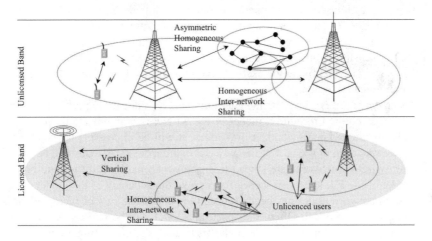

Fig. 3.4. Various types of coexistence.

hole, no restrictions on the transmission of the secondary users are imposed. Adaptation to spectrum dynamics is only limited by the granularity for defining the holes in time, space, or frequency. Also, this approach fits well with existing spectrum allocation. Legacy systems continue their operation without being affected by the secondary users. As a result, this promising concept was picked up by the DARPA XG program [24] who introduced the term opportunistic spectrum access (OSA). The first worldwide standard based on the cognitive radio technology uses this definition of cognitive radio [25]. This IEEE 802.22 project was started in 2004 and targets the definition of a cognitive wireless regional area network for operation in the TV bands, as will be discussed in Sect. 3.5.

3.4 Dynamic Sharing and the Cognition Cycle

Irrespective of the type of coexistence considered, or the level of reconfigurability or spectrum awareness, it is possible to identify three important steps in the dynamic sharing process. In this section, an overview of these steps that together represent a cognition cycle for dynamic sharing is presented.

3.4.1 Basic Cognition Cycle for Dynamic Sharing

Crucial in the design of dynamic sharing techniques is the cognitive capability of the radio systems. Such cognitive capability allows the fast interaction with the environment to dynamically determine the best communication or spectrum access strategy. The main cognitive tasks required to achieve dynamic sharing are depicted in Fig. 3.5, and are referred to as *cognitive cycle*. The three main steps [26] of this cycle are *spectrum sensing*, *spectrum analysis* and *spectrum decision*:

- *Spectrum sensing:* To be able to adapt to the dynamics present in the wireless communication scene, it is imperative to monitor it in detail. Typically, spectrum sensing is defined as monitoring the available spectrum bands to detect spectrum holes. The main challenge is to do this energy efficiently. Also, the hardware cost for this sensing should be taken into account. Next to sensing the spectrum, it will be shown later that it can be useful to monitor other information in the wireless communication scene.
- *Spectrum analysis:* Based on the measurements obtained through spectrum sensing, it is required to build a model of the wireless communication scene. Due to hardware and energy budget limitations, it will be impossible to monitor the whole spectrum or scene continuously and in great detail. Also, as will be shown later, wireless networks are spread in space which makes it difficult to build a model of the full spectrum scene. To improve the accuracy of the model, cooperation between nodes has been proposed in [27], at the cost of increased communication overhead. Often, the spectrum model will be built on local partial information. Techniques for local spectrum analysis are an important research objective for this task.

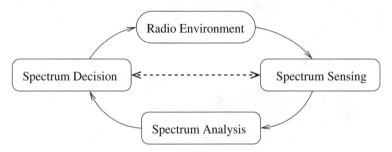

Fig. 3.5. Cognitive cycle for dynamic sharing.

- *Spectrum decision:* The spectrum decision is about whether and how to access the spectrum. The more reconfigurability present in the cognitive radio, the more optimization options are available. The optimal spectrum access option is the one that maximizes the application or user requirements given the environment or spectrum policy constraints. The spectrum decision is framed as an optimization problem using the model built during spectrum analysis. This optimization problem can have a local or a global optimization goal. Next, the spectrum decision should be communicated across users, which is often done through a common channel. Since the availability of such channel cannot be relied on in the context of dynamic opportunistic spectrum access, the major spectrum decision challenge relates to the development of spectrum coordination techniques that do not rely on such a common channel.

3.4.2 Spectrum Sensing

Cognitive capability relies to a large extent on the awareness of the environment dynamics. For dynamic spectrum sharing, important information is related to the spectrum use by other nodes in the network. These other users can be primary users that should not be interfered with. Alternatively, they can be users of equal regulatory status that occupy the channel. These other users are not necessarily equipped with cognitive capability, and more importantly, they can be much more powerful. In all cases, it is necessary to become aware of the presence of the different other users.

Wireless communication typically involves a transmitter and one or more receivers. Depending on the type of the other users, and depending on the optimization goal of the cognitive or sensing user, it is required to get information on other transmitters or alternatively other receivers. In case the other users are primary users that should not be interfered with, the sensing user should obtain information on primary receivers involved in communication. When communicating, the cognitive radio will cause harmful interference to receivers in its interference range (Fig. 3.6). As a result, the sensing node should be able to detect whether receivers are present in this range. Alternatively, when the cognitive radio is mainly interested in avoiding interference by other users in the network, it has to determine if there are harmful transmitters in the neighborhood. If the cognitive sensing radio is currently not in a transmitter

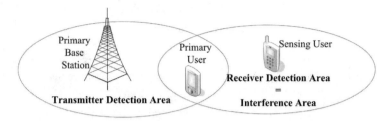

Fig. 3.6. Transmitter detection versus receiver detection.

detection area (Fig. 3.6), it can communicate without interference. Finally, next to listening to the spectrum itself, a cognitive radio can capture important information by listening to its peer nodes. Sensing information can be exchanged in order to improve the accuracy of the spectrum awareness, or alternatively information can be exchanged to improve and harmonize the decisions of cognitive users. The more environment knowledge a cognitive radio can rely on, the better it can optimize the performance of its transmissions. In the next subsections, we will discuss transmitter detection, receiver detection and network monitoring (Fig. 3.7).

3.4.2.1 Sense for Other Transmitters

Three approaches exist for transmitter detection, based on the sensing users' knowledge on the transmitted signals. A matched filter is the most powerful approach, but it, however, relies on synchronization and knowledge of the primary users' signaling. Energy detection is a non-coherent detection method that needs only basic information on the signals. The drawback of this energy detection is that it is less accurate than matched filter detection, for a given number of samples or sensing time. The third method, cyclostationary feature detection, improves the performance of the energy detection by exploiting the inherent periodicity present in communication signals [28].

The main problem is the energy and hardware cost of wide-band sensing. Wideband analog systems are difficult to design, and fast analog-to-digital converters typically consume a lot of energy. In [29] approaches are proposed to sense the full

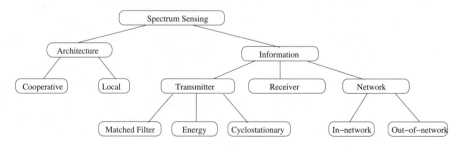

Fig. 3.7. Spectrum sensing options.

spectrum in two steps: a first inaccurate wide-band step and a second focused step to refine the sensing where appropriate. Alternatively, it is possible to couple the sensing strategy to the higher layers operation to determine the minimal frequency band and time period for sensing.

As illustrated in Fig. 3.6, sensing for other transmitters is not sufficient to avoid interference caused to other (primary) users in the network. This is the hidden node problem that can be solved by an additional control message exchange prior to each communication (e.g., the Request to Send (RTS) and Clear to Send (CTS) messages in 802.11 networks). However, in the case of dynamic spectrum access, cooperation and communication between heterogeneous users cannot be relied on. As a result, it is not possible to gather sufficient information by local transmitter detection only. In [27], it is proposed that sensing information is exchanged across the network. When sensing nodes collaborate, the information they can potentially gather is much more detailed and a better view of the wireless communication scene can be established. This approach is also followed in the first standard for cognitive radio [7] as will be shown in Sect. 3.5.

In [30], two distinct networks are deployed separately. A first sensor network is deployed for cooperative spectrum sensing only and an operational network is used for data transmission. The sensor network is deployed in the target sensing area. A central controller processes the information gathered from the sensors and makes the spectrum occupancy map for the operational network.

3.4.2.2 Sense for Other Receivers

As explained before, to detect the presence of other communication that should not be interfered with, it is necessary to be able to detect the receivers involved in that communication. The easiest technique for receiver detection is relying on cooperation from those receivers. This is the case in the RTS/CTS exchange proposed in 802.11 to solve the hidden node problem: the receiver itself helps in avoiding the problem by transmitting a sequence, i.e., the CTS message in this case. Typically, it is, however, not possible to rely on this active cooperation from the receiver. This is mainly because this receiver is typically a device without cognitive capability, and hence not aware of the other devices present in the network.

A first approach for receiver detection that does not rely on receiver cooperation is proposed in [31], for the application of secondary wireless networks operating in the TV bands. It is based on detecting the local oscillator (LO) leakage power in the receiver. However, this approach suffers from a long detection time and a very short detection range. In [31], they propose to deploy large sensing networks dedicated to this spectrum opportunity detection task, with sensors close to each receiver.

Another approach models the total interference temperature at each location in the network. This interference temperature results from summing the contributions from each transmission in the cognitive network. Communication is allowed when the interference is below a certain threshold at each possible moment and location, i.e., low enough not to harm receivers. This is in fact the spectrum underlay sharing approach. As mentioned before, the drawback of this approach is that it does not adapt or take advantage of spectrum holes and often maintains very pessimistic interference temperature levels.

3.4.2.3 Sense for Cross-Layer Information

Next to information about the use of the spectrum or the physical channel, it can be useful to gather information related to the higher layers of the communication protocol stack. For instance, information about the application that is currently used can help to establish the appropriate model of the spectrum use dynamics. Listening to routing protocol messages can help to establish a model of the mobility of the users in the network. The beacons sent by the medium access (MAC) protocol, typically contains useful information about the transmission and sleep schedule of the network. If the signals of the primary or competing users can be detected, they can contain a lot of useful information. This information detection is called out-of-network monitoring.

It is, however, also useful to monitor information from other, cooperative, cognitive users, which is referred to as in-network monitoring. Indeed, wireless communication involves typically a transmitter and a receiver. In order to find a proper frequency band for communication, it is required that the transmitter could harmonize its local decision with the receiver. For that purpose, listening to information of the intended receiver can help the transmitter choose a channel that is optimal for both. It will be shown in Sect. 3.6 that listening to beacons can help improve the connectivity and robustness of the network.

3.4.3 Spectrum Analysis

By analyzing the information measured, it is now possible to identify spectrum opportunities. Intuitively, a spectrum opportunity can be considered to be a spectrum unit that is currently not used by another user. A spectrum unit is defined in time, frequency, or space. Whether a spectrum unit is an opportunity depends on the sharing type considered, as mentioned before and as illustrated in Fig. 3.6. Indeed, if the main goal of the sharing is to avoid causing interference to primary or other users present in the area, an opportunity is present if there are no receivers active in a certain time, space or frequency spot. Alternatively, if the main goal of the sharing is to avoid interference caused by other users, then an opportunity is present if there are no transmitters active in the considered spectrum unit.

Typically, the information gathered by a cognitive or sensing node is not complete or very accurate. This is because wide-band sensing is very hardware and energy costly. Also, because of the spatial distribution of communication nodes in a network, local sensing or data gathering does not give the complete information. Finally, because of the dynamic nature of wireless communication, it is often difficult to react timely, and also estimates of future spectrum access changes are needed. In any case, the spectrum model will be probabilistic in nature. If the dynamics of the wireless scene are not varying too fast, the model can be made more accurate by taking into account a history of spectrum information. In [32], this is modeled as a partially observable markov decision process (POMDP). In [1], how machine learning techniques can help in this spectrum analysis phase is explored. This will be explained in more detail in Sect. 3.6.

The spectrum analysis can be carried out locally in each node, or alternatively it is possible to carry out the analysis centrally by a more powerful device (e.g., in the IEEE 802.22 standard).

The spectrum model built can finally be classified as follows (Fig. 3.8). Some spectrum models base their analysis on the current spectrum sensing results only. If the spectrum is idle, it can be used. Alternatively, if another user is detected, the channel is vacated. Other models keep track of the spectrum usage history, assuming that the spectrum use is rather slowly varying. This allows to save energy or hardware for spectrum sensing. Alternatively, it is possible to embed the variation probabilities in the spectrum model.

3.4.4 Spectrum Decision

The spectrum decision is about whether and how to access the spectrum. This decision is based on the model derived after spectrum analysis. Given the stochastic na-

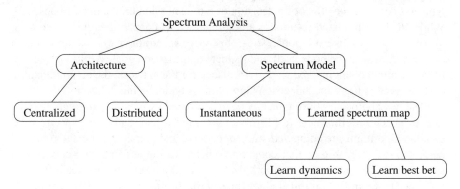

Fig. 3.8. Spectrum analysis classification.

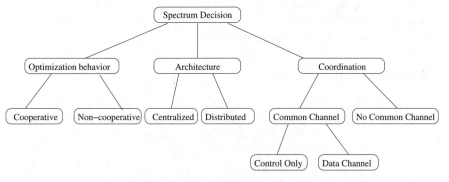

Fig. 3.9. Spectrum decision classification.

ture of that model, a spectrum decision is sometimes better described as a spectrum bet. The goal of the spectrum decision or bet is to best meet the user communication requirements, while satisfying a set of constraints, e.g., the acceptable interference that can be caused to other users in the spectrum. The optimization goal, i.e., the outcome that best meets the user requirements, can be a local or a global criterion. Next, it is important that the spectrum decision is coordinated across cognitive users in the network (Fig. 3.9). Typically, approaches rely on a common control channel (CCC) to achieve this. Relying on the availability of such channel is dangerous, and alternative approaches will be addressed below. Finally, in this section, it is shown that the spectrum decision can impact the future spectrum sensing and hence the amount of "learning" the wireless communication scene. This effect results in the additional arrow connecting spectrum sensing and decision in Fig. 3.5.

3.4.4.1 Local Versus Global Optimization

Two types of behaviors are possible to decide on the spectrum use. More specifically, the optimization objective can be a local one, in which case the decision is taken in a non-cooperative way. Alternatively, cognitive users can aim to optimize a global performance or cost, which results in cooperative behavior. Cooperative decisions typically consider the effect of a decision or a transmission on the other nodes [32]. While all the centralized solutions can be regarded as cooperative, there also exist distributed cooperative solutions. Non-cooperative spectrum sharing is more selfish and only considers the node itself and its local optimization goal. While such selfish behavior might result in a reduced performance for the network, the communication cost involved in those local decision algorithms is typically much lower.

The two approaches have typically been compared by means of throughput, fairness or spectrum efficiency. In [33], both centralized and distributed spectrum allocation algorithms are compared, and cooperative and non-cooperative approaches are compared. It is shown that cooperative behavior outperforms non-cooperative spectrum allocation. Distributed optimization techniques closely approach the performance of centrally controlled algorithms while minimizing the control overhead.

Finally, it should be clear that the optimal decision for communication also depends on the receiver. As mentioned earlier, if information about the receiver is overheard by in-network monitoring, such information can already be considered in the spectrum model. This will be illustrated in Sect. 3.6.

3.4.4.2 Coordination Mechanisms

After the best spectrum opportunity is selected, the decision should be communicated through the network. If the spectrum decision was taken in a centralized way, both receiver(s) and transmitter should be informed. In case of distributed decision taking, only the intended receiver(s) should be updated with the new access strategy. Typical approaches tend to rely on the availability of a common control channel (CCC), which is a channel that is guaranteed to be available to all the cognitive nodes and can be used for control information exchange.[4] This is the approach taken in the first cognitive radio standard 802.22, and in many other projects such as the European DRiVE and OverDRiVE [16]. The advantage is clearly that the spectrum decision can be fully coordinated.

A first main drawback of this approach is that it is typically not possible to rely on a predefined common control channel in a dynamic cognitive radio context. Even if the common channel could be selected adaptively, it is typically even not possible to guarantee the availability of a single common channel for a large network. The availability of spectrum holes varies often drastically over space, which can be due to a varying availability of networking technologies but also due to small-scale fading effects. As a result, a channel that is available in one spot might be unavailable in another part of the network. A second limitation is that relying on a common control channel for the network limits the scalability of the approach. When the network size increases, the common channel often gets congested [34].

Alternatively, approaches exist that do not rely on a common control channel. Stations can change their frequency according to a predetermined pseudo-random pattern. This pseudo-random sequence can be completely specified by knowing two parameters: current channel number and the seed. This method assumes every station is assigned a seed that is different from all other stations. It also assumes that stations are aware of each others' hopping pattern. Transmitters migrate to the current channel of the intended receiver, until the current channel is found to be free and communication is possible. The disadvantage of this approach is the potential long delay before a free channel is found.

In [17], a different approach is taken. They propose a distributed channel selection algorithm to optimize the next channel decision. They assume that transmitter and receiver synchronize, and then follow the same distributed channel selection algorithm and hence keep synchronized. The initial handshake (where they synchronize to a point from which they can start hopping around together) is receiver based. Each receiver has a set of channels on which it should regularly receive to see if a transmitter is looking for it. Transmitters announce pending communication through a handshake message on the subset of the receiver channels that is currently free. The disadvantage of this approach is, however, that receiver and transmitter need to be synchronized in time accurately, to be able to keep hopping together.

3.4.4.3 Spectrum Decision Trade-Off

As shown in Fig. 3.5, the spectrum decision potentially affects future spectrum sensing. Due to hardware limitations and energy cost of spectrum monitoring, users cannot sense the full spectrum continuously. As mentioned also in [17], an optimal sensing decision is to catch a spectrum opportunity for immediate access and obtain statistical information on spectrum occupancy so that more rewarding decisions can be made in the future. A trade-off has to be reached between these two often conflicting

---

[4] Some common control channels are reserved for control information exchange only, while others can be used for data transmission too.

objectives. This trade-off between exploration and exploitation will be illustrated in detail in Sect. 3.6.

3.5 IEEE 802.22 Proposed Approaches for Spectrum Sharing

In this section, some proposed techniques from the IEEE 802.22 WG are discussed, as an illustration and instantiation of the many possible techniques for spectrum sharing. First, the application domain for devices operating under this standard is addressed. Next, spectrum sensing, spectrum analysis and spectrum decision are discussed in more detail. For a more detailed overview of the IEEE 802.22, we refer to [7, 25, 35].

3.5.1 Overview of 802.22

The IEEE 802.22 Working Group (WG) was formed in November 2004, after the FCC released its Notice of Proposed Rule Making (NPRM) for the TV bands in May 2004. This WG specifies an air interface (including PHY and MAC specifications) for wireless regional area networks (WRAN) to coexist with legacy TV transmission relying on cognitive capability. First, the application domain is discussed briefly, to better understand the ultimate goal of the standard and hence the decisions made for spectrum sensing, analysis and decision. Next, a short system overview is given.

3.5.1.1 Application

The main application target for 802.22 systems is wireless broadband access in rural and remote areas. Typical broadband access involves data, voice and QoS support. The use of the lower frequency bands are particularly useful for rural access because of the favorable propagation conditions encountered for those lower frequencies. Although the population density is often very small in rural areas, large coverage areas might render the deployment of 802.22 base stations (BSs) a profitable business. These lower frequency bands are licensed for TV broadcasting and wireless microphones. However, many TV channels are largely unoccupied in many parts of the United States, and often TV is delivered through cable access or satellite. As a result, opening up those bands for WRAN systems makes a good case, both from business and technical points of view. Next to the main WRAN application domain, 802.22 networks can also be used for smaller markets such as small businesses or home offices.

3.5.1.2 System Architecture

An example of a deployed 802.22 network is given in Fig. 3.10. The 802.22 networks operate in a fixed point-to-multi-point topology where a BS controls a cell consisting of a number of consumer premise equipments (CPEs). The BS is an entity installed by an operator and controls the cell strictly. Next to more traditional medium access control, that addresses when to transmit, it decides on how CPEs should access the spectrum. Moreover, the BS maintains control of a distributed sensing strategy to keep track of potential primary users (TV or wireless microphone signals). Clearly, it is possible to have multiple 802.22 cells that interfere. This is aggravated because of the very large transmission area of those systems. Coexistence issues of 802.22 cells are hence also addressed in the 802.22 standard.

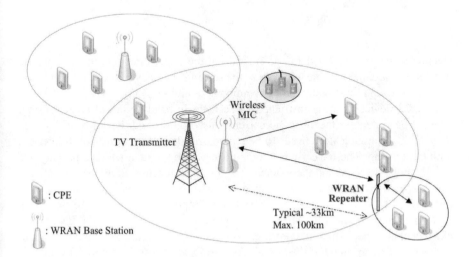

Fig. 3.10. 802.22 deployment scenario.

3.5.2 Spectrum Sensing

One of the important components of the 802.22 draft MAC to achieve the required cognitive capability is related to spectrum measurements. The spectrum measurement in 802.22 is primarily based on transmitter detection. In order to check the presence of primary signals, 802.22 devices need to be able to detect signals at very low signal-to-noise ratio (SNR) levels. Since the detection is done at low SNR, it is assumed that the detection of TV signals is done in a non-coherent manner, which means that no synchronization is needed [35].

The required accuracy of the spectrum sensing, the frequency band and time period, is determined in a centralized way by the BS. Using the local measurements, the BS can establish a spectrum occupancy map. The BS does not require the same sensing accuracy of each CPE, and algorithms to optimize or distribute the sensing load across CPEs can be used.

To optimize the sensing, 802.22 devices are supposed to be equipped with a dedicated omnidirectional antenna for sensing. This is in addition to a directional

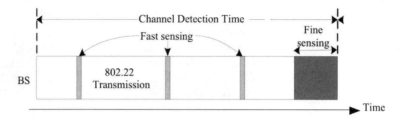

Fig. 3.11. The two-phase in-band sensing.

antenna which is used for data transmission in the target direction, minimizing the interference area. To be able to optimize the sensing accuracy of the omnidirectional antenna, it would most likely have to be mounted outdoors [35].

802.22 devices can be instructed to perform in-band or out-of-band sensing, where a band denotes the TV band currently used by the cell. For in-band sensing, the 802.22 communication needs to be temporarily halted, in order not to interfere with the sensing. There clearly is a trade-off between speed at which a primary TV signal can be detected and the efficiency or throughput achieved by the 802.22 cell. To avoid too frequent long connectivity halts, a two-phase sensing mechanism is proposed (Fig. 3.11). Fast sensing, i.e., based on a simple and fast sensing technique, is performed more frequently. After one (or more) fast sensing period the BS can decide whether to perform a fine sensing. This fine sensing takes more time but should in fact only be carried out if the fast sensing results are not sufficient to draw conclusions. Given the fact that TV signals do not come on the air frequently, this two-phase sensing method proves highly effective [35].

If multiple 802.22 cells operate in the same area, it is required that their sensing strategy is synchronized (i.e., they should halt communication when other cells sense). Since coexistence among different 802.22 cells is an important issue, such synchronization is embedded in the 802.22 standard.

Contrary to the TV signals detection, sensing of wireless microphone transmissions is much harder as these transmit at a much lower power and occupy much lower bandwidths. Therefore, in addition to transmitter detection, a second sensing option is enabled in the 802.22 standards. This second option relies on the transmission of beacons by the microphones themselves or a special device carried by microphone operators. This primary network information monitoring is embedded in the 802.22 MAC.

3.5.3 Spectrum Analysis

As mentioned before, transmitter detection is not the optimal sensing strategy, since to avoid causing interference to primary communication, it is in fact required to detect primary receivers. It might be the case that a node is outside the transmission range of a primary transmitter, but still capable of interfering with a primary receiver (Fig. 3.6). As a result, to optimize the probability of transmitter detection, it is possible to combine measurements from different CPEs and hence from different locations into a single spectrum occupancy map. The BS might use techniques such as data fusion and referendums over all measured data to obtain a reliable spectrum occupancy figure [35].

The BS vacates a channel if licensed signals are detected above certain well-defined thresholds at any receiver [35]. Depending on the type of primary signals (analog or digital TV or wireless microphones), the thresholds vary. Next, the BS can compute a keep-out region based on the information gathered from local measurements. In [35] it is shown that a WRAN station should typically protect an area of 155 km around a TV transmitter when transmitting at maximum EIRP. Depending on the actual distance to the transmitter, or depending on the current propagation conditions, the transmit power can be adjusted.

The spectrum occupancy map can be used to update the spectrum usage table. This table classifies channels as per availability, which can be occupied (by a primary user or another 802.22 user), free for use or prohibited (cannot be used at all by 802.22) [35]. Entries in this table might also be filled in by a system operator (e.g., setting some channels prohibited).

3.5.4 Spectrum Decision

Based on the spectrum occupancy map, the BS decides on spectrum access for each of the CPEs. Spectrum availability might vary over time and frequency, and effective use of the spectrum hence relies on the availability of a very flexible physical layer. In this section, the control options of the 802.22 PHY are given. Next, the MAC is introduced.

3.5.4.1 How to Send

The 802.22 PHY is based on multi-carrier modulation which makes it possible to define time and frequency slots in a very flexible way depending on the interference constraints and user requirements. More specifically, the 802.22 PHY is based on OFDMA enhanced with channel bonding. This channel bonding makes it possible to use different TV bands for a single transmission. Next, the modulation and coding is adaptive, resulting in a variable throughput or SNR requirement. Clearly, a lot of configuration options are available to the BS, that should be equipped with powerful optimization schemes.

3.5.4.2 Coordination

Communication between BS and CPEs follows a well-defined structure as illustrated in Fig. 3.12. Each frame consists or a preamble for synchronization and a separated downlink (DL) and uplink (UL) slot. In the downlink part, the BS sends control information and data to each of the associated CPEs. In the uplink phase, data transmission from CPEs to base station is scheduled. Next to the scheduled uplink slots, a contention interval can be used for initialization, bandwidth request and urgent coexistence situation notification. In the uplink slot, the BS can also schedule the broadcast of synchronization messages to synchronize different 802.22 cells for co-existence.

The above-presented frame structure, however, assumes that each CPE is associated, and hence synchronized, with the BS. Initialization or association of a new CPE with a BS, is however, a difficult problem. Indeed, the new CPE cannot know in which channel to look for the BS. Moreover, because of the channel bonding, the new CPE does not even know exactly how the channel is defined. To facilitate network entry and initialization, a superframe structure has been proposed (Fig. 3.13) [35]. At the beginning of each superframe, the BS sends a special preamble and super-frame control header (SCH) containing information on the BS channel selection. This preamble and SCH is sent on any of the TV channels that are free. After a CPE has decided on the locally free channels, it then scans each of those channels for the duration of a superframe period. Once a preamble and SCH is detected, the CPE has sufficient information to associate with the BS and get synchronized with the communication of frames as depicted in Fig. 3.12.

3.5.4.3 Spectrum Decision Trade-Off

As mentioned earlier, a decision to sense often impacts the current operation, so there is a trade-off between exploration and exploitation. In case of 802.22 networks, this has been solved to a large extent by the introduction of the two-phase sensing scheme. Fast sensing limits the impact on performance, unless when more fine sensing really makes sense.

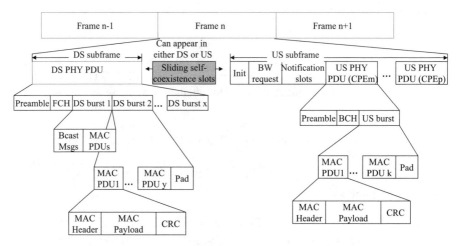

Fig. 3.12. 802.22 MAC frame structure proposal. From [2]; *[©2005] IEEE*. All rights reserved.

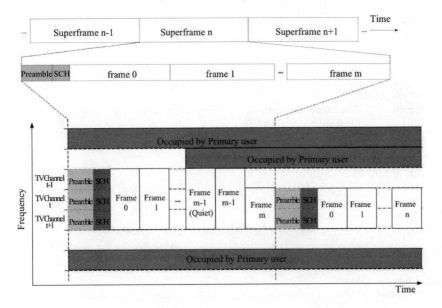

Fig. 3.13. 802.22 MAC superframe structure proposal. From [2]; *[©2005] IEEE.* All rights reserved.

3.5.5 Summary

The spectrum sensing, spectrum analysis and spectrum decision for 802.22 are summarized in Table 3.1. 802.22 dynamic spectrum sharing is essentially based on cooperative sensing, a centralized analysis of the sensed information and a decision coordination relying on a common channel. In the next section a solutions for coexistence of 802.15.4 with 802.11 will be introduced that was proposed in [1]. This solution implements a local spectrum sensing, local analysis and does not rely on communication for the decision coordination.

3.6 802.15.4 and 802.11 Coexistence

3.6.1 Introduction

In this section we *focus* on the coexistence between two major wireless standards that operate in the 2.4 GHz ISM band, namely 802.11g wireless LAN [12] and 802.15.4 sensor networks [36]. Their overlapping frequency channels are shown in Fig. 3.14. This special case of coexistence has started to receive attention since it has been shown in [37] that the impact of 802.11 on the sensor network is large and leads to above 92% of packet loss. An adaptive scheme using multiple radios has been proposed in [38] to overcome this, but this solution assumes that communication between nodes on a channel is possible, even after interference has been detected

Table 3.1. Spectrum sensing, analysis and decision for 802.22 networks and for the 802.15.4 coexistence schemes proposed in Sect. 3.6.

	802.22	802.15.4 distributed coexistence [1]
Spectrum sensing		
Information	Transmitter detection	Transmitter detection
	Optional out-of-network beacon detection for microphone detection	In-network beacon detection
Architecture	Distributed/cooperative sensing	Local sensing
Spectrum analysis		
Architecture	Centralized analysis Local analysis by CPE at initialization	Local sensing with local analysis
Model	Instantaneous clearage of channel upon incumbent detection Spectrum usage table for history	Comparison of techniques with and without history learning
Spectrum decision		
Optimization	Cooperative decision by central BS	Cooperative local decision
Architecture	Centralized	Distributed, local
Coordination	Common channel – data and control on same channel	No common channel

on that channel. This is not robust to the extreme interference patterns which are encountered in this context.

The characteristics of both networks are very different, resulting in a problem that is asymmetric in nature. Indeed, the output power of 802.15.4 devices is typically as low as 0 dBm [39], whereas the output power of 802.11g devices is typically 15 dBm or above. Also, 802.15.4 sensor networks are designed to monitor the environment or buildings, and can be very large, while 802.11 networks are mostly local hotspots organized around an access point (AP). Finally, sensor network applications are not demanding in terms of throughput, but, however, require a high reliability and robustness against attacks or unknown events. They should also be self-organizing since it is impossible to maintain such large networks efficiently. In comparison, 802.11 networks are typically used by a limited number of throughput-intensive applications. There is in fact only one common requirement: both 802.15.4 and 802.11 devices are battery-powered so that energy consumption is a major design criterion. Any algorithm for those networks should take the energy cost into account, including the non-negligible hardware power contribution associated with idle mode operation, scanning and receive processing.

In this context, distributed channel selection algorithms to optimize the 802.15.4 performance under varying 802.11 interference patterns have been proposed in [1]. The considered algorithms are fully distributed to improve scalability (since sensor networks are large), robustness (which is an important requirement for sensor net-

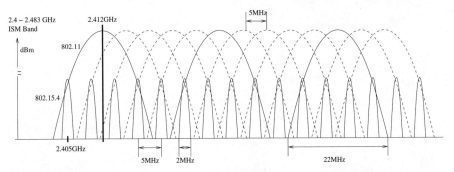

Fig. 3.14. 802.11 and 802.15.4 channels in the 2.4 GHz ISM band.

work applications) and adaptability. Next, the energy cost of the sensing algorithms are considered in the problem statement.

In this section, the algorithms presented in [1] are discussed. A first reason to choose this case study is that it is one of the few examples discussed in the literature discussing asymmetric coexistence in the ISM band. More importantly, it is one of the first examples proposing a fully distributed and statistical coordination that does not rely on any common channel establishment. Next, the algorithms proposed take into account the energy cost of the sensing. Also, a comparison is given between different spectrum analysis models (i.e., with and without spectrum learning). And finally, the case study allows for a nice illustration of the coupling between spectrum exploration and exploitation.

First the system model is discussed. Next, the spectrum sensing, analysis and decision are detailed. Finally some results are given.

3.6.2 System Model

In this section, we give a detailed overview of the models used for the sensor network and for the wireless LAN interference pattern as defined in [1]. The considered energy and performance metrics that are relevant for the investigated scenario are introduced next. These metrics will be used to evaluate the proposed distributed adaptation algorithms in [1].

3.6.2.1 802.15.4 Network Model

In [1] the 802.15.4 network is represented by a large number of nodes N that are arranged in a string or a rectangular topology. An $N \times N$ connectivity matrix $\mathbf{C}_{(N,N)}$ is used to denote which sensors can overhear each others. It is assumed that this is the case for all sensors in a range Rd from each other, where d is the inter-node distance and R is a parameter. In Fig. 3.15 a simple string topology is presented with $R = 2$. Each of the sensor nodes operates in one frequency channel among F possible ones, with $F = 16$ for 802.15.4 networks operating in the 2.4 GHz ISM band. A N-dimensional vector $\mathbf{t}_N$ keeps track of the current frequency $f_i(i \in [1, \dots, F])$, which

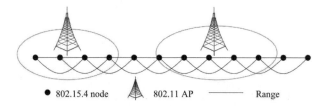

Fig. 3.15. Considered 802.15.4 network affected by 802.11 interference.

the different nodes are using to transmit. Although transmission is only possible on one frequency, nodes can be designed to scan or receive on multiple channels at the cost of increased scanning energy consumption and hardware.

The 802.15.4 medium access control can operate in different modes, depending on the use or absence of beacons. In [1] solutions are proposed that operate in the beacon-enabled mode, which results in a so-called superframe structure [36]. Such a superframe typically consists of a beacon, a contention access period and a contention-free period. More importantly, it is typically followed by an inactive period, during which the involved nodes are put asleep. It is assumed that sensors can change their channel without extra energy cost, when they wake-up from sleep mode. As a result, it is assumed that the sensors potentially swap frequency every interbeacon period, which ranges from 15 ms to above 4 min according to the standard. The frequency swapping strategy will be determined by the distributed adaptation algorithms discussed below.

On top of this medium access, three types of networks can be formed: star, peer-to-peer and cluster-tree. For the algorithms introduced in [1], the 802.15.4 network is assumed to consist of nodes of equal importance that are connected in a mesh network through peer-to-peer connections. These connections are reflected in the connectivity matrix **C** introduced earlier in this section. Each of those connections in such type of networks is maintained by scanning for the beacons sent by the peer.

3.6.2.2 802.11 Interference Model

For the performance analysis in [1], they use a large 802.15.4 network, which is affected by 802.11 interference. As it will be described hereafter, this 802.11 interference can vary over frequency, time and space.

The variations in frequency result from the fact that different 802.11 networks can operate using different channels (Fig. 3.9). Dynamic frequency selection (DFS) is a new functionality currently added to most of those 802.11 networks [40]. It is developed to optimize the frequency allocation of 802.11 networks that are subject to interference. As a result of this DFS, 802.11 interference is varying over time. It can be seen in Fig. 3.9 that each 802.11 network covers four consecutive 802.15.4 channels. The power distribution over these channel is more or less flat, especially for the OFDM-based 802.11g which is getting widely used thanks to its throughput improvement w.r.t. conventional 802.11b networks [12].

The variation in time results from the fact that the use of 802.11 networks is varying over time since user activities vary over time. Different timescales can be considered for these variations, where the smallest timescale considers packet bursts. In [1], only 802.11 traffic variations of minutes to hours are taken into account. These variations result from 802.11 ad hoc networks that are opportunistically established to transfer a large file, or from 802.11 hotspots that become operational when users join the network, as it was analyzed in [41]. It can thus reasonably be assumed that the 802.11 interference stays constant for several minutes to hours, and hence several (thousands) of 802.15.4 inter-beacon periods.

The spatial variations considered in [1] result from the fact that 802.15.4 networks are typically deployed over a large area (e.g., for monitoring purpose). As a result, the 802.15.4 networks are expected to be large both in terms of the area they cover and the number of nodes they consist of. The theoretical transmission, and hence also interference, range of 802.11 networks is 100 m to even 250 m. Although this is a significant range, sensor networks can cover larger areas since they consist of a large number of nodes in a mesh topology. Also, due to obstacles and walls, the effective range of 802.11 networks can be considerably smaller than this theoretical range. As a result, the 802.11 interference is assumed to affect large geographical subsets of the 802.15.4 nodes (Fig. 3.15).

For simplicity, it is assumed in [1] that an active or interfering 802.11 network will always be detected by the 802.15.4 nodes. In real systems, however, some noise should be considered in this detection of 802.11 interferers, since the 802.15.4 nodes could be scanning during a short inactive burst in between 802.11 packets. As the 802.15.4 beacon period is typically large compared to the 802.11 beacon or packet burst periods, this noise can be considerably reduced under the assumption that the 802.15.4 nodes scan during the whole period.

The 802.11 interference can thus finally be modeled as a $N \times F$ matrix $\mathbf{I}_{(N,F)}$. Each interfering network i then corresponds to a submatrix of dimensions $N_i \times 4$, where N_i denotes the number of nodes that are in the range of network i (depending on its output power), and where it is taken into account that every 802.11 interference pattern has a width of four 802.15.4 channels. Networks can swap frequency over time, disappear or appear, but this time variation is assumed to be slow compared to the 802.15.4 frequency adaptation.

3.6.2.3 Performance and Energy Measures

The relevant performance metric used in [1] is delay since throughput requirements in sensor networks are typically low. More precisely, assuming that sensors monitor a variable that should be communicated to a central sink the delay is the average number of periods required to forward a measurement to a fixed central sink. This average is computed over time and over the nodes in the network. The more the network is affected by interference, the more periods will be required on average to reach the sink. For the delay computation, in [1], it is assumed that every packet is forwarded only once during each period, to the node closest to the sink that can be reached during that period. As a result, packets travel the largest possible distance each period.

The energy cost is fixed per period, and hence independent of the actual packets sent, received or beacons overheard. This is a valid assumption since throughput is very low in sensor network applications, and moreover the system transmit power cost P_{Tx} is typically lower than the receive power cost P_{Rx} [39]. Moreover, since the full receive chain is required to be on for scanning, the power consumption in that mode is the same as the power cost in the receive mode. During every superframe, each node is awake to listen at least to its current frequency channel. The quality of a frequency channel can be assessed by counting the number of overheard beacons of neighbors. If no beacons are heard, energy detection, which is part of the 802.15.4 specifications, can be used to detect interference on the channel. As a result, in [1], the energy consumption only varies with the number of channels that are scanned (or listened to) in parallel.

$$E_{\text{tot}} = f_{\text{scan}} \times T_{\text{active}} \times P_{\text{Rx}} \qquad (3.1)$$

where f_{scan} is the number of frequencies considered and T_{active} is the active period per superframe.

3.6.3 Distributed Spectrum Sharing

In this section, the distributed channel selection algorithms to improve the 802.15.4 performance and robustness in presence of 802.11 interference, that are introduced in [1], are discussed. The algorithms do not rely on any coordination between the nodes (which would otherwise require a dedicated interference-free channel). Indeed, they should be robust against virtually any interference, since in the 2.4 GHz ISM band such dedicated interference-free bands do not exist. The algorithms are discussed in terms of the cognition cycle given in Fig. 3.5.

3.6.3.1 Spectrum Sensing

The main goal of the 802.15.4 sensing is to get informed about harmful 802.11 transmissions. As a result, transmitter detection will be carried out. Energy detection as it is enabled in the 802.15.4 standard [36] will be used for this purpose.

Secondly, the goal of the 802.15.4 nodes is to send their data through the network. For that purpose, they want to connect to other 802.15.4 nodes. Assuming that all nodes send beacons, it is possible to detect the current channel of other 802.15.4 nodes by listening to their beacons.

3.6.3.2 Spectrum Analysis

Spectrum Quality Metric

The quality metric is computed based on one hand on the output of the built-in 802.15.4 energy detector [36] that enables to capture the presence of 802.11 interference and on the other hand on the number of beacons heard in the scanned channel.

It is assumed that no beacons can be heard in the presence of 802.11 interference. The metric G is defined as:

$$G = \begin{cases} \sum \text{beacons heard} + 1, & \text{if no energy detected} \\ 0, & \text{if energy detected.} \end{cases} \tag{3.2}$$

When 802.11 interference is present, the channel quality is assumed to be equal to 0 (worst case). When no 802.11 interference is present, the channel quality is assumed to be proportional to the number of heard beacons, augmented by one to distinguish from the aforementioned worst case.

Spectrum Quality Predictions

Next, in [1], spectrum analysis techniques are proposed that learn the quality of each channel. The goal is to derive spectrum decisions based on experience rather than on expensive scanning.

In the considered spectrum learning model, a policy is represented by a two-dimensional lookup table indexed by states and actions. In the considered problem statement, both a current state and action correspond to a channel frequency: f_s and f_a. The reward function (Q^*) represents for each state and action the expected rewards when taking that action:

$$Q^*(f_s, f_a) = G(f_a) - G(f_s). \tag{3.3}$$

This means that the expected reward, which is expressed by the quality function G, is the expected quality improvement by moving to the new state f_a. The problem is now that the Q^* function should be approximated (*learned*) online by an estimate $\hat{Q}^*$. To do so, for every possible action (channel selection f_a), the available estimate $\hat{Q}^*$ should be updated as:

$$\hat{Q}^*(f_a) = (1 - \alpha)\hat{Q}^* + \alpha G(f_a) \tag{3.4}$$

where α is a learning parameter.

To conclude, this model learns the expected quality of a given channel f_a. Alternatively, it is possible to actually scan the channel and determine the channel quality G directly.

3.6.3.3 Spectrum Decision

Various spectrum decision algorithms are proposed in [1]. All are distributed, but they differ in the amount of scanning required, or in the model that is used for the local decision. It is important to note that because of the local sensing and decision taking, the techniques scale well for large networks, which is very useful for sensor networks.

Random Frequency Selection

The simplest distributed frequency selection solution is a scheme where nodes randomly (following a uniform distribution) pick a channel every period. Packets are forwarded to any other node closer to the sink within communication range that happened to pick the same channel. It can of course be expected that the average delay in this scheme will be large. However, since it does not rely on any coordination between the nodes and does not rely on an environment model, it can adapt to any possible event.

Simulated Annealing

It is possible to outperform the random frequency selection algorithm described above, since the considered 802.11 interference does not vary every 802.15.4 period (once an interference-free channel is found for the whole network, the nodes should indeed continue using that channel until adaptation is required.) In [1] a new approach relying on the simulated annealing optimization method is proposed. This technique can be elegantly implemented in the considered network setup. Simulated annealing is a very effective heuristic optimization strategy for finding a global optimum, developed by Metropolis et al. [42]. The basic idea of the method is to sample the search space using a Gaussian distribution, and to *anneal* this distribution as the optimum is approached.

Applied to the present context (i.e., optimizing the frequency allocation over a large sensor network affected by dynamic interference), nodes have to keep looking for another channel (i.e., sampling the search space). Since the 802.11 interference probability over the 802.15.4 channels is uniformly distributed, this search space sampling can be done uniformly. Every period, next to the node i's current frequency channel f_i, another channel f_{random} is considered and its performance is assessed. This is done according to the given channel quality metric G.

In simulated annealing, exploration is embedded in the algorithm to allow the system to *jump* out of a local optimum. This means that a new channel f_{random} can be accepted even if it is measured to be worse than the current channel f_i according to the quality metric G. This happens with a certain probability that should be decreased (i.e., annealed) when the system converges to its optimal solution. In this dynamic context, no annealing is, however, used. The probability to select the current frequency even if it is scanned to be worse depends on the quality metric G. In the algorithm proposed in [1], f_{random} is accepted with probability:

$$\exp(-G(f_i)/A) \times \{G(f_{\text{random}}) > 0\} \tag{3.5}$$

where A is the annealing temperature and where the second condition ($G(f_{\text{random}}) > 0$)) avoids the system to swap to a new channel when 802.11 interference is present (corresponding to $G = 0$). Further exploring a channel that is known to be bad is indeed clearly a waste of resources.

As far as energy is concerned, the considered algorithm requires to scan the current channel f_i and an extra channel f_{random}, so that the energy cost is doubled with respect to the random frequency selection algorithm. Obviously, it is possible to increase the number of channels to sample simultaneously in the proposed algorithm, and continue the basic algorithm with the best one in terms of G. In [1], the case where two random channels are selected at every iteration of the algorithm, at the cost of an increased energy consumption, is also covered.

Learning-Based Distributed Approaches

Alternatively, in [1], distributed channel selection schemes are proposed that are based on a learned estimate of the channel quality, rather than an estimate based on instantaneous sensing. This estimate of the quality implicitly defines a greedy policy that selects the action f_a with the largest expected reward:

$$f_a = \max \hat{Q}^*(f_a). \tag{3.6}$$

It is important to note that the learning algorithm updates the estimate for each action, but in fact does not specify what actions should be taken. The learning allows arbitrary experimentation while at the same time preserving the current best estimate of states' values. This is an important property in a time-varying environment and allows decoupling the learning phase from the decision policy.

In [1] the authors proposed the following algorithm to select the next channel. When experimentation (i.e., *learning*) is allowed, a random frequency f_a is selected with a probability similar to that in (3.5) used for the simulated annealing algorithm. Since the frequency f_a was not scanned before selecting it, the second factor in (3.5) cannot be included in this exploration factor. This probability writes thus:

$$\exp(-G(f_i)/A). \tag{3.7}$$

When no experimentation is required (i.e., *experience*), the greedy policy defined in (3.6) is used.

To sum up, the learning algorithm selects a frequency f_a for the next period that is expected to maximize the reward or quality G. This optimal policy is learned online and some exploration is allowed to adapt to varying interference patterns. It is clear that during every superframe period only one frequency is scanned, so that the energy consumption is similar to that of the random selection algorithm. However, it is possible that a channel is chosen with interference since the decision is taken before the channel is actually scanned. In simulated annealing, no decisions are made before the channel is scanned. This costs scanning energy but allows to avoid interference more proactively. In the results section we will investigate how the predictions based on learning compare with the more costly approaches based on scanning.

3.6.4 Simulation Results

Simulation results are given for the proposed distributed channel selection schemes. Networks of different size ($N \in [50, 100, 200]$) in a simple string topology are considered, with varying average interference (25 or 50% of affected nodes) and different time variations. Traffic is generated randomly in the sensors and forwarded to the

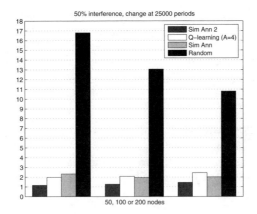

Fig. 3.16. Normalized delay increase compared to ideal channel allocation. 802.11 interference remains static for 25,000 802.15.4 superframe periods.

sink node located at the end of the string topology. The connectivity range of each node is assumed to be $R = 10$. The performance measure is the average number of periods it takes for each packet to reach the sink, compared to the expected delay in case of ideal channel allocation (which could only be achieved by a central entity that can monitor the whole network interference). For the considered R and N, this ideal average number of periods can be shown to be of 3, 5.5 and 10.5 for N respectively equal to 50, 100 and 200. The algorithm parameters are set to $A = 4$ and $\alpha = 0.1$.

The simulation results in [1] showed that it is indeed possible to design fully distributed algorithms, that do not rely on a common control channel, and achieve a performance close to the optimum. Fig. 3.16 shows that both the algorithms based on scanning, and those based on an estimate of the spectrum, achieve a performance within a factor 2 of the optimum. Surprisingly, the very simple learning scheme proposed in [1] achieves a performance that is very close to the scanning based approach. The learning approach results in energy and hardware savings of 50%.

3.6.5 Summary

In this section, distributed channel selection algorithms have been discussed for 802.15.4 networks subject to interference from 802.11 networks. These distributed channel selection algorithms operate without the use of a common control channel, which is a very important characteristic in the context of dynamic spectrum sharing. Because of the lack of a common communication channel, the sensing is carried out locally. Next, different spectrum analysis methods were considered and compared. Finally, in this section, it has been illustrated how the spectrum decision affects spectrum sensing.

Conclusion

In this chapter, the use of cognitive radio for dynamic spectrum sharing has been introduced. First, it is shown that dynamic use of the spectrum is required to meet the increasing quality requirements of applications. However, it is also shown that recent trends in hardware development and protocol design have the potential to evolve toward solutions for such cognitive radios. Following the motivation for dynamic spectrum sharing, a taxonomy and classification has been given for coexistence and dynamic sharing scenarios. Next, the cognitive capability required to implement dynamic sharing has been introduced: *spectrum sensing*, *spectrum analysis* and *spectrum decision*. The main challenges related to each of the subtasks have been identified, and solutions in the literature are listed. Finally, two relevant cases for coexistence are discussed, to illustrate the introduced concepts in more detail. First, the coexistence of wireless regional area networks with TV transmitters is discussed, which is an example of vertical coexistence between primary licensed users and secondary unlicensed users. In this case, which is covered by the IEEE 802.22 standard proposal, spectrum sensing is carried out in a distributed way through transmitter detection. Spectrum analysis is mainly done centrally. Finally, for the spectrum decision coordination, a common channel is considered. Next, the relevant problem of coexistence of IEEE 802.15.4 with 802.11 in the ISM band is assessed. For this horizontal coexistence between systems of equal regulatory status, algorithms are discussed that do not rely on a common channel for the decision coordination. Spectrum sensing is carried out locally and different spectrum analysis models are discussed. These two considered case studies cover many of the important issues encountered in coexistence or dynamic sharing problems.

Acknowledgments

The author would like to thank Mustafa Ergen, Michael Timmers, Bill Hodge and Antoine Dejonghe for their help with parts of this chapter.

References

1. S. Pollin, M. Ergen, M. Timmers, A. Dejonghe, L. Van der Perre, I. Moerman, F. Catthoor, and A. Bahai, "Distributed cognitive coexistence of 802.15.4 with 802.11," in *Proc. Intl. Conf. on Cognitive Radio Oriented Wireless Networks Commun.*, June 8–10, 2006.
2. D. Birru, V. Gaddam, C. Cordeiro, K. Challapali, M. Bellec, P. Pirat, L. Escobar, and D. Callonnec, "A cognitive PHY/MAC proposal for IEEE 802.22 WRAN systems Part 1: The cognitive PHY," IEEE 802.22 Working Group doc.: IEEE 802.22 – 05/0103r0, 2005.
3. IEEE 802.11a, Part 11, Amendment 1, "High-speed physical layer in the 5 GHz band," 1999.
4. F. Horlin, F. Petre, E. Lopez-Estraviz, F. Naessens, and L. Van der Perre, "Flexible transmission scheme for 4G wireless systems with multple antennas," *EURASIP J. Wireless Commun. Netw.*, vol. 3, pp. 308–322, Aug. 2005.

5. J. Mitola, "Cognitive radio: An integrated agent architecture for software-defined radio," PhD Thesis, Royal Institute of Technology, 2000.

6. R. Brodersen, W. Wolisz, D. Cabri, S. M. Mishra, and D. Willkomm, "A cognitive radio approach for usage of virtual unlicensed spectrum," CORVUS White Paper, 2004.

7. IEEE 802.22 WG on Wireless Regional Area Networks.

8. C. Blanch, S. Pollin, G. Lafruit, and W. Eberle, "Channel adaptive rate control," in *Proc. International Packet Video Workshop* (Hangzhou, China), 2006.

9. N. Khaled, "Transmit and receive optimization for MIMO/OFDM-based high-throughput wireless local area networks," PhD Thesis, K.U.Leuven, Belgium, 2005.

10. C. U. Saraydar, N. B. Mandayam, and D. J. Goodman, "Efficient power control via pricing in wireless data networks," *IEEE. Trans. Commun.*, vol. 50, no. 2, pp. 291–303, Feb. 2002.

11. B. Bougard and B. Debaillie, "Energy-scalable OFDM transmitter design and control," in *Proc. 43rd Annual Conference on Design Automation*, pp. 536–541, 2006.

12. IEEE 802.11g, Part 11, Amendment 4, "Further higher-speed physical layer extension in the 2.4 GHz band," 2003.

13. IEEE 802.11e, Part 11, Amendment 8, "Medium access control enhancements for quality of service," 2005.

14. R. Coase, "The Federal Communication Commission," *J. Law Econ.*, pp. 1–49, 1959.

15. D. Hatfield and P. Weiser, "Property rights in spectrum: Taking the next step," in *Proc. IEEE DySPAN 2005*, pp. 43–55, Nov. 2005.

16. L. Xu, R. Tnjes, T. Paila, W. Hansmann, M. Frank, and M. Albrecht, "DRiVE-ing to the Internet: Dynamic radio for IP services in vehicular environments," in *Proc. IEEE Conference on Local Computer Networks*, pp. 281–289, 2000.

17. Q. Zhao and B. M. Sadler, "Dynamic spectrum access: Signal processing, networking and regulatory policy," to appear in *IEEE Signal Process. Mag.*, vol. 55, no. 5, pp. 2294–2309, May 2007.

18. Y. Benkler, "Overcoming agoraphobia: Building the commons of the digitally networked environment," *Harvard J. Law Technol.* vol. 287, 1998.

19. W. Lehr and J. Crowncroft, "Managing shared access to a spectrum commons," in *Proc. IEEE DySPAN 2005*, pp. 420–444, Nov. 2005.

20. G. Marias, "Spectrum scheduling and brokering based on QoS demands of competing WISPs," in *Proc. IEEE DySPAN 2005*, pp. 684–687, Nov. 2005.

21. O. Ileri, D. Samarzija, T. Sizer, and N. B. Mandayam, "Demand responsive pricing and competitive spectrum allocation via a spectrum server," in *Proc. IEEE DySPAN 2005*, pp. 194–202, Nov. 2005.

22. X. Jing and D. Raychaudhuri, "Spectrum co-existence of IEEE 802.11b and 802.16a networks using CSCC etiquette protocol," in *Proc. IEEE DySPAN 2005*, pp. 243–250, Nov. 2005.

23. J. Mitola, "Cognitive radio for flexible mobile multimedia communications," *IEEE Mobile Multimedia Conference*, pp. 3–10, 1999.

24. "DARPA: The Next Generation (XG) Program." http://www.darpa.mil/ato/programs/xg/index.htm.

25. C. Cordeiro, K. Challapali, D. Birru, and N. S. Shankar, "IEEE 802.22: The first worldwide wireless standard based on cognitive radios," in *Proc. IEEE DySPAN 2005*, pp. 328–337, Nov. 2005.

26. I. F. Akyldiz, W. Y. Lee, M. C. Vuran, and S. Mohanty, "NeXt generation/dynamic spectrum access/cognitive radio wireless networks: A survey," *Comput. Netw. J. (Elsevier)*, vol. 50, pp. 2127–2159, Sept. 2006.

27. A. Sahai, N. Hoven, and R. Tandra, "Some fundamental limits on cognitive radio," in *Allerton Conf. Communication, Control, and Computing*, Oct. 2004.
28. W. A. Gardner, "Signal interception: A unifying theoretical framework for feature detection," *IEEE. Trans. Commun.*, vol. 36, no. 8, pp. 897–906, Aug. 1988.
29. Z. Tian and G. B. Giannakis, "A wavelet approach to wideband spectrum sensing for cognitive radios," in *Proc. Intl. Conf. on Cognitive Radio Oriented Wireless Networks Comnun.*, June 8–10, 2006.
30. S. Shankar, "Spectrum agile radios: Utilization and sensing architecture," in *Proc. IEEE DySPAN 2005*, Nov. 2005.
31. B. Wild and K. Ramchandran, "Detecting primary receivers for cognitive radio applications," in *Proc. IEEE DySPAN 2005*, pp. 124–130, Nov. 2005.
32. Q. Zhao, L. Tong, and A. Swami, "Decentralized cognitive MAC for dynamic spectrum access," in *Proc. IEEE DySPAN 2005*, pp. 224–232, Nov. 2005.
33. C. Peng, H. Zheng, and B. Y. Zhao, "Utilization and fairness in spectrum assignment for opportunistic spectrum access," *Mobile Netw. Appl.*, vol. 11, issue 4, pp. 555–576, Aug. 2006.
34. W. So, J. Mo, and J. Walrand, "Comparison of multi-channel MAC protocols," in *Proc. 8th ACM/IEEE International Symposium on Modeling, Analysis and Simulation of Wireless and Mobile Systems*, 2005.
35. C. Cordeiro, K. Challapali, D. Birru, and N. S. Shankar, "IEEE 802.22: The first worldwide wireless standard based on cognitive radios," *J. Commun. (JCM)*, pp. 38–47, Apr. 2006.
36. IEEE 802.15.4, "Wireless medium access control (MAC) and physical layer (PHY) specifications for low-rate wireless personal area networks (LR-WPANs)," 2003.
37. Steibeis-Transfer Centre, "Compatibility of IEEE 802.15.4 (Zigbee) with IEEE802.11 (WLAN), Bluetooth, and Microwave Ovens in 2.4 GHz ISM-Band." http://www.baloerrach.de.
38. C. Won, J.-H. Youn, H. Ali, H. Sharif, and J. Deogun, "Adaptive radio channel allocation for supporting coexistnce of 802.15.4 and 802.11b," in *Proc. IEEE Vehicular Tech. Conf. Fall*, pp. 2522–2526, 2005.
39. http://www.chipcon.com/files/CC2420 Data Sheet 1 2.pdf.
40. focus.ti.com/pdfs/bcg/tnetw1130 prod bulletin.pdf.
41. D. Tang and M. Baker, "Analysis of a local-area wireless network," in *Proc. ACM Mobicom 2000*, pp. 110, Aug. 2000.
42. N. Metropolis, A. W. Rosenbluth, M. N. Rosenbluth, A. H. Teller, and E. Teller, "Equations of state calculations by fast computing machines," *J. Chem. Phys.*, vol. 21, pp. 1087–1091, 1953.

4

Cooperative Spectrum Sensing

Khaled Ben Letaief and Wei Zhang

Department of Electronic & computer Engineering,
Hong Kong University of Science & Technology,
Clear Water Bay, Kowloon,
Hong Kong
{eekhaled,eewzhang}@ece.ust.hk

4.1 Introduction

With the rapid growth of wireless applications and services in the recent decade, spectrum resources are facing huge demands. The radio spectrum is a limited resource and is regulated by government agencies such as the Federal Communications Commission (FCC) in the United States. Within the current spectrum regulatory framework, all of the frequency bands are exclusively allocated to specific services and no violation from unlicensed users is allowed. The spectrum scarcity problem is getting worse due to the emergence of new wireless services. Fortunately, the worries about spectrum scarcity are being shattered by a recent survey made by a Spectrum Policy Task Force (SPTF) within FCC. It indicates that the actual licensed spectrum is largely under-utilized in vast temporal and geographic dimensions [1]. For instance, a field spectrum measurement, which is taken in New York City, has shown that the maximum total spectrum occupancy is only 13.1% from 30 MHz to 3 GHz [2]. The exciting findings shed light on the problem of spectrum scarcity and motivate a new direction to solve the conflicts between spectrum scarcity and spectrum under-utilization.

A remedy to spectrum scarcity is to improve spectrum utilization by allowing secondary users to access under-utilized licensed bands dynamically when/where licensed users are absent. Recently, FCC has issued a Notice of Proposed Rule Making (NPRM-FCC 03-322 [3]) advocating cognitive radio technology as a candidate to implement opportunistic spectrum sharing. Meanwhile, IEEE has also endeavored to formulate a novel wireless air interface standard based on cognitive radios: the IEEE 802.22 working group. The IEEE 802.22 WG aims to develop wireless regional area network physical (PHY) and medium access control (MAC) layers for use by unlicensed devices in the spectrum allocated to TV bands [4].

Cognitive radio is a novel technology which improves the spectrum utilization by allowing secondary networks (users) to borrow unused radio spectrum from primary licensed networks (users) or to share the spectrum with the primary networks (users) [5–7]. As an intelligent wireless communication system, cognitive radio is aware of the radio frequency environment, selects the communication parameters (such as carrier frequency, bandwidth and transmission power) to optimize the spectrum usage and adapts its transmission and reception accordingly. One of most critical components of cognitive radio technology is spectrum sensing. By sensing and adapting to the environment, a cognitive radio is able to fill in spectrum holes and serve its users without causing harmful interference to the licensed user. To do so, the cognitive radio must continuously sense the spectrum it is using in order to detect the re-appearance of the primary user. Once the primary user is detected, the cognitive radio should withdraw from the spectrum instantly so as to minimize the interference it may possibly incur. This is a very difficult task as the various primary users will be employing different modulation schemes, data rates and transmission powers in the presence of variable propagation environments and interference generated by other secondary users. Another great challenge of implementing spectrum sensing is the hidden terminal problem, which occurs when the cognitive radio is shadowed, in severe multipath fading or inside buildings with high penetration loss while a primary user is operating in the vicinity [8]. Due to the hidden terminal problem, a cognitive radio fails to see the presence of the primary user and then will access the licensed channel and cause interference to the licensed users. In order to deal with the hidden terminal problem in cognitive radio networks, multiple cognitive users can cooperate to conduct spectrum sensing.

Cooperative communications has been recently recognized as a powerful solution that can overcome the limitation of wireless systems [9]. The basic idea behind cooperative transmission rests on the observation that in a wireless environment, the signal transmitted or broadcast by a source to a destination node, each employing a single antenna, is also received by other terminals, which are often referred to as relays or partners. The relays process and retransmit the signals they receive. The destination node then combines the signals coming from the source and the partners, thereby creating spatial diversity and taking advantage of the multiple receptions of the same data at the various terminals and transmission paths. In addition, the interference among terminals can be dramatically suppressed by distributed spatial processing technology. By allowing multiple cognitive radios to cooperate in spectrum sensing, the hidden terminal problem can be addressed [10, 11].

Cooperative spectrum sensing in cognitive radio networks has an analogy to a distributed decision in wireless sensor networks, where each sensor makes a local decision and those decision results are reported to a fusion center to give a final decision according to some fusion rule [12]. The main difference between these two applications lies in the wireless environment. Compared to wireless sensor networks, cognitive radios and the fusion center (or common receiver) are distributed over a larger geographic area. This difference brings out a much more challenging problem to cooperative spectrum sensing because sensing channels (from the primary user to cognitive radios) and reporting channels (from cognitive radios to the fusion center or common receiver) are normally subject to fading or heavy shadowing.

In this chapter, a survey of cooperative spectrum sensing for cognitive radios is given. We shall also review some well-known spectrum sensing techniques and introduce the concept and principle of cooperative spectrum sensing. The performance analysis of cooperative spectrum sensing over realistic fading channels is given. Several robust cooperative spectrum sensing techniques are are also proposed.

4.2 Spectrum Sensing

Spectrum sensing is a key element in cognitive radio communications as it should be firstly performed before allowing unlicensed users to access a vacant licensed channel. The essence of spectrum sensing is a binary hypothesis-testing problem:

H_0 : Primary user is absent.

H_1 : Primary user is in operation.

The key metric in spectrum sensing are the probability of correct detection, probability of false alarm and probability of miss, which are given by respectively,

$$P_d = \text{Prob}\{\text{Decision} = H_1|H_1\} \tag{4.1}$$

$$P_f = \text{Prob}\{\text{Decision} = H_0|H_0\} \quad \text{and} \tag{4.2}$$

$$P_m = \text{Prob}\{\text{Decision} = H_0|H_1\}. \tag{4.3}$$

4.2.1 Spectrum Sensing Techniques

To enhance the detection probability, many signal detection techniques can be used in spectrum sensing. In the following, we give an overview of some well-known spectrum sensing techniques. For further details, interested readers are referred to [10, 13–17].

4.2.1.1 Energy Detection

The energy detection method is optimal for detecting any unknown zero-mean constellation signals [13]. In the energy detection approach, the radio frequency energy in the channel or the received signal strength indicator (RSSI) is measured to determine whether the channel is occupied or not. The energy detection implementation for spectrum sensing is shown in Fig. 4.1a. The received signals $x(t)$ sampled in a time window are first passed through an FFT device to get the spectrum $X(f)$. The peak of the spectrum is then located. After windowing the peak in the spectrum of $x(t)$, we get $Y(f)$. The signal energy is then collected in the frequency domain. Finally, the following binary decision is made,

$$\begin{cases} H_1, \text{ if } \sum |Y(f)|^2 \geq \lambda \\ H_0, \quad \text{otherwise.} \end{cases} \tag{4.4}$$

Although the energy detection approach can be implemented without any prior knowledge of the primary user signal, it still has some drawbacks. The first problem is that it can only detect the signal of the primary user if the detected energy is above a threshold. Another challenging issue is that the energy approach cannot distinguish between other secondary users sharing the same channel and the primary user [14]. The threshold selection for energy detection is also problematic since it is highly susceptible to the changing background noise and interference level.

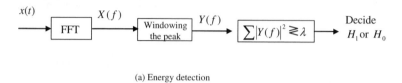

(a) Energy detection

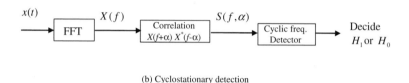

(b) Cyclostationary detection

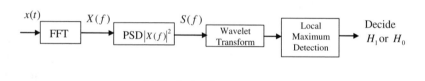

(c) Wavelet detection

Fig. 4.1. Implementation of various detection approaches for spectrum sensing. $x(t)$ and $X(f)$ denote the time domain and frequency domain of the observed signal, respectively. $S(f)$ denotes the power spectral density of $x(t)$. $S(f, \alpha)$ represents the cyclic spectrum.

4.2.1.2 Matched Filter

A matched filter is an optimal detection method as it maximizes the signal-to-noise ratio (SNR) of the received signal in the presence of additive Gaussian noise. A matched filter is obtained by correlating a known signal, or template, with an unknown signal to detect the presence of the template in the unknown signal. This is equivalent to convolving the unknown signal with a time-reversed version of the template. Matched filters are commonly used in radar transmission. In the cognitive radio scenario, however, the use of the matched filter can be severely limited since the information of the primary user signal is hardly available at the cognitive radios. If partial information of primary user signal such as pilots or preambles is known, the use of matched filter is still possible for coherent detection [10]. For example, in order to detect the presence of a digital television (DTV) signal, we may detect its pilot tone by passing the DTV signal through a delay-multiply circuit. If the squared magnitude of the output signal is larger than a threshold, the presence of the DTV signal can be detected. The detailed matched filter implementation for spectrum sensing is shown in Fig. 4.1b.

4.2.1.3 Cyclostationary Detection

If the signal of the primary user exhibits strong cyclostationary properties, it can be detected at very low SNR values. A signal is said to be cyclostationary (in the wide sense) if its autocorrelation is a periodic function of time t with some period. The cyclostationary detection can be performed as follows [15]. Firstly, one can calculate the cyclic autocorrelation function (CAF) of the observed signal $x(t)$, $R_x(\tau)$, as

$$R_x(\tau) = E[x(t+\tau)x^*(t-\tau)e^{-j2\pi\alpha t}]$$

where $E[\cdot]$ denotes the statistical expectation operation and α is called *cyclic frequency*. The discrete Fourier transformation of the CAF can then be computed to obtain the spectral correlation function (SCF), $S(f, \alpha)$, also called cyclic spectrum, which is a two-dimensional function in terms of frequency and cyclic frequency. Finally, the detection is completed by searching for the *unique cyclic frequency* corresponding to the peak in the SCF plane. The detailed cyclostationary detection implementation for spectrum sensing is shown in Fig. 4.1c. This detection approach is robust to random noise and interference from other modulated signals, because the noise has only a peak of SCF at the zero cyclic frequency and the different modulated signals have different unique cyclic frequencies.

4.2.1.4 Wavelet Detection

For signal detection over wideband channels, the wavelet approach offers advantages in terms of both implementation cost and flexibility in adapting to the dynamic spectrum as opposed to conventional use of multiple narrowband bandpass filters (BPF) [16]. In order to identify the locations of vacant frequency bands, the entire wideband is modeled as a train of consecutive frequency sub-bands where the power spectral characteristic is smooth within each sub-band but changes abruptly on the border of two neighboring sub-bands. By employing a wavelet transform of the power spectral density (PSD) of the observed signal $x(t)$, the singularities of the PSD $S(f)$ can be located and thus the vacant frequency bands can be found. The detailed wavelet detection implementation for spectrum sensing is shown in Fig. 4.1d.

Table 4.1. Advantages and Disadvantages of Spectrum Sensing Techniques.

Spectrum sensing approach	Advantages	Disadvantages
Energy detection	Does not need any prior information low computational cost	Cannot work in low SNR cannot distinguish users sharing the same channel
Matched filter	Optimal detection performance low computational cost	Requires a prior knowledge of the primary user
Cyclostationary detection	Robust in low SNR robust to interference	Requires partial information of the primary user high computational cost
Wavelet detection	Effective for wideband signal	Does not work for spread spectrum signals; high computational cost

One critical challenge of implementing the wavelet approach in practice is the high sampling rates for characterizing the large bandwidth.

The advantages and disadvantages of the aforementioned spectrum sensing techniques are summarized and compared in Table 4.1.

4.2.2 Performance of Spectrum Sensing

It has been found that the optimal detector for detecting a weak unknown signal from a known zero-mean constellation is the energy detector, even though there are some fundamental limits when SNR is below a certain threshold [13]. The energy detection is performed by measuring the energy of the received signal in a fixed bandwidth W over an observation time window T. The performance analysis of the energy detector has been studied for AWGN channels in [18, 19] and for Rayleigh fading channels in [20–22]. In the following, we briefly present the main results that describe the performance of the energy detector over Rayleigh fading channels. The details of the proof are omitted here and can be found in [20, 21].

We assume that each cognitive radio performs local spectrum sensing independently. For simplicity, we consider the ith cognitive radio ($1 \leq i \leq K$) only to see how the energy detector works. The local spectrum sensing is to decide between the following two hypotheses,

$$x_i(t) = \begin{cases} n_i(t), & H_0 \\ h_i s(t) + n_i(t), & H_1 \end{cases} \tag{4.5}$$

where $x(t)$ is the observed signal at the ith cognitive radio and $s(t)$ is the signal transmitted from the primary user, $n_i(t)$ is the additive white Gaussian noise (AWGN) and h_i is the complex channel gain of the sensing channel between the primary user and the ith cognitive radio.

As shown in Fig. 4.1a, the energy collected in the frequency domain is $D_i = \sum |Y(f)|^2$ which serves as a decision statistic with the following distribution

$$D_i \sim \begin{cases} \chi^2_{2u}, & H_0 \\ \chi^2_{2u}(2\gamma_i), & H_1 \end{cases} \tag{4.6}$$

where χ^2_{2u} denotes a central chi-square distribution with $2u$ degrees of freedom and $\chi^2_{2u}(2\gamma_i)$ denotes a non-central chi-square distribution with $2u$ degrees of freedom and a non-centrality parameter $2\gamma_i$, respectively. γ_i is the instantaneous SNR of the received signal at the ith cognitive radio and $u = TW$.

For the ith cognitive radio with the energy detector, the average probability of false alarm, the average probability of detection, and the average probability of miss over Rayleigh fading channels are given by, respectively,

$$P_{f,i} = \mathbf{E}_{\gamma_i} [\text{Prob}\{D_i > \lambda | H_0\}]$$
$$= \frac{\Gamma(u, \frac{\lambda_i}{2})}{\Gamma(u)}, \tag{4.7}$$

$$P_{d,i} = \mathbf{E}_{\gamma_i} [\text{Prob}\{D_i > \lambda | H_1\}]$$
$$= e^{-\frac{\lambda}{2}} \sum_{n=0}^{u-2} \frac{1}{n!} \left(\frac{\lambda_i}{2}\right)^n + \left(\frac{1+\bar{\gamma}_i}{\bar{\gamma}_i}\right)^{u-1}$$
$$\times \left[e^{-\frac{\lambda_i}{2(1+\bar{\gamma}_i)}} - e^{-\frac{\lambda_i}{2}} \sum_{n=0}^{u-2} \frac{1}{n!} \left(\frac{\lambda_i \bar{\gamma}_i}{2(1+\bar{\gamma}_i)}\right)^n \right] \tag{4.8}$$

and

$$P_{m,i} = 1 - P_{d,i} \tag{4.9}$$

where λ_i and $\bar{\gamma}_i$ denote the energy threshold and the average SNR at the ith cognitive radio, respectively. $\mathbf{E}_{\gamma_i}[\cdot]$ represents the expectation over the random variable γ_i. Likewise, $\Gamma(\cdot, \cdot)$ is the incomplete gamma function and $\Gamma(\cdot)$ is the gamma function.

In Fig. 4.2, complementary receiver operating characteristic (ROC) curves (P_m versus P_f) of the energy detection for one cognitive radio are plotted for a variety of SNR values according to (4.7) and (4.9). It shows that the energy detection performance of one cognitive radio becomes worse when SNR decreases.

4.3 Cooperative Spectrum Sensing

One of the most critical issues of spectrum sensing is the hidden terminal problem, which happens when the cognitive radio is shadowed. In Fig. 4.3, cognitive radio 1 is shown to be shadowed by a high building over the sensing channel. In this case, the cognitive radio cannot reliably sense the presence of the primary user due to the very low SNR of the received signal. Then, this cognitive radio assumes that the observed channel is vacant and begins to access this channel while the primary user is

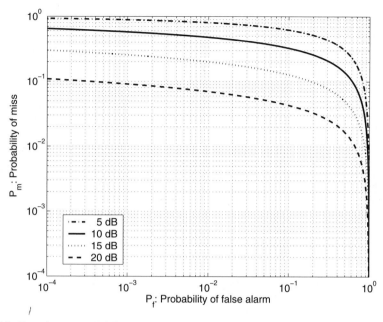

Fig. 4.2. Complementary ROC curves in a Rayleigh fading channel with SNR $\bar{\gamma} = 5, 10, 15$ and 20 dB for one cognitive radio.

still in operation. To address this issue, multiple cognitive radios can be coordinated to performance spectrum sensing cooperatively. Several recent works have shown that cooperative spectrum sensing can greatly increase the probability of detection in fading channels [22]. In general, cooperative spectrum sensing is performed as follows:

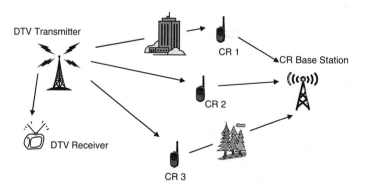

Fig. 4.3. Cooperative spectrum sensing in cognitive radio (CR) networks; CR 1 is shadowed over the sensing channel and CR 3 is shadowed over the reporting channel.

- *Step 1*: Every cognitive radio performs local spectrum measurements independently and then makes a binary decision.
- *Step 2*: All the cognitive radios forward their binary decisions to a common receiver which is an access point (AP) in a wireless LAN or a base station (BS) in a cellular network.
- *Step 3*: The common receiver combines those binary decisions and makes a final decision to infer the absence or presence of the primary user in the observed band.

In the above cooperative spectrum sensing algorithm, each cooperative partner makes a binary decision based on its local observation and then forward one bit of the decision to the common receiver. At the common receiver, all one-bit decisions are fused together according to an "OR" logic. This cooperative sensing algorithm is referred to as *decision fusion*. An alternative form of cooperative spectrum sensing can be performed as follows. Instead of transmitting the one-bit decision to the common receiver in Step 2 of the above algorithm, each cognitive radio can just send its observation value directly to the common receiver [23]. This alternative approach can then be seen as *data fusion* for cooperative networks. Obviously, the one-bit decision needs a low bandwidth control channel. Moreover, it has been recently found that a hard decision approach can perform almost as well as that of the soft decision one in terms of detection performance [24].

4.3.1 Cooperative Spectrum Sensing Performance

In cooperative spectrum sensing, all cognitive radios measure the licensed spectrum and make the decisions independently. If the decision in one cognitive radio is H_0, then a symbol $\{-1\}$ will be transmitted to the BS. If H_1 is true, then $\{1\}$ is forwarded to the BS. The transmission of the decisions from all cognitive radios to the BS can be seen as a multiuser access protocol, which can be based on TDMA or FDMA. The BS collects all K decisions and makes the final decision using an OR rule. Let Z denote the decision statistic in the BS, then it can be described as

$$
Z \sim \begin{cases} \{H_0^{\mathrm{BS},1}, \cdots, H_0^{\mathrm{BS},K}\}, & \mathcal{H}_0 \text{ (signal is absent)} \\ \text{otherwise}, & \mathcal{H}_1 \text{ (signal is present)} \end{cases} \tag{4.10}
$$

where $H_0^{\mathrm{BS},i}$ denotes the decision H_0 received from the ith cognitive radio at the BS for $i = 1, \cdots, K$. The expression (4.10) demonstrates that the BS decides the signal is absent only if all cognitive radios decide the absence of the signal. On the other hand, the BS assumes that the primary user is present if there exists at least one cognitive radio which assumes the presence of the primary user signal. Therefore, the false alarm probability of the cooperative spectrum sensing is given by

$$
\begin{aligned}
Q_{\mathrm{f}} &= \mathrm{Prob}\{\mathcal{H}_1 | H_0\} \\
&= 1 - \mathrm{Prob}\{\mathcal{H}_0 | H_0\} \\
&= 1 - \prod_{i=1}^{K} (1 - P_{\mathrm{f},i})
\end{aligned} \tag{4.11}
$$

where $P_{f,i}$ denotes the false alarm probability of the ith cognitive radio in its local spectrum sensing.

The miss probability of cooperative spectrum sensing is given by

$$Q_m = \text{Prob}\{\mathcal{H}_0|H_1\}$$
$$= \prod_{i=1}^{K} P_{m,i} \qquad (4.12)$$

where $P_{m,i}$ denotes the miss probability of the ith cognitive radio in its local spectrum sensing.

Assume that every cognitive radio achieves the same P_f and P_m in the local spectrum sensing, the false alarm probability and the miss probability of cooperative spectrum sensing over Rayleigh fading channels are then given by

$$Q_f = 1 - (1 - P_f)^K$$
$$Q_m = (P_m)^K. \qquad (4.13)$$

Figure 4.4 shows that the performance curves of cooperative spectrum sensing for different number of cognitive radios over Rayleigh fading channels with the average SNR $\bar{\gamma} = 10$ dB. It is obvious that the probability of miss is greatly reduced with a larger value K for a given probability of false alarm. As such, we may refer to K as the *sensing diversity gain* of the cooperative spectrum sensing.

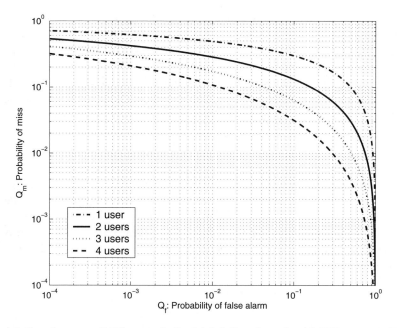

Fig. 4.4. Complementary ROC curves in Rayleigh fading channels with SNR $\bar{\gamma} = 10$ dB for different number of cognitive radios (CRs), $K = 1, 2, 3, 4$.

It can be seen that cooperative spectrum sensing will go through two successive channels: (1) *sensing channel* (from the primary user to cognitive radios); and (2) *reporting channel* (from the cognitive radios to the common receiver). The merit of cooperative spectrum sensing primarily lies in the achievable space diversity brought by the independent sensing channels, namely *sensing diversity gain*, provided by the multiple cognitive radios. Even though one cognitive radio may fail to detect the signal of the primary user, there are still many chances for other cognitive radios to detect it. With the increase of the number of cooperative cognitive radios, the probability of missed detection for all the users will be extremely small. Another merit of cooperative spectrum sensing is the mutual benefit by communicating with each other to improve the sensing performance [25]. When one cognitive radio is far away from the primary user, the received signal may be too weak to detect. However, by employing a cognitive radio who is located nearby the primary user as a relay, the signal of the primary user can be detected reliably by the far user.

4.3.2 Performance in Realistic Fading Channels

In practice, the reporting channels between the cognitive radios and the common receiver will also experience fading and shadowing such as cognitive radio 3 in Fig. 4.3. This will typically deteriorate the transmission reliability of the sensing results reported from the cognitive radios to the common receiver. For example, when one cognitive radio reports a sensing result $\{1\}$ (denoting the presence of the primary user) to the common receiver through a realistic fading channel, the common receiver will most likely detect it to be the opposite result $\{0\}$ (denoting the absence of the primary user) because of the disturbance from the random complex channel coefficient and random noise. Eventually, the performance of cooperative spectrum sensing will be degraded by the imperfect reporting channels.

Definition 4.1. *The probability of reporting errors of the ith cognitive radio, denoted by $P_{e,i}$, is defined as the error probability of signal transmission over the reporting channels between the ith cognitive radio and the common receiver.*

Theorem 4.1. *Let Q_f and Q_m denote the probability of false alarm and probability of miss of cooperative spectrum sensing, respectively. Then,*

$$Q_f = 1 - \prod_{i=1}^{K} [(1 - P_{f,i})(1 - P_{e,i}) + P_{f,i}P_{e,i}] \tag{4.14}$$

$$Q_m = \prod_{i=1}^{K} [P_{m,i}(1 - P_{e,i}) + (1 - P_{m,i})P_{e,i}] \tag{4.15}$$

where $P_{f,i}$ and $P_{m,i}$ are the false alarm probability and miss probability of the local spectrum sensing of the ith cognitive radio, respectively.

Corollary 4.1. *Suppose that the local spectrum sensing conducted by cognitive radio i results in $P_{f,i} = P_f$ and $P_{m,i} = P_m$, for all $i = 1, \cdots, K$, and that the probabilities of reporting errors are identical for all cognitive radios, then*

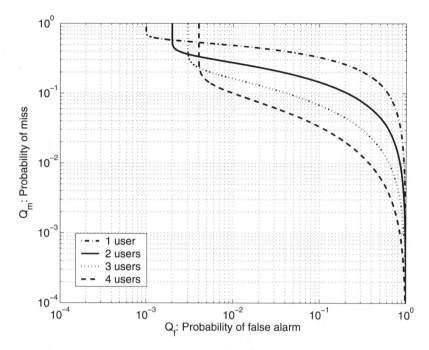

Fig. 4.5. Cooperative spectrum sensing performance (probability of missed detection versus probability of false alarm) for different number of cognitive radios.

$$\begin{cases} Q_f = 1 - [(1 - P_f)(1 - P_e) + P_f P_e]^K \\ Q_m = [P_m(1 - P_e) + (1 - P_m)P_e]^K. \end{cases} \quad (4.16)$$

Furthermore, Q_f is bounded by

$$Q_f \geq \bar{Q}_f \triangleq \lim_{P_f \to 0} Q_f = 1 - (1 - P_e)^K \approx K P_e. \quad (4.17)$$

Figure 4.5 shows the ROC curves (probability of miss, Q_m versus probability of false alarm, Q_f) of cooperative spectrum sensing under imperfect reporting scenarios for different number of cognitive radios. Both sensing channels and reporting channels are simulated as flat Rayleigh fading. The SNR of the sensing channels and the reporting channels are taken as 10 and 25 dB, respectively. Energy detection is used for local spectrum sensing at each cognitive radio and the decision fusion is employed for reporting the sensing results to the common receiver. Obviously, when the number of cognitive radios increases, the miss probability becomes smaller for any given false alarm probability. However, it can be seen in Fig. 4.5 that each curve is chopped by a vertical line, which is called the *false alarm wall*, denoted by $\bar{Q}_f$. This implies that the false alarm probability cannot be sufficiently small due to the bound. It can be shown from (4.17) that the false alarm wall results from the *reporting error probability*, P_e, which characterizes the error probability when the sensing result is transmitted from one cognitive radio to the common receiver over an imperfect reporting channel. Furthermore, it can be seen that the false alarm wall $\bar{Q}_f$ becomes higher when the number of cognitive radios increases. Therefore, for the case that the desired false alarm probability is smaller than $\bar{Q}_f$, cooperative spectrum sensing will be completely invalid.

4.4 Robust Cooperative Spectrum Sensing Techniques

It has been shown that the use of multiple cognitive radios to perform spectrum sensing cooperatively may improve the detection probability but the performance is limited by the realistic reporting channels. To alleviate the performance degradation resulting from the imperfect reporting channels, in the following, we propose several robust cooperative spectrum sensing techniques.

4.4.1 Cooperative Diversity for Cooperative Spectrum Sensing

Multiple antennas technology has been shown as an efficient way to provide superior reception performance due to the potential high-space diversity [26]. In cognitive radio networks, implementing multiple antennas at each cognitive radio is not practical due to the increasing cost and hardware complexity. However, a virtual antenna array can be formed by allowing multiple cognitive radios to cooperate. Hence, the classical space–time coding approaches [27] which have been widely used in multiple-input multiple-output (MIMO) systems can be used in cognitive radio networks so as to achieve a high cooperative diversity. Consider the case when two-located cognitive users cooperate on spectrum sensing, as illustrated in Fig. 4.6. Since the two users are close, the channels between two users can be assumed to be ideal. Firstly, the two users perform local spectrum sensing independently and obtain the sensing results D_1 and D_2 for user 1 and user 2, respectively. Then, they exchange their decisions and send them alternatively in two time slots, with user 1 transmitting $\{D_1, D_2\}$ and user 2 transmitting $\{-D_2, D_1\}$. By doing so, each decision is reported to the common receiver through two independent fading channels. This gives rise to a space

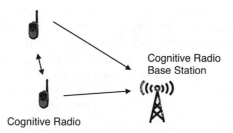

Fig. 4.6. Cooperative diversity technique for cooperative spectrum sensing. The two co-located cognitive radios exchange their local decisions and form a distributed antenna array over the reporting channels.

diversity gain of 2. When the number of cognitive radios in cooperative spectrum sensing is K, it can be expected that a diversity gain of K will be achieved.

For high data rate wireless communications, channel frequency selectivity becomes a critical challenging issue that can significantly affect the system performance. Orthogonal frequency division multiplexing (OFDM) is a powerful tool that can deal with the detrimental effects of multipath fading [28] and has been adopted in many wireless standards such as DTV and wireless LAN. In particular, an OFDM-based cognitive radio system structure is considered by the IEEE 802.22 working group for wireless regional area networks (WRAN). For OFDM-based cognitive radios, cooperative diversity technique can be performed as follows. The two users exchange their local spectrum sensing decisions. Then, the decisions will be sent through two separated sub-channels from each user to the common receiver. By doing so, a frequency diversity gain of 2 can be achieved over frequency-selective fading channels. Therefore, by exploiting a cooperative diversity among co-located cognitive radios, we can reduce the reporting error probability and then enhance the cooperative spectrum sensing performance.

4.4.2 Relay Diversity for Cooperative Spectrum Sensing

When the reporting channels of some cognitive radios experience heavy shadowing, the local decisions in these cognitive radios cannot be forwarded to the BS. Then, the maximum cooperative diversity gain of cooperative spectrum sensing will be reduced. Assume that cognitive radio i fails to send its decision X_i to the BS due to heavy shadowing in its reporting channel. This is the case when the received signal power is so weak that it is merged into the noise. In this case, BS has to make a random decision between H_1 and H_0 if it incorporates such an unreliable cognitive radio into the cooperative decision. Hence, using some unreliable cognitive radios cannot improve the cooperative spectrum sensing performance. To address this issue, BS could censor the SNR of the received signal to check whether or not this cognitive radio is reliable enough before counting it into the cooperative decision. If the SNR of the received signal from the cognitive radio i is lower than a predesigned threshold, then the cognitive radio i will be labeled as an unreliable one. Under the supervision of the BS, the unreliable one can relay its local spectrum sensing result to other cognitive radios which are in enough good channel state, as shown in Fig. 4.7.

With the relay technique, we see that cooperative spectrum sensing achieves the full cooperation among cognitive radios by avoiding transmission of local sensing results over bad reporting channels. Suppose that M out of K cognitive radios experience heavy shadowing. Without any relay, the diversity gain of the cooperative spectrum sensing is only $(K - M)$. However, with the help of other relay cognitive radios, it is demonstrated that the maximum cooperative diversity gain K can be achieved. Although relay-based cooperative spectrum sensing can exploit the full cooperation of all cognitive radios in the case of heavy shadowing, the bound $\bar{Q}_f$ will also increase with an increase in the number of cooperative cognitive radios, as can be seen from (4.17). To decrease the bound $\bar{Q}_f$ while maintaining the maximum cooperative diversity, it is of interest to explore a coding approach combined with relay

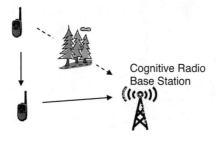

Fig. 4.7. Relay diversity technique for cooperative spectrum sensing. The cognitive radio in heavy shadowing relays its local decision to a neighboring cognitive radio.

diversity. Here, we employ an algebraic coding approach [29] to achieve signal space diversity for relay cognitive radios. Assume that cognitive radio i experiences heavy shadowing and cognitive radio j experiences Rayleigh fading. In order to achieve the maximum cooperative diversity, cognitive radio i will relay its decision X_i to cognitive radio j. Then, the two decisions X_i and X_j which are BPSK symbols are encoded as

$$[C_i \ C_j]^{\mathrm{T}} = \Theta [X_i \ X_j]^{\mathrm{T}}$$

where Θ is a 2×2 rotation matrix and the superscript $\mathcal{T}$ denotes the transpose of a vector or matrix. Subsequently, C_i and C_j are sent through orthogonal channels $H_j(m_i)$ and $H_j(m_j)$, respectively. At the common receiver, the received symbols will be jointly decoded and then forwarded to perform a joint decision. A 2×2 matrix Θ can be given by

$$\Theta = \frac{1}{\sqrt{2}} \begin{pmatrix} 1 & e^{\mathrm{j}\pi/4} \\ 1 & e^{\mathrm{j}5\pi/4} \end{pmatrix} \tag{4.18}$$

which can guarantee a diversity gain of 2 over Rayleigh fading channels [29].

Figure 4.8 shows the performance of the cooperative spectrum sensing with the proposed relay diversity and algebraic coding for two cognitive radios. The average SNRs of the sensing channels of the two cognitive radios are both $\bar{\gamma} = 15$ dB and the average SNR of the reporting channel of the first cognitive radio is $\bar{\eta} = 14$ dB. The second cognitive radio experiences heavy shadowing in its reporting channel and cannot forward the decision to the BS. For comparison, we have also plotted the complementary ROC curve without the use of relays. It can be seen that the curve without relay has the worse performance among the three examined cases at a large value of Q_{f}. This implies that without relay, the *sensing diversity order* of cooperative spectrum sensing is lost under this scenario. Meanwhile the other two curves have a similar performance when Q_{f} is larger than their lower bounds. This indicates that the diversity gain of cooperative spectrum sensing can be retrieved by relaying the decision of the second cognitive radio to the first cognitive radio.

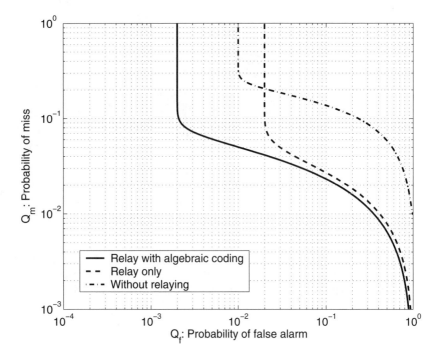

Fig. 4.8. Cooperative spectrum sensing performance with relay diversity and algebraic coding for two cognitive radios. Both sensing channels have an average SNR $\bar{\gamma} = 15$ dB. The reporting channel of the first cognitive radio has average SNR $\bar{\eta} = 14$ dB. The reporting channel of the second cognitive radio experiences heavy shadowing.

However, it can be also observed that the lower bound $\bar{Q}_f$ in the case of relay is larger than that in the case of without relay. This substantiates (4.17) and indicates that cooperative spectrum sensing with many cognitive radios will induce an increase of the lower bound of Q_f. The curve of relay with algebraic coding has the best performance in Fig. 4.8 and achieves both cooperative diversity and lower bound $\bar{Q}_f$. This is because relay diversity allows us to achieve the maximum diversity and algebraic coding results in lowering the bound of Q_f.

4.4.3 Multiuser Diversity for Cooperative Spectrum Sensing

In order to reduce the reporting error probability, we may take advantage of multiuser diversity in cooperative spectrum sensing. Multiuser diversity is a form of selection diversity in which the user with the highest SNR is chosen as the only physical transmission link [30]. In cognitive radio networks, cognitive radios are scattered. This results in different distances between the cognitive radios and the common receiver. Thus, the SNR of the reporting channels between the cognitive radios and the common receiver are varied and are also independently changing due to the independent fading. By taking advantage of these independent fading channels, multiuser diver-

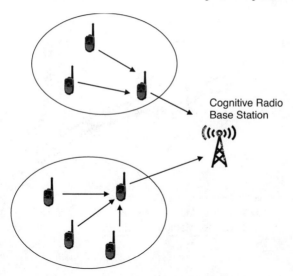

Fig. 4.9. Multiuser diversity technique for cooperative spectrum sensing. Cognitive radios are separated into a few clusters and only the cluster head (with the highest SNR of the reporting channel) participates in the reporting process.

sity can be exploited in cooperative spectrum sensing. Figure 4.9 shows a cognitive radio network with a two-layer hierarchy in implementing the multiuser diversity technique. In the first layer, all cognitive radios are configured into few clusters according to some distributed clustering method. Then, a cluster head is chosen in each cluster according to the highest SNR of the reporting channels. Once every cognitive radio in the same cluster finishes the local spectrum sensing, the sensing results will be reported to the cluster head which will then make a preliminary cooperative decision according to an "OR" logic rule. In the second layer, only cluster heads are required to report to the common receiver with their preliminary cooperative decisions and based on these decisions, the common receiver will make a final decision according to an "OR" logic rule. The advantages of this cluster-based cooperative spectrum sensing are twofolds [31]: firstly, only the user with the highest SNR is chosen as the cluster head to report the decisions to the common receiver. By doing so, it produces a selection diversity gain to reduce the reporting error probability. Secondly, the total amount of sensing bits reported to the common receiver can be greatly reduced since the work of reporting has been taken by the cluster heads, not all cognitive radios, thereby facilitating a low bandwidth control channel.

Figure 4.10 demonstrates the cooperative spectrum sensing performance using cluster-based method, where energy fusion and decision fusion are both considered at the cluster head. For comparison, the conventional method without clustering is simulated. In the simulation, seven cognitive radios are considered and separated into two clusters. The SNR of the sensing channels for two clusters are 10 and 5 dB, respectively, and the SNR of the reporting channels are 10 dB for both clusters. It can be observed that the sensing performance are improved by using cluster-based

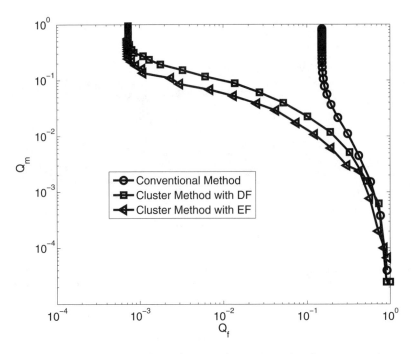

Fig. 4.10. Performance comparison of cooperative spectrum sensing among using conventional method, cluster-based method with energy fusion (EF) and decision fusion (DF).

method. This results from the fact that the multiuser diversity can reduce the reporting error probability with the best user selection.

4.4.4 Censored Decision for Cooperative Spectrum Sensing

For a cognitive radio network with a large number of cognitive radios, the total number of sensing bits transmitted to the common receiver tends to be very large and this will require a high demand in terms of control channel bandwidth and also result in a long sensing time. Note that since the local decision $D \in \{0, 1\}$ is obtained by comparing the local observation O with a predesigned threshold λ, the observation values in the vicinity of the detection threshold are not reliable enough due to the noise disturbance. To exclude the ambiguous detection region around the threshold, a censored decision approach can be used in cooperative spectrum sensing. By carefully setting the ambiguous detection region as the interval $[\lambda_1, \lambda_2]$, only the cognitive radios having the observation values out of this region are required to report to the common receiver. Specifically, the cognitive radio will report a local decision D:

$$D = \begin{cases} 0, & 0 \leq O \leq \lambda_1 \\ 1, & O \geq \lambda_2. \end{cases} \tag{4.19}$$

But if $\lambda_1 < O < \lambda_2$, the cognitive radio will not report anything to the common receiver. The probability of the event that one cognitive radio participates in the reporting process can be calculated by

$$\bar{K} = 1 - \text{Prob}\{\lambda_1 < O < \lambda_2\}. \tag{4.20}$$

Therefore, the average number of sensing bits (one bit for one active cognitive radio in reporting) is $\bar{K}K$. The normalized amount of sensing bits over total number of cognitive radios is $\bar{K}$. It can be expected that by using the proposed censored decision approach, the average transmitted sensing bits will be greatly reduced without much affecting the sensing performance much. This is because those unreliable decisions are censored and excluded from the final decision.

Figure 4.11 shows the miss probability in terms of the normalized amounts of the sensing bits under several given false alarm probabilities. It can be seen that with increase of $\bar{K}$, the miss probability is reduced. However, when $\bar{K}$ is above 0.5, the miss probability will change slightly. It implies that employing half of total number of cognitive radios for cooperative spectrum sensing will not necessarily lead to the

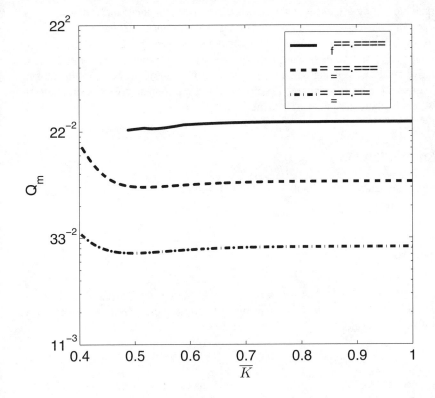

Fig. 4.11. Performance of cooperative spectrum sensing using censored decision. $\bar{K}$ denotes the normalized amounts of sensing bits.

loss of performance. This is because the other half of total number of cognitive radios has a local decision in ambiguous region which will be much unreliable and cannot improve the sensing performance.

Conclusion

Cognitive radio is an agile radio technology that can efficiently utilize the spectrum holes of the licensed channels in different locations and times. To detect the spectrum holes accurately and quickly, spectrum sensing is a critical component in cognitive radio systems. In this chapter, a survey of spectrum sensing techniques for cognitive radios has been presented. The conventional spectrum sensing methods have firstly been introduced and their advantages and disadvantages have been discussed. In order to deal with the hidden terminal problem, which is commonly seen in wireless networks, cooperative spectrum sensing has been considered. By allowing a number of cognitive radios to perform local spectrum sensing independently and fusing their local decision results together at common receiver, the spectrum sensing performance is greatly enhanced. Cooperative spectrum sensing has also been considered for realistic fading scenarios, where both the sensing channels and reporting channels are subject to fading and/or shadowing. Performance analysis of cooperative spectrum sensing under realistic fading channels has been given and a limitation of the cooperative spectrum sensing has been observed. To address this and other cooperative spectrum sensing challenges, several robust cooperative spectrum sensing techniques have been proposed. Further research on cooperative spectrum sensing can be envisioned on wideband sensing.

Acknowledgment

The authors would like to thank Miss Sunny Chunhua Sun for her help in generating Figs. 4.10 and 4.11.

Appendix

Proof of Theorem 4.1: In the following, we assume that the common receiver is a base station (BS). Then, the whole sensing and reporting process can be described as the following flowchart:

$$H_0/H_1(\text{PU}) \longrightarrow H_0/H_1(\text{CR}) \longrightarrow H_0/H_1(\text{BS}). \tag{4.21}$$

In the first stage, one cognitive radio performs local spectrum sensing to get the local decision H_0^{CR} or H_1^{CR}. Then, the local decision is reported to the BS through reporting channels which are usually subject to fading and hence the local decision will be contaminated by the fading channels and additive noise. After signal recovery at the BS, H_0^{BS} or H_1^{BS} is decoded.

Let $\mathrm{Prob}\{H_0^{\mathrm{BS}}|H_0, H_0^{\mathrm{CR}}\}$ denote the probability of the event:

$$H_0(\mathrm{PU}) \longrightarrow H_0(\mathrm{CR}) \longrightarrow H_0(\mathrm{BS}).$$

Then,

$$\mathrm{Prob}\{H_0^{\mathrm{BS}}|H_0, H_0^{\mathrm{CR}}\} = (1 - P_{\mathrm{f}})(1 - P_{\mathrm{e}}). \tag{4.22}$$

Let $\mathrm{Prob}\{H_0^{\mathrm{BS}}|H_0, H_1^{\mathrm{CR}}\}$ denote the probability of the event:

$$H_0(\mathrm{PU}) \longrightarrow H_1(\mathrm{CR}) \longrightarrow H_0(\mathrm{BS}).$$

Then,

$$\mathrm{Prob}\{H_0^{\mathrm{BS}}|H_0, H_1^{\mathrm{CR}}\} = P_{\mathrm{f}}P_{\mathrm{e}}. \tag{4.23}$$

Therefore,

$$\mathrm{Prob}\{H_0^{\mathrm{BS}}|H_0\} = (1 - P_{\mathrm{f}})(1 - P_{\mathrm{e}}) + P_{\mathrm{f}}P_{\mathrm{e}}. \tag{4.24}$$

Likewise, we can obtain

$$\mathrm{Prob}\{H_0^{\mathrm{BS}}|H_1\} = P_{\mathrm{m}}(1 - P_{\mathrm{e}}) + (1 - P_{\mathrm{m}})P_{\mathrm{e}}. \tag{4.25}$$

Next, consider a cognitive radio network with K cognitive users. The decision statistic for the cooperative spectrum sensing can be given by (4.10). Hence,

$$\begin{aligned}
Q_{\mathrm{f}} &= \mathrm{Prob}\{\mathcal{H}_1|H_0\} \\
&= 1 - \mathrm{Prob}\{H_0^{\mathrm{BS},1}, \cdots, H_0^{\mathrm{BS},K}|H_0\} \\
&= 1 - \prod_{i=1}^{K} \mathrm{Prob}\{H_0^{\mathrm{BS},i}|H_0\}.
\end{aligned} \tag{4.26}$$

By substituting (4.24) into (4.26) results in (4.14). Likewise,

$$\begin{aligned}
Q_{\mathrm{m}} &= \mathrm{Prob}\{\mathcal{H}_0|H_1\} \\
&= \mathrm{Prob}\{H_0^{\mathrm{BS},1}, \cdots, H_0^{\mathrm{BS},K}|H_1\} \\
&= \prod_{i=1}^{K} \mathrm{Prob}\{H_0^{\mathrm{BS},i}|H_1\}.
\end{aligned} \tag{4.27}$$

By substituting (4.25) into (4.27) results in (4.15). ∎

Proof of Corollary 4.1: By substituting $P_{\mathrm{f},i} = P_{\mathrm{f}}$ and $P_{\mathrm{m},i} = P_{\mathrm{m}}$ for all $i = 1, \cdots, K$ into (4.14) and (4.15), immediately we get (4.16).

For given P_e and K, it can be seen from (4.16) that the minimization of Q_f is equivalent to the maximization of

$$(1 - P_f)(1 - P_e) + P_f P_e = (1 - P_e) - P_f(1 - 2P_e).$$

Because we always have $P_e < 0.5$, we obtain

$$\min Q_f = \lim_{P_f \to 0} Q_f$$
$$= 1 - (1 - P_e)^K$$
$$\approx K P_e. \tag{4.28}$$

∎

References

1. Federal Communications Commission, "Spectrum Policy Task Force," Report ET Docket no. 02-135, Nov. 2002.
2. M. A. McHenry, "NSF spectrum occupancy measurements project summary," Shared Spectrum Company Report, Aug. 2005.
3. Federal Communications Commission, ET Docket No. 03-322. Notice of Proposed Rule Making and Order, Dec. 2003.
4. C. Cordeiro, K. Challapali, D. Birru, and N. Sai Shankar, "IEEE 802.22: The first world-wide wireless standard based on cognitive radios," in *Proc. IEEE Symp. New Frontiers in Dynamic Spectrum Access Networks* (Baltimore, USA), pp. 328–337, Nov. 8–11, 2005.
5. N. Devroye, P. Mitran, and V. Tarokh, "Limits on communications in a cognitive radio channel," *IEEE Commun. Mag.*, vol. 44, pp. 44–49, June 2006.
6. J. Mitola and G. Q. Maguire, "Cognitive radio: Making software radios more personal," *IEEE Pers. Commun.*, vol. 6, pp. 13–18, Aug. 1999.
7. D. Cabric, I. D. O'Donnell, M. S.-W. Chen, and R. W. Brodersen, "Spectrum sharing radios," *IEEE Circuits Syst. Mag.*, vol. 6, no. 2, pp. 30–45, 2006.
8. S. Haykin, "Cognitive radio: Brain-empowered wireless communications," *IEEE J. Select. Areas Commun.*, vol. 23, pp. 201–220, Feb. 2005.
9. A. Sendonaris, E. Erkip, and B. Aazhang, "User cooperation diversity – part I: System description," *IEEE Trans. Commun.*, vol. 51, pp. 1927–1938, Nov. 2003.
10. D. Cabric, S. M. Mishra, and R. W. Brodersen, "Implementation issues in spectrum sensing for cognitive radios," in *Proc. Asilomar Conf. on Signals, Systems, and Computers*, vol. 1, pp. 772–776, Nov. 7–10, 2004.
11. T. Weiss, J. Hillenbrand, and F. Jondral, "A diversity approach for the detection of idle spectral resources in spectrum pooling systems," in *Proc. 48th Int. Sci. Colloquium* (Ilmenau, Germany), Sept. 2003.
12. P. K. Varshney and C. S. Burrus, *Distributed detection and data fusion*. New York: Springer, 1997.
13. A. Sahai, N. Hoven, and R. Tandra, "Some fundamental limits on cognitive radio," in *Proc. Allerton Conf. on Communications, Control, and Computing* (Monticello), Oct. 2004.

14. N. Sai Shankar, C. Cordeiro, and K. Challapali, "Spectrum agile radios: Utilization and sensing architectures," in *Proc. IEEE Symp. New Frontiers in Dynamic Spectrum Access Networks* (Baltimore, USA), pp. 160–169, Nov. 8–11, 2005.

15. W. A. Gardner and C. M. Spooner, "Signal interception: Performance advantages of cyclic-feature detectors," *IEEE Trans. Commun.*, vol. 40, pp. 149–159, Jan. 1992.

16. Z. Tian and G. B. Giannakis, "A wavelet approach to wideband spectrum sensing for cognitive radios," in *Proc. Int. Conf. on Cognitive Radio Oriented Wireless Networks and Communications* (Greece), June 8–10, 2006.

17. Y. Hur, J. Park, W. Woo, K. Lim, C.-H. Lee, H. S. Kim, and J. Laskar, "A wideband analog multi-resolution spectrum sensing (MRSS) technique for cognitive radio (CR) systems," in *Proc. IEEE Int. Symp. Circuit and System*, pp. 4090–4093, May 21–24, 2006.

18. H. Urkowitz, "Energy detection of unknown deterministic signals," *Proc. IEEE*, vol. 55, pp. 523–531, Apr. 1967.

19. J. Hillenbrand, T. Weiss, and F. K. Jondral, "Calculation of detection and false alarm probabilities in spectrum pooling systems," *IEEE Commun. Lett.*, vol. 9, pp. 349–351, Apr. 2005.

20. V. I. Kostylev, "Energy detection of a signal with random amplitude," in *Proc. IEEE Int. Conf. Commun.* (New York, NY), pp. 1606–1610, Apr. 28–May 2, 2002.

21. F. F. Digham, M.-S. Alouini, and M. K. Simon, "On the energy detection of unknown signals over fading channels," in *Proc. IEEE Int. Conf. Commun.* (Anchorage, AK, USA), pp. 3575–3579, May 11–15, 2003.

22. A. Ghasemi and E. S. Sousa, "Collaborative spectrum sensing for opportunistic access in fading environments," in *Proc. IEEE Symp. New Frontiers in Dynamic Spectrum Access Networks* (Baltimore, USA), pp. 131–136, Nov. 8–11, 2005.

23. E. Visotsky, S. Kuffner, and R. Peterson, "On collaborative detection of TV transmissions in support of dynamic spectrum sensing," in *Proc. IEEE Symp. New Frontiers in Dynamic Spectrum Access Networks* (Baltimore, USA), pp. 338–345, Nov. 8–11, 2005.

24. S. M. Mishra, A. Sahai, and R. Brodersen, "Cooperative sensing among cognitive radios," in *Proc. IEEE Int. Conf. Commun.* (Turkey), June 2006.

25. G. Ganesan and Y. G. Li, "Agility improvement through cooperation diversity in cognitive radio," in *Proc. IEEE Global Communications Conference* (St Louis, Missouri, USA), vol. 5, pp. 2505–2509, Nov. 28–Dec. 2, 2005.

26. R. D. Murch and K. B. Letaief, "Antenna systems for broadband wireless access," *IEEE Commun. Mag.*, vol. 40, no. 4, pp. 76–83, Apr. 2002.

27. S. M. Alamouti, "A simple transmit diversity technique for wireless communication," *IEEE J. Select. Areas Commun.*, vol. 16, pp. 1451–1458, Oct. 1998.

28. D. Huang, K. B. Letaief, and J. Lu, "A receive space diversity architecture for OFDM systems using orthogonal designs," *IEEE Trans. Wireless Commun.*, vol. 3, pp. 992–1002, May 2004.

29. J. Boutros, and E. Viterbo, "Signal space diversity: A power and bandwidth efficient diversity technique for the Rayleigh fading channel," *IEEE Trans. Inf. Theory*, vol. 44, pp. 1453–1467, July 1998.

30. K. B. Letaief and Y. Zhang, "Dynamic multiuser resource allocation and adaptation for wireless systems," *IEEE Wireless Commun.*, vol. 13, no. 4, pp. 38–47, Aug. 2006.

31. C. Sun, W. Zhang, and K. B. Letaief, "Cooperative spectrum sensing for cognitive radios under bandwidth constraints," in *Proc. IEEE Int. Wireless Commun. and Networking Conf.* (Hong Kong), Mar. 11–15, 2007.

Additional Reading

1. I. F. Akyildiz, W.-Y. Lee, M. C. Vuran, and S. Mohanty, "Next generation/dynamic spectrum access/cognitive radio wireless networks: A survey," *Comput Netw*, vol. 50, pp. 2127–2159, 2006.
2. M. Ghozzi, M. Dohler, F. Marx, and J. Palicot, "Cognitive radio: Methods for the detection of free band," *CR Phys.*, *Elsevier*, vol. 7, pp. 794–804, Sept. 2006.
3. N. han, S. Shon, J. H. Chung, and J. M. Kim, "Spectral correlation based signal detection method for spectrum sensing in IEEE 802.22 WRAN systems," in *Proc. Int. Conf. Adv. Commun. Technol.* (Phoenix Park, Korea), vol. 3, pp. 1765–1770, Feb. 20–22, 2006.
4. C. Sun, W. Zhang, and K. B. Letaief, "Cluster-based cooperative spectrum sensing for cognitive radio systems," in *Proc. IEEE Int. Conf. Commun.* (Glasgow, Scotland, UK), June 24–28, 2007.
5. S. Bandyopadhyay and E. Coyle, "An energy-efficient hierarchical clustering algorithm for wireless sensor networks," in *Proc. IEEE INFOCOM* (San Francisco, CA, USA), pp. 1713–1723, Apr. 2003.
6. O. Younis and S. Fahmy, "Distributed clustering in ad hoc sensor networks: A hybrid, energy-efficient approach," in *Proc. IEEE INFOCOM* (Hong Kong, China), pp. 629–640, Mar. 2004.

5

A Protocol Suite for Cognitive Radios in Dynamic Spectrum Access Networks

Michael B. Pursley and Thomas C. Royster IV

Department of Electrical and Computer Engineering
Clemson University
Clemson, SC, 29634, USA
{pursley,troyste}@ces.clemson.edu

5.1 Introduction

In this chapter, a protocol suite is presented for initiating and controlling transmissions among cognitive radios in dynamic spectrum access networks. A framework is provided for the selection of the initial modulation to be used in a session after a frequency band has been selected. During the first few packet transmissions in a new session, a power-adjustment protocol compensates for uncertainties in the interference and propagation characteristics in the designated frequency band. Throughout the session, an adaptive transmission protocol compensates for variations in the communications environment. Because increases in transmitter power can disrupt other sessions that are underway in the network, adaptation of modulation and coding is the preferred mechanism for responding to increased interference or propagation loss. For a wide range of modulation techniques and channel models, performance results for our protocols are compared with performance results for ideal protocols that are furnished perfect channel-state information.

A suite of protocols is required for cognitive packet radios that wish to initiate and maintain reliable communications in a wireless ad hoc dynamic spectrum access network. The session initiation process begins when one wireless communications device, referred to as the *source*, wishes to set up a session to send a sequence of packets to another wireless communications device, the *destination*. For the packet sizes that we consider, a session for the transfer of a 1 MB file requires the delivery of approximately 2000–8000 packets, depending on the rate of the error-control code. Other wireless communications devices that are within range of the source are referred to as *unintended receivers*. These devices may receive interference from the source's transmissions if they are operating in the same frequency band.

In this chapter, it is assumed that the source and destination are within range of each other; relaying of packets is not required. The wireless communications devices employ half-duplex packet transmission, so simultaneous two-way communication is not possible. The only feedback information is provided in acknowledgment packets, so the goal of the adaptive transmission protocol is to respond to such channel variations as those due to changes in shadow loss or propagation distance between the source and destination.

Once an available frequency band is identified for the session, a *modulation-selection protocol* must choose a modulation technique according to the capabilities of the radios, the established etiquette for transmission in the network, and the quality-of-service (QoS) priorities for the session. Because of the uncertainties in the propagation characteristics in the frequency band that is selected for the session, a *power-adjustment protocol* must adjust the transmitter power during the first few packets. In this chapter, a *packet* is an information packet unless specified otherwise (e.g., an acknowledgment packet). The adjusted power level must be high enough to provide dependable delivery of packets to the destination but not so high that the transmissions waste energy and cause unnecessary interference to other radios in the network. As the session continues, the communications environment may change, so an *adaptive transmission protocol* must modify the modulation and coding as needed to maintain reliable communications without increasing interference to other radios. At the end of the session, the adaptive transmission protocol can supply information to the modulation-selection and power-adjustment protocols if there is to be another session in the same frequency band involving the same source and destination.

The radios in the network are cognitive radios (e.g., [1] or [2]) that extract information about their environments. The information they extract permits other radios to adapt their transmission parameters to changes in propagation loss or interference. For our protocols, the cognitive radios are not required to employ channel estimation techniques or even make measurements of received power. Instead, the destination obtains simple statistics from its demodulator and decoder, chooses the modulation format and code for the next packet, and communicates these choices to the source in the acknowledgment packet. We avoid the use of complex channel measurements so that the protocols can be implemented with modest complexity and employed in half-duplex packet radios using current and near-term future technology.

We provide a framework for modulation selection that accounts for the three primary spectrum etiquette parameters (time, bandwidth, and power) for each of the modulation formats. We define and evaluate a power-adjustment protocol that converges within the first few packets of a session, and we describe a family of adaptive transmission protocols and provide performance evaluations for static channels with unknown parameters and dynamic channels with time-varying parameters. Our protocols are applicable to any modulation format. For illustrative purposes, we give performance results for a wide range of modulation formats, including quadrature-amplitude modulation (QAM), phase-shift keying (PSK), binary orthogonal modulation, nonbinary biorthogonal modulation, and complementary code keying (CCK). Each of the modulation techniques can be employed with direct-sequence (DS) or frequency-hop (FH) spread spectrum.

Because of the nature of dynamic spectrum access networks, we do not consider adaptive transmission protocols that increase transmitter power as the primary response to deteriorating channel conditions. The increased interference that results from ramping up the transmitter power degrades frequency reuse in the network and it may disrupt ongoing sessions and prevent the initiation of new ones. The preferred alternative is to use adaptive coding and modulation to respond to an increase in propagation loss or interference. The desire to minimize interference to other radios and the need for energy conservation, particularly for hand-held mobile communications devices, lead us to design protocols that increase the transmitter power only as a last resort.

5.2 Modulation Formats

The combinations of error-control coding and data modulation that we selected to illustrate the performance of the protocols are forms of bit-interleaved coded modulation [3]. The encoding, modulation, demodulation, and decoding are depicted in Fig. 5.1. Spread-spectrum techniques [4] may be applied to the coded modulation. As shown in Fig. 5.1, a signature sequence (e.g., a pseudo-random sequence) can be applied at the modulator output to give DS spread spectrum. If DS spread spectrum is not employed, the system of Fig. 5.1 is modified by removing the signature sequence generators and multipliers or by letting the signature sequence be a sequence of 1 s so that it has no effect. For FH spread spectrum, the sequence generator and multiplier are replaced by a frequency hopper in the transmitter and a frequency dehopper in the receiver. The modulation techniques can also be used with various forms of multi-carrier modulation [5] such as orthogonal frequency-division multiplexing [6] or multi-carrier DS spread spectrum (e.g., see [7] or [8]), and it is easy to modify the protocols that we provide to accommodate multicarrier transmission.

We evaluate the performance of our adaptive transmission protocols for static channels with unknown parameters and for dynamic channels with time-varying parameters whose variations are modeled by finite-state Markov chains. Our performance results are for turbo product codes, S-random [9] or helical [10] interleaving, and soft-decision iterative decoding. The turbo product codes are available commercially on a single chip [11]. Similar results would be obtained for standard convolutional codes, except that higher power levels are required. Hypothetical ideal protocols with perfect channel-state information are used to provide performance bounds for dynamic channels.

The rows of an $N \times N$ Hadamard matrix provide an N-orthogonal signal set. If $M = 2N$, then the orthogonal signals and their complements give an M-biorthogonal signal set. QAM has both inphase and quadrature modulation, as does QPSK. If the inphase and quadrature signals are M-biorthogonal, then we get I–Q biorthogonal modulation with M^2 signals in the set. M^2 I–Q biorthogonal modulation requires half the bandwidth of standard M-biorthogonal modulation for the same information rate. The 11 Mb/s CCK signal set for IEEE 802.11b [12] has 256 signals, so we denote it by 256-CCK. For the same information rate, 256-CCK has the same bandwidth as 256 I–Q biorthogonal modulation. For the turbo product code of rate approximately $1/2$, frames that represent approximately 2000 bits, and an additive white Gaussian noise (AWGN) channel, 256 I–Q biorthogonal modulation gives better performance than 256-CCK by more than 1.5 dB. For the NASA-standard convolutional code of constraint length $K = 7$ and rate $1/2$ with soft-decision Viterbi

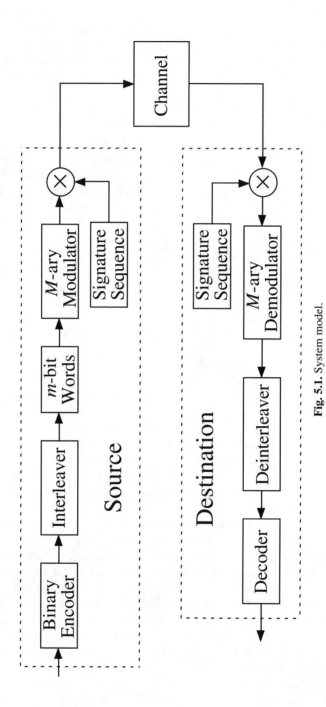

Fig. 5.1. System model.

decoding, the performance advantage of 256 I–Q biorthogonal modulation over 256-CCK is more than 2 dB.

5.3 Code and Modulation Parameters

We denote the set of modulation techniques by $\{\mathcal{M}_j : 1 \leq j \leq n_m\}$ and the set of codes by $\{C_i : 1 \leq i \leq n_c\}$, where the codes are indexed in the order of increasing rates (i.e., $r_1 < r_2 < \cdots < r_{n_c}$). It is convenient to index code–modulation combinations with a single subscript, so we let $n = n_c\, n_m$ and denote the set of combinations by $\{\mathcal{D}_k : 1 \leq k \leq n\}$ for an arbitrary one-to-one mapping between the pairs (i, j) and the indices k. If a definite mapping is desired, we can use $k = i + n_c(j-1)$.

A signature sequence can be applied to some modulation formats to provide *spread-spectrum multiple-access capability* [4], which permits multiple simultaneous sessions to be accommodated in the same frequency band. Frequency hopping can be applied for the same purpose. For modulation format $\mathcal{M}_j$, let η_j denote the number of signature sequence chips per modulation chip and let q_j denote the number of frequency slots over which the signal is hopped. If a signature sequence is not used, then $\eta_j = 1$; if there is no frequency hopping, then $q_j = 1$. The *modulation chip* is the elemental pulse used by the data modulation waveform. For example, a set of M-orthogonal modulation symbols of duration $T = MT_c$ is obtained if each modulation symbol is a sequence of M rectangular pulses of duration T_c and the sequences of pulse amplitudes for different symbols correspond to different rows of an $M \times M$ Hadamard matrix (e.g., see [4] or [13]). Let m_j denote the number of binary symbols per modulation symbol and let L_j denote the number of modulation chips per modulation symbol. For example, if $\mathcal{M}_j$ is standard QPSK with no spread-spectrum modulation, then $m_j = 2$ and $L_j = \eta_j = q_j = 1$, but if $\mathcal{M}_j$ is standard BPSK without spread spectrum, then $m_j = L_j = \eta_j = q_j = 1$. If $\mathcal{M}_j$ is the modulation format for the reverse link of the TIA-95 cellular CDMA system [14], then $\eta_j = 4$, $m_j = 6$, $L_j = 64$, and $q_j = 1$. If $\mathcal{M}_j$ is 256-CCK modulation, then $m_j = L_j = 8$ and neither a signature sequence nor frequency hopping can be used if the signals must conform to the IEEE 802.11b standard [12].

In Table 5.1, several modulation formats are listed along with their null-to-null bandwidths for transmission of uncoded data at an information rate of 1 b/s with no spread spectrum (i.e., $\eta_j = q_j = 1$). Nonbinary orthogonal modulation formats are not listed in Table 5.1 because M-orthogonal modulation has twice the bandwidth of M-biorthogonal modulation for the same information rate. Binary orthogonal modulation (e.g., BFSK) is also not included in the table. Its approximate bandwidth is in the range 3.0–4.0 Hz, depending on the signal design. Although QAM gives the best spectral efficiency among the modulation formats in Table 5.1, it has no multiple-access capability so it is not suitable for a shared frequency band. On the other hand, the orthogonal and biorthogonal modulation formats can tolerate multiple-access or multipath interference, especially when a signature sequence is applied. Even for the AWGN channel, orthogonal and biorthogonal signals require less energy per bit than QAM, BPSK, or QPSK. Orthogonal modulation, biorthogonal modulation, and PSK

Table 5.1. Bandwidths for several modulation formats (1 b/s, no spread spectrum).

Modulation	Bandwidth (HZ)
64-Biorthogonal	10.67
32-Biorthogonal	6.40
16-Biorthogonal	4.00
256 I–Q Biorthogonal	2.00
256-CCK	2.00
BPSK	2.00
QPSK/4-QAM	1.00
16-QAM	0.50
64-QAM	0.33

do not have the amplitude fluctuations of QAM and they do not require amplitude reference levels for their demodulation.

For our numerical results, the codes in the set $\{C_i : 1 \leq i \leq 5\}$ are turbo product codes. We consider fixed-length packets with $n_b = 4096$ binary code symbols per packet. If a packet is encoded with code C_i, then it represents $k_i = r_i n_b$ bits of information. The code block length is 4096 for each of the three *primary codes* C_2, C_3, and C_5, which have 1331, 2028, and 3249 information bits per block, respectively. The corresponding approximate rates are 0.325, 0.495, and 0.793. Code C_1 has block length 2048 with 484 information bits per block, which gives an approximate rate of 0.236. The block length for code C_4 is 1024, and it has 676 information bits per block for an approximate rate of 0.660. There is one code word per packet for each of the primary codes, two code words per packet for C_1 and four code words for packet for C_4. Because the encoders and iterative decoders for all five codes are available on a single chip [11], the set $\{C_i : 1 \leq i \leq 5\}$ is very attractive for use in adaptive-rate coding.

For systems that demodulate coherently, the log-likelihood-ratio (LLR) metric (e.g., see [15–17]) is used for all modulation formats except QAM. A simpler distance metric [16] with approximately the same performance as the LLR metric is employed for QAM. For noncoherent demodulation of binary orthogonal signals, we employ the log-ratio metric, which is the logarithm of the ratio of the outputs of the two noncoherent detectors (e.g., envelope detectors).

For a channel with no fading, the energy per information bit is denoted by $\mathcal{E}_b$ and the energy per binary code symbol is denoted by $\mathcal{E}_s$. If the error-control code has rate r, then the two energies are related by $\mathcal{E}_s = r\mathcal{E}_b$. For QAM, these energies are not the same for all points in the constellation, so $\mathcal{E}_b$ and $\mathcal{E}_s$ denote the average energies for the information bits and binary code symbols, respectively. The receiver's thermal noise is modeled as white Gaussian noise with one-sided power spectral density N_0. Although each of our channel models includes additive white Gaussian noise (AWGN), we reserve the phrase *AWGN channel* for a channel that has no fading and no other noise or interference. Power and energy ratios are expressed in dB by defining such parameters as $\mathrm{ENR} = 10\log_{10}(\mathcal{E}_b/N_0)$ and $\mathrm{SENR} = 10\log_{10}(\mathcal{E}_s/N_0)$.

Table 5.2. Required ENR for 10^{-2} packet error probability for five turbo product codes (AWGN channel, coherent BPSK).

Code	Rate (r_i)	Block length	Capacity limit(dB)	ENR	Difference (dB)
C_1	0.236	2048	−0.8	1.5 dB	2.3
C_2	0.325	4096	−0.5	1.1 dB	1.6
C_3	0.495	4096	0.2	1.7 dB	1.5
C_4	0.660	1024	1.0	2.9 dB	1.9
C_5	0.793	4096	2.0	2.9 dB	0.9

The energy requirements for the five turbo product codes when used with BPSK and coherent demodulation on an AWGN channel are listed in Table 5.2. The values of ENR required by five turbo product codes are compared with the capacity limits for the five rates. The three primary codes are within 1.6 dB of their respective capacity limits. Table 5.2 is also valid for QPSK.

For the Rician fading channel, the energy per binary code symbol for the specular (unfaded) component is $\mathcal{E}_s$, the same as for the AWGN channel. The average energy per binary symbol in the diffuse (Rayleigh faded) component is denoted by $\mathcal{E}_d$ and the average total energy per binary symbol for the received signal is denoted by $\mathcal{E}_a$. Numerical results are presented in terms of $\text{SPENR} = 10\log_{10}(\mathcal{E}_s/N_0)$, $\text{DENR} = 10\log_{10}(\mathcal{E}_d/N_0)$, and $\text{SENR} = 10\log_{10}(\mathcal{E}_a/N_0)$. The *Rician fading parameter* γ is the square-root of the ratio of the average energy in the diffuse component to the energy in the specular component; that is, $\gamma^2 = \mathcal{E}_d/\mathcal{E}_s$. The AWGN channel ($\gamma^2 = 0$) and the Rayleigh fading channel ($\gamma^2 = \infty$) are obtained as special cases.

5.4 Time-Bandwidth Product: A Basis for Modulation Selection

The source and destination are part of a dynamic spectrum access network, so their first step is to select (or be assigned) an available frequency band that meets the requirements for the session. Next, they must identify modulation and coding techniques that are usable by both radios and satisfy any constraints imposed by the frequency band that was chosen. For example, it may be that the destination cannot demodulate some of the source's modulation techniques, or it may be that some modulation techniques that are usable by both devices exceed the available bandwidth at the chosen frequency. Within the set of modulation and coding methods that are compatible with both radios, the source must choose methods that conform to established etiquette for the spectrum access network and accommodate the QoS priorities for the session. For example, subject to constraints imposed by spectrum etiquette, it may be desirable to maximize throughput or minimize delay for the session traffic.

We introduce a spectrum-etiquette measure for use in the selection of the initial modulation and coding for a new session. The spectrum-etiquette measure is a function of the three etiquette parameters that have the greatest impact on other users of the spectrum access network: bandwidth, transmission time, and power. The measure of transmission time that we use is the session's *data-transmission time*, which is the total time that the source is actively transmitting data during the session. The influence of bandwidth and time are clear. If a session uses more bandwidth for a longer time, then there are fewer opportunities for other sessions. The power level used by the source is equally important, because it determines the number of unintended receivers that experience interference during the session.

For simplicity, let each packet contain the same number n_b of binary code symbols, but of course the number $k_i = r_i n_b$ of information bits that a packet represents depends on the code C_i that is used for the packet. The spectral occupancy parameter that we use is normalized relative to uncoded BPSK. For code C_i and modulation format $\mathcal{M}_j$, the spectral occupancy parameter is defined as

$$\lambda_{i,j} = \frac{\eta_j q_j L_j n_b}{m_j k_i}. \tag{5.1}$$

Notice that $\lambda_{i,j} = 1$ for uncoded BPSK. The spectral occupancy parameter is related to the null-to-null bandwidth as follows: For a rectangular chip waveform and an information rate of R_b b/s, the null-to-null bandwidth for modulation format $\mathcal{M}_j$ and code C_i is $B_{i,j} = 2\lambda_{i,j} R_b$ Hz. If the modulation incorporates spread spectrum, then $B_{i,j}$ is the spread bandwidth for the signal. For a FH system, $q_j > 1$ and we can view $B_{i,j}/q_j$ as the null-to-null bandwidth for each frequency slot. If the chip waveform is not rectangular, then the factor of 2 is replaced by an appropriate constant; for example, the constant is 3 for the half-sinewave pulse used in minimum-shift keying (MSK). Similarly, measures of bandwidth other than null-to-null bandwidth are accommodated by adjusting the constant appropriately, but the choice of constant is relatively unimportant because comparisons between different combinations of coding and modulation are dependent only on the ratios of their bandwidths. Thus, if the information rates are the same for two combinations, then the spectral occupancy parameter $\lambda_{i,j}$ contains all the information that is needed to determine the best combination. If the null-to-null bandwidth is fixed at B Hz, then the information rate that can be accommodated with modulation format $\mathcal{M}_j$ and code C_i is $R_{i,j} = B/(2\lambda_{i,j})$ b/s.

If a session that employs modulation format $\mathcal{M}_j$ and code C_i is required to deliver N_b bits of information, then it must deliver $N_i = \lceil N_b/k_i \rceil \approx N_b/k_i$ packets. The expected number of packet transmissions required to complete the session, including retransmissions of any failed packets, is $N_i/Q_{i,j}$, where $Q_{i,j} = 1 - P_{i,j}$ is the packet success probability and $P_{i,j}$ is the packet error probability. If the frequency band for the session provides null-to-null bandwidth $B_{i,j}$, then the average data-transmission time per session is closely approximated by

$$\mathcal{T}_{i,j} \approx \frac{1}{k_i Q_{i,j}} \left[\frac{2\eta_j q_j L_j n_b N_b}{m_j B_{i,j}} \right]. \tag{5.2}$$

The approximation is exact if N_b/k_i is an integer. From (5.2) we obtain the very important conclusion that the average data-transmission time for a given modulation format $\mathcal{M}_j$ and a fixed bandwidth $B_{i,j} = B$ is minimized by the code that maximizes the expected throughput $S_{i,j} = k_i Q_{i,j}$, which is the approach that we use for adaptive coding in Sect. 5.8. The time required by packet headers or preambles (e.g., for synchronization) is not included in (5.2), but it is straightforward to incorporate such overhead times into the analysis [18].

An important spectral etiquette measure is the product of the bandwidth and the session's data-transmission time. The *time-bandwidth product* $\Psi_{i,j}$ for a session that employs code C_i and modulation format $\mathcal{M}_j$ is $\Psi_{i,j} = T_{i,j}B_{i,j}$. From (5.2) we see that

$$\Psi_{i,j} \approx \frac{1}{k_i Q_{i,j}} \left[\frac{2\eta_j q_j L_j n_b N_b}{m_j} \right] \tag{5.3}$$

and we observe there is no dependence on i among the terms inside the brackets of (5.3). Thus, for modulation format $\mathcal{M}_j$, the code with the largest expected throughput provides the smallest time-bandwidth product. If we divide the time-bandwidth product by the constant $2N_b$, then we obtain

$$\tau_{i,j} = \frac{\Psi_{i,j}}{2N_b} \approx \frac{1}{k_i Q_{i,j}} \left[\frac{\eta_j q_j L_j n_b}{m_j} \right] \tag{5.4}$$

which we refer to as the *normalized time-bandwidth product*. It follows from (5.4) that $\tau_{i,j}$ is approximately equal to the average number of transmitted chips per delivered information bit. If N_b/k_i is an integer, then the approximations in (5.3) and (5.4) are exact.

5.5 Resource Consumption

Assume the source, which is randomly placed in the plane, wishes to initiate a session, and the unintended receivers are randomly and independently located according to a uniform distribution in a region whose boundaries are far beyond the transmission range of the source. It is desired that the packet success probability for the session be no less than Q. The source has a set $\{\mathcal{D}_k : 1 \le k \le n\}$ of combinations of codes and modulation formats from which it selects one combination $\mathcal{D}$ to use at the beginning of the session. For the remainder of this section, the code-modulation combination $\mathcal{D}$ is fixed, and the requirement for the packet success probability is Q; consequently, the subscripts i and j are omitted for the symbols that appear in (5.1)–(5.4).

The *average power density* for a transmission is the average power divided by the bandwidth B. If the source uses code-modulation combination $\mathcal{D}$ and average power density ζ in a frequency band of width B Hz for an average of T seconds per session, then the frequency band available for receptions by the $\mathcal{N}$ unintended receivers that are within range of the source's transmission is reduced by B for a time period whose average duration is T. The average power density ζ for the transmission is determined by the power-adjustment protocol. For a perfect power-adjustment protocol, ζ is the minimum average power density that provides packet success probability Q at the destination. In Sect. 5.7, it is shown that the power-adjustment protocol can set the average power density close to this minimum.

If the source's transmission prevents one radio from receiving in a band of width 1 Hz for a time period of 1 s, then we say that one unit of resource has been consumed by the transmission. For code-modulation combination $\mathcal{D}$, the source's transmission at power density level ζ prevents $\mathcal{N}$ radios from receiving in a band of width B Hz for an average of $\mathcal{T}$ seconds, so the average resource consumption for the session is

$$\text{RC} = \mathcal{T} B \mathcal{N}. \tag{5.5}$$

We adopt a simple threshold model for the acceptable level of interference in each radio. A radio that receives interference with an average power density greater than the threshold is said to be in the *interference region* for the source's transmission, and the *interference area* is the area of the interference region. Let $\mathcal{N}$ be the average number of unintended receivers that are in the interference region when the source employs code-modulation combination $\mathcal{D}$. Without loss of generality, we let the threshold level be one, so the power-density unit that is used to specify ζ is equal to the interference threshold for the unintended receivers. If unintended receivers do not have the same sensitivity to interference, then we use the minimum tolerable interference among the unintended receivers as the threshold. Because we are interested only in comparisons of resource consumption for different code-modulation combinations, multiplicative constants are unimportant. For simplicity, we set each such constant equal to one as it arises in our development of an expression for the resource consumption.

If the unintended receivers are uniformly distributed, then the average number of them that receive interference power density greater than one unit is proportional to the interference area, and this area is a function of the transmission's range. For transmissions using code-modulation combination $\mathcal{D}$, the relationship between the range ρ and the transmitted power density ζ is obtained from $\zeta \rho^{-\alpha} = 1$, where α is the propagation loss exponent [19]. The propagation loss exponent is typically in the interval $2 \le \alpha \le 6$, and normally $\alpha < 2$ only for short-range, indoor, line-of-sight communications [20]. For an arbitrary exponent, the range is

$$\rho = \zeta^{1/\alpha}. \tag{5.6}$$

The interference area A is proportional to the square of the range, regardless of the beamwidth of the source's antenna. Because the constant of proportionality is unimportant, we let

$$A = \rho^2. \tag{5.7}$$

From (5.6) and (5.7), we see that $A = \zeta^{2/\alpha}$. For most communication environments, the maximum area for the interference region is $A = \zeta$, which occurs if $\alpha = 2$ (free-space propagation). Because the location of the source is arbitrary and the terrain around it is unknown, we use a small value for the propagation loss exponent in the development of a measure for resource consumption. As a consequence, it is very unlikely that the source's transmission produces an interference density greater than the interference threshold for any radio outside the interference region. The average number of unintended receivers that are in the interference region is proportional to the area, so we conclude that

$$\mathcal{N} = \zeta . \tag{5.8}$$

It follows from (5.5) and (5.8) that the resource consumption is

$$\text{RC} = T B \zeta \tag{5.9}$$

for a session that uses bandwidth B, has average data-transmission time T, and transmits average power density ζ. The transmitted power is $\mathcal{P}_t = B\zeta$. If the transmitted power is held constant after the initial power adjustment, then the average transmitted energy per session is $\mathcal{E}_{\text{session}} \approx \mathcal{P}_t T$, so the resource consumption is approximately equal to the average transmitted energy for the session. The approximation is accurate because the power-adjustment protocol converges within the first few packets in a session that typically requires the delivery of a few hundred to several thousand packets.

The units for power, time, and bandwidth are unimportant, because we are interested only in comparisons of the resource consumption for different code-modulation combinations. Consequently, we can use any convenient normalization. The *normalized resource consumption* is

$$\mathcal{R} = \frac{G}{N_b N_0} \text{RC} = \frac{T B \xi}{N_b N_0} = \frac{\mathcal{P}_r T}{N_b N_0} \tag{5.10}$$

where G is the channel gain from the source to the destination, $\xi = G\zeta$ is the power density for the received signal at the destination, and $\mathcal{P}_r = G\mathcal{P}_t$ is the received power at the destination. It follows from (5.10) that $\mathcal{R} = \mathcal{E}_b/N_0$, where $\mathcal{E}_b$ is the average received energy per information bit. For convenience, we assume that N_b is a multiple of k, the number of information bits per packet, so the approximations in (5.2)–(5.4) are exact. It follows from (5.4) that $T B = 2 N_b \tau$, so (5.10) implies

$$\mathcal{R} = \frac{2\tau\xi}{N_0} = P\tau \tag{5.11}$$

where τ is the normalized time-bandwidth product and $P = 2\xi/N_0$ is proportional to the received power density at the destination.

Values for the normalized resource consumption $\mathcal{R}$ are given in Table 5.3 for several combinations of turbo product codes and modulation formats. The values for $\mathcal{R}$ are determined from $\mathcal{R} = \mathcal{E}_b/N_0$, and the values for $\mathcal{E}_b/N_0$ are determined from simulations of the iterative decoding of packets at a success probability of $Q = 0.99$. The values for the normalized time-bandwidth product τ are computed analytically from (5.4). From the table, we see that smaller values for the resource consumption are obtained for combinations that compromise between the power density and the time-bandwidth product. In particular, each combination with a very small time-bandwidth product has a relatively large resource consumption. In [21], we determine the Shannon bounds on resource consumption, and we show that these bounds also

Table 5.3. Resource consumption for a 10^{-2} packet error probability.

Modulation	Code, rate	P	τ	$\mathcal{R}$
64-Biorth	$C_1, 0.236$	0.10	22.83	2.3
64-Biorth	$C_2, 0.325$	0.12	16.58	2.0
64-Biorth	$C_3, 0.495$	0.15	10.88	1.6
64-Biorth	$C_5, 0.793$	0.24	6.79	1.6
256 I–Q Biorth	$C_3, 0.495$	0.81	2.04	1.7
256-CCK	$C_3, 0.495$	1.22	2.04	2.5
QPSK	$C_2, 0.325$	0.83	1.55	1.3
QPSK	$C_3, 0.495$	1.45	1.02	1.5
QPSK	$C_5, 0.793$	3.07	0.64	2.0
64-QAM	$C_2, 0.325$	6.71	0.52	3.5
64-QAM	$C_3, 0.495$	14.73	0.34	5.0
64-QAM	$C_5, 0.793$	45.02	0.21	9.5

predict that the best modulation formats are compromises between low power density and small time-bandwidth product. The Shannon bounds give accurate guidelines if good error-control codes are used (i.e., codes that give performance near the capacity limit for the modulation format that is used). If the bandwidth is held constant, the duration of the session is proportional to τ and the transmitted power is proportional to P. Thus, the tradeoff we obtain from Table 5.3 when the code-modulation combinations are required to have the same bandwidth is between the amount of power that is required to achieve $Q = 0.99$ and the amount of time required to complete the session. The lowest resource consumption is obtained by compromising between these two performance measures.

We envision a wide range of applications for the concept of resource consumption. As we have indicated, one application is to provide a quantitative basis for the selection of the initial modulation and coding for a session. The protocol for selection of modulation and coding might minimize $\mathcal{R}$, perhaps subject to QoS constraints (e.g., delay) or limitations on resources (time, bandwidth, power). However, for wireless communications, a communications channel can be arbitrarily poor for arbitrarily long periods of time, so it may not be feasible to impose firm constraints. A more reasonable approach is to employ resource consumption in conjunction with QoS *priorities* to perform tradeoffs between session duration, bandwidth requirements, energy consumption, and interference effects. If the radios in a dynamic spectrum access network must pay for the use of spectrum, then the resource consumption could be the basis for determining the fee. For an ad hoc network in which packets must be relayed between the source and destination, the resource consumption is a suitable link resistance metric for use in least-resistance routing (e.g., see [22] or [23]).

5.6 Statistics for Power Adjustment and Adaptive Transmission

Each packet has a header and a payload. The payload carries the information bits for the packet, and the header includes a few bits that specify the modulation and error-control code that are used for the payload. For the packet formats in this chapter, the payload represents many more bits than the header, and the transmission time for the header is negligible compared to the transmission time for the payload. The number of binary code symbols represented by the payload is constant, independent of the code rate, so the number of information bits per packet varies with the code rate. We assume the radios employ half-duplex packet transmission, which implies that feedback information from the destination cannot be received during the transmission of a packet. However, the header of each acknowledgment packet includes a field that carries a few bits of feedback information from the destination. The use of the header in the acknowledgment packet may be different for the power-adjustment period than for the longer-term adaptive-transmission period.

Our protocols are not given any channel-state information; instead, power adjustment and adaptive transmission are driven by statistics that are derived in the receiver. The feedback may consist of the receiver statistics or, equivalently, it may consist of commands that are derived from the statistics (e.g., a command to increase the code rate or change the modulation format). Because of the variability in the amount of time between consecutive packet transmissions and the desire to minimize complexity, we do not require channel estimation techniques or measurements of received power. Instead, the destination extracts simple statistics from the demodulator and decoder and sends to the source either a representation of these statistics or the commands that are derived from them. Depending on whether the feedback consists of the statistics or adaptation commands, the source or the destination will choose the modulation format and code for the next packet. In our preferred mode of operation, the destination sends an acknowledgment packet in response to each packet that it receives from the source. If an acknowledgment is not received, then the source retransmits the packet, perhaps using a more powerful code or modulation format. In one alternative mode, which is not evaluated in this chapter, a single acknowledgment packet is sent in response to a specified number of consecutive packets from the source.

Examples of statistics that are derived in the receiver are shown in Fig. 5.2. The *iteration count for a received word* is the number of iterations performed by the decoder for the word, and the *iteration count for a packet* is the average of the iteration counts for the words in the packet. The *error count* for a packet is the number of errors in the binary symbols that are derived from hard decisions made at the demodulator output. As illustrated in Fig. 5.2, the error count can be determined for a correctly decoded packet (e.g., as verified by a CRC code) by encoding the information symbols at the output of the soft-decision decoder and comparing the resulting code symbols with the hard-decision demodulated binary symbols. If there are multiple code words per packet, then the error count for the packet is the sum of the error counts for the received words. The type of *demodulator statistic* that is employed for adaptation depends on the modulation format, and it may also depend

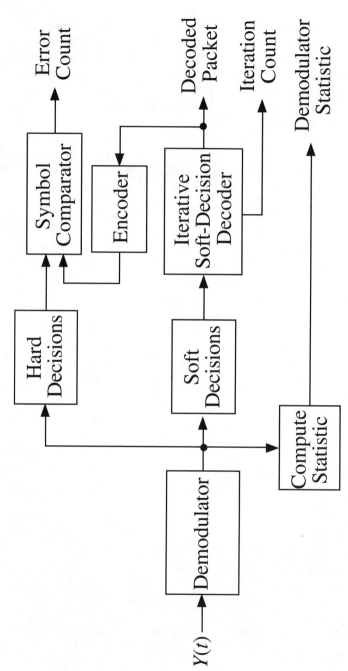

Fig. 5.2. Extraction of statistics for use in power adjustment and adaptive transmission.

on the soft-decision metric that is used for decoding. For each modulation symbol in a packet that uses M-biorthogonal modulation, the magnitudes of the $M/2$ correlator outputs are examined and the ratio of the largest output to the second largest is computed. This measure of the quality of a demodulated symbol was originally devised by Viterbi [24] for anti-jam communications. Applications to soft-decision decoding are given in [17]. The average $\bar{r}$ of the ratios for the modulation symbols in the packet is determined, and the demodulator statistic for the packet is $1 - \bar{r}$. For QPSK and QAM, the Euclidean distance between each received symbol and its closest point in the signal constellation is determined, and the demodulator statistic for each packet is the average Euclidean distance for the modulation symbols in the packet.

5.7 Power Adjustment

When a new session between the source and destination is to be conducted in a different frequency band than any of their recent sessions, the propagation loss is very difficult to predict. Several empirical formulas for outdoor propagation are reviewed in [6] and [20], and each formula includes modifications and correction factors that depend on terrain, antenna height, etc. In the Hata model [25] for the frequency band 100–1500 MHz, the correction factors differ by several decibels, even for a suburban area as compared with an open area. For indoor communications, a common model includes a Gaussian random variable that has a standard deviation as large as 10 dB for certain types of buildings (e.g., see [20]). Shadow loss in urban areas can vary by 20 dB or more [09].

Because the source typically does not know the propagation loss in the assigned frequency band, the initial power may be too low or too high. It is especially detrimental to the network if the initial power level greatly exceeds the power needed to obtain a satisfactory packet error rate. Equipment designers may even include a positive bias in the initial power setting to increase the chances that the session's first packets are received and acknowledged, because the acknowledgments provide important feedback information for subsequent transmissions. It has minimal impact on other sessions to transmit excessive power for ten to twenty packets if the sessions involve thousands of packets, but it is very disruptive to other sessions to continue emitting excess power that is perhaps 10 dB or more greater than what is required. Eliminating the excess power also reduces energy consumption at the source. Of course, the problem is that the source does not know how much of the power is excess power, so a protocol is required to obtain the necessary statistics in the destination receiver and use them efficiently to adjust the power to a satisfactory level within the first few packets of a new session.

Our power-adjustment protocols use one or more adaptation statistics that are derived in the destination's demodulator and decoder, as shown in Fig. 5.2. If only a single statistic is used for power adjustment, then a demodulator statistic gives the best performance. For each modulation format, a simple interval test is used to select the power level for the next packet, and a stopping condition is applied to

Table 5.4. Accuracies for the power-adjustment protocol.

Modulation	Statistic	Δ_1 (dB)	Δ_2 (dB)
64-Biorth.	Ratio	0.3	1.1
QPSK	Distance	0.1	1.2
16-QAM	Distance	0.3	1.8

the sequence of demodulator statistics from consecutive packets to determine when to terminate the power adjustment. The feedback, which consists of the statistics or a decision based on the statistics, is included in the acknowledgment packet. If an acknowledgment is not received for a packet that is sent during the power-adjustment phase of a session, then the source increases the power by a fixed amount for the next packet. For our numerical results, a power increase of 5 dB was employed in response to unacknowledged packet transmissions during the power-adjustment period.

For each code-modulation combination $\mathcal{D}$ there is a received power level $P_{\min}$ (in dB) that is the minimum possible received power that gives a packet success probability of Q or larger when $\mathcal{D}$ is used for the packet. We set the target power level at $P_{\text{tar}} = P_{\min} + 0.5$ to give a margin of 0.5 dB. We conducted simulation tests to verify the convergence and accuracy of the power-adjustment protocol when it is employed with the appropriate demodulator statistics. Each test consists of a sequence of 10,000 sessions. The random initial power for each session has a uniform distribution on the interval from 15 dB below the target to 15 dB above the target. The initial power levels for different sessions are independent. Each of the three modulation formats in Table 5.4 was employed with the turbo product code of rate 0.793. For each session in the simulation of 10,000 sessions, the received power level when the power-adjustment protocol was stopped was in the range from $P_{\min} + \Delta_1$ to $P_{\min} + \Delta_2$, and the number of packets transmitted when the demodulator statistic triggered the stopping condition was never more than eight. In each of the 10,000 sessions for each modulation format, the received power level at the time the power adjustment was stopped was above $P_{\min}$.

5.8 Adaptation of the Code Rate

After the power adjustment is completed (usually within the first seven or eight packets of a session), the system applies adaptive protocols to change the coding and modulation as required by any changes that occur in the communications environment during the session, such as an increase in the propagation loss, perhaps due to shadowing. The system could respond to such an increase by transmitting more power, but that would raise the interference level in the network. It is much better to increase the transmitter power only if the propagation loss increases so much that it cannot be offset by changes in modulation and coding. We begin by considering the adaptation of the error-control code. Some results on the adaptation of the modulation are given in Sect. 5.11.

In the adaptive-coding protocol, the feedback from a packet reception is used to adapt the rate of the error-control code for the next packet that the source will send. If the source transmits a packet but an acknowledgment is not received, then the source retransmits the packet, perhaps using a code of lower rate. According to our results in Sect. 5.4, we should choose the code that maximizes the expected throughput. The expected throughput is the number of information bits per packet times the probability of success for the packet (i.e., $k_i Q_{i,j}$ in the notation of Sect. 5.4). The corresponding performance measure for a session is its *average throughput*, which is defined as the total number of information bits in packets that are decoded correctly at the destination divided by the total number of packet transmissions that are made by the source (including retransmissions). An information bit is not counted in the numerator of the throughput expression unless the entire packet is decoded correctly. All packet transmissions, whether they are decoded correctly or not, are counted in the denominator, so the protocol is penalized for failed packets.

The protocol strives to achieve the best balance between a high-rate code, which represents a larger number of information bits per packet but may have a lower success probability, and a low-rate code, which may have a higher success probability but carries fewer information bits in each packet. The best choice depends on the channel state, which is unknown to the protocol. The protocol selects the code for the next packet transmission according to one or more adaptation statistics from the previous packet transmission. In addition to the turbo product code, a high-rate CRC code is employed to verify the decoder output is correct before the error count or iteration count is determined. If the previous transmission was not decoded correctly, then the code rate is reduced by one step if possible (i.e., if the code of lowest rate was not used); otherwise, the code rate is unchanged for the next transmission. If the packet is decoded correctly, then the choice of code rate for the next packet is based on comparisons of the adaptation statistics with a fixed set of adaptation parameters.

5.9 Adaptive-Coding Protocol Performance for Static Channels

The first performance results for our adaptive-coding protocol are given in Fig. 5.3 for coherent demodulation of 64-QAM on an AWGN channel with fixed but unknown propagation loss, so the value of SENR is unknown to the protocol. The throughput curves for the adaptive-coding protocols are shown along with the five throughput curves for fixed-rate coding with the same five turbo product codes that are used in the adaptive-coding system. The statistic for one adaptive-coding protocol is the error count and the statistic for the other is the iteration count. In each case, a simple interval test is applied to the statistic to decide which code to use for the next packet. Each curve that is labeled by a code rate is for fixed-rate coding with the corresponding turbo product code. The upper envelope of these curves represents the performance of an ideal protocol that is told the exact value of SENR and uses the code that maximizes the throughput for that value. The performance of our adaptive protocol, which is given only the error count or iteration count for the previous

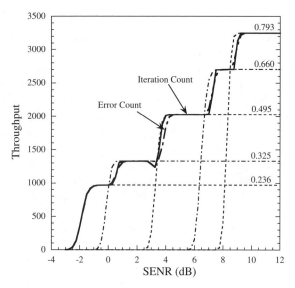

Fig. 5.3. 64-QAM with coherent demodulation for an AWGN channel.

packet, is almost as good as the performance of an ideal protocol that is given perfect channel-state information (i.e., the exact value of SENR).

The performance results shown in Fig. 5.4 are for a Rician fading channel whose parameters γ^2 and SENR are fixed but unknown to both the transmitter and receiver. These results also represent the steady-state performance of our protocol following a change in the channel parameters. Our protocol is given no information about the values of γ^2 and SENR; instead, it uses only the error count from the most recent packet transmission to select the code rate for the next packet transmission. Also shown in Fig. 5.4 are the individual throughput curves for each of the five turbo product codes. The upper envelope of the five dashed curves in Fig. 5.4 represents the performance of an ideal protocol that is told the exact values for γ^2 and SENR and uses the code that maximizes the throughput for those values. The performance of our adaptive protocol, which is given only the error count for the previous packet, is almost as good as the performance of the ideal protocol that is given perfect channel-state information. The comparison between the upper envelope and the throughput curve for our adaptive-coding protocol implies that there is very little to be gained from making channel measurements or using other receiver statistics in addition to the error count. For the same modulation and channel model, we also evaluated the adaptive-coding protocol with the iteration count, and we found that its throughput curve is approximately the same as the throughput curve in Fig. 5.4. Additional results and more discussion of the system and channel models are given in [26].

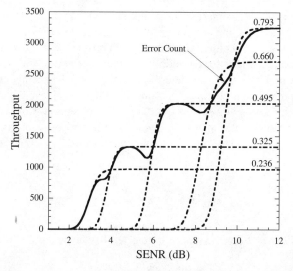

Fig. 5.4. Throughput for binary orthogonal modulation, noncoherent demodulation, frequency hopping, and a static Rician-fading channel with $\gamma^2 = 0.3$.

5.10 Adaptive Coding Results for Dynamic Channels

Each time-varying parameter for a dynamic channel is modeled by the Markov chain illustrated in Fig. 5.5. The state is fixed for the duration of a packet, but it can change from one packet to the next. If the time-varying parameter is the propagation loss, then the K states correspond to excess propagation losses $L_1, L_2, \ldots, L_K$, in increasing order. The *excess propagation loss* is the amount in dB by which the actual propagation loss exceeds some reference level. The reference level corresponds to state 1, so the excess path loss for state 1 is always $L_1 = 0$ dB. The results presented here are for a four-state Markov model with $p = 0.1$ and $L_k = (k-1)\Delta$ for $2 \le k \le 4$ and $\Delta = 1.5$ dB. We have also investigated six-state models with $\Delta = 2$ dB.

In order to obtain benchmarks against which to compare our protocols for dynamic channels, we evaluate two hypothetical protocols in which perfect information is supplied to the protocol about the past (previous state) or the future (next state). *Perfect previous-state information* consists of the exact value of the path loss on the channel for the previous packet transmission, and *perfect next-state information* consists of the exact value of the path loss on the channel for the next packet

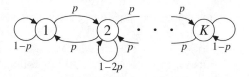

Fig. 5.5. Markov model for changes in channel parameters.

transmission (i.e., the transmission for which the code is being selected by the protocol). In either case, the hypothetical protocol selects the code that maximizes the conditional expected throughput given the perfect channel-state information.

From simulation results for the individual fixed-rate codes, such as those shown as dashed lines in Fig. 5.4, we can determine analytically the average throughput for each of the hypothetical protocols. The throughput that is achieved by code C_i when the channel is in state k is denoted by $s(i|k)$. The transition probability $p(k|j)$ is the probability that the next state is k given that the previous state is j. First, consider the protocol with perfect previous-state information. The conditional expected throughput for code C_i is

$$\bar{s}(i|j) = \sum_{k=1}^{K} s(i|k)\, p(k|j).$$

(5.12)

When the previous state is j, code C_{i_j} is selected for the next transmission if

$$\bar{s}(i_j|j) = \max\{\bar{s}(i|j) : 1 \le i \le n_c\}.$$

(5.13)

If π_j denotes the steady-state probability for state j in the Markov chain, then the average throughput for the protocol with perfect previous-state information is

$$\bar{\mathcal{S}}_1 = \sum_{j=1}^{K} \pi_j\, \bar{s}(i_j|j).$$

(5.14)

Now, consider the protocol with perfect next-state information. The conditional expected throughput for code C_i given that the next state is k is $s(i|k)$. When the next state is k, code C_{i_k} is selected for the next transmission if

$$s(i_k|k) = \max\{s(i|k) : 1 \le i \le n_c\}$$

(5.15)

and the resulting average throughput for the protocol with perfect next-state information is

$$\bar{\mathcal{S}}_2 = \sum_{k=1}^{K} \pi_k\, s(i_k|k).$$

(5.16)

The analytical performance results for the protocol with perfect next-state information represents an upper bound on the throughput for any protocol that uses the five codes listed in Table 5.2. The five turbo product codes are also employed with the protocol that has perfect previous-state information to give a more realistic benchmark for our protocols. It is unrealistic to assume that the protocol will have perfect knowledge of the future channel state, so the best we can hope for is to use statistics that provide adequate information about the previous channel state.

The performance results for Fig. 5.6 are for adaptive coding for 16-QAM on a channel with time-varying propagation loss that is modeled as a four-state Markov chain with $\Delta = 1.5$ dB and transition probability $p = 0.1$. The nominal value

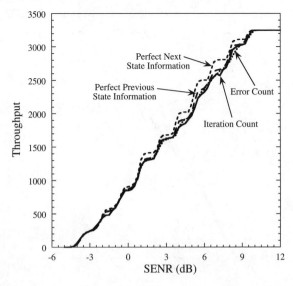

Fig. 5.6. Throughput for 16-QAM with adaptive coding on a channel with time-varying propagation loss.

of the binary code symbol energy to noise density ratio for the AWGN channel is denoted by SENR, and the actual value when the channel is in state k is $\text{SENR}_k = \text{SENR} - L_k$. Our protocols do not know the value of SENR or the state of the channel. In Fig. 5.6, the throughput graphs for the error count and the iteration count are compared with the throughput graphs for the two ideal protocols that have perfect channel-state information and use the same five turbo product codes that our protocol uses. We see from Fig. 5.6 the protocol that uses only the error count and the protocol that uses only the iteration count have nearly the same average throughput as the protocol that is given perfect previous-state information, including the exact value of SENR. Our results show that more complex methods of estimating the previous channel state are not needed and will not give better average throughput than we achieve with the iteration count or the error count. We have found that the error count gives good performance with other coding systems as well (e.g., convolutional codes with Viterbi decoding or Reed–Solomon codes with bounded-distance decoding).

We next consider 64-biorthogonal modulation that employs a signature sequence with one sequence chip per modulation chip (i.e., $\eta_j = 1$). The performance of the adaptive-coding protocol is evaluated for a time-varying multipath channel in which there are two specular components with a separation (i.e., differential delay) of 32 chips, which corresponds to the 64-biorthogonal symbol duration. The four states in the Markov chain represent four relative strengths for the two multipath components. We refer to the first-arriving multipath component as the *primary* component and the second-arriving component as the *secondary* component. The receiver is based on standard matched filters or correlators; it is not a rake receiver. As a result, the secondary component acts as an interference signal. The *multipath power ratio* is the

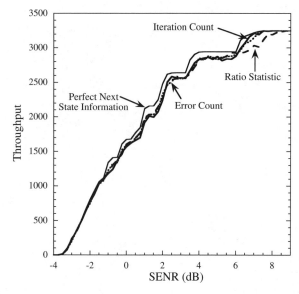

Fig. 5.7. Throughput for 64-biorthogonal modulation with adaptive coding on a dynamic two-component specular multipath channel.

ratio of the power in the primary component to the power in the secondary component. The four states correspond to multipath power ratios of -2 dB, 0 dB, 3 dB, and ∞; the latter is a channel with no secondary component. The transition probability is $p = 0.1$.

The throughput results for the adaptive-coding protocol with three different adaptation statistics are shown in Fig. 5.7 as a function of SENR, the ratio (in dB) of the binary symbol energy for the primary component to the noise density. The three adaptation statistics are the error count, the iteration count, and the ratio statistic. Included for comparison is the throughput graph for the ideal protocol with perfect next-state information. The throughput graphs for our three protocols are nearly the same as the throughput graph for the ideal protocol, which again indicates that the adaptation statistics provide all the necessary information for adaptive coding. Additional statistics or channel measurements would be of no value. In [27], comparisons are made with the upper bounds on throughput that correspond to the ideal protocol with perfect next-state information and five capacity-achieving codes of the same rates as the turbo product codes.

5.11 Adaptive Modulation for Dynamic Channels

For adaptive M-biorthogonal modulation, the value of M is adapted as the channel changes, but we hold the chip rate constant. By keeping the chip rate constant, we maintain a constant bandwidth as the modulation is adapted. Because the number

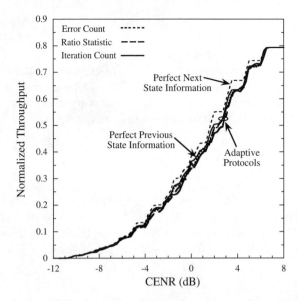

Fig. 5.8. Adaptive M-biorthogonal modulation for a time-varying propagation loss.

of binary code symbols per packet is fixed ($n_b = 4096$), the number of chips per packet changes as M changes. For fair comparisons among the different values of M, the throughput measure that is used for adaptive M-biorthogonal modulation is normalized by dividing by the number of chips per packet. Thus, the *normalized throughput* shown in Fig. 5.8 is the throughput per transmitted chip. The parameter CENR is the ratio of the energy per chip to the one-sided noise density.

The results in Fig. 5.8 are for a channel with a time-varying propagation loss that is modeled by a four-state Markov chain with $\Delta = 1.5$ dB and transition probability $p = 0.1$. The iteration count, error count, and ratio statistic give nearly equal throughput for each value of CENR, and their throughput graphs are very close to the throughput graphs for ideal protocols with perfect channel-state information. Although we initially considered all powers of 2 in the range $4 \le M \le 64$, we found that only three modulation formats are needed for good results: $M = 4$, $M = 8$, and $M = 64$. Similarly, we found that only four of the five codes are needed for good results: C_1 (rate 0.236), C_2 (rate 0.325), C_3 (rate 0.495), and C_5 (rate 0.793).

Conclusion

We have provided a framework for modulation selection and we have presented and evaluated protocols for initial power adjustment, adaptive coding, and adaptive modulation in cognitive radios. Protocols for the adaptation of coding and modulation are important for dynamic spectrum access networks, because they permit the radios to respond to changes in propagation and interference without increasing power, which would increase interference to other users of the network. The new protocols require no channel measurements; instead, they rely only on a few bits of information that can be supplied in acknowledgment packets. The adaptation statistics used by the protocols are very simple and are obtained easily from the demodulator and decoder. Performance results for static channels with unknown parameters and for dynamic channels with time-varying parameters show that our protocols perform almost as well as ideal protocols with perfect channel-state information.

Acknowledgments

The performance results for frequency-hop transmission over Rician fading channels were obtained by Jason Skinner and Michael Masse. The preparation of this chapter was supported by Grant N00014-04-1-0563 from the US Office of Naval Research. Thomas Royster was also supported by a fellowship from the US National Science Foundation.

References

1. S. Haykin, "Cognitive radio: Brain-empowered wireless communications," *IEEE J. Select. Areas Commun.*, vol. 23, no. 2, pp. 201–220, Feb. 2005.
2. J. Mitola III and G. Q. Maguire Jr., "Cognitive radio: Making software radios more personal." *IEEE Personal Commun.*, vol. 6, no. 4, pp. 13–18, Aug. 1999.
3. G. Caire, G. Taricco, and E. Biglieri, "Bit-interleaved coded modulation," *IEEE Trans. Inf. Theory*, vol. 44, no. 3, pp. 927–946, May 1998.
4. M. B. Pursley, *Introduction to Digital Communications*. Upper Saddle River, NJ: Prentice Hall, 2005.
5. J. A. C. Bingham, "Multicarrier modulation for data transmission: An idea whose time has come," *IEEE Commun. Mag.*, vol. 28, pp. 5–14, May 1990.
6. A. Goldsmith, *Wireless Communications*, Cambridge University Press, New York, 2005.
7. S. Kondo and L. B. Milstein, "Performance of multicarrier DS-CDMA systems," *IEEE Trans. Commun.*, vol. 44, pp. 238–246, Feb. 1996.
8. E. Sourour and M. Nakagawa, "Performance of orthogonal multicarrier CDMA in a multipath fading channel," *IEEE Trans. Commun.*, vol. 44, pp. 356–367, Mar. 1996.
9. S. Dolinar and D. Divsalar, "Weight distributions for turbo codes using random and non-random permutations," JPL TDA Progress Report 42-122, pp. 56–65, Aug. 1995.
10. W. W. Wu, D Haccoun., R. Peile, and Y. Hirata, "Coding for satellite communication," *IEEE J. Select. Areas Commun.*, vol. SAC-5, pp. 724–748, May 1987.
11. Advanced Hardware Architectures, Inc., Product Specification for AHA4501 Astro 36 Mbits/sec Turbo Product Code Encoder/Decoder. http://www.aha.com.
12. Institute of Electrical and Electronics Engineers, Standard 802.11b-1999, Part 11: Wireless LAN Medium Access Control (MAC) and Physical Layer (PHY) Specifications: Higher-Speed Physical Layer Extension in the 2.4 GHz Band, http://grouper.ieee.org/groups/802/11/, Jan. 2000.
13. M. K. Simon, S. M. Hinedi, and W. C. Lindsey, *Digital Communication Techniques*. Upper Saddle River, NJ: Prentice Hall, 1995.

14. TIA/EIA Mobile Station-Base Station Compatibility Standard for Wideband Spread Spectrum Cellular Systems, ANSI/TIA/EIA-95-B-99. Washington, DC: Telecommunications Industry Association, Mar. 1999.

15. S. Le Goff, A. Glaviuex, and C. Berrou, "Turbo-codes and high spectral efficiency modulation," in *Proc. of the 1994 IEEE International Conference on Communications*, pp. 645-649, 1994.

16. W. G. Phoel, J. A. Pursley, M. B. Pursley, and J. S. Skinner, "Frequency-hop spread spectrum with quadrature amplitude modulation and error-control coding," in *Proc. of the 2004 IEEE Military Communications Conference* (Monterey, CA), Nov. 2004.

17. M. B. Pursley and T. C. Royster IV, "High-rate direct-sequence spread spectrum with error-control coding," *IEEE Trans. Commun.*, vol. 54, no. 9, pp. 1693–1702, Sept. 2006.

18. M. B. Pursley, T. C. Royster IV, and J. S. Skinner, "Protocols for the selection, adjustment, and adaptation of transmission parameters in dynamic spectrum access networks," in *Proc. of the 2005 IEEE Int. Symp. on New Frontiers in Dynamic Spectr. Access Networks* (Baltimore), pp. 649–657, Nov. 2005.

19. W. C. Jakes, editor, *Microwave Mobile Communications*, Psicataway, NJ: IEEE Press, 1974.

20. T. S. Rappaport, *Wireless Communications: Principles and Practice*, 2nd ed. Upper Saddle River, NJ: Prentice Hall PTR, 2002.

21. M. B. Pursley and T. C. Royster IV, "Resource consumption in dynamic spectrum access networks: Applications and Shannon limits," in *Proc. of the Workshop on Information Theory and Its Applications* (LaJolla, CA), Jan. 2007.

22. M. B. Pursley and H. B. Russell, "Network protocols for frequency-hop packet radios with decoder side information," *IEEE J. Select. Areas in Commun.*, vol. 12, pp. 612–621, May 1994.

23. M. B. Pursley, H. B. Russell, and J. S. Wysocarski, "Energy-efficient routing of multimedia traffic in frequency-hop packet radio networks," *Int. J. Wireless Inf. Networks*, vol. 13, no. 3, pp. 193–205, July 2006.

24. A. J. Viterbi, "A robust ratio-threshold technique to mitigate tone and partial band jamming in coded MFSK systems," in *Proc. of the IEEE Military Communications Conference*, vol. 2, (Boston), pp. 22.4.1–22.4.5, Oct. 1982.

25. M. Hata, "Emperical formula for propagation loss in land mobile radio services," *IEEE Trans. Vehicular. Technol.*, vol. VT-29, no. 3, pp. 317–325, Aug. 1980.

26. M. R. Masse, M. B. Pursley, T. C. Royster IV, and J. S. Skinner, "Adaptive coding for wireless spread-spectrum communication systems," in *Proc. of the 2006 International Conference on Communications, Circuits, and Systems*, vol. 1, (Guilin, China), pp. 1321–1326, June 2006.

27. M. B. Pursley, T. C. Royster IV, and J. S. Skinner, "Adaptive transmission for mobile packet-radio networks: Protocol performance vs. capacity limits," http://ita.ucsd.edu/workshop/06/talks/papers/123.pdf, in *Proc. of the Workshop on Information Theory and Its Applications* (LaJolla, CA), Feb. 2006

6

OFDM-Based Cognitive Radios for Dynamic Spectrum Access Networks

Rakesh Rajbanshi, Alexander M. Wyglinski, and Gary J. Minden

The University of Kansas, USA
{rajbansh,alex,gminden}@ittc.ku.edu

6.1 Introduction

With the advent of new wireless applications, as well as growth of existing wireless services, demand for additional bandwidth is rapidly increasing. As a result, the possibility of spectrum scarcity becomes more of a reality. Existing "command-and-control" spectrum allocations defined by government regulatory agencies prohibit unlicensed access to licensed spectrum, constraining them instead to several heavily populated, interference-prone frequency bands. This spectrum scarcity is apparent since it has been shown that the spectrum is not utilized efficiently. For instance, measurement studies have shown that many licensed bands are relatively unused across time and frequency [1]. To make better use of radio spectrum resources, government regulatory agencies such as the Federal Communications Commission (FCC) are currently working on the concept of unlicensed users "borrowing" spectrum from incumbent license holders. This concept is called dynamic spectrum access (DSA) [2, 3]. Wireless communication technology needs to be sufficiently agile in order to perform DSA such that spectrum utilization is improved while not interfering with incumbent user transmissions.

The development of software-defined radio (SDR) technology has made modern wireless transceivers more versatile, powerful, and portable, by performing baseband processing, such as modulation/demodulation and equalization, entirely in software and digital logic. With the ease and speed of programming baseband operations, SDR technology is a prime candidate for DSA networks. In addition to the agility of the SDR technology, the radio needs to be spectrally aware as well as autonomous in order to dynamically utilize spectrum. A radio that can adapt its transmitter parameters based on interaction with the environment[1] in which it operates is known as *cognitive radios* [4]. With recent developments in cognitive radio technology, it is now possible for these systems to simultaneously respect the rights of incumbent license holders while providing additional flexibility and access to spectrum.

---

[1] These changing environments can be at the physical, network, and/or application layers of the system.

Research and development of cognitive radios involves experts from various disciplines, including but not limited to the following categories:

- Spectrum sharing policy: Regulatory agencies assign radio spectrum to license holders, which maintain exclusive rights to a finite bandwidth. However, since radio transmissions propagate throughout space, it is necessary to define enforceable rules and regulations that guarantees the rights of the incumbent license holders [5]. Simultaneously, it is possible to grant access to the unlicensed users to enable the secondary spectrum utilization.
- Artificial intelligence (AI): The radio should autonomously and dynamically determine the appropriate radio parameters without intervention from the user in order to enable the efficient secondary spectrum utilization [6].
- Cognitive network protocols: The coordination of cognitive radios require the sharing of information in order to agree upon communication parameters dynamically. Dissemination of control traffic among the users is quite important for effective and efficient spectrum sharing [7].
- Reconfigurable hardware: Adaptation to dynamically changing operating parameters require the cognitive radio hardware to be rapidly reconfigurable. Therefore, software defined radios and FPGA-based techniques are prime candidates to build a cognitive radio [8, 9].
- Agile physical layer transmission techniques: As with any wireless communications system, including cognitive radios, the choice of a physical layer transmission technique is an important design decision [10]. The primary goal of a transmission technique employed by the cognitive radio unit would be to achieve sufficient agility enabling unlicensed users to transmit in a licensed band while not interfering with the incumbent users.

Therefore, cognitive radio research is highly interdisciplinary and various issues need to be addressed to meet the regulatory requirements before becoming a reality.

In this chapter, we will focus on the design and implementation of agile physical layer transmission techniques for cognitive radios. To support throughput-intensive applications, multi-carrier modulation (MCM) techniques can be employed due to its ability for handling distortions introduced by frequency selective channels [11–13]. Moreover, cognitive radio transceivers based on MCM can readily enable DSA networks by employing spectrum pooling, where secondary users may temporarily rent spectral resources during the idle periods of licensed users [14].

One form of MCM that possesses an efficient implementation is orthogonal frequency division multiplexing (OFDM). One variant of OFDM that is capable of deactivating subcarriers across its transmission bandwidth which could potentially interfere with transmissions from other users, is non-contiguous OFDM (NC-OFDM) [15–20]. NC-OFDM is designed to support a high aggregate data rate with the remaining active subcarriers while simultaneously transmitting in the proximity of other users in the same region of frequency spectrum.

Despite the advantages of NC-OFDM, there exist several issues with this technique. First, all OFDM-based implementations employ the fast Fourier transform (FFT) at the core of its design. The FFT blocks are used for modulating and demodulating the individual subcarriers to different center frequencies. When a significant number of subcarriers are deactivated, the computation time of FFT-based modulation can be optimized by reducing the total number of multiply/add operations. Second, OFDM-based systems suffer from a high peak-to-average power ratio phenomenon, requiring the components, such as power amplifier (PA), digital-to-analog (D/A) converters, and analog-to-digital (A/D) converters, to have large dynamic ranges in order to avoid any clipping or non-linear distortions of the signal [21]. Third, OFDM subcarrier parameters can be individually altered in order to enhance overall system performance. However, the process by which to optimize these parameters must be both fast and efficient. In this chapter, solutions to these issues that have been proposed in the literature will be investigated.

The rest of this chapter is organized as follows: In Sect. 6.2, an overview of DSA techniques is presented. Section 6.3 provides a brief introduction to the NC-OFDM framework, as well as FFT pruning technique used to improve the efficiency of the FFT computation. Moreover, the error performance of the NC-OFDM technique is analyzed. Section 6.4 describes the peak-to-average power ratio problem in an NC-OFDM system. Non-uniform bit allocation employed to improve the performance of the NC-OFDM technique is presented in Sect. 6.5, and several concluding remarks are presented.

6.2 Dynamic Spectrum Access Techniques

Spectrum sharing can mitigate the apparent spectrum scarcity problem and improve spectrum efficiency. Several spectrum sharing strategies have been proposed in literature, which can be broadly categorized based on [22]:

1. Network architecture
   (i) Centralized approach
   (ii) Distributed approach
2. Spectrum allocation behavior
   (i) Cooperative approach
   (ii) Non-cooperative approach
3. Spectrum access technique
   (i) Underlay approach
   (ii) Overlay approach

The following sections describe these categories in greater detail.

6.2.1 Centralized and Distributed Spectrum Sharing Approaches

In a *centralized spectrum sharing* approach, a centralized entity coordinates with arbitrary wireless technologies and manages access to arbitrary radio spectra by issuing clients temporary leases for parts of the radio spectrum [23]. In this model, a centralized server collects information from a collaborating group of secondary users, which learn about the primary user transmission characteristics, along with primary

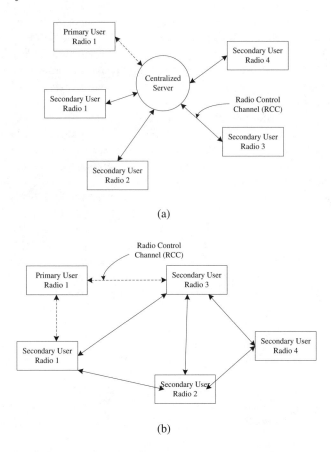

(a)

(b)

Fig. 6.1. **a** Centralized and **b** distributed spectrum sharing.

user cooperation, if possible,[2] and manages a database for the spectrum access and availability information. The users communicate with the centralized server using a pre-assigned dedicated radio control channel (RCC). A basic framework for a centralized spectrum sharing model is shown in Fig. 6.1(a). In the figure, the dashed RCC link between the primary user and the centralized server implies that the primary user may or may not choose to cooperate, whereas the solid RCC link between the secondary user and the centralized server implies that they must cooperate with each other. This form of spectrum management offers simpler and coordinated spectrum access, which enables efficient spectrum sharing and utilization in wireless environments.

---

[2] Since spectrum sharing techniques generate additional interference to the primary user, and the process of cooperation implies the addition of overhead for the incumbent license holders, primary user cooperation would not be easily anticipated without providing significant benefit to the incumbent users.

Even though a centralized server can optimize across network-wide information, there are two serious limitations [24]:

- The spectrum server and all secondary users need to communicate using a pre-assigned dedicated RCC. As the network grows in density, a pre-defined control channel can limit the bandwidth available for data communication.
- As the number of users grows, the server processing complexity will scale at least polynomially [24]. Therefore, any central spectrum server can quickly become a computational bottleneck in the system.

Several centralized spectrum sharing approaches have been proposed in the literature, including the dynamic intelligent management of spectrum for ubiquitous mobile access networks (DIMSUMnet) [25], which can be employed as a regional spectrum broker, and the dynamic spectrum access protocol (DSAP) [26], which can be used as a spectrum broker for heavily-used, densely-populated localized areas where lease updates can occur frequently.

In a *distributed spectrum sharing* approach, each node is responsible for its own spectrum allocation and access based on primary user transmissions in its vicinity and policies [27, 28]. In this model, since secondary users can sense and share the local spectrum access information among themselves, primary user contributions need not be enforced. Therefore, this model poses an advantage for the primary license holders, since there would be no overhead involved with the incumbent users. A basic framework for a distributed spectrum sharing model is shown in Fig. 6.1(b). In the figure, the dashed RCC link between the primary user and other secondary users implies that the primary user may or may not choose to cooperate, whereas the solid RCC link among the secondary users implies that they must cooperate with each other. Since individual nodes are responsible for maintaining the correct information about current spectrum usage, distributed spectrum sharing results in increased overhead communications among the secondary users. However, cooperative distributed algorithm can produce effects similar to global optimization through cooperative local actions distributed throughout the system [24]. One of the serious drawbacks of the distributed spectrum sharing approach can be a *hidden node problem*, where the secondary users fail to detect incumbent users[3] and result inadvertently interfering with the incumbent user transmissions [29]. Moreover, large amounts of measurement information gathered by the secondary users terminals during the detection cycle need to be transmitted to the other users, which can be a significant overhead in the system.

6.2.2 Cooperative and Non-cooperative Spectrum Sharing Approaches

Spectrum sharing techniques can be classified into *cooperative* and *non-cooperative* spectrum sharing based on the spectrum allocation behavior. In cooperative spectrum sharing, the primary and secondary users can cooperate and share spectrum

---

[3] The secondary user may fail to detect incumbent user because of its low power, inactivity, distance, or the poor channel conditions.

occupancy information with each other to improve the spectral usage. The model can either use centralized server sharing [26], where a centralized entity maintains the database of the spectrum usage and coordinates the spectrum access information among the users, or distributed sharing [27, 28], where each user maintains the information about the local spectrum usage and share its knowledge with other nearby users to improve spectrum utilization efficiency. Even though the cooperative approach seems to be the most straightforward method, the primary user must be involved for efficient sharing of spectrum access information among the secondary users, which is often an unwanted burden on the part of the primary users. On the other hand, secondary users may cooperate with each other without any involvement of primary users and share information to detect the presence of a primary user to achieve significant performance enhancements on spectrum utilization and interference avoidance [30].

Cooperative approaches may lead to results that closely approximate the optimal spectrum allocation among the users. However, a cooperative approach model may heavily depend on the communication resources of the DSA networks. As a result, this communication overhead might limit the available spectrum for data communications. Since the ultimate goal of the cooperative approach is to achieve acceptable overall spectrum utilization, the users must be somewhat selfless, occasionally sacrificing local performance to improve overall system utility [31].

In a non-cooperative spectrum sharing approach, information exchange among the users is kept to a minimum, such that the secondary users independently interpret the spectrum usage and availability, while not interacting with the primary users [32, 33]. The non-cooperative approaches result in minimal communication requirements among the nodes at the expense of poor spectrum utilization efficiency. The non-cooperative approaches may act in a selfish, greedy, or rational way [34].

6.2.3 Underlay and Overlay Spectrum Sharing Approaches

Spectrum sharing techniques can be classified into *underlay* and *overlay* spectrum sharing based on the spectrum access techniques. *Underlay systems* use spread spectrum techniques, such as ultrawide band (UWB) [35, 36] and code division multiple access (CDMA) [37], to transmit the signal below the noise floor of the spectrum [38]. An example of the time and frequency domain information of an underlay spectrum sharing system is shown in Fig. 6.2(a). In the figure, we see that the underlay systems use wide band low power signals for transmissions. However, this technique can increase the overall noise temperature and thereby worsening error robustness of the primary users as compared to the case without underlay systems. To avoid any interference to the primary users, underlay system can use interference avoidance techniques, such as notching [39] and waveform adaptation [40].

To improve the spectral efficiency, *overlay systems* utilize the unused portions of the spectrum. The spectrum holes[4] filled in by secondary transmissions in an overlay system is shown in Fig. 6.2(b). As shown in the figure, the overlay systems use the

---

[4] A spectrum hole is an unused portion of the licensed spectrum [41].

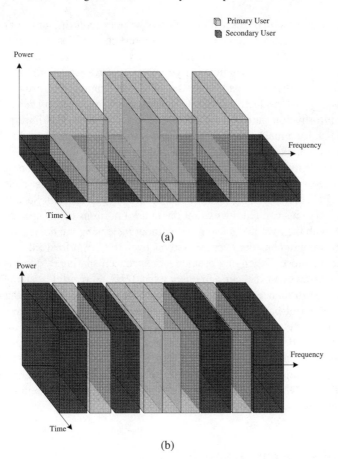

Primary User
Secondary User

Fig. 6.2. a Overlay and b underlay spectrum sharing.

unoccupied portions of the spectrum with a reasonable amount of guard intervals for secondary transmissions keeping the interference to the primary users to a minimum. Since the licensed system has privileged access to the spectrum, it must not be disturbed by any secondary transmissions. This results in two main design goals for an overlay system [42]:

- Minimum interference to licensed transmissions
- Maximum exploitation of the gaps in the time-frequency domain.

In order to achieve these goals, the overlay system needs information about the spectrum allocation of the licensed systems by regularly performing spectrum measurements. When interference among the users is high, it has been shown that frequency division multiplexing is an optimal technique [34].

To enhance spectral efficiency, an approach called *spectrum pooling* is proposed, which enables secondary access to licensed frequency bands by filling in the spectrum holes with secondary user transmissions without requiring any changes to the primary licensed systems. Spectral pooling represents the idea of merging spectral ranges from different spectrum owners (military, trunked radio, etc.) into a common pool, where users may temporarily rent spectral resources during idle periods of licensed users, thereby enabling the secondary utilization of already licensed frequency bands [14]. In spectrum pooling system, a centralized entity can collect measurement information gathered by the secondary user terminals during the detection cycle, and maintain the spectrum usage information. The centralized entity is responsible for making decisions on granting portions of the spectrum to the secondary users. With the use of a centralized entity, the information management of a spectrum access network would be relatively simple. However, this same entity can also easily be a bottleneck for the network, as already explained in Sect. 6.2.1. Since the overlay systems can readily exploit the unused portions of the spectrum without interfering with the incumbent users and without increasing the noise temperature of the system, we will consider overlay systems from this point forward.

One of the most challenging problems of spectrum sharing systems is their successful co-existence in the same frequency band, i.e. an overlay system should not degrade the performance of systems already 1working in the target frequency band. For instance, out-of-band radiation has to be reduced in order to enable co-existence. The transmitter spectral mask is a measure of the transmitter spectral profile in order to verify that the device is not transmitting excessive amounts of energy outside its assigned channel bandwidth. Several approaches have been proposed in literature for suppressing the sidelobe levels, such as the deactivation of subcarriers lying at the borders of an OFDM spectrum [43], windowing [44], subcarrier weighting [45], and insertion of cancellation carriers [46].

6.3 Non-Contiguous Transmission

As mentioned previously, MCM is highly suited for high-speed data transmissions, due to its ability to efficiently handle the distortion introduced by frequency selective channels [11]. Moreover, MCM techniques, such as OFDM, can provide the necessary agile spectrum usage, when portions of the target licensed spectrum are occupied by both primary and secondary users. This is achieved by deactivating (i.e. nulling) subcarriers that can potentially interfere with other users. This form of an OFDM, where the implementation achieves the high data rates via collective usage of a large number of non-contiguous subcarriers, is called NC-OFDM. A frequency spectra for 16-subcarrier NC-OFDM with nine active and seven deactivated subcarriers is shown in Fig. 6.3, where the subcarriers are orthogonally overlapped. The subcarriers corresponding to the spectrum occupied by incumbent user transmissions, which are determined from the spectrum sensing measurements, are deactivated.[5] In this section, we present a brief overview of the NC-OFDM framework, efficient

---

[5] The deactivated subcarriers implies that no information is transmitted over these subcarriers.

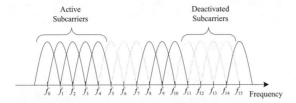

Fig. 6.3. Frequency spectra of NC-OFDM subcarriers.

implementations of NC-OFDM transceivers, and conduct a performance analysis of the system.

6.3.1 Non-contiguous OFDM Framework

A general schematic of an NC-OFDM transceiver is shown in Fig. 6.4. Without loss of generality, a high-speed data stream, $x(n)$, is modulated using either M-ary phase shift keying (PSK) or M-ary quadrature amplitude modulation (QAM). The modulated data stream is then split into N slower data streams using a serial-to-parallel (S/P) converter. Note that the subcarriers in the NC-OFDM transceiver do not need to be all active as in conventional OFDM transmission. Moreover, the active subcarriers are located in the unoccupied spectrum bands, which are determined by dynamic spectrum sensing and channel estimation techniques [15, 34, 47]. The inverse fast Fourier transform (IFFT) is then applied to these streams, modulating them to different subcarrier center frequencies. The output of the IFFT block for the mth NC-OFDM symbol is given by

$$Y_{m,n} = \frac{1}{\sqrt{N}} \sum_{k=0}^{N-1} X_{m,k} \exp(j2\pi kn/N) \quad n = 0, 1, \ldots, N-1 \qquad (6.1)$$

where $X_{m,k}$ is the symbol of the kth subcarrier,[6] and $j = \sqrt{-1}$. The symbol over the kth deactivated subcarrier is $X_{m,k} = 0$.

Prior to transmission, a guard interval of length greater than the channel delay spread is added to each NC-OFDM symbol, known as a cyclic prefix (CP). This block is used to mitigate the effects of intersymbol interference (ISI). Following parallel-to-serial (P/S) conversion, the baseband NC-OFDM signal, $s(n)$, is then passed through the transmitter radio frequency (RF) chain, which amplifies the signal and upconverts it to the desired center frequency.

The receiver performs the reverse operation of the transmitter, mixing the RF signal to baseband for processing, yielding the signal $r(n)$. Then, the signal is converted into parallel streams using S/P converter, the CP is removed, and the fast Fourier transform (FFT) is applied to transform the time domain data into the frequency domain. After compensating for distortion introduced by the channel using

---

[6] For example, $X_{m,k} \in \{1, -1\}$ for BPSK signaling and $X_{m,k} \in \{1, -1, j, -j\}$ for QPSK signaling.

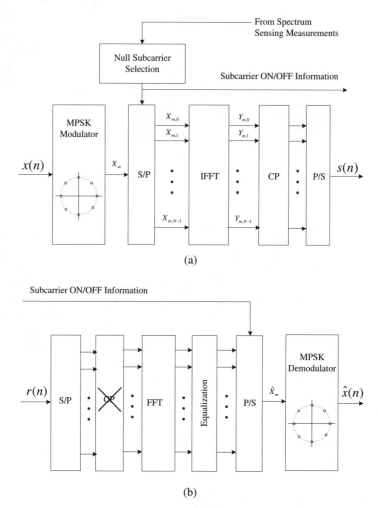

Fig. 6.4. Schematic of an NC-OFDM transceiver. **a** NC-OFDM Transmitter. **b** NC-OFDM Receiver.

equalization techniques, the data in the active subcarriers is multiplexed using a P/S converter, and demodulated into a reconstructed version of the original high-speed input, $\hat{x}(n)$.

The "null carrier selection" block at the NC-OFDM transmitter periodically collects information about the spectrum occupancy from the spectrum sensing measurements. Then, the subcarriers corresponding to the incumbent user transmissions are deactivated at the transmitter for avoiding any interference to the primary license holders. This information can be transmitted to the receiver via control channel before any data communication process begins so that the data over the active subcarriers are demodulated correctly.

6.3.2 Efficient Modulation

As we have seen in the previous section, OFDM-based transceivers employ IFFT and FFT blocks. The FFT algorithm make modulation and demodulation of the subcarriers highly efficient in terms of hardware and computational complexity [48]. However, an NC-OFDM transceiver may have several subcarriers that are deactivated in order to avoid any interference to the incumbent user transmissions. These deactivated subcarriers would result in zero-valued inputs to the IFFT and FFT blocks. Thus, the hardware resources of the FFT are not fully exploited since the computations involving zeroes are unnecessary. Therefore, an approach is needed to efficiently implement the FFT blocks when several subcarriers are deactivated.

It has been shown that for situations in which the relative number of zero-valued inputs is quite large, significant time savings can be obtained by "pruning" the FFT algorithm[7] [49]. Several algorithms have been proposed in the literature for enhancing the efficiency of the FFT algorithm based on the decimation-in-time (DIT) and the decimation-in-frequency (DIF) algorithms [50–57]. However, most of these algorithms are suitable only for systems with specific zero-input pattern distributions. Nevertheless, there exists several algorithms in the literature that prune the FFT for any zero-input pattern, yielding an efficient implementation with respect to computational time [58, 59].

In a wide-band communication system, a large portion of frequency channels may be occupied by primary or other secondary transmissions. As a result, these frequencies are considered occupied and an NC-OFDM transceiver must deactivate subcarriers in the vicinity of these other transmissions. For highly sparse unoccupied spectrum, the number of zero-valued inputs in the FFT may be significant relative to the total number of the usable subcarriers. When the relative number of zero-valued inputs is quite large, significant time savings can be obtained by pruning the FFT algorithm [49].

6.3.2.1 An FFT Pruning Example

A 16-point DIF FFT butterfly structure is shown in Fig. 6.5, where a_i represents the ith input signal to the FFT block. Suppose incumbent users are located at subcarriers $a_2, a_3, a_5, a_7, a_{10}, a_{11}, a_{13}$ and a_{15}. As a result, these subcarriers must be nulled in order to avoid interference with the existing signals. Thus, some of the multiplies and adds associated with the nulled subcarriers can be removed. For a conventional FFT algorithm, the total number of multiplications and additions would be $N \log_2 N$. However, with an FFT pruning algorithm, the unnecessary multiplications and addition operations at the stages $b_2, b_3, b_5, b_7, b_{10}, b_{11}, b_{13}, b_{15}, c_3, c_7, c_{11}$, and c_{15} can be pruned as their values will always be zeros. Moreover, multiplications and additions at nodes $c_1, c_2, c_5, c_6, c_9, c_{10}, c_{13}, c_{14}, d_1, d_3, d_5, d_7, d_9, d_{11}, d_{13}$, and

---

[7] *FFT pruning* refers to the procedure for improving the efficiency of the fast Fourier transform by removing operations on input values which are zeros, and on output values which are not required [49].

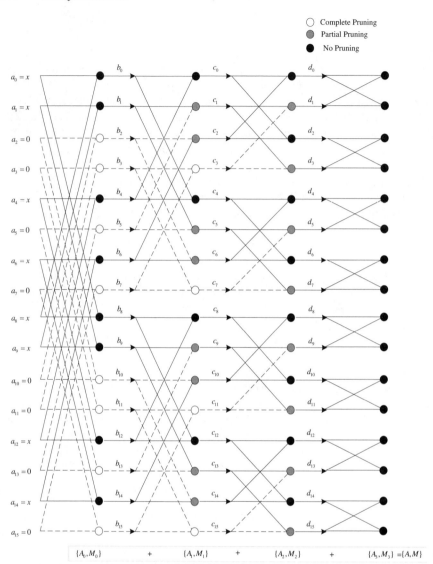

Fig. 6.5. FFT butterfly structure. A value of "0" denotes a zero-valued subcarrier and "x" denotes a data bearing subcarrier. The dotted lines represent the computations that can be pruned.

d_{15} can be replaced with simple "copy" operation, whereas addition operations in nodes $c_1, c_3, c_5,$ and c_7 can be pruned to save the FFT computation time. Therefore, the FFT computation time can be significantly improved with partial and complete pruning.

In wideband communication systems, the channel conditions and incumbent spectral occupancy[8] (ISO) varies over time. Thus, the FFT pruning algorithm should be able to design an efficient FFT implementation every time the channel condition and/or ISO changes [58].

In [58], an FFT pruning algorithm designed for NC-OFDM transceivers was presented that can quickly design an efficient FFT implementation for any zero-input pattern. Consider a radix-2 FFT algorithm with N levels. First, the algorithm generates a matrix M, with N columns and 2^N rows, where each element of the matrix corresponds to an addition/multiplication node of the FFT flow graph. The node needs to be computed if the corresponding element in the matrix M is non-zero and vice-versa. Second, the FFT pruning algorithm uses the information provided by the M to prune unnecessary computations at the corresponding nodes, hence reducing the execution time for the FFT computation and/or reducing the hardware components, such as addition/multiplication blocks. From the simulation results, it was demonstrated that pruning the FFT yields an implementation possessing a faster execution time. Given that cognitive radio units employing NC-OFDM would need to quickly adapt to the changing operating environment, with hardware resources of a small form factor cognitive radios being limited, such an algorithm would be very beneficial for its ability to reduce the computation time and/or the hardware components.

6.3.3 Performance Evaluation

NC-OFDM is sufficiently agile with respect to spectrum usage, "filling in" the available spectral gaps within a transmission bandwidth partially occupied by other users (incumbent and other unlicensed) while not sacrificing its error robustness [18, 19]. Since power of the nulled subcarriers can be redistributed to the active subcarriers to improve signal-to-noise ratio (SNR) in NC-OFDM systems, their bit error rate (BER) performance can be improved as compared to conventional OFDM.

The SNR is defined as the ratio of the desired signal power to the noise power [59]. The SNR indicates reliability of transmission link between the transmitter and receiver, and is accepted as a standard measure of signal quality.

Assuming a wide sense stationary uncorrelated scattering (WSSUS) channel [60], the instantaneous SNR of an NC-OFDM signal is given by

$$\gamma = \frac{|X \cdot H|^2}{|\tilde{n}|^2} = \frac{|X|^2 \cdot |H|^2}{|\tilde{n}|^2}. \tag{6.2}$$

where X is a vector of symbols from the N subcarriers, H is a vector of the channel frequency response for the N subcarriers, and $\tilde{n} = \text{FFT}(n)$, with n being a vector of zero-mean complex Gaussian independent random variables across N subcarriers.

Therefore, the mean SNR can be given by [61]

---

[8] Incumbent spectral occupancy (ISO) is defined as the fraction of the intended transmission bandwidth occupied by incumbent user transmissions.

$$E(\gamma) = \frac{E(|\boldsymbol{X}|^2 \cdot |\boldsymbol{H}|^2)}{E(|\tilde{\boldsymbol{n}}|^2)} = \frac{E(|\boldsymbol{X}|^2) \cdot E(|\boldsymbol{H}|^2)}{E(|\tilde{\boldsymbol{n}}|^2)} \tag{6.3}$$

where $E(\cdot)$ denotes an expectation operator.

In the following two paragraphs, we present the SNR analysis for the NC-OFDM system over additive white Gaussian noise (AWGN) and Rayleigh fading channels [18, 19].

6.3.3.1 AWGN Channel

Consider an AWGN channel with noise spectral density N_0 and bandwidth B, the noise power is given by

$$E(|\tilde{\boldsymbol{n}}|^2) = \sigma_N^2 = N_0 B \tag{6.4}$$

while the SNR is given by

$$E(\gamma) = 10 \log_{10} \left(\frac{E(|\boldsymbol{X}|^2)}{\sigma_N^2} \right) = 10 \log_{10} \left(\frac{E(|\boldsymbol{X}|^2)}{N_0 B} \right). \tag{6.5}$$

Suppose the incumbent spectral occupancy[9] (ISO) is α, then the total available bandwidth would be $(1 - \alpha)B$. Since the channel response is assumed to be flat, the signal power would remain constant, irrespective of the available bandwidth. However, the effective noise power would be

$$\sigma_N^2 = N_0(1 - \alpha)B \tag{6.6}$$

with the SNR given by

$$E(\gamma) = 10 \log_{10} \left(\frac{E(|\boldsymbol{X}|^2)}{\sigma_N^2} \right) = 10 \log_{10} \left(\frac{E(|\boldsymbol{X}|^2)}{N_0(1 - \alpha)B} \right). \tag{6.7}$$

Therefore, the SNR gain is

$$\text{SNR}_{\text{gain}} = -10 \log_{10} (1 - \alpha). \tag{6.8}$$

However, the total throughput would also be reduced to $(1 - \alpha)NR_b$, where R_b represents the bit rate over an individual subcarrier.

6.3.3.2 Rayleigh Fading Channel

Suppose we consider a frequency non-selective slow fading channel, i.e. flat channel response, where the channel magnitude response $E(|H_i|^2)$ is flat over the spectrum band. The deactivation of subcarriers due to incumbent users will result in a non-zero ISO. This would also filter out a portion of the channel magnitude response, which

---

[9] Incumbent spectral occupancy (ISO) is defined as the fraction of the intended transmission bandwidth occupied by incumbent user transmissions.

results in an increase in the magnitude of $E(|H_i|^2)$. As a result, the SNR gain is given by

$$\text{SNR}_{\text{gain}} = 10\log_{10}\left(\frac{E(|\boldsymbol{X}|^2) \cdot E(|\boldsymbol{H}|^2)/(1-\alpha)}{N_0(1-\alpha)B}\right)$$
$$- 10\log_{10}\left(\frac{E(|\boldsymbol{X}|^2) \cdot E(|\boldsymbol{H}|^2)}{N_0 B}\right)$$
$$= -10\log_{10}(1-\alpha)^2. \tag{6.9}$$

In case of a frequency selective multipath channel, the channel magnitude response is not flat over the spectrum. Thus, deactivating a portion of the spectrum would also flatten a portion of the channel magnitude response, which results in an increase in $E(|\boldsymbol{H}|^2)$. Therefore, the SNR gain would not be linear as in the case with a flat AWGN channel.

6.4 Peak-to-Average Power Ratio Reduction for NC-OFDM

An OFDM signal consists of a sum of independent signals modulated over several orthogonal subcarriers of equal bandwidth. Therefore, when added up coherently, the OFDM signal may exhibit large peaks, while the mean power remains relatively low. Being a variant of OFDM, NC-OFDM signals also suffer from this same problem. By definition, the peak-to-average power ratio (PAPR)[10] is the ratio of the peak instantaneous power to the average power of a given signal, which characterizes the envelope variations of the signal in time domain, namely [21]:

$$\text{PAPR}(s(t)) = \frac{\max\limits_{0\leq t\leq T}|s(t)|^2}{E\{|s(t)|^2\}} \tag{6.10}$$

where $E\{.\}$ denotes the expectation operator. Without loss of generality, we can safely neglect the cyclic extension from the analysis since it does not contribute to the PAPR problem. The continuous time PAPR of $s(t)$ can be approximated using the discrete time PAPR, which is obtained using samples of the NC-OFDM signal, $s(n)$. It has been shown that an oversampling factor of four is sufficient to estimate the continuous PAPR of a BPSK system [62, 63].

When signals with the same phase are added together, the highest PAPR occurs. When high PAPR occurs, the digital-to-analog (D/A) converter and power amplifier of the transmitter would require a large dynamic range in order to avoid amplitude clipping, thus increasing both power consumption and component cost of the transceiver.

Numerous techniques have been proposed in the literature that attempt to reduce PAPR of an OFDM signal. These solutions include *clipping/filtering* [64, 65], *error*

---

[10] Several authors refer to Crest factor (CF) as a measure of envelope variations in the time domain, where CF is given by the square root of the PAPR.

control coding [66, 67], and *constellation shaping techniques* (phase, power, or both) [68–70]. The PAPR reduction techniques can achieve a decrease in PAPR, but at the cost of increased system complexity, reduced information rate, or degraded BER performance. Moreover, due to non-contiguous subcarriers, the PAPR reduction techniques proposed for the OFDM signals may need to be modified for reducing the values of PAPR for NC-OFDM signals. Design requirements of the PAPR reduction techniques for NC-OFDM signals will be addressed later.

6.4.1 Motivations for Reducing PAPR

When the PAPR of an NC-OFDM transmission is high, the D/A converters and power amplifiers require a large dynamic range to avoid clipping of the given signal mitigating undesirable consequences, such as signal distortion and spectral spillage. Moreover, a large dynamic range implies increased complexity, reduced efficiency and increased cost of the components. The motivations for reducing PAPR will be elaborated in the following two sections.

6.4.1.1 Dynamic Range of D/A Converters and Power Amplifiers

Amplitude clipping of an NC-OFDM signal causes several undesirable outcomes, such as signal distortion and spectral regrowth [64]. It also causes in-band noise, which results in bit error rate (BER) performance degradation, and higher-order harmonics that spill over into out-of-band spectrum. It has been shown that filtering after the power amplifier to remove this spectral leakage is very inefficient with respect to power usage, thus making it an undesirable solution [71]. Therefore, the dynamic range of D/A converters should be large enough to accommodate large peaks of signals, i.e. high PAPR. A high-precision D/A converter supports high PAPR with reasonable quantization noise, but may be very expensive for a given sampling rate of the system. On the other hand, low-precision D/A converter would be cheaper, but quantization noise will be significant which reduces signal SNR, when the dynamic range of the D/A converter is increased to support high PAPR. Otherwise, the D/A converter will saturate and clipping will occur [21].

Similarly, the dynamic range of the power amplifiers should also be large to accommodate large PAPR. Otherwise, power amplifiers may saturate, resulting in amplitude clipping. The component cost of the D/A converters and power amplifiers increase with the increase in the dynamic range.

6.4.1.2 Power Savings

Power amplifiers with a high dynamic range exhibit poor power efficiency, which is the ratio of power delivered to the load and total power consumed. For a given NC-OFDM signal, the average input power needs to be adjusted such that the peaks of the signal are rarely clipped. Therefore, the efficiency of the power amplifiers is inversely proportional to the PAPR. Therefore, the net power savings is directly proportional to the desired average output power and is highly dependent upon the clipping probability level. Therefore, PAPR reduction leads to significant power savings, making it highly desirable [72].

NC-OFDM system utilizes non-contiguous blocks of subcarriers for high data rate transmissions. When the number of deactivated subcarriers is large, the common assumption of the input symbols being identically and independently distributed (i.i.d.) may not hold. This results in a different statistical properties of the PAPR for NC-OFDM signals as compared to that for OFDM signals.

6.4.2 Statistical Analysis

The complementary cumulative distribution function (CCDF) of the PAPR denotes the probability of an NC-OFDM signal exceeds a given threshold [21]. It is the most frequently used parameter to characterize PAPR and also as performance measures for PAPR reduction techniques.

In Fig. 6.6(a), we present the CCDF of PAPR with the fixed number of active subcarriers and different numbers of deactivated subcarriers. From the results, we observe that the probability of occurrence of high PAPR increases with the increase the total number of subcarriers, for a given number of active subcarriers. In Fig. 6.6(b), we show the CCDF of PAPR for various number of active subcarriers, while the total number of subcarriers are kept constant. The CCDF of PAPR shows the probability of getting high PAPR increases with the increase in the number of active subcarriers, even though the total number of subcarriers is kept constant.

6.4.3 Design Requirements

The conventional PAPR reduction techniques for OFDM systems inherently assume a contiguous set of subcarriers. Therefore, PAPR reduction techniques proposed for OFDM systems would need to be adapted to a system employing NC-OFDM. In spectrum opportunistic systems, the active subcarriers are located in proximity to the incumbent user transmissions. As a result, both intersymbol interference (ISI) and intercarrier interference (ICI) may cause distortion in the primary user transmissions. Therefore, time-domain-based or distortion-based techniques, such as clipping and filtering [73], and frequency domain-based techniques assuming contiguous subcarriers, such as coding [66], cannot be used for reducing the PAPR of NC-OFDM signals. However, frequency-domain PAPR reduction techniques are better suited, since it is easier to sort out the nulled subcarriers avoiding any interference to existing user transmissions. The techniques, such as interleaving [74], selected mapping (SLM) [75], and partial transmit sequences (PTS) [76], need to be aware of the locations of the active subcarriers. Moreover, in a dynamic spectrum access network, the total number of active subcarriers and their locations might change continuously and the PAPR reduction techniques should be able to adapt to these changes.

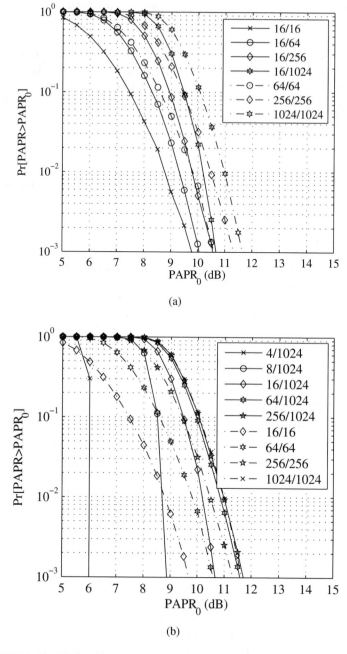

Fig. 6.6. CCDF of PAPR for random location distributions of active subcarriers, where c_a/c_t is ratio of the number of active subcarriers over the total number of subcarriers. **a** Fixed number of active subcarriers. **b** Fixed number of total subcarriers.

6.5 Non-uniform Bit Allocation

To exploit the flexibility offered by cognitive radio transceivers and NC-OFDM, *bit allocation* can be employed to enhance system performance. The process of bit allocation involves tailoring the subcarrier signal constellations to the prevailing channel conditions in order to meet a specified objective (e.g., enhanced error robustness, increased data throughput). Moreover, since the distortion affecting the subcarriers across the channel may not be uniform, the best-possible choice for a signal constellation can vary across the subcarriers. As a result, there exist several algorithms in the literature designed to solve for a bit allocation given an objective function and constraints (see [77] and references therein).

Mathematically, the process of performing bit allocation in order to increase the overall throughput of the system while ensuring the mean BER, $\bar{P}$, is below a specified mean BER limit, P_T, can be defined by the following optimization problem:

$$\max_{b_i} \sum_{i=0}^{N-1} b_i \tag{6.11}$$

subject to

$$\bar{P} = \left(\sum_{i=0}^{N-1} b_i P_i \right) \Big/ \left(\sum_{i=0}^{N-1} b_i \right) \leq P_T \tag{6.12}$$

where b_i is the number of bits per symbol for subcarrier i, N is the number of subcarriers, and P_i is the BER for subcarrier i, which is computed from the subcarrier SNR, γ_i, via closed form expressions [59].[11]

Although bit allocation offers the potential for improved throughput or error robustness, its main disadvantage is the amount of overhead information generated. To reduce the overhead information, one solution is to perform *uniform bit allocation*. As opposed to *non-uniform bit allocation*, where the subcarrier signal constellations can vary [77], uniform bit allocation imposes the additional constraint of

$$b_0 = b_1 = \ldots = b_{N-1} \tag{6.13}$$

when trying to solve for the objective function of (6.11). Another solution that employs some of the flexibility offered by multicarrier modulation is to assign a signal constellation to a block of B subcarriers. The bit allocation process would assess the average SNR of each block of subcarriers, and then select an appropriate signal constellation for each block, insuring that the BER constraint of (6.12) is satisfied while attempting to increase the system throughput in (6.11).

The throughput performance of a cognitive radio transceiver employing NC-OFDM employing either uniform or non-uniform bit allocation was examined in [20]. To reduce the overhead information and the bit allocation algorithm complexity, the transceiver was implemented to assign the same signal constellation and activity

---

[11] In a practical implementation, the BER values would be stored in a look-up table.

level to blocks of subcarriers. The simulation results showed that for low spectral occupancy by the incumbent users, the cost of using blocks of subcarriers to reduce overhead was acceptable relative to the throughput penalty incurred by using blocks. However, as the incumbent spectral occupancy increases, the benefits in reduced overhead relative to the throughput penalty diminished very quickly. Therefore, it was recommended that one adaptable parameter to be included is an algorithm that decides on a value for the subcarrier block size, which is a function of the incumbent spectral occupancy.

Conclusion

DSA techniques can enable the secondary utilization of the spectrum, thereby improving the spectrum efficiency and mitigate the apparent spectrum scarcity. In this chapter, we presented non-contiguous OFDM (NC-OFDM) as a viable transmission technology for cognitive radio transceivers operating in DSA networks. While operating in an DSA network, it was shown that NC-OFDM can be optimized with respect to computational complexity through FFT pruning, PAPR reduction, and overall data throughput via bit allocation.

Acknowledgments

This work was generously supported by the National Science Foundation (NSF) through grants ANI-0230786 and ANI-0335272.

References

1. Federal Communications Commission, "Spectrum policy task force report," ET Docket No. 02–135, 2002.
2. M. J. Marcus, "Unlicensed cognitive sharing of TV spectrum: The controversy at the federal communications commission," *IEEE Commun. Mag.*, vol. 43, pp. 24–25, May 2005.
3. Federal Communications Commission, "Unlicensed operation in the TV broadcast bands," ET Docket No. 04-186, 2004.
4. J. Mitola III, "Cognitive radio for flexible mobile multimedia communications," in *Proc. IEEE Int. Wksp. Mobile Multimedia Commun.*, vol. 1, (San Diego, CA, USA), pp. 3–10, Nov. 1999.
5. F. N. Hatfield and P. J. Weiser, "Property rights in spectrum: Taking the next step," in *Proc. IEEE Int. Symp. New Frontiers Dynamic Spectr. Access Networks*, vol. 1, (Baltimore, MD, USA), pp. 43–55, Nov. 2005.
6. T. R. Newman, B. A. Barker, A. M. Wyglinski, A. Agah, J. B. Evans, and G. J. Minden, "Cognitive engine implementation for wireless multicarrier transceivers," *Wireless Commun. Mobile Comput.*, vol. 6, in press.

7. J. Zhao, H. Zheng, and G.-H. Yang, "Distributed coordination in dynamic spectrum allocation networks," in *Proc. IEEE Int. Symp. New Frontiers Dynamic Spectr. Access Networks*, vol. 1, (Baltimore, MD, USA), pp. 259–268, Nov. 2005.

8. T. W. Rondeau, B. Le, D. Maldonado, D. Scaperoth, and C. W. Bostian, "Cognitive radio formulation and implementation," in *Proc. 1st Int. Conf. on Cognitive Radio Oriented Wireless Networks and Commun.* (Mykonos, Greece), June 2006.

9. G. J. Minden, J. B. Evans, L. Searl, D. DePardo, R. Rajbanshi, J. Guffey, Q. Chen, T. Newman, V. R. Petty, F. Weidling, M. Lehnherr, B. Cordill, D. Datla, B. Barker, and A. Agah, "An agile radio for wireless innovation," submitted to *IEEE Commun. Mag.*, Oct. 2006.

10. A. J. Viterbi, "Wireless digital communication: A view based on three lessons learned," *IEEE Commun. Mag.*, vol. 29, pp. 33–36, Sept. 1991.

11. J. A. C. Bingham, "Multicarrier modulation for data transmission: An idea whose time has come," *IEEE Commun. Mag.*, vol. 28, pp. 5–14, May 1990.

12. B. R. Saltzberg, "Comparison of single-carrier and multitone digital modulation for ADSL applications," *IEEE Commun. Mag.*, vol. 36, pp. 114–121, Nov. 1998.

13. L. J. Cimini Jr., "Analysis and simulation of a digital mobile channel using orthogonal frequency division multiplexing," *IEEE Trans. Commun.*, vol. 33, pp. 665–675, July 1985.

14. T. A. Weiss and F. K. Jondral, "Spectrum pooling: An innovative strategy for the enhancement of spectrum efficiency," *IEEE Commun. Mag.*, vol. 43, pp. S8–14, Mar. 2004.

15. H. Tang, "Some physical layer issues of wide-band cognitive radio systems," in *Proc. IEEE Int. Symp. New Frontiers Dynamic Spectr. Access Networks*, vol. 1, (Baltimore, MD, USA), pp. 151–159, Nov. 2005.

16. J. D. Poston and W. D. Horne, "Discontiguous OFDM considerations for dynamic spectrum access in idle TV channels," in *Proc. IEEE Int. Symp. New Frontiers Dynamic Spectr. Access Networks*, vol. 1, (Baltimore, MD, USA), pp. 607–610, Nov. 2005.

17. M. P. Wylie-Green, "Dynamic spectrum sensing by multiband OFDM radio for interference mitigation," in *Proc. IEEE Int. Symp. New Frontiers Dynamic Spectr. Access Networks*, vol. 1, (Baltimore, MD, USA), pp. 619–625, Nov. 2005.

18. R. Rajbanshi, Q. Chen, A. M. Wyglinski, J. B. Evans, and G. J. Minden, "Comparative study of frequency agile data transmission schemes for cognitive radio transceivers," in *Proc. 2nd Annual International Wireless Internet Conference – Int. Wksp. on Technol. and Policy for Accessing Spectrum* (Boston, MA, USA), Aug. 2006.

19. R. Rajbanshi, Q. Chen, A. M. Wyglinski, G. J. Minden, and J. B. Evans, "Quantitative comparison of agile modulation techniques for cognitive radio transceivers," in *Proc. IEEE Consumer Commun. and Networking Conf. – Workshop on Cognitive Radio Networks* (Las Vegas, NV, USA), Jan. 2007.

20. A. M. Wyglinski, "Effects of bit allocation on non-contiguous multicarrier-based cognitive radio transceivers," in *Proc. 64th IEEE Veh. Technol. Conf. – Fall*, (Montreal, Canada), Sept. 2006.

21. J. Tellado, *Multicarrier Modulation with low PAR: Applications to DSL and Wireless*. Massachusetts, USA: Kluwer Academic Publishers, 2000.

22. I. F. Akyildiz, W. Y. Lee, M. C. Vuran, and S. Mohanty, "NeXt generation/dynamic spectrum access/cognitive radio wireless networks: A survey," *Computer Networks Journal* (Elsevier), vol. 50, pp. 2127–2159, Sept. 2006.

23. C. Raman, R. D. Yates, and N. B. MAndayam, "Scheduling variable rate links via a spectrum server," in *Proc. IEEE Int. Symp. New Frontiers Dynamic Spectr. Access Networks*, vol. 1, (Baltimore, MD, USA), pp. 110–118, Nov. 2005.

24. C. Peng, H. Zheng, and B. Y. Zhao, "Utilization and fairness in spectrum assignment for opportunistic spectrum access," *Mobile Networks Appl.*, vol. 11, pp. 555–576, Aug. 2006.

25. M. M. Buddhikot, P. Kolodzy, S. Miller, K. Ryan, and J. Evans, "DIMSUMnet: New directions in wireless networking using coordinated dynamic spectrum," in *Proc. IEEE Int. Symp. World of Wireless Mobile Multimedia Networks* (Taormina, Italy), pp. 78–85, June 2005.

26. V. Brik, E. Rozner, S. Banerjee, and P. Bahl, "DSAP: A protocol for coordinated spectrum access," in *Proc. IEEE Int. Symp. New Frontiers Dynamic Spectr. Access Networks*, vol. 1, (Baltimore, MD, USA), pp. 611–614, Nov. 2005.

27. L. Cao and H. Zheng, "Distributed spectrum allocation via local bargaining," in *Proc. IEEE Sensor and Ad Hoc Commun. and Networks* (Santa Clara, CA, USA), pp. 475–486, Sept. 2005.

28. J. Huang, R. A. Berry, and M. L. Honig, "Spectrum sharing with distributed interference compensation," in *Proc. IEEE Int. Symp. New Frontiers Dynamic Spectr. Access Networks*, vol. 1, (Baltimore, MD, USA), pp. 88–93, Nov. 2005.

29. W. Krenik and A. Batra, "Cognitive radio techniques for wide area networks," in *42nd Design Automation Conf.* (Anaheim, CA, USA), pp. 409–412, June 2005.

30. A. Ghasemi and E. S. Sousa, "Collaborative spectrum sensing for opportunistic access in fading environments," in *Proc. IEEE Int. Symp. New Frontiers Dynamic Spectr. Access Networks*, vol. 1, (Baltimore, MD, USA), pp. 131–136, Nov. 2005.

31. H. Zheng and L. Cao, "Device centric spectrum management," in *Proc. IEEE Int. Symp. New Frontiers Dynamic Spectr. Access Networks* (Baltimore, MD, USA), pp. 56–65, Sept. 2005.

32. S. Sankaranarayanan, P. Papadimitratos, A. Mishra, and S. hershey, "A bandwidth sharing approach to improve licensed spectrum utilization," in *Proc. IEEE Int. Symp. New Frontiers Dynamic Spectr. Access Networks*, vol. 1, (Baltimore, MD, USA), pp. 611–614, Nov. 2005.

33. Q. Zhao, L. Tong, and A. Swami, "Decentralized cognitive MAC for dynamic spectrum access," in *Proc. IEEE Int. Symp. New Frontiers Dynamic Spectr. Access Networks*, vol. 1, (Baltimore, MD, USA), pp. 224–232, Nov. 2005.

34. R. Etkin, A. Parekh, and D. Tse, "Spectrum sharing for unlicensed bands," in *Proc. IEEE Int. Symp. New Frontiers Dynamic Spectr. Access Networks*, vol. 1, (Baltimore, MD, USA), pp. 251–258, Nov. 2005.

35. M. Z. Win and R. A. Scholtz, "Impulse radio: How it works," in *IEEE Communications Letters*, vol. 2, No. 2, pp. 36–38, Feb. 1998.

36. Federal Communications Commission, "FCC first report and order: Revision of part 15 of the commissions's rules regarding ultra-wideband transmission systems," ET Docket No. 98-153, Apr. 2002.

37. K. S. Gilhousen, I. M. Jacobs, R. Padovani, A. J. Viterbi, L. A. Weaver Jr., and C. E. Wheatley III, "On the capacity of a cellular CDMA system," *IEEE Trans. Veh. Technol.*, vol. 40, pp. 303–312, May 1991.

38. R. Menon, R. M. Buehrer, and J. H. Reed, "Outage probability based comparison of underlay and overlay spectrum sharing techniques," in *Proc. IEEE Int. Symp. New Frontiers Dynamic Spectr. Access Networks*, vol. 1, (Baltimore, MD, USA), pp. 101–109, Nov. 2005.

39. J. Wang, "Narrowband interference suppression in time hopping impulse radio," in *Proc. 60th IEEE Veh. Technol. Conf.*, vol. 3, (Los Angeles, CA, USA), pp. 2138–2142, Sept. 2005.

40. C. Rose, S. Ulukus, and R. D. Yates, "Wireless systems and interference avoidance," *IEEE Trans. Wireless Commun.*, vol. 1, pp. 415–428, July 2002.

41. S. Haykin, "Cognitive radio: Brain-empowered wireless communications," *IEEE J. Select. Areas Commun.*, vol. 23, pp. 201–220, Feb. 2005.

42. U. Berthold and F. K. Jondral, "Guidelines for designing OFDM overlay systems," in *Proc. IEEE Int. Symp. New Frontiers Dynamic Spectr. Access Networks*, vol. 1, (Baltimore, MD, USA), pp. 626–629, Nov. 2005.

43. T. Weiss, J. Hillenbrand, A. Krohn, and F. K. Jondral, "Mutual interference in OFDM-based spectrum pooling systems," in *Proc. 59th IEEE Veh. Technol. Conf. – Spring*, vol. 4, (Milan, Italy), pp. 1873–1877, May 2004.

44. S. Kapoor and S. Nedic, "Interference suppression in DMT receivers using windowing," in *Proc. IEEE Int. Conf. Commun.*, vol. 2, (New Orleans, LA, USA), pp. 778–782, June 2000.

45. I. Cosvic, S. Brandes, and M. Schnell, "A technique for sidelobe suppression in OFDM systems," in *Proc. IEEE Global Telecommun. Conf.*, vol. 1, (St. Louis, MO, USA), pp. 204–208, Nov. 2005.

46. S. Brandes, I. Cosvic, and M. Schnell, "Sidelobe suppression in OFDM systems by insertion of cancellation carriers," in *Proc. 62nd IEEE Veh. Technol. Conf. – Fall*, vol. 1, (Dallas, TX, USA), pp. 152–156, Sept. 2005.

47. F. Weidling, D. Datla, V. Petty, P. Krishnan, and G. J. Minden, "A framework for RF spectrum measurements and analysis," in *Proc. IEEE Int. Symp. New Frontiers Dynamic Spectr. Access Networks*, vol. 1, (Baltimore, MD, USA), pp. 573–576, Nov. 2005.

48. S. B. Weinstein and P. M. Ebert, "Data transmission by frequency division multiplexing using the discretefourier transform," *IEEE Trans. Commun. Technol.*, vol. 19, pp. 628–634, Oct. 1971.

49. J. D. Markel, "FFT Pruning," *IEEE Trans. Audio Electroacoust.*, vol. 19, pp. 305–311, Dec. 1971.

50. R. G. Alves, P. L. Osorio, and M. N. S. Swamy, "General FFT pruning algorithm," in *Proc. 43rd IEEE Midwest Symp. Circuits and Systems*, vol. 3, (Lansing, MI, USA), pp. 1192–1195, Aug. 2000.

51. S. Holm, "FFT pruning applied to time domain interpolation and peak localization," in *Proc. IEEE Int. Conf. Acoust., Speech, Signal Process.*, vol. 35, (Dallas, TX, USA), pp. 1776–1778, Dec. 1987.

52. Z. Hu and H. Wan, "A novel generic fast Fourier transform pruning technique and complexity analysis," *IEEE Trans. Signal Process.*, vol. 53, pp. 274–282, Jan. 2005.

53. L. P. Jaroslavski, "Comments on 'FFT algorithm for both input and output pruning'," in *Proc. IEEE Int. Conf. Acoust., Speech, Signal Process.*, vol. 29, (Atlanta, GA, USA), pp. 448–449, June 1981.

54. J. Schoukens, R. Pintelon, and H. Van Hamme, "The interpolated fast Fourier transform: A comparative study," *IEEE Trans. Instrum. Meas.*, vol. 41, pp. 226–232, Apr. 1992.

55. D. P. Skinner, "Pruning the Decimation in time FFT algorithm," in *Proc. IEEE Int. Conf. Acoust., Speech, Signal Process.*, vol. 24, (Philadelphia, PA, USA), pp. 193–194, Apr. 1976.

56. T. V. Sreenivas and P. V. S. Rao, "High-resolution narrow band spectra by FFT pruning," in *Proc. IEEE Int. Conf. Acoust., Speech, Signal Process.*, vol. 28, (Denver, CO, USA), pp. 254–257, Apr. 1980.

57. T. V. Sreenivas and P. Rao, "FFT algorithm for both input and output pruning," in *Proc. IEEE Int. Conf. Acoust., Speech, Signal Process.*, vol. 27, (Washington, DC, USA), pp. 291–292, June 1979.

58. R. Rajbanshi, A. M. Wyglinski, and G. J. Minden, "An efficient implementation of NCOFDM transceivers for cognitive radios," in *Proc. 1st Int. Conf. on Cogn. Radio Oriented Wireless Networks and Commun.* (Mykonos, Greece), June 2006.

59. J. G. Proakis, *Digital Communications,* 4th ed. New York, NY, USA: McGraw-Hill, 2001.

60. T. S. Rappaport, *Wireless Communications: Principles and Practice*. Upper Saddle River, NJ, USA: Prentice Hall, 1996.

61. K. S. Shanmugan and A. M. Breipohl, *Random Signals: Detection, Estimation and Data Analysis*. New York, NY, USA: Wiley, 1988.

62. C. Tellambura, "Computation of the continuous time PAR of an OFDMsignal with BPSK subcarriers," *IEEE Commun. Lett.*, vol. 5, pp. 185–187, May 2001.

63. H. Yu and G. Wei, "Computation of the continuous time PAR of an OFDM signal," in *Proc. IEEE Int. Conf. Acoust., Speech, Signal Process.*, vol. 2003, (Hong Kong), pp. 529–531, Apr. 2003.

64. X. Li and L. J. Cimini, "Effects of clipping and filtering on the performance of OFDM," *IEEE Commun. Lett.*, vol. 2, pp. 131–133, May 1998.

65. L. Wang and C. Tellambura, "A simplified clipping and filtering technique for PAR reduction in OFDM systems," *IEEE Signal Process. Lett.*, vol. 12, pp. 453–456, June 2005.

66. H. Ahn, Y. M. Shin, and S. Im, "A block coding scheme for peak to average power ratio reduction in an orthogonal frequency division multiplexing system," in *Proc. 51st IEEE Veh. Technol. Conf. – Spring*, vol. 1, (Tokyo, Japan), pp. 56–60, May 2000.

67. A. Jones, T. Wilkinson, and S. Barton, "Block coding scheme for reduction of peak to mean envelope power ratio of multicarrier transmission schemes," *Electron. Lett.*, vol. 30, pp. 2098–2099, Dec. 1994.

68. S. Sezginer and H. Sari, "Peak power reduction in OFDM systems using dynamic constellation shaping," in *Proc. 13th Eur. Signal Process. Conf.* (Antalya, Turkey), Sept. 2005.

69. B. S. Krongold and D. L. Jones, "PAR reduction in OFDM via active constellation extension," *IEEE Trans. Broadcast.*, vol. 49, pp. 258–268, Sept. 2003.

70. S. H. Han and J. H. Lee, "Peak-to-average power ratio reduction of an OFDM signal by signal set expansion," in *Proc. IEEE Int. Conf. Commun.*, vol. 2, (Paris, France), pp. 867–971, June 2004.

71. B. S. Krongold, "New techniques for multicarrier communication systems," PhD Thesis, University of Illinois at Urbana-Champaign, Urbana, Illinois, 2003.

72. R. J. Baxley and G. T. Zhou, "Power savings analysis of peak-to-average power ratio reduction in OFDM," *IEEE Trans. Consumer Electron.*, vol. 50, pp. 792–798, Aug. 2004.

73. J. Armstrong, "Peak-to-average power reduction for OFDM by repeated clipping and frequency domain filtering," *Electron. Lett.*, vol. 38, pp. 246–247, Feb. 2002.

74. A. D. S. Jayalath and C. Tellambura, "The use of interleaving to reduce the peak to average power ratio of an OFDM signal," in *Proc. IEEE Global Telecommun. Conf.*, vol. 1, (San Francisco, CA, USA), pp. 82–86, Nov. 2000.

75. R. W. Bauml, R. F. H. Fischer, and J. B. Huber, "Reducing the peak to average power ratio of multicarrier modulation by selective mapping," *Electron. Lett.*, vol. 32, pp. 2056–2057, Oct. 1996.

76. S. H. Muller and J. B. Huber, "OFDM with reduced peak to average power ratio by optimum combination of partial transmit sequences," *Electron. Lett.*, vol. 33, pp. 368–369, Feb. 1997.

77. A. M. Wyglinski, F. Labeau, and P. Kabal, "Bit loading with ber-constraint for multicarrier systems," *IEEE Trans. Wireless Commun.*, vol. 4, pp. 1383–1387, July 2005.

Link Adaptation in OFDM-Based Cognitive Radio Systems

Gaurav Bansal, Md. Jahangir Hossain, and Vijay K. Bhargava

Department of Electrical and Computer Engineering,
The University of British Columbia, Canada
{gauravbs,jahangir,vijayb}@ece.ubc.ca

7.1 Introduction

Radio spectrum is one of the most scarce and valuable resources in wireless communications. Given this fact, new insights into the use of spectrum have challenged the traditional *static* spectrum allocation approach to spectrum management. Actual measurements have shown that most of the allocated spectrum is largely underutilized [1]. The Spectrum-Policy Task Force appointed by the Federal Communications Commission (FCC) drew a similar conclusion. Specifically, FCC reported vast temporal and geographic variations in the usage of allocated spectrum, with utilization ranging from 15 to 85% [2].

Spectrum utilization can be significantly improved by giving *opportunistic* access to the frequency bands instead of employing *static* spectrum allocation. According to the opportunistic spectrum access policy, a group of potential users may use a frequency or spectrum band for wireless communications provided that the legacy users of this band are not deprived of their priority right to use the band. On the other hand, development of *software-defined radio* (SDR) technology [3] has enabled radio transceivers to perform baseband processing functionalities, e.g., modulation and demodulation, using software and digital logic. Software-defined radio, therefore, becomes the promising technology in developing versatile wireless transceivers that will have the capability of accessing different radio networks with different technologies. In order to facilitate opportunistic spectrum access, this versatile transceiver needs to be spectrally aware, which motivates the design of *cognitive radio (CR)* technology [4]. Cognitive radio technology is an innovative radio design philosophy that involves smartly sensing the swaths of spectrum and then determining the transmission characteristics (e.g., symbol rate, power, bandwidth, latency) of a group of secondary users (also referred to as CR users)[1] based on the behavior of the users to whom the spectrum has been licensed [5, 6]. As such cognitive radio has been proposed as a way to improve spectrum utilization by exploiting unused spectrum in a dynamically changing environment.

---

[1] Throughout this chapter, we use the terms *secondary users* and *CR users* interchangeably.

However, in order to fully exploit the CR paradigm, adaptive access technologies must be developed for CR systems. Therefore, the current main focus of the wireless communication research community is to research and develop such enabling adaptive radio access technologies. Before CR systems become a reality, extensive research in the following two major areas is required.

- *Spectrum sensing:* In order to identify and access a suitable portion of spectrum with a minimum interference to the legacy users, i.e., the primary users, the first critical design challenge is to monitor the activity levels of the legacy users. This monitoring or sensing is critical in the sense that it needs to process a very wide bandwidth and reliably detect the presence of primary users. Therefore, spectrum sensing techniques should have a very high sensitivity, linearity and dynamic range of circuitry in the radio frequency front-end. In pursuit of these goals, various digital signal processing techniques, for example, matched filtering, energy detection and cyclostationary feature detection, have already been studied in the literature [7]. In order to develop an effective spectrum sensing algorithm, the computational complexity, storage requirements, total search time as well as the knowledge the CR has regarding the primary user signal characteristics must be considered. The burden on the signal processing techniques can be alleviated to a large extent by using cooperative diversity between CR spectrum sensors [8]. Few CR spectrum sensors under independent fades can help in reducing individual sensitivity requirements.

- *Efficient spectrum utilization:* Based on the available spectrum information as determined by the sensing algorithms, the next challenging task is for the secondary users to utilize it in an efficient fashion. As such the transmission capacity of the secondary users is maximized while the interference introduced to the primary users remains within the tolerable range. Once an unused or suitable portion of the licensed spectrum is identified by the secondary users, a number of challenging questions arise. For example, what would be the physical layer transmission parameters, e.g., transmission power and rates of the secondary users? Due to the great flexibility in dynamically allocating the unused spectrum among the secondary users as well as the ease of analysis of the spectral activity of the primary users [9], orthogonal frequency division multiplexing (OFDM) has already been recognized as a potential transmission technology for CR systems.

This chapter focuses on exploring some of the research challenges involved in the design of adaptive power and bit loading algorithms for an OFDM-based CR system where the secondary users access unused portions of the spectrum using an OFDM technique. Our objectives are threefold. First, we explore some of the research challenges involved in the design of link adaptation algorithms, i.e., power and bit loading algorithms, for an orthogonal frequency division (OFDM)-based cognitive radio (CR) system. In such a system, the secondary users (also referred to as CR users) are considered to co-exist with the primary users by filling the unused portions of the frequency band and using OFDM modulation at the air interface. Second, we provide some solutions to these challenging problems. More specifically, we study interference versus capacity performance of the existing power and bit loading algorithms when they are employed in an OFDM-based CR system. An optimal power and bit loading algorithm for such a scenario is devised by formulating the loading problem as a constrained optimization problem. In order to minimize the level of interference to the primary users' band, a suboptimal loading algorithm for discrete bit loading is proposed and the well-known discrete bit loading algorithms, which were proposed earlier for conventional wireless networks, are modified. Third, the effect of subcarriers' nulling on system performance is presented.

The rest of the chapter is organized as follows. In Sect. 7.2, we present an overview of opportunistic spectrum access architecture with a specific focus on *spectrum pooling*. Section 7.3 describes the research challenges involved in designing adaptive power and bit loading algorithms for an OFDM-based cognitive radio system. Specifically, an optimal loading algorithm for the continuous rate variation case is studied. The performance of the optimal scheme is compared with a classical loading algorithm. In this section, the effect of subcarrier nulling mechanism is also presented. Section 7.4 examines the interference versus transmission rate performance of well-known existing discrete bit loading algorithms that have been proposed previously for *conventional* wireless networks. In order to minimize the interference to the primary user's band, a suboptimal scheme is presented, and the existing schemes are modified. The effect of subcarrier nulling mechanism on different integer bit loading schemes is also presented in this section. Finally, we conclude the chapter.

7.2 Opportunistic Spectrum Access Strategy: Overview

One of the most challenging problems in opportunistic spectrum sharing is the successful co-existence of primary and secondary users in the same frequency band. Several strategies have been proposed in the literature for opportunistic spectrum access. Examples of these strategies have been surveyed in [10] and include *spectrum pooling* [9], the CR approach to usage of the virtual unlicensed spectrum (CORVUS) [1], DARPA's neXt Generation (XG) program [11, 12], IEEE 802.22 [13], dynamic intelligent management of spectrum for ubiquitous mobile network (DIMSUMnet) [14], the OFDM-based cognitive radio (OCRA) network [15] and European dynamic radio for IP services in vehicular environments (DRiVE) [16]. In spectrum pooling architecture, the CR system is highly flexible, since in this manner, the spectrum bands that are left idle by the licensed users can be efficiently filled. The goal of this architecture is to overlay secondary users on the existing licensed users without requiring any changes to the licensed system, and thereby increase spectrum utilization.

7.2.1 Spectrum Pooling

According to the *spectrum pooling* strategy of opportunistic spectrum sharing, secondary users access licensed frequency bands by filling the unused portion of the spectrum without making any changes to the primary users' system. The notion of spectrum pooling was first introduced in [17]. Basically, spectrum pooling involves merging spectral ranges from different licensed owners (GPRS, UMTS, military, emergency services, TV band, etc.) into a common pool. Unused portions of the spectrum can then be assigned to the cognitive users from this common pool. The spectrum pooling strategy shown in Fig. 7.1 depicts secondary users co-existing in the same band with primary users by filling the unused or idle portions of the primary users' bands.

According to spectrum pooling strategy, both secondary and primary users co-exist side by side in the same band but may have different access technologies. Therefore, mutual interference is the limiting factor for the performance of both networks. Specifically, in [17] the authors have shown that using OFDM modulation causes mutual interference between the primary and CR users due to the non-orthogonality of the transmitted signals. The amount of interference introduced to the primary user's band by a CR user's subcarrier depends on the power allocated to that subcarrier as well as the spectral distance between the subcarrier and the primary user's band. The study also showed that the subcarrier's nulling mechanism reduces interference in the primary user's band.

The model presented in Fig. 7.1 is a generalized picture of co-existence of both types of users according to a spectrum pooling strategy. The interference model presented in [17], indicates that a secondary user's transmission in a given unused portion of the spectrum produces higher interference to the adjacent primary user's band. In other words, the interference introduced into a primary user's band is dominated by the adjacent secondary users' transmission. Interference from the distant secondary users decays as distance increases. Therefore, if we consider only two dominant interferers, two possible co-existence scenarios, which we designate scenario 1 and scenario 2, can be imagined, as shown in Fig. 7.2. Scenario 1 shows that secondary user(s) may be located in the middle of the primary users. In scenario 2, secondary users access the left and right side of the unused portion of the primary

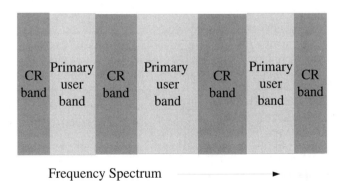

Frequency Spectrum ————————▶

Fig. 7.1. Co-existence of primary and cognitive users according to a spectrum pooling strategy.

users' band. In this chapter, we limit our study to scenario 2. Scenario 1 may be studied using a similar methodology but it will result in a different loading profile.

In scenario 2, it is assumed that frequency band B, which has been occupied by the primary user(s), is known and is located in the middle (see Fig. 2(b)). The middle band can be occupied by more than one primary users. Since we are considering overall interference in the primary user band, without loss of generality we assume that only one primary user is using the middle band. Hence interference introduced to this primary user is the limiting factor for the successful co-existence of the primary and secondary users in the same band. Since in this chapter we do not consider the subcarrier allocation problem among the secondary users, we assume that all of the unused spectrum is used by a single cognitive user employing the OFDM modulation format at the air interface. One of the main advantages of using OFDM is that the specific subcarrier can be deactivated by feeding it with zero power. The other advantage of using OFDM is that the M-fast Fourier transform (FFT) used in the OFDM transmission can also be used to analyze the spectral activity of the licensed users. The available bandwidth for CR transmission is divided into N subcarriers, $N/2$ on each side, and each with a bandwidth of Δf. Further, it is assumed that the

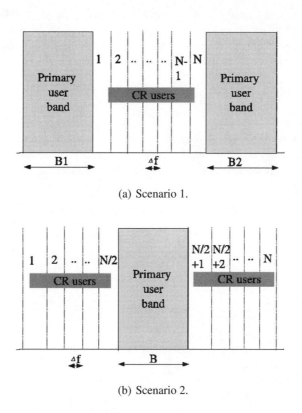

(a) Scenario 1.

(b) Scenario 2.

Fig. 7.2. Possible co-existence scenarios.

cognitive user does not have any knowledge of the primary users' access method, whether it is also OFDM or not. If the primary user also uses OFDM modulation and the secondary user has knowledge of it, their transmission could be made orthogonal. However, in practice the primary user might not be using OFDM, even if it is, it would be very difficult for the cognitive radio user to know the required parameters of the primary user in order to maintain orthogonality. Due to the co-existence of primary and secondary users in this fashion, there are two types of interference in the system [17]. One is introduced by the primary user into the CR user band and the other is introduced by the cognitive user into the primary user band as described below.

7.2.1.1 Interference Introduced by the Secondary User's Signal

The power density spectrum of the ith subcarrier in the CR user's band can be written as [17]

$$\phi_i(f) = P_i T_s \left(\frac{\sin \pi f T_s}{\pi f T_s} \right)^2 \tag{7.1}$$

where P_i is the total transmit power emitted by the ith subcarrier in the CR user's band and T_s is the symbol duration. The interference introduced by the ith subcarrier to the primary user's band is the integration of the power density spectrum of the ith subcarrier across the primary user's band, and can be written as

$$I_i(d_i, P_i) = P_i T_s \int_{d_i - B/2}^{d_i + B/2} \left(\frac{\sin \pi f T_s}{\pi f T_s} \right)^2 \mathrm{d}f \tag{7.2}$$

where d_i represents the spectral distance between the ith subcarrier of the CR user's band and the primary user's band. $I_i(d_i, P_i)$ represents the interference introduced by the ith subcarrier for a transmit power, P_i, into the primary user's band.

7.2.1.2 Interference introduced by the primary user's signal

The power density spectrum of the primary user's signal after the M-fast Fourier transform (FFT) processing can be expressed by the following expected value of the periodogram [17]:

$$E\{I_N(w)\} = \frac{1}{2\pi M} \int_{-\pi}^{\pi} \phi_{\mathrm{PU}}(\mathrm{e}^{\mathrm{j}w}) \left(\frac{\sin(w - \psi)M/2}{\sin(w - \psi)/2} \right)^2 \mathrm{d}\psi \tag{7.3}$$

where w represents the frequency normalized to the sampling frequency and $\phi_{\mathrm{PU}}(\mathrm{e}^{\mathrm{j}w})$ is the power density spectrum of the primary user's signal. The primary user's signal has been taken as an elliptically filtered white noise process with amplitude P_{PU} [17]. The interference introduced by the primary user's signal to the ith subcarrier will be the integration of the power density spectrum of the primary user's signal across the ith subcarrier, and can be written as

$$J_i(d_i, P_{\mathrm{PU}}) = \int_{d_i-\Delta f/2}^{d_i+\Delta f/2} E\{I_N(w)\}dw \tag{7.4}$$

where $J_i(d_i, P_{\mathrm{PU}})$ represents the interference introduced by the primary user's signal into the ith subcarrier of the CR user's band. In the study presented in this chapter, we assume this interference to be additive white noise to the secondary user.

7.3 Adaptive Power and Bit Loading

Since different subcarriers in an OFDM system may have different fading gains in a given channel access, use of the same modulation order in all subcarriers leads to inefficient utilization of the overall spectrum [18]. Assuming that the channel state information (CSI) is available at the transmitter, different power, bit or both power and bit loading schemes have been proposed in literature. These loading schemes exploit the time varying nature of fading gains across the OFDM subcarriers in order to improve overall system performance. Different loading algorithms have different end goals [19]. One broad class of bit loading algorithms minimizes the transmit power while attaining a fixed transmission rate as well as a given target bit error rate (BER) (see, for example, [20, 21]). In another version of bit loading algorithms, instantaneous capacity is maximized at a fixed transmit power. All these algorithms maximize the transmission capacity of OFDM-based systems and are useful for *conventional* wireless networks where there is only one group of users i.e., primary users. As mentioned earlier in this section, there is mutual interference between the primary and secondary users when both types of users co-exist. Therefore, use of the classical loading algorithms, e.g., the uniform power but variable rate algorithm and the water-filling algorithm for the secondary user's transmission may result in higher mutual interference to the primary user's band. Throughout this chapter, we consider a CR downlink scenario where interference introduced to the primary user, rather than the transmit power is limiting factor. In fact, such an interference-limited scenario limits the transmit power as well as the achievable transmission capacity of the secondary users. Hence, the design problem is given an interference threshold prescribed by the primary users to determine how much power and how many bits should be loaded into each subcarrier such that the overall transmission capacity of the CR user is maximized.

According to the classical power and bit loading schemes, e.g., the water-filling algorithm,[2] more power and bits should be loaded into the subcarrier that has the higher channel gain. However, the amount of interference introduced by allowing transmission in a secondary user's subcarrier depends on the location of the subcarrier with respect to the primary user's spectrum. From the interference point of view, more power should be loaded into a distant subcarrier. Therefore, a judicious loading policy is required that considers not only the fading gains of the subcarriers

---

[2] Although we are assuming no constraint on the transmit power for the downlink CR scenario, later on we will compare the performance of the water-filling algorithm in detail.

but also the spectral distance of the subcarriers from the primary user's band. The authors in [22] have proposed an unequal bit loading algorithm for a non-contiguous (NC)-OFDM-based CR system. However, in this scheme uniform power allocation among the OFDM subcarriers is used. Later in this chapter, we will see that the use of uniform transmit power in each subcarrier can significantly reduce the transmission capacity of the secondary user. In what follows, we formulate the power and bit loading algorithm for an OFDM-based CR system as a constrained optimization problem. This optimization problem maximizes the transmission capacity of the secondary subcarriers while keeping the interference introduced to the primary user below a specified threshold.

In the formulation, it is assumed that each subcarrier goes under frequency flat fading and the instantaneous fading gains are perfectly known at the transmitter. The transmit power and bits are adaptively loaded in each subcarrier. With an ideal coding scheme, the transmission rate at the ith subcarrier, R_i for transmit power, P_i and channel fading power gain h_i is connected via the Shannon capacity formula, and is given by

$$R_i(P_i, h_i) = \Delta f \log_2 \left(1 + \frac{h_i P_i}{\sigma^2 + J_i} \right) \tag{7.5}$$

where σ^2 denotes the single-sided noise power spectral density and J_i denotes the interference introduced by the primary user into the ith subcarrier. If the system can tolerate some transmission error instead of having an error-free transmission, a practical modulation and coding scheme can be used. In this case the transmission rate, R_i^{prac} corresponding to the transmit power P_i and fading power gain h_i can be approximated using the well-known SNR gap approximation [23]

$$R_i^{\text{prac}}(P_i, h_i) = \Delta f \log_2 \left(1 + \frac{h_i P_i}{\Theta(\sigma^2 + J_i)} \right) \tag{7.6}$$

where Θ is a coding and modulation dependent parameter. For example, with uncoded M-ary quadrature modulation (M-QAM), Θ can be calculated as [23]

$$\Theta = \frac{1}{3} \left[Q^{-1} \left(\frac{\text{BER}_0}{4} \right) \right]^2 \tag{7.7}$$

where $Q^{-1}(\cdot)$ is the inverse standard Gaussian Q-function and BER_0 is the target BER. In the following formulation, we assume an ideal error-free transmission scenario. The extension for a given target BER is quite straightforward.

7.3.1 Optimal Scheme: Continuous Case

Since the design goal is to maximize the *instantaneous* transmission capacity of the CR user while keeping the *instantaneous*[3] interference introduced to the primary

---

[3] It is a system requirement that the interference introduced to the primary user should not go beyond some threshold at any instant. Thus, *constraint* is used in an instantaneous sense here.

users below a certain threshold, it can be formulated as the following constrained optimization problem

$$C = \max_{P_i} \sum_{i=1}^{N} R_i(P_i, h_i) \tag{7.8}$$

$$s.t. \ : \ \sum_{i=1}^{N} I_i(d_i, P_i) \leq I_{th} \tag{7.9}$$

where C denotes the total transmission capacity of the CR user and I_{th} denotes the interference threshold prescribed by the primary user. As we will see that this interference threshold limits the transmit power as well as the achievable transmission capacity of the CR user. The optimization problem in (7.8) and (7.9) can be solved using the Lagrange method [24]. Skipping all the details of this solution, the optimal transmit power, P_i^* in the ith subcarrier can be written as

$$P_i^* = \frac{1}{\lambda K_i} - \frac{\sigma^2 + J_i}{h_i} \tag{7.10}$$

where K_i is a constant defined as

$$K_i = T_s \int_{d_i - B/2}^{d_i + B/2} \left(\frac{\sin \pi f T_s}{\pi f T_s} \right)^2 \mathrm{d}f. \tag{7.11}$$

The parameter λ in (7.10) is the Lagrange multiplier which can be obtained from the following equation:

$$\sum_{i=1}^{N} I_i(d_i, P_i^*) = I_{th} \tag{7.12}$$

where $I_i(d_i, P_i^*)$ is the interference introduced into the primary user's band for transmit power P_i^* in the ith subcarrier and is given in (7.2). It should be noted that power can come out to be negative for some subcarriers using (7.10) and (7.12). In this case, zero power is assigned to that subcarrier whose power has the highest negative value. The whole scheme is then reiterated for the remaining subcarriers. Hence, by using the above scheme the optimal power allocation policy, which maximizes the transmission capacity of the secondary user while keeping the interference introduced to the primary user below the specified threshold, can be obtained.

7.3.2 Comparison with Uniform Power Loading/Water-Filling Schemes

As mentioned earlier, a number of loading algorithms exist that can improve the performance of conventional OFDM-based systems. In situations where there is a power constraint, the capacity maximizing power allocation policy is the well-known water-filling policy in the frequency domain [25]. This policy suggests that more power should be allocated to the subcarriers that have relatively better channel quality and that less power should be allocated to those with poor channel quality. Due to the implementation complexity of water-filling policy, uniform power loading was proposed later on. According to uniform power loading policy, the total transmit power is allocated equally among all the subcarriers. However, the transmission rate should be adjusted according to the subcarriers' fading gains. This scheme is similar to a fixed power but variable rate transmission scheme and the degradation in capacity due to varying only rate is very negligible [26].

It is obvious that both uniform power loading and water-filling schemes are suboptimal for the interference-limited CR system as they do not have any constraint on interference. In order to make a fair comparison with the optimal scheme presented in Sect. 3.1, all the schemes considered should maintain a given interference threshold. Then it would be interesting to observe which scheme offers a higher transmission rate for the CR user. Intuitively, different schemes may require different amounts of transmit power for a given interference threshold as this threshold limits the amount of power that can be transmitted by the CR user. In other words, power as well as the transmission capacity of each scheme is determined by the interference threshold. Therefore, for a fair comparison, at first we determine how much power can be transmitted using a uniform power allocation scheme for a given interference threshold. Then this total power is divided equally among the subcarriers. According to the uniform power allocation policy, the transmit power, P_i^{U} per subcarrier for the given interference threshold I_{th} can be written as

$$P_i^{\mathrm{U}} = \frac{I_{\mathrm{th}}}{\sum_{i=1}^{N} K_i}. \tag{7.13}$$

For distributing power according to the water-filling policy, we first determine the total power used by the uniform scheme for a given interference threshold. Using total power as the instantaneous power constraint, we determine power for each subcarrier using the water-filling algorithm. Then for this distribution of power, we calculate the capacity achieved by the CR users and the interference introduced to the primary user's band.

7.3.3 Numerical Results: Continuous Case

In the numerical results presented in this section, we use values for T_s, Δf, and B of $4\,\mu$ s, 0.3125 MHz, and 0.3125 MHz, respectively. Additive white Gaussian noise (AWGN) variance of 10^{-3} is assumed. The channel power gain h_i is assumed to be Rayleigh fading with an average channel power gain of 10 dB. The value of amplitude P_{PU} is assumed to be 10 mW. Further, we assume that there are 10 subcarriers for the CR user, five on each side of the primary user's band. Since the channel fading power gains for different realizations of h_i can be different, an average transmission capacity of 100,000 independent simulation runs is considered.

The sample average transmission rate of CR user versus interference introduced to the primary user's band for the optimal scheme is plotted in Fig. 7.3. In order to compare the performances of the uniform power but variable rate transmission

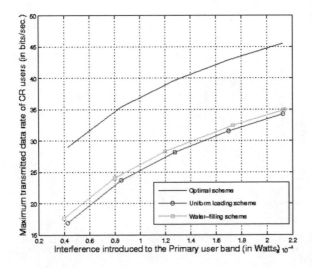

Fig. 7.3. Transmission capacity of CR user versus interference threshold.

scheme and the water-filling scheme, the interference versus capacity curves of these schemes are also plotted in Fig. 7.3. From this figure, it can be seen that for a given interference threshold, the uniform loading and water-filling schemes degrade the capacity of the CR user. The optimal scheme, on the other hand, increases transmission capacity as it allows more judicious transmission of power. In Fig. 7.4, the transmit power of the CR user is plotted for different values of the interference threshold for the schemes under consideration. It is observed from Fig. 7.4 that the optimal scheme allows more power to be transmitted than the other schemes do for a given interference threshold. The uniform power loading and water-filling schemes load less power as they do not judiciously take interference into account in their loading policy for a given interference threshold.

For a given realization of subcarriers' fading power gains, power profiles are plotted in Fig. 7.5 for various schemes.

7.3.4 Effect of Subcarrier Nulling Mechanism: Continuous Case

In [17], the authors studied the effect of subcarrier nulling mechanism. They showed that the interference introduced to the primary user's band can be reduced by nulling the adjacent subcarriers since the adjacent subcarriers produce higher interference than the subcarriers that are located far apart from the primary user's band. Nulling more than one adjacent carrier was also shown to decrease the interference introduced to the primary user's band assuming uniform power loading in each subcarrier. For example, if one adjacent subcarrier is nulled, there is a significant reduction in interference for the same amount of total transmit power. If two adjacent subcarriers are nulled, the reduction in interference to the primary user's band is not as

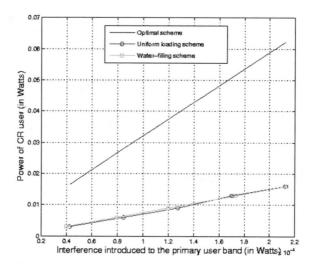

Fig. 7.4. Transmit power of CR user versus interference threshold.

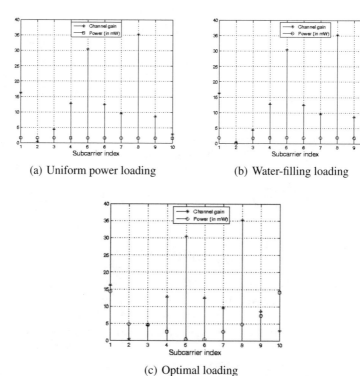

Fig. 7.5. Power profile of different loading schemes.

high as it is when one subcarrier is nulled. Although this study is very interesting from the interference reduction point of view, it does not explore the reduction in capacity for different nulling cases. Since the adjacent subcarrier is the dominant interfering subcarrier, it may be expected that higher transmission capacity can be achieved by nulling adjacent subcarriers. This is because one may expect that for a given interference threshold, nulling dominant interfering subcarriers may allow the CR user to transmit more power to the remaining active subcarriers in order to achieve higher transmission capacity. However, when nulling adjacent subcarriers, the time varying nature of their fading gains cannot be exploited. In other words, the adjacent subcarrier is always assigned zero power, even when it has very good channel gain. Therefore, nulling creates a trade-off between the amount of interference that can be reduced and the achievable transmission capacity. In this section, we discuss the effect of the nulling mechanism on the uniform power loading and water-filling schemes.

The transmission capacity of the CR user versus interference introduced to the primary user's band is plotted for uniform power loading schemes for various nulling scenarios in Fig. 7.6. Similar plot has been shown for water-filling case in Fig. 7.7. Here, *one nulling* means that one adjacent subcarrier from each side of the primary user's band has been assigned zero power. Similarly, for the two nulling case, two subcarriers from each side have been assigned zero power. For the sake of comparison we have also plotted the transmission capacity of the optimal scheme in Figs. 7.6 and 7.7. It can be observed from Figs. 7.6 and 7.7 that after nulling, the performance of the uniform power loading and water-filling scheme improves compared with the no nulling case. However, the optimal scheme still achieves the highest capacity for a given interference threshold. Interestingly, it can also be observed from Figs. 7.6 and 7.7 that both uniform power loading and water-filling schemes degrade the performance for the two nulling case compared to the one nulling case. This is because of the tradeoff between the interference that can be reduced by nulling additional subcarriers and the capacity that can be achieved by keeping them active. From this selected numerical results it can be concluded that nulling additional subcarriers does not always help to improve overall system performance. We did not consider further nulling as the associated performance degradation has been checked via simulation.

In Fig. 7.8, we present the curves for the transmit power of the CR user versus interference introduced to the primary user's band for the uniform power loading scheme, for various nulling cases. Similar plot has been shown for water-filling case in Fig. 7.9. In order to make a comparison, the transmit power of the CR user for the optimal scheme is also plotted in Figs. 7.8 and 7.9. The interesting observation from these figures is that for some interference threshold, the uniform power loading and water-filling schemes can load more power than the optimal scheme for the two nulling case. However, the transmission capacity is always lower than for the optimal scheme because it completely loses the opportunity of using these two subcarriers even if they have superior channel qualities.

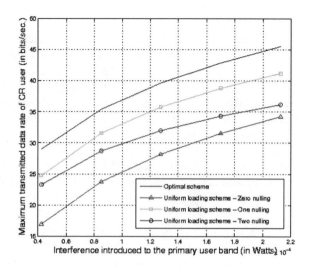

Fig. 7.6. Transmission capacity of CR user versus interference to the primary user for various nulling using uniform loading scheme.

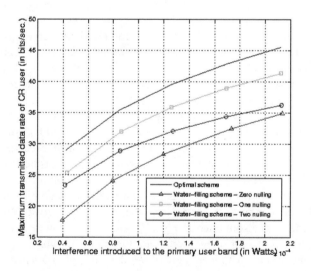

Fig. 7.7. Transmission capacity of CR user versus interference to the primary user for various nulling using water-filling scheme.

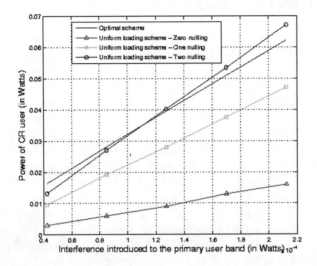

Fig. 7.8. Transmit power of CR user versus interference introduced to the primary user's band for various nulling scenarios using uniform loading scheme.

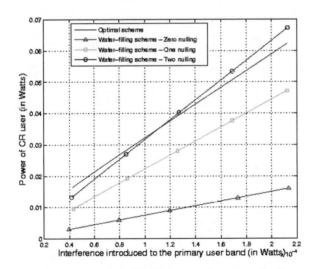

Fig. 7.9. Transmit power of CR user versus interference introduced to the primary user's band for various nulling scenarios using water-filling scheme.

7.4 Discrete Bit Loading Algorithms

In Sect. 7.3, an optimal power and bit loading algorithm has been formulated and studied. It has also been shown that the uniform power loading scheme and water-filling policy both degrade the transmission capacity of the OFDM-based CR system. The study was based on the operating assumption that the transmission rate can be varied continuously. Most of the coding and modulation schemes that are used in practice provide a discrete or integer transmission rate which has led the researchers to develop discrete bit loading algorithms for OFDM-based systems. In this context, a number of algorithms have been proposed in the literature. Examples of well-known algorithms include the Hughes–Hartogs [27] and the Chow et. al. [25] algorithms. These algorithms are not directly applicable to an OFDM-based CR system as the interference they introduce to the adjacent primary user's band may increase significantly. In this section, our focus is to modify these algorithms so that they will be applicable to the OFDM-based CR system.

First we present a suboptimal scheme for an integer bit loading case. The algorithm is suboptimal in the sense that it approximates the optimal continuous rate value to the nearest integer value. Then the we modify Hughes–Hartogs and the Chow et al. algorithms to minimize the interference to the primary user's band.

Scheme A. The goal is to minimize the instantaneous interference to the adjacent primary user's band while transmitting at a fixed data rate, R_{spec} and a specified BER, $\mathrm{BER}_{\mathrm{spec}}$. This scheme can be written mathematically as follows:

$$\min_{R_i \in \mathcal{R}} \quad \sum_{i=1}^{N} I_i(d_i, P_i(R_i)) \tag{7.14}$$

$$s.t. \; : \; \sum_{i=1}^{N} R_i = R_{\mathrm{spec}} \tag{7.15}$$

$$\mathrm{BER}_i(P_i, R_i, h_i) \le \mathrm{BER}_{\mathrm{spec}} \tag{7.16}$$

where $\mathcal{R} = \{0, 1, 2, \cdots\}$ represents the integer transmission rate and BER_i denotes the BER for the ith subcarrier for a transmission rate R_i specified as [26]

$$\mathrm{BER}_i = 0.2 \exp\left[\frac{-1.6 P_i h_i}{(\sigma^2 + J_i)(2^{R_i} - 1)}\right]. \tag{7.17}$$

The nature of the optimization problem in (7.14), (7.15) and (7.16) is *combinatorial* and is difficult to solve. Therefore, we assume that the rate R_i can have continuous value and then the optimal transmission rate is rounded to the nearest integer value. Hence, the scheme is suboptimal for the discrete rate adaptation case. Using the Lagrange formulation and assuming continuous rate variation, the optimal solution is derived as follows:

$$R_i = \frac{R_{\text{spec}}}{N} + \frac{1}{N}\sum_{i=1}^{N}\log_2\left[\frac{-K_i\ln(5\text{BER}_{\text{spec}})(\sigma^2 + J_i)\ln(2)}{1.6 \times h_i}\right]$$

$$+ \log_2\left[\frac{-1.6h_i}{K_i\ln(5\text{BER}_{\text{spec}})(\sigma^2 + J_i)\ln(2)}\right] \tag{7.18}$$

where K_i and J_i have been defined in (7.11) and (7.4), respectively. If R_i in (7.18) is negative for some subcarriers, zero bit is assigned to them and we reiterate the whole scheme for the remaining subcarriers. Since the rates R_i can only be integers, we round R_i to the nearest integer R_{qi} and determine the round-off error as follows: $\Delta R_i = R_i - R_{qi}$. The next part of the scheme is adopted from [25]. The sum $\sum_{i=1}^{N} R_{qi}$ is calculated. If it is larger (smaller) than the R_{spec}, then the rate of the channel with the largest (smallest) ΔR_i is incremented (decremented). The algorithm stops when $\sum_{i=1}^{N} R_{qi} = R_{\text{spec}}$.

7.4.1 Modifications of the Existing Schemes

Scheme B. The Hughes-Hartogs algorithm [27] incrementally allocates an integer number of bits at the cost of high computational complexity. The algorithm adds bits successively to the subcarrier that will require the least amount of power for the specified BER. But for the CR scenario, the goal is to minimize the interference introduced to the primary user's band. Hence, we allocate bits successively to the subcarrier that will introduce the least amount of interference to the primary user's band. The modified algorithm works as follows:

1. The subcarrier is searched that will introduce the least interference to the primary user's band for the specified BER in assigning one more bit.
2. The bit is assigned to the subcarrier that introduces the minimum interference.
3. The above steps are repeated until the given data rate is achieved.

It is obvious that the above algorithm requires extensive searching and hence is slow. However, it is optimal as the bits are loaded in a manner that minimizes total interference. Although Scheme B is optimal, it is very slow for practical applications compared with Scheme A which has a closed-form expression.

Scheme C. The Chow et al. algorithm [25] omits intensive sorting as it allocates rate among the subcarriers according to channel capacity approximation. The goal of the algorithm is to minimize the transmitted power or to maximize the noise margin given a data rate and a target BER. Noise margin (γ_{margin}) is defined as the amount of additional noise that can be tolerated, while still achieving the specified BER. The algorithm can be described as follows [25]:

1. Initialize $\gamma_{\text{margin}} = 0$, IterateCount(number of iterations) = 0, UsedCar(total number of subcarriers) = N and $\epsilon(i)$(energy of a particular subcarrier) = 1.
2. For $\forall i$ calculate

$$R(i) = \log_2\left(1 + \frac{\text{SNR}(i)}{\tau + \gamma_{\text{margin}}}\right) \tag{7.19}$$

where τ is the SNR gap in the gap approximation [23].

$$\hat{R}(i) = \text{round}[R(i)] \tag{7.20}$$

where $\hat{R}(i)$ is an integer number of bits that are assigned to a particular subcarrier.

$$\text{diff}(i) = R(i) - \hat{R}(i). \tag{7.21}$$

$$\text{If} \quad \hat{R}(i) = 0, UsedCar = UsedCar - 1. \tag{7.22}$$

Now, if $R_{\text{total}} = \sum_{i=1}^{N} \hat{R}(i) = 0$, the algorithm is stopped.

3. If $R_{\text{total}} \neq 0$, the new γ_{margin} is calculated according to:

$$\gamma_{\text{margin}} = \gamma_{\text{margin}} + 10 \log_{10}(2^{\frac{R_{\text{total}} - R_{\text{spec}}}{UsedCar}}) \tag{7.23}$$

where R_{spec} is the given data rate. Increment IterateCount by 1.

4. If $R_{\text{total}} \neq R_{\text{spec}}$ and IterateCount $< MaxCount$ (maximum number of allowed iterations), let UsedCar $= N$ and go to step 2, else go to step 5. It should be noted that if the algorithm does not converge after $MaxCount$ iterations, convergence is forced using step 5.

5. If $R_{\text{total}} > R_{\text{spec}}$, then one bit is subtracted from the subcarrier that has the minimum diff(i) and it is repeated until R_{total} becomes equal to R_{spec}. On the other hand, if $R_{\text{total}} < R_{\text{spec}}$, then one bit is added to the subcarrier which has the maximum diff(i) and this is repeated until R_{total} becomes equal to R_{spec}.

6. The power of each subcarrier is adjusted such that the BER of each subcarrier (here, we exclude the subcarriers that have been assigned 0 bits) is equal to the specified BER for the given bit allocation $\hat{R}(i)$.

Basically, the algorithm first finds the optimal system performance margin (in steps 1–4), and then if the algorithm does not converges in $MaxCount$ iterations, forced convergence is imposed (in step 5). Finally, the input energy distribution is adjusted.

For the CR scenario, we change Eq. (7.19) so that the bits are allocated more to the subcarriers that are far from the primary user's band, as they introduce less interference. We define $\text{const}(i)$ as follows:

$$\text{const}(i) = (1/k(i))/\sum_{i=1}^{N}(1/k(i)). \tag{7.24}$$

It should be noted that the subcarriers which are near to the primary user's band introduce more interference and hence, has a higher value of $k(i)$ as compared to the subcarriers which are far from the primary user band. For the CR scenario, we modify (7.19) as follows:

$$R(i) = \log_2\left(1 + \frac{\text{const}(i)\text{SNR}(i)}{\tau + \gamma_{\text{margin}}}\right). \tag{7.25}$$

Hence, by introducing $\text{const}(i)$, we are allocating less bits to subcarriers which are near to the primary user band as they produce more interference and more bits are given to the subcarriers that are far from the primary user's band as they produce less interference. The rest of the algorithm remains same.

7.4.2 Numerical Results: Discrete Case

In the numerical results presented in this section, we use values for T_s, Δf and B of $4\,\mu$ s, 0.3125 MHz, and 0.3125 MHz, respectively. Noise variance of 10^{-6} is assumed. The channel fading power gain h_i is assumed to be Rayleigh faded with an average channel power gain of 5 dB. A specified BER of 10^{-7} is used and the value of τ is taken to be 9.8 dB as in [25]. The amplitude P_{PU} is assumed to be 1 mW. $MaxCount$ is taken to be 10.

In Fig. 7.10, we present the interference introduced to the primary user's band versus given data rate plots for the proposed schemes A, B, and C, as well as the conventional Hughes-Hartogs scheme and the Chow et al. scheme. The plotted data rates represent an average of 100,000 independent simulation runs. From Fig. 7.10, we observe that for a given data rate, the existing Hughes-Hartogs and Chow et al. schemes introduce more interference to the primary user's band compared with schemes A, B, and C. Scheme B is optimal and introduces the minimum interference to the primary user's band. Scheme A is suboptimal and introduces less interference than does the suboptimal scheme C and the same as optimal scheme B. It should be

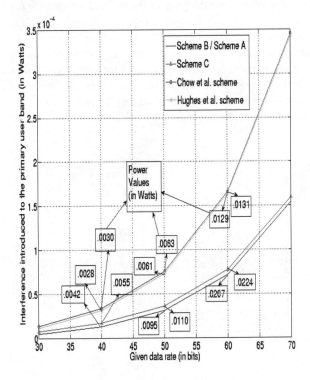

Fig. 7.10. Interference introduced to the primary user's band versus given data rate for discrete case (BER = 10^{-7}).

noted that the suboptimal scheme A has similar performance to the optimal scheme B because the quantization error does not produce any visual degradation in performance. Further, in Fig. 7.10 we have marked the power values for some selected data rates. We observe from these values that the conventional optimal Hughes-Hartogs scheme, which minimizes the power, requires the least power for a given data rate. Our proposed schemes A, B, and C require higher transmit power than does the Hughes-Hartogs scheme as they assign more power to the distant subcarriers than to the nearest subcarriers in order to reduce interference to the primary user's band.

7.4.3 Effect of Subcarrier Nulling Mechanism: Discrete Case

In this section, we study the effect of nulling on various discrete bit loading algorithms proposed in Sect. 7.4. Since scheme B performs optimally, we do not study the effect of nulling for this scheme and so is for scheme A.

In Fig. 7.11, the interference versus data rate for the proposed scheme C is plotted under various nulling scenarios. Similar curves are plotted in Figs. 7.12 and 7.13 for the Hughes-Hartogs scheme and the Chow et al. scheme, respectively. In these figures, we have also plotted the data rate of the optimal scheme for the sake of comparison. From Figs. 7.11–7.13, we observe that the performances of Scheme C, the Hughes-Hartogs scheme and the Chow et al. scheme are improved for several data rates after nulling compared with the no nulling case. However, the optimal scheme achieves the highest transmission rate for a given interference threshold. We did not consider additional nulling as we know from simulation that this causes the performance to degrade. These figures also indicate that the two nulling case performs the worst for most of the data rates. In Figs. 7.11–7.13, we have also identified the power

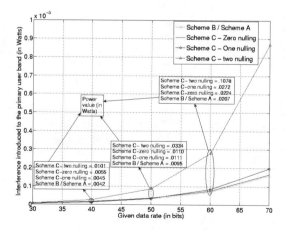

Fig. 7.11. Interference introduced to the primary user's band versus data rate (BER = 10^{-7}) for Scheme C under various nulling scenarios.

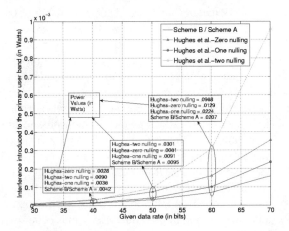

Fig. 7.12. Interference introduced to the primary user's band versus data rate (BER = 10^{-7}) for Hughes–Hartogs scheme under various nulling scenarios.

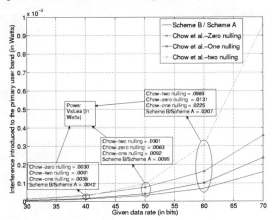

Fig. 7.13. Interference introduced to the primary user's band versus data rate (BER = 10^{-7}) for Chow et al. scheme under various nulling scenarios.

values that are required by different schemes under various nulling scenarios. These values can be used as important parameters in designing an OFDM-based cognitive radio system.

Conclusion

In this chapter, we have explored some of the challenges involved in the design of adaptive power and bit loading algorithms for an OFDM-based CR system. For a given interference constraint, we presented a downlink transmission capacity maximization power and bit loading algorithm. The performance of the existing loading algorithms used in conventional OFDM-based systems are compared with the optimal scheme. For the continuous rate variation case, the results show that for a given interference constraint, the uniform power but variable bit loading algorithm has a high-capacity degradation compared with the optimal scheme. In other words, the optimal scheme allows more power to be loaded to achieve a higher transmission capacity for a given interference threshold. For the integer rate adaptation case, a suboptimal bit loading algorithm is presented as well as the well-known algorithms are modified in order to minimize interference to the primary user's band. The results in this case showed that suboptimal and modified algorithms can significantly reduce interference. We also studied the effect of nulling mechanism in both the continuous and discrete cases. Selected numerical results showed that nulling degrades the transmission capacity for a given interference threshold compared with the optimal scheme. We found that nulling of one subcarrier yields better performance than nulling of two or zero subcarriers.

Acknowledgment

This research was supported by the Natural Sciences and Engineering Research Council (NSERC), Canada, under a Strategic Project Grant.

References

1. R. W. Broderson, A. Wolisz, D. Cabric, S. M. Mishra, and D. Willkomm, "CORVUS: A cognitive radio approach for usage of virtual unlicensed spectrum," White Paper submitted at the University of Berkeley, CA, July. 2004.
2. Federal Communications Commission, "Spectrum Policy Task Force," Report ET Docket no. 02-135, Nov. 2002.
3. J. Mitola, "The software radio architecture," *IEEE Commun. Mag.*, vol. 33, no. 5, pp. 26–38, May 1995.
4. J. Mitola III, "Cognitive radio for flexible mobile multimedia communications," in *Proc. 6th International Workshop on Mobile Multimedia Commun. (MoMuC'99)* (San Diego, CA), pp. 310, Nov. 1999.
5. J. Mitola III and G. Q. Maguire Jr., "Cognitive radios: Making software radios more personal," *IEEE Pers. Commun. Mag.*, vol. 6, pp. 13–18, Aug. 1999.
6. S. Haykin, "Cognitive radio: Brain-empowered wireless communications," *IEEE J. Select. Areas Commun.*, vol. 23, no. 2, pp. 201–220, Feb. 2005.
7. D. Cabric, S. M. Mishra, and R. W. Brodersen, "Implementation issues in spectrum sensing for cognitive radios," in *Proc. Asilomar Conf. on Signals, Systems, and Computers* (Pacific Grove, CA), pp. 772–776, Nov. 2004.
8. S. M. Mishra, A. Sahai, and R. Brodersen, "Cooperative sensing among cognitive radios," in *Proc. IEEE Int. Conf. Commun. (ICC'06)* (Istanbul, Turkey), June 2006.
9. T. Weiss and F. K. Jondral, "Spectrum pooling: An innovative strategy for the enhancement of spectrum efficiency," *IEEE Commun. Mag.*, vol. 43, no. 3, pp. S8–S14, Mar. 2004.

10. I. F. Akyildiz, W. Y. Lee, M. C. Vuran, and S. Mohanty, "NeXt generation/dynamic spectrum access/cognitive radio wireless networks: A survey," *Comput Netw (Elsevier)*, vol. 50, pp. 2127–2159, Sept. 2006.

11. Darpa XG Working group, "The xg architectural framework," rfc v1.0 2003.

12. Darpa XG Working group, "The xg vision," rfc v1.0 2003.

13. C. Cordeiro, K. Challapali, D. Birru, and S. Shankar, "IEEE 802.22: The first worldwide wireless standard based on cognitive radios," in *Proc. IEEE Int. Symposium on Dynamic Spectrum Access Networks (DySPAN'05)*, pp. 328–337, Nov. 2005.

14. M. M. Buddhikot, P. Kolody, S. Miller, K. Ryan, and J. Evans, "DIMSUMNet: New directions in wireless networking using coordinated dynamic spectrum access," in *Proc. IEEE Int. Symposium on a World of Wireless, Mobile and Multimedia Networks (WoWMoM' 05)*, pp. 78–85, June 2005.

15. I. F. Akyildiz and Y. Li, "OCRA: OFDM-based cognitive radio networks," Technical report, Broadband and Wireless Networking Laboratory, Georgia Institute of Technology, Mar. 2006.

16. L. Xu, R. Tonjes, T. Paila, W. Hansmann, M. Frank, and M. Albrecht, "DRiVE-ing to the internet: Dynamic radio for IP services in vehicular environments," in *Proc. 25th Annual IEEE Conference on Local Computer Networks*, pp. 281–289, Nov. 2000.

17. T. Weiss, J. Hillenbrand, A. Krohn, and F. K. Jondral, "Mutual interference in OFDM-based spectrum pooling systems, " in *Proc. IEEE Vehicular Technol. Conf. (VTC'04)*, vol. 4, pp. 1873–1877, May 2004.

18. T. Keller and L. Hanzo, "Multicarrier modulation: A convenient framework for time-frequency processing in wireless communications," *Proc. IEEE*, vol. 88, no. 5, pp. 611–640, May 2000.

19. A. T. Toyserkani, J. Ayan, S. Naik, Y. Made, and O. Al-Askary, "Sub-carrier based adaptive modulation in HIPERLAN/2 system," in *Proc. IEEE Int. Conf. Commun. (Paris, France)*, pp. 3460–3464, June 2004.

20. L. Goldfeld and V. Lyandres, "Capacity of the multicarrier channel with frequency-selective Nakagami fading," *IEICE Trans. Commun.*, vol. E83-B, no. 3, pp. 697–702, Mar. 2000.

21. C. Y. Wong, R. S. Cheng, K. B. Letaief, and R. D. Murch, "Multiuser OFDM with adaptive subcarrier, bit, power, and power allocation," *IEEE J. Select. Areas Commun.*, vol. 17, no. 10, pp. 1747–1758, Oct. 1999.

22. A. M. Wyglinski, "Effects of bit allocation on non-contiguous multicarrier-based cognitive radio transceivers," in *Proc. 64th IEEE Veh. Technol. Conf. – Fall* (Montreal, Canada), Sept. 2006.

23. J. M. Cioffi, "A multicarrier primer," *ANSI T1E1.4 Committee Contribution*, pp. 91–157, Nov. 1991.

24. A. Antonio and W.-S. Lu, *Optimization methods, algorithms, and applications.* Kluwer Academic, 2005.

25. P. Chow, J. Cioffi, and J. Bingham, "A practical discrete multitone transceiver loading algorithm for data transmission over spectrally shaped channels," *IEEE Trans. Commun.*, vol. 43, no. 2/3/4, pp. 773–775, Feb./Mar./Apr. 1995.

26. S. T. Chung and A. J. Goldsmith, " Degrees of freedom in adaptive modulation: A unified view, " *IEEE Trans. Commun.*, vol. 49, no. 9, pp. 1561–1571, Sept. 2001.

27. J. Bingham, "Multicarrier modulation for data transmission: An idea whose time has come," *IEEE Commun. Mag.*, vol. 28, no. 5, pp. 5–14, May 1990.

UWB-Based Cognitive Radio Networks

Hüseyin Arslan and Mustafa E. Şahin

University of South Florida, Tampa, FL, USA
{arslan,mesahin}@eng.usf.edu

8.1 Introduction

As wireless communication systems are making the transition from wireless telephony to interactive internet data and multi-media type of applications, the desire for higher data rate transmission is increasing tremendously. As more and more devices go wireless, it is not hard to imagine that future technologies will face spectral crowding, and coexistence of wireless devices will be a major issue. Considering the limited bandwidth availability, accommodating the demand for higher capacity and data rates is a challenging task, requiring innovative technologies that can offer new ways of exploiting the available radio spectrum. Ultra-wideband (UWB) and cognitive radio are two exciting technologies that offer new approaches to the spectrum usage.

Ignited by the earlier work of Mitola [1], cognitive radio is a novel concept for future wireless communications, and it has been gaining significant interest among the academia, industry, and regulatory bodies [2]. Cognitive radio provides a tempting solution to spectral crowding problem by introducing the opportunistic usage of frequency bands that are not heavily occupied by their licensed users. Cognitive radio concept proposes to furnish the radio systems with the abilities to measure and be aware of parameters related to the radio channel characteristics, availability of spectrum and power, interference and noise temperature, available networks, nodes, and infrastructures, as well as local policies and other operating restrictions. The primary advantage targeted with these features is to enable the cognitive systems to utilize the available spectrum in the most efficient way. An interconnected set of cognitive radio devices that share information is defined as a cognitive network. Cognitive networks aim at performing the cognitive operations such as sensing the spectrum, managing available resources, and making user-independent, intelligent decisions based on cooperation of multiple cognitive nodes. In order to be able to achieve the goals of the cognitive radio concept, cognitive networks need a suitable wireless technology that will facilitate collaboration of the nodes.

Ultra-wideband is defined as any wireless technology that has a bandwidth greater than 500 MHz or a fractional bandwidth[1] greater than 0.2. Ultra-wideband systems have been attracting an intense attention from both the industry and academic world since 2002, when the US Federal Communications Commission (FCC) released a spectral mask officially allowing the unlicensed usage of UWB. There are two commonly proposed means of implementing UWB. These two technologies are the Orthogonal Frequency Division Multiplexing based UWB (UWB-OFDM) and the impulse radio based UWB (IR-UWB).

Under the current FCC regulation, UWB is a promising technology for future short- and medium-range wireless communication networks with a variety of throughput options including very high data rates. UWB's most significant property is that it can coexist in the same temporal, spatial, and spectral domains with other licensed/unlicensed radios because it is an underlay system. Other tempting features of UWB include that it has a multi-dimensional flexibility involving adaptable pulse shape, bandwidth, data rate, and transmit power. On top of these, UWB has a low power consumption, and it allows significantly low complexity transceivers leading to a limited system cost. Another very important feature of UWB is providing secure communications. It is very hard to detect UWB transmission as the power spectrum is embedded into the noise floor. This feature introduces very secure transmission in addition to other higher layer encryption techniques.

When the wireless systems that are potential candidates for cognitive radio are considered, UWB seems to be one of the tempting choices because it has an inherent potential to fulfill some of the key cognitive radio requirements. Along with the inherent UWB attributes mentioned, especially IR-UWB offers some extraordinary uses that can add a number of extra intellective features to cognitive systems. These special uses are brought by the high multipath resolution property, which enables UWB to act as an accurate radar, ranging, and positioning system. Examples of specific UWB features include sensing the physical environment to enable situation awareness, and providing geographical location information.

Owing to all its distinctive properties mentioned, in this chapter, UWB is considered as one of the enabling technologies of cognitive radio networks. The flow of this chapter is as follows. Cognitive radio and cognitive networks are described in Sect. 8.2. The basics of UWB and its suitability for cognitive networks are addressed in Sect. 8.3. Finally, in Sect. 8.4, various UWB cognitive networks related issues are discussed in detail.

8.2 Cognitive Networks and Cognitive Radio

When we look at the evolution of wireless standards and technologies, it can be seen that the adaptive features and intelligent network capabilities are gradually adapted as the hardware and software technologies improve. Especially, with the recent trend

---

[1] Fractional bandwidth = $2 \cdot \frac{F_H - F_L}{F_H + F_L}$, where F_H and F_L are the upper and lower edge frequencies, respectively.

and interest in software defined radio based architectures, cognitive radio and cognitive networks attracted more interest. In addition to these, the increasing demand for wireless access along with the scarcity of the wireless resources (specifically the spectrum) bring about the desire for new approaches in wireless communications. Therefore, even though cognitive networks and cognitive radio terms have recently become popular, it is actually a natural evolution of the wireless technologies. With the emergence of cognitive radio and cognitive network concepts, this evolution process has been more formalized and structured. Also, with these new concepts the perception of adaptation and optimization of wireless communication systems gained new dimensions and perspectives. Especially, the emergence of cognitive networks (with cooperative functions and cognitive engine concepts) is a promising solution for the barrier that arises from the flaws of the conventional layered design architecture.

The term "cognitive radio" defines the wireless systems that can sense, be aware of, learn, and adapt to the surrounding environment according to inner and outer stimuli. Overall cognition cycle can be seen as an instance of artificial intelligence, since it encompasses observing, learning, reasoning, and adaptation. Adaptation itself in the cognition cycle is a complex problem, because cognitive radio needs to take into account several inputs at the same time including its own past observations as a result of learning property. Although the adaptation of wireless networks is not a new concept, the previous standards and technologies strive to obtain an adaptive wireless communication network from a narrower perspective (commonly focused on a single-layer adaptation with a single objective function) as compared to that of cognitive radio, which considers a global adaptation that includes multiple layers and goal functions.

For many researchers and engineers, the cognitive radio concept is not limited to a single intelligent radio, but it also includes the networking functionalities. However, within this chapter, we will use the term of cognitive networks to define the networking functionalities of the cognition cycle. Hence, cognitive networks can be defined as intelligent networks that can automatically sense the environment (individually and collaboratively) and current network conditions, and adapt the communication parameters accordingly. Comparing the cognitive radio and cognitive networks definition, it can be seen that the definitions are similar, except cognitive networks have more broader perspective that also include all the network elements.

Cognitive networks are expected to shape the future wireless networks with important applications in dynamic spectrum access, and co-existence and interoperability of different wireless networks. Among the special features of cognitive networks, the leading ones are advanced interference management strategies, efficient use of wireless resources, safe and secure wireless access methodologies, and excellent Quality of Service (QoS). In spite of all these great features and possibilities, being a new concept, the cognitive radio network poses many new technical challenges. As it will be described in the subsequent sections, such networks have requirements in dynamic spectrum management, power and hardware efficiency, complexity and size, spectrum sensing and interference identification, environment awareness, user awareness, location awareness, new distributed algorithm design, distributed spectrum measurements, QoS guarantees, and security. Addressing these requirements is very critical for the success of these networks in wireless communication market, and the authors of this chapter believe that ultra-wideband technology and networking has the capability to accommodate some of these key requirements as it will be discussed throughout this chapter.

8.3 UWB Basics and UWB's Suitability for Cognitive Networks

The two main techniques considered for UWB physical layer are the impulse radio and OFDM. In this section, first, the fundamentals of both of these technologies will be given to provide a technical background. Then, by providing a one-by-one matching between various cognitive radio needs and UWB properties it will be explained how suitable UWB is for cognitive networks.

8.3.1 Background on UWB

According to the current FCC regulations in the USA, UWB systems are allowed to operate in the 3.1–10.6 GHz band without a license requirement. However, the transmit power of these systems is strictly limited. Both in indoors and outdoors, UWB systems are not permitted to transmit more than −42 dBm/MHz in the specified band. This limitation ensures that the UWB systems do not affect the licensed operators that use various frequency bands in the UWB band. However, it should also be kept in mind that it is not unlikely that revisions can be made in the UWB-related FCC regulations, especially regarding the transmit power limits. In the near future, if the UWB radios are provided with cognitive properties that allow them to sense the spectrum to determine the occupancy of their target bands and to ensure the absence of licensed users, it is possible that regulatory agencies may consider to offer more freedom to UWB.

Impulse radio based implementation of UWB is carried out by transmitting extremely short low-power pulses that are on the order of nanoseconds [3,4] as illustrated in Fig. 8.1. Impulse radio UWB is advantageous in that it enables to employ

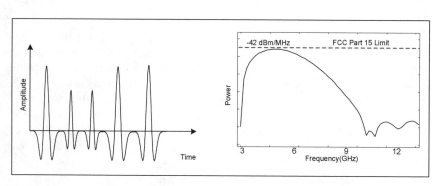

Fig. 8.1. Impulse radio based UWB pulses and the spectrum of a single pulse.

various types of modulations, including on-off keying (OOK), pulse amplitude modulation (PAM), pulse shape modulation (PSM), pulse interval modulation (PIM), pulse position modulation (PPM), and phase shift keying (PSK) [5].

For multi-user access, IR-UWB systems employ time hopping (TH) codes that are specific to each user [6]. These specific pseudo-random noise (PN) codes enable the UWB system to provide access to multiple users conveniently. The multi-user parameters can be adaptively modified according to the change in number of users. To enable more users to communicate, for example, the UWB system can increase the number of chips in each frame at the expense of decreasing each user's data rate.

Different types of receivers can be utilized for IR-UWB communications which include coherent receivers (such as Rake and correlator receivers) as well as non-coherent ones such as energy detector and transmitted reference receivers. Along with the flexibility in modulation methods and receiver types, IR-UWB also offers a variety of options regarding the shapes of the transmitted pulses. Various analog and digital methods to implement pulse shaping for impulse radio can be found among others in [1,3–39].

Besides being a communication system, IR-UWB is a precise radar technology as well as a highly accurate ranging and positioning system. These extra features are owed to the fact that IR-UWB systems have an excellent multipath resolving capability because of the extremely wide frequency band that they occupy.

In OFDM-based UWB, orthogonal subcarriers are employed to modulate the transmitted data. Figure 8.2 shows a typical OFDM waveform in frequency domain. As long as the total occupied bandwidth is not less than 500 MHz, the number of subcarriers and the subcarrier spacing may be assigned various values according to the needs. In the current multi-band OFDM planning, which divides the entire UWB band into 14 subbands, each subband is considered to be 528 MHz and contain 128 subcarriers. The subcarrier spacing is usually chosen to be less than the channel coherence bandwidth. This enables that each subcarrier goes through a flat fading channel. Hence, the UWB-OFDM receiver needs a simple equalizer implementation to recover the originally transmitted signal. One of the most tempting properties of

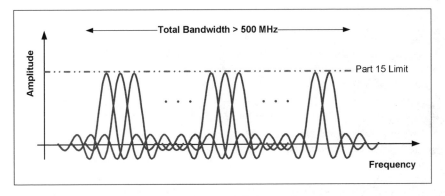

Fig. 8.2. OFDM based UWB waveform.

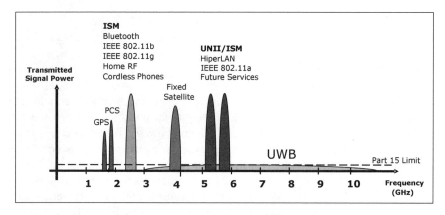

Fig. 8.3. Narrowband systems and UWB spectrum.

UWB-OFDM is the easiness of avoiding interference to licensed systems. A UWB-OFDM transmitter can avoid jamming a licensed signal by simply turning off the subcarriers that overlap with the spectra of the licensed system.

The final significant feature of UWB is its interference immunity. UWB has a considerable resistance against the multi-user access interference (MAI), which is investigated in detail in [4, 35–38]. UWB systems are immune to not only MAI, but also against narrowband interference (NBI), which is caused by the licensed and unlicensed systems that exist in the frequency band occupied by the UWB system [5, 17–37], which are illustrated in Fig. 8.3.

8.3.2 Cognitive Radio Requirements vs. UWB Features

One of the main goals targeted with cognitive radio is to utilize the existing radio resources in the most efficient way. To ensure the optimum utilization, cognitive radio requires a number of conditions to be satisfied. A wireless system that is a potential candidate for cognitive radio applications is expected to fulfill some of these conditions.

The primary cognitive radio requirements include

- negligible interference to licensed systems,
- capability to adapt itself to various link qualities,
- ability to sense and measure critical parameters about the environment, channel, etc.
- ability to exploit variety of spectral opportunity,
- flexible pulse shape and bandwidth,
- adjustable data rate, adaptive transmit power, information security, and limited cost.

At this point, if the main properties of UWB are considered, it is seen that there is a strong match between what the cognitive radio requires and what UWB offers.

In the following, the primary features of UWB will be investigated from the point of satisfying the requirements of cognitive radio.

Cognitive radios aim at an opportunistic usage of frequency bands that are owned by their licensed users. Therefore, one of the most significant requirements of cognitive radio is that the interference caused by cognitive devices to licensed users remains at a negligible level.

UWB offers the possibility of being implemented both in *underlay* and *overlay* modes. The difference between the two modes is the amount of transmitted power. In the underlay mode, UWB has a considerably restricted power, which is spread over a wide frequency band. In this mode, it complies with the corresponding regulations of the FCC in the USA. When a UWB system is operating in the underlay mode, it is quite unlikely that any coexisting licensed system is affected from it. On top of this, underlay UWB can employ various narrowband interference avoidance methods.[2] In the overlay mode the transmitted power can be much higher. However, this mode is only applicable if the UWB transmitter ensures that the targeted spectrum is completely free of signals of other systems, and, of course, if the regulations allow this mode of operation. If these conditions are met, the transmitted UWB power can be increased to a certain level that is comparable to the power of licensed systems. UWB can also operate in both underlay and overlay modes simultaneously. Depending on the spectrum opportunities, the signaling and the spectrum of the transmitted signal can be shaped in such a way that part of the spectrum is occupied in an underlay mode and some other parts are occupied in an overlay mode.

Apparently, in any mode of operation, UWB causes negligible interference to other communication systems, if it does at all. This special feature of UWB makes it very tempting for the realization of cognitive radio.

One of the main features of the cognitive radio concept is that the targeted frequency spectrum is scanned periodically in order to check its availability for opportunistic usage. According to the results of this spectrum scan, the bands that will be utilized for cognitive communication are determined. Since at different times and locations the available bands can vary, cognitive radio is expected to have a high flexibility in determining the spectrum it occupies.

Flexible spectrum shaping is a part of UWB's nature. In IR-UWB, since the communication is basically realized via the transmission of short pulses, varying the duration or the form of the pulses directly alters the occupied spectrum. In UWB-OFDM, on the other hand, spectrum shaping can be conveniently accomplished by turning some subcarriers on or off according to the spectral conditions.

The availability of unused bands is of vital importance for the continuity of communications in cognitive radio. Any increase in the utilization of the bands by the licensed systems directly results in narrowed freedom for the cognitive radio, which can force it to decrease its data rate and QoS, or even to terminate its communication. Therefore, cognitive radio systems are expected to be able to adjust their throughput

---

[2] For a detailed discussion of narrowband interference avoidance and cancelation methods in UWB systems, the readers are referred to [37].

according to the available bandwidth. They should also provide a solution for the cases when the available bandwidth is so limited that the communication cannot be continued.

UWB systems are able to make abrupt changes in their data rates. An IR-UWB system responds to a decrease in available bandwidth by switching to a different pulse shape that is wider in shape. If there is more band to use, it can respond by doing the opposite. The adjustment of the occupied bandwidth in UWB-OFDM is much simpler. The subcarriers that overlap with the occupied bands are turned off, and this way, the data rate is decreased.

On top of its flexible data rate property, UWB provides an exceptional solution regarding the dropped calls. As mentioned earlier, UWB can be performed both in underlay and overlay modes. Assuming that the normal operation mode is overlay, in cases when it becomes impossible to perpetuate the communication, UWB can switch to the underlay mode. Since the licensed systems are not affected by UWB when it is in the underlay mode, this gives the UWB the opportunity to maintain the communication link even though it is at a low quality.

The existence of licensed systems and other unlicensed users is not the only limitation regarding the secondary usage of the spectrum. The spectral masks that are imposed by the regulatory agencies (such as the FCC in the USA) are also determinative in spectrum usage in that they set a limit to the transmit power of wireless systems. UWB offers a satisfactory solution to the adaptable transmit power requirement of cognitive radio. Both UWB-OFDM and IR-UWB systems can comply with any set of spectral rules mandated upon the cognitive radio system by adapting their transmit power.

Since the cognitive radio concept includes free utilization of unused frequency bands, there will be a number of users willing to make use of the same spectrum opportunities at the same time. Therefore, cognitive radio networks should be able to provide access to multiple users simultaneously. During the operation of a cognitive radio, changes may occur in the overall spectrum occupancy, or the signal quality observed by each user can fluctuate because of various factors. These changes may require the cognitive radio to modify its multiple access parameters accordingly.

UWB is very flexible in terms of multiple access. In IR-UWB, by modifying the number of chips in a frame, the number of users can be determined. In UWB-OFDM, on the other hand, the subcarriers assigned to each user can be decreased in order to allow more users to communicate. Therefore, also from the point of adaptive multiple access, UWB proves to be a proper candidate for cognitive radio applications.

The primary objectives targeted with cognitive radio include preserving the privacy of information. UWB is one of the systems that have information security in their nature. If a UWB system is working in the underlay mode, because of the very low power level, it is impossible for unwanted users to detect even the existence of the UWB signals. Therefore, underlay UWB is a highly secure means of exchanging information. Overlay mode UWB, on the other hand, can also be considered a safe communication method. In overlay IR-UWB, multiple accessing is enabled either by time hopping or by direct sequencing. Therefore, receiving a user's information is only possible if the user's time hopping or spreading code is known. UWB-OFDM also provides security by assigning different subcarriers to different users. The level of security can be increased by periodically changing these subcarrier assignments. Apparently, UWB is a secure way of communicating in both its underlay and overlay modes. Hence, UWB can be considered a strong candidate for cognitive radio applications in terms of information security.

Being a future wireless concept, cognitive radio targets at a low cost for each of its components. This is necessary for the system to be able to reflect the profit earned by using the spectrum in an opportunistic way (rather than purchasing a license) to its subscribers. UWB signals can be generated and processed by inexpensive transceiver circuitries. The RF front-end required to send and capture UWB signals are also quite uncomplicated and inexpensive. Therefore, UWB communication can be accomplished by employing very low cost transmitter and receivers. This property of UWB makes it very attractive for cognitive radio, which aims at limited infrastructure and transceiver costs.

8.4 Cognitive UWB Network Related Issues

As it is pointed out throughout the previous section, UWB is highly competent in satisfying many basic requirements of cognitive radio. Therefore, employing UWB in cognitive radio networks could be very instrumental for the successful penetration of cognitive radio into the wireless world. Nevertheless, since today's spectrum regulations prohibit employing UWB systems in the overlay mode, UWB based implementation of cognitive radio might not become a reality in the near future. However, besides being a strong candidate for practical cognitive radio implementation, UWB can be considered as a supplement to cognitive radio systems that are realized by means of other wireless technologies. Therefore, it can be concluded that this way or the other, UWB will be an inseparable part of cognitive radio applications.

UWB can offer various kinds of support to cognitive radio network. These include sharing the spectrum sensing information via UWB, locating the cognitive nodes in a cognitive network by means of IR-UWB, and sensing the physical environment/channel with IR-Radar. In the following, various cognitive UWB networks related issues including these supplementary uses of UWB will be discussed.

8.4.1 Spectrum Sensing Information Exchange in Cognitive Networks

In order to be able to opportunistically utilize the available licensed frequency bands, cognitive radio systems periodically scan their target spectrum and detect the spectrum opportunities. In cognitive networks, it is mandatory that all nodes agree on the spectral opportunities to be utilized. Therefore, it is a major issue for a cognitive node to share the spectrum sensing information with the other nodes. In some works in the literature, it is considered to have an allocated control channel to transmit this information [33]. In some other works, it is proposed to have a centralized controller that gathers this information, decides for spectrum availability, and allocates distinct bands to different cognitive users [8, 10]. An alternative to these methods is to transmit spectrum sensing results via low power UWB signaling that complies with the FCC regulations [34]. Since this transmission will be accomplished in an underlay manner, it can be done simultaneously with the real data communication without affecting it regardless of the wireless technology employed to realize the cognitive radio itself. Considering the relatively low throughput needed to transmit the sensing information as well as the low cost transceiver requirement, it turns out to be a proper option to use an uncomplicated non-coherent receiver such as an energy detector, and to employ on-off keying (OOK) modulation.[3] By using a proper mapping scheme (from sensing information to binary codewords), coding, and OOK modulation, spectrum information can be conveniently shared between the nodes.

A cognitive network (see Fig. 8.4) can be realized by allowing its nodes to communicate with each other using UWB to exchange spectrum information. One of the aims of cognitive radio is to increase the range of communication as much as possible, and at the first glance, UWB signaling may not seem to be very appropriate for this purpose because of the limited range of underlay UWB. The answer to this question can be obtained by looking at the bit error rate (BER) expression for OOK modulated UWB signals. This BER expression can be stated as

$$\text{BER} = Q\left(\sqrt{\frac{N_s A E_p}{2N_0}}\right) \tag{8.1}$$

where N_s is the number of pulses per symbol, A is the pulse amplitude, E_p is the normalized pulse energy, and the additive white Gaussian noise (AWGN) has a double sided spectrum of $\frac{N_0}{2}$. In this expression, it is seen that increasing the number of

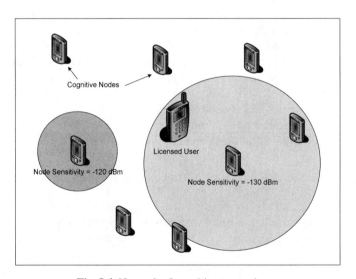

Fig. 8.4. Network of cognitive transceivers.

---

[3] The implementation issues regarding the OOK based energy detector receivers such as estimating the optimal threshold and determining the optimum integration interval can be found in [24].

pulses per symbol results in lower BER. Increasing N_s requires a repeated transmission of data, i.e. processing gain. By applying the necessary amount of processing gain, it can be made possible that the farthest nodes in a cognitive network can share the spectrum sensing information. Although this comes at the expense of lowered throughput, it is not a limiting factor in this case because a quite low data rate is enough to transmit the spectrum sensing information. By enabling all the nodes in a cognitive network to talk to each other via UWB, there is no need

- either to allocate a separate channel for sharing the sensing information,
- or to employ a centralized controller that collects this information, processes it, and sends it to the cognitive users in the network.

Spectrum information can also be shared in an ad-hoc multi-hopping scheme that uses UWB. This way, long range transmission is not needed. Multiple nodes collaboratively share the information and route this information to other nodes using low power, low cost UWB technology. In essence, a UWB based sensor network with the collaboration of multiple radios is formed.

8.4.2 Receiving Sensitivity of Cognitive Nodes and Size of Cognitive Networks

In cognitive radio networks, in order to make sure that the intended frequency spectrum is being used by its licensed user, all nodes involved in communication have to scan the spectrum and inform each other about the spectral conditions. It is not hard to imagine that there should not be a physical gap between the sensing ranges of the nodes. If the sensing ranges are not at least partially overlapping, there is always a risk that a licensed user located inside the gap between the sensing ranges is not detected, which would cause the cognitive nodes to jam the licensed user's signal. Therefore, the receiving sensitivity of the nodes in the network has an integral role in determining the range of communication. As an example, assume a rather high sensitivity around -120 dBm to -130 dBm and consider free space propagation, in which the transmitted power (P_{tx}) and received power (P_{rx}) are related to each other by the Friis equation (ignoring the system loss and antenna gains)

$$P_{rx} = \frac{P_{tx}\,\lambda^2}{(4\pi)^2\,d^2} \tag{8.2}$$

where λ is the wavelength and d is the distance. With these assumptions, it is seen that the distance between two cognitive nodes can go up to 50–150 m, getting the cognitive network classified as a medium-sized network according to its coverage area.

The sensing information received from all the other nodes in the network can be combined in each node, and pulse design can be done according to the common white (unused) spaces. Increasing the network size results in an increased probability of overlapping with licensed systems. This fact sets a practical limit to the size of the cognitive network, because continuing to enlarge the network, the common available spectra become less and less, and after some point their amount becomes insufficient to ensure the minimum QoS. For the details of how the common white bands are going to be shared by the cognitive nodes in the network, the readers are referred to [9, 11] and [22].

8.4.3 Locating the Cognitive Nodes via IR-UWB

Owing to the extremely wide band they occupy, IR-UWB systems have an advanced multipath resolving capability. This desirable feature enables these systems to be considered as a means of highly accurate (centimeter range) positioning besides being communication systems [16]. Because of this reason, IR-UWB is the primary candidate for the IEEE 802.11.4a standardization group, which aims at determining a new physical layer for very low power, low data rate communications with a special emphasis on accurate location finding.

The positioning capability can make IR-UWB systems an excellent supplement for small sized cognitive networks. Since such networks aim at not interfering with other radios in their physical environment, it can be very beneficial for them to be able to determine the locations of the nodes in the network closely. Having information about the precise locations of the nodes in a cognitive network, accurate and high efficiency beamforming [20] can be achieved towards the direction of the target nodes. Also, spatial nulls can be generated towards undesired receivers/signal sources to avoid interference. Beamforming can be accomplished by planar antenna arrays, which can be put onto very small areas for high-frequency systems (such as 60 GHz radios), and these arrays can be employed even by wireless nodes that are smaller than a hand palm in size.

The accurate positioning capability of IR-UWB can also be utilized to determine the transmit power adaptively. Using the positioning data, the distance between the transmitting and receiving nodes can be found, and based on the distance information the radiated power can be set. Such an implementation would not only optimize the power consumption but also help to ensure the link quality between the distant nodes.

Another nice utilization of the positioning capability can be tracking cognitive nodes or devices that are mobile. Updating the corresponding positioning information in a frequent manner, a cognitive node can be tracked in space. This way, any communication link directed to it would not be lost although its location is changing continuously.

Examples of using the positioning feature to augment the cognitive communication quality can be increased. All these examples lead to the idea that IR-UWB can leverage cognitive radio networks by providing a very strong support through its accurate positioning capability.

8.4.4 Sensing the Physical Environment of Cognitive Radio Network with Impulse Radar

Among the various impulse radio UWB applications, impulse radar is one of the oldest, and it has been used especially for military purposes [29, 40]. Practical implementations of impulse radar have been addressed in [4, 31–41]. As in the case of the other IR-UWB applications mentioned so far, impulse radar can improve cognitive communications from a number of aspects when combined with cognitive radio systems.

One of the potential uses of impulse radar can be to determine objects and walls in the indoor environments. Determining the objects can yield a rough estimation of the directions of multipath components, which can improve the channel estimation. Determining the walls, on the other hand, yields information about the physical borders of an indoor network, which may be very useful when establishing a cognitive network.

In mobile applications, impulse radar can allow to estimate the speed of the mobile users, it can enable a cognitive mobile device to measure its own speed. Such a capability would result in being able to estimate the Doppler spread and the channel coherence time, which are important parameters to know in mobile communications.

Impulse radar can also be used to detect the movement of human beings in the wireless channel, which can be very effective on the link quality between cognitive nodes especially for extremely high-frequency systems such as the 60 GHz radios [15].

8.4.5 A Cognitive Network Case Study

In order to provide a case study, computer analysis and simulations are performed regarding the practical implementation of a cognitive radio network. These simulations are related to the transmission of spectrum sensing results via UWB, the range of a cognitive network, and the capability of a cognitive network to detect a licensed system. In the simulations regarding the UWB signaling, the channel model $CM3$ in [13], which corresponds to an office environment with line-of-sight (LOS), is utilized. All parameters used in these simulations are listed in Table 8.1.

Table 8.1. List of simulation parameters for UWB signaling.

Parameter	Value
-10 dB Bandwidth	500 MHz
Freq. range	3.1–3.6 GHz
Geometric center freq.	3.34 GHz
Channel model	Office LOS (CM3)
Reference path loss	35.4 dB
Path loss exponent (n)	1.63
Rx antenna noise fig.	17 dB
Implementation loss	3 dB
Throughput (R_b)	20 Mbps
Integration interval	30 ns

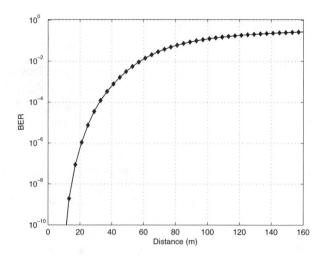

Fig. 8.5. BER vs. distance between the nodes for UWB signaling.

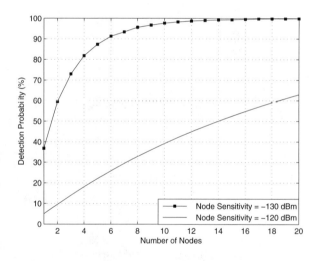

Fig. 8.6. Probability of the licensed transmitter being detected by the cognitive network.

A theoretical analysis is performed to investigate the performance of OOK modulated UWB data transmission, which is used to share the spectrum sensing results in a cognitive network, depending on the distance between a cognitive transmitter–receiver pair. According to [13] the path loss assumed can be shown as

$$L(d) = L_0 + 10n \, \log_{10} \left(\frac{d}{d_0} \right) \tag{8.3}$$

where the reference distance (d_0) is set as 1 m, L_0 is the path loss at d_0, and n is the path loss exponent. The average noise power per bit is

$$N = -174 + 10 \log_{10}(R_b) \tag{8.4}$$

where R_b is the throughput. In Fig. 8.5, the effect of distance on the probability of error is demonstrated. The results show that the BERs obtained for up to 40 m are still acceptable. For further distances, however, some processing gain is definitely needed.

Another simulation is done to investigate the effect of the number of nodes on the probability of a licensed system being detected by the cognitive network. Figure 8.4 demonstrates a network composed of cognitive radio devices. The nodes in the network are randomly distributed in a 200 m × 200 m area inside a building. It is assumed that there is a licensed transmitter, which is a $GSM\,900$ cell phone transmitting at -60 dBm, whose location is random, as well. Depending on the level of the node sensitivity, the number of nodes required to make a reliable detection might vary. The results of this simulation are demonstrated in Fig. 8.6. It is seen that an increase in the number of nodes certainly increases the detection probability. However, a low node sensitivity may lead to a considerably high number of nodes to be employed.

Conclusion

In this chapter, the attractiveness of the UWB technology for purposes of implementing cognitive radio networks is investigated from two main approaches. The first one considers UWB as a direct means of practical cognitive radio realization. Under this approach, the UWB features such as negligible interference to licensed systems, dynamically adjustable bandwidth and data rate, and adaptive transmit power and multiple access are discussed emphasizing their closeness to the cognitive radio requirements. UWB is shown to be a proper candidate for implementing the cognitive radio networks. The concern regarding UWB's being the technology of cognitive radio is that the overlay mode operation of UWB is currently not allowed by regulatory agencies. Therefore, this option may have to be deferred until it is proven that licensed systems can co-exist with specifically designed overlay UWB systems that have advanced sensing and spectrum shaping capabilities.

The second approach considers UWB as a source of supplementary uses for cognitive radio networks. Among the numerous uses that will enhance cognitive communications, some significant ones such as sharing the spectrum sensing information via UWB, locating the cognitive nodes using UWB, and providing awareness via impulse radar are addressed in this chapter.

It should be emphasized that even in the case that the impulse radio UWB is not accepted as the means of implementing the cognitive radio networks, its supplementary uses are so beneficial that UWB cannot be separated from cognitive radio systems of future.

References

1. J. Mitola, "Cognitive radio for flexible mobile multimedia communications," in *Proc. Mobile Multimedia Commun. (MoMuC '99)*, pp. 3–10, Nov. 1999.
2. W. D. Horne, "Adaptive spectrum access: Using the full spectrum space," in *Proc. 31st Annual Telecommun. Policy Res. Conf. (TPRC 03)*, Oct. 2003.
3. M. Z. Win and R. A. Scholtz, "Impulse radio: How it works," *IEEE Commun. Lett.*, vol. 2, pp. 36–38, Feb. 1998.
4. M. Win and R. A. Scholtz, "Ultra-wide bandwidth time-hopping spread-spectrum impulse radio for wireless multiple-access communications," *IEEE Trans. Commun.*, vol. 48, pp. 679–689, Apr. 2000.
5. I. Guvenc and H. Arslan, "On the modulation options for UWB systems," in *Proc. IEEE Military Commun. Conf.*, vol. 2, (Boston, MA), pp. 892–897, Oct. 2003.
6. R. Scholtz, "Multiple access with time-hopping impulse modulation," in *Proc. IEEE Military Commun. Conf.*, vol. 2, (Boston, MA), pp. 447–450, Oct. 1993.
7. G. Lu, P. Spasojevic, and L. Greenstein, "Antenna and pulse designs for meeting UWB spectrum density requirements," in *Proc. IEEE Ultrawideband Syst. Technol. (UWBST)*, (Reston, VA), pp. 162–166, Nov. 2003.
8. J. Hillenbrand, T. Weiss, and F. Jondral, "Calculation of detection and false alarm probabilities in spectrum pooling systems," *IEEE Comm. Lett.*, vol. 9, pp. 349–351, Apr. 2005.
9. R. Thomas, L. DaSilva, and A. MacKenzie, "Cognitive networks," in *Proc. IEEE Int. Symp. Dynamic Spectrum Access Networks (DySPAN) 2005*, (Baltimore, MD), pp. 352–360, Nov. 2005.
10. G. Ganesan and Y. Li, "Cooperative spectrum sensing in cognitive radio networks," in *Proc. IEEE Int. Symp. Dynamic Spectrum Access Networks (DySPAN) 2005*, (Baltimore, MD), pp. 137–143, Nov. 2005.
11. H. Zheng and L. Cao, "Device-centric spectrum management," in *Proc. IEEE Int. Symp. Dynamic Spectrum Access Networks (DySPAN) 2005*, (Baltimore, MD), pp. 56–65, Nov. 2005.
12. D. Sostanovsky and A. Boryssenko, "Experimental UWB sensing and communication system," *IEEE Aerospace Electron. Syst. Mag.*, vol. 21, pp. 27–29, Feb. 2006.
13. A. Molisch, K. Balakrishnan, C. C. Chong, S. Emami, A. Fort, J. Karedal, J. Kunisch, H. Schantz, U. Schuster, and K. Siwiak, "IEEE 802.15.4a channel model – final report," Sep. 2004.
14. Y. Wu, A. Molisch, S.-Y. Kung, and J. Zhang, "Impulse radio pulse shaping for ultra-wide bandwidth (UWB) systems," in *Proc. IEEE Int. Symp. on Personal, Indoor and Mobile Radio Commun. PIMRC 2003*, vol. 1, (Beijing, China), pp. 877–881, Sep. 2003.
15. S. Collonge, G. Zaharia, and G. Zein, "Influence of the human activity on wideband characteristics of the 60 GHz indoor radio channel," *IEEE Trans. Wireless Commun.*, vol. 3, pp. 2389–2406, Nov. 2004.
16. S. Gezici, Z. Tian, G. Giannakis, H. Kobayashi, A. Molisch, H. Poor, and Z. Sahinoglu, "Localization via ultra-wideband radios: A look at positioning aspects for future sensor networks," *IEEE Signal Proc. Mag.*, vol. 22, pp. 70–84, July 2005.
17. I. Bergel, E. Fishler, and H. Messer, "Narrowband interference mitigation in impulse radio," *IEEE Trans. Commun.*, vol. 53, pp. 1278–1282, Aug. 2005.
18. R. Dilmaghani, M. Ghavami, B. Allen, and H. Aghvami, "Novel UWB pulse shaping using prolate spheroidal wave functions," in *Proc. IEEE Int. Symp. on Personal, Indoor and Mobile Radio Commun. PIMRC 2003*, vol. 1, (Beijing, China), pp. 602–606, Sept. 2003.

19. A. Taha and K. Chugg, "On designing the optimal template waveform for UWB impulse radio in the presence of multipath," in *Proc. IEEE Ultrawideband Syst. Technol. (UWBST)*, (Baltimore, MD), pp. 41–45, May 2002.

20. S. K. Yong, M. E. Sahin, and Y. H. Kim, "On the effects of misalignment and angular spread on the beamforming performance," in *Proc. IEEE Consumer Commun. and Networking Conf. (CCNC)*, (Las Vegas, NV), Jan. 2007.

21. R. Tesi, M. Hamalainen, J. Iinatti, J. Oppermann, and V. Hovinen, "On the multi-user interference study for ultra wideband communication systems in AWGN and modified Saleh-Valenzuela channel," in *Proc. Ultra Wideband Systems, 2004. Joint with Conference on Ultrawideband Systems and Technologies. Joint UWBST & IWUWBS. 2004 International Workshop on*, pp. 91–95, May 2004.

22. S. Mangold, A. Jarosch, and C. Monney, "Operator assisted cognitive radio and dynamic spectrum assignment with dual beacons – detailed evaluation," in *Proc. Commun. Syst. Software and Middleware (Comsware 2006)*, pp. 1–6, Jan. 2006.

23. X. Wu, Z. Tian, T. Davidson, and G. Giannakis, "Optimal waveform design for UWB radios," *IEEE Trans. on Signal Proc.*, vol. 54, pp. 2009–2021, June 2006.

24. M. E. Sahin, I. Guvenc, and H. Arslan, "Optimization of energy detector receivers for UWB systems," in *Proc. IEEE Vehic. Technol. Conf.*, vol. 2, (Stockholm, Sweden), pp. 1386–1390, May 2005.

25. L. Piazzo and F. Ameli, "Performance analysis for impulse radio and direct-sequence impulse radio in narrowband interference," *IEEE Trans. Commun.*, vol. 53, pp. 1571–1580, Sep. 2005.

26. K.Wallace, B. Parr, B. Cho, and Z. Ding, "Performance analysis of a spectrally compliant ultra-wideband pulse design," *IEEE Trans. Wireless Commun.*, vol. 4, pp. 2172–2181, Sept. 2005.

27. S. Gezici, H. Kobayashi, H. Poor, and A. Molisch, "Performance evaluation of impulse radio UWB systems with pulse-based polarity randomization," *IEEE Trans. Signal Process.*, vol. 53, pp. 2537–2549, Jul. 2005.

28. L. Zhao and A. Haimovich, "Performance of ultra-wideband communications in the presence of interference," *IEEE J. Select. Areas Commun.*, vol. 20, no. 9, pp. 1684–1691, 2002.

29. R. Fontana, "Recent system applications of short-pulse ultra-wideband (UWB) technology," *IEEE Trans. Microwave Theory Tech.*, vol. 52, pp. 2087–2104, Sep. 2004.

30. M. Mahfouz, A. Fathy, Y. Yang, E. Ali, and A. Badawi, "See-through-wall imaging using ultra wideband pulse systems," in *Proc. 34th Applied Imagery Pattern Recognition Workshop*, (Washington, DC), Oct. 2005.

31. I. Immoreev, S. Samkov, and T.-H. Tao, "Short-distance ultra wideband radars," *IEEE Aerospace Electron. Syst. Mag.*, vol. 20, pp. 9–14, June 2005.

32. Y.-P. Nakache and A. Molisch, "Spectral shaping of UWB signals for time-hopping impulse radio," *IEEE J. Select. Areas Commun.*, vol. 24, pp. 738–744, Apr. 2006.

33. X. Jing and D. Raychaudhuri, "Spectrum co-existence of IEEE 802.11b and 802.16a networks using the CSCC etiquette protocol," in *Proc. IEEE Int. Symp. Dynamic Spectrum Access Networks (DySPAN) 2005*, (Baltimore, MD), pp. 243–250, Nov. 2005.

34. M. E. Sahin and H. Arslan, "System design for cognitive radio communications," in *Proc. Cognitive Radio Oriented Wireless Networks and Commun. (CrownCom)*, (Mykonos Island, Greece), June 2006.

35. J. Foerster, "The performance of a direct-sequence spread ultrawideband system in the presence of multipath, narrowband interference, and multiuser interference," in *Proc. Ultra Wideband Systems and Technologies, IEEE Conf. on*, pp. 87–91, 2002.

36. M. G. Di Benedetto and L. De Nardis, "Tuning UWB signals by pulse shaping," *Special Issue on Signal Proc. in UWB Commun., Eurasip Journal on Signal Proc.*, Elsevier Publishers, 2005.
37. H. Arslan and M. E. Sahin, *Ultra Wideband Wireless Communication, ch. Narrowband Interference Issues in Ultrawideband Systems*. Hoboken, NJ: Wiley, Sept. 2006.
38. M. Win and R. Scholtz, "Ultra-wide bandwidth time-hopping spread-spectrum impulse radio forwireless multiple-access communications," *IEEE Trans. Commun.*, vol. 48, no. 4, pp. 679–689, 2000.
39. S. Gezici, Z. Sahinoglu, H. Kobayashi, and H. Poor, "Ultra-wideband impulse radio systems with multiple pulse types," *IEEE J. Select. Areas Commun.*, vol. 24, pp. 892–898, Apr. 2006.
40. I. Immoreev and J. Taylor, "Ultrawideband radar special features & terminology," *IEEE Aerospace Electron. Syst. Mag.*, vol. 20, pp. 13–15, Mar. 2005.
41. A. Yarovoy, L. Ligthart, J. Matuzas, and B. Levitas, "UWB radar for human being detection," *IEEE Aerospace Electron. Syst. Mag.*, vol. 21, pp. 10–14, Mar. 2006.

9

Degrees of Cooperation in Dynamic Spectrum Access for Distributed Cognitive Radios *

Zhu Han

Electrical and Computer Engineering Department,
Boise State University, Boise, ID 83725, USA
zhuhan@boisestate.edu

9.1 Introduction

The Federal Radio Act under Federal Communications Commission (FCC) allows predetermined users the right to transmit at a given frequency. Non-licensed users are regarded as "harmful interference" and not allowed to transmit in a certain frequency bands. As the demands for wireless communication become more and more pervasive, the wireless devices must find a way for the right to transmit at frequencies in the extremely limited radio band. However, there exist a large number of frequency bands that have considerable, and sometimes periodic, dormant time intervals. In the literature, those frequency bands refer to spectrum holes [1,2]. So there is a dilemma that on one hand the mobile users have no spectrum to transmit, while on the other hand some spectrums are not fully utilized.

In order to cope with the dilemma, the FCC has recently investigated the efficient spectrum usage for cognitive radios, which is a novel paradigm that improves the spectrum utilization by allowing secondary networks (users) to borrow unused radio spectrum from primary licensed networks (users) or to share the spectrum with the primary networks (users). As an intelligent wireless communication system, cognitive radios are aware of the radio frequency environment, select the communication parameters (such as carrier frequency, bandwidth and transmission power) to optimize the spectrum usage and adapt the transmission and reception accordingly. Cognitive radios can bring a variety of benefits: for a regulator, cognitive radios can significantly increase in spectrum availability for new and existing applications. For a license holder, cognitive radios can reduce the complexity of frequency planning, facilitate the secondary spectrum market agreements, increase system capacity through access to more spectrum and avoid interference. For equipment manufacturers, cognitive radios can increase demands for wireless devices. Finally, for a user, cognitive radios can bring more capacity per user, enhance inter-operability

---

* Table 9.2 and Figs. 9.1–9.7 reprinted, with permission, from [3–7] ©[2004], [2005], [2007] IEEE.

and bandwidth-on-demand and provide ubiquitous mobility with a single user device across disparate spectrum access environments.

The process for spectrum access is first to sense what the available spectrum is, then to get access to some of the available spectrum, next to use the available spectrum and finally to release the used spectrum. Significant research is necessary to investigate how to dynamically access the spectrum, which enables the opportunistic management of radio resources within a single access system or between different radio access systems. As a result, dynamic spectrum access can improve spectral efficiency, increased capacity and improve ease of access to the spectrum. In the literature, much work [8,9] has been done for dynamic spectrum access.

In this chapter, we classify some of the dynamic spectrum access techniques for cognitive radios, according to the degrees of cooperation. The relations between distributed cognitive radios ranges from complete autonomy and non-cooperation, to full obeyance to the centralized controller. Specifically, we will discuss the following techniques for different degrees of cooperation:

1. Non-cooperative competition (Sect. 9.2)
2. Learning for better equilibria (Sect. 9.3)
3. Referee mediation (Sect. 9.4)
4. Threat and punishment from repeated interactions (Sect. 9.5)
5. Spectrum auction (Sect. 9.6)
6. Mutual benefits via bargaining (Sect. 9.7)
7. Contract using cooperative game (Sect. 9.8)
8. centralized scheme (Sect. 9.9)

There are some tradeoffs for different types of approaches. For example, for non-cooperative competition, the transceiver is simple but the performance can be inferior due to the extensive non-cooperation. On the other hand, the centralized scheme can achieve the optimal solution, but it is necessary for extensive measuring of channels and signaling to exchange channel information. Our goals are to investigate those different approaches with different degrees of cooperation, study in which network scenarios the approaches fit most and understand the underlying tradeoffs for the wireless cognitive network design.

9.2 Non-cooperative Competition

In cognitive wireless networks, it is hard for an individual cognitive user to know the channel conditions of the other users. The cognitive users cannot cooperate with each other for spectrum usage. They act selfishly to maximize their own performances in a distributive fashion. Such a fact motivates us to adopt game theory. Dynamic spectrum access can be modeled as a game that deals largely with how rational and intelligent individuals interact with each other in an effort to achieve their own goals. In this game, each cognitive user is self-interested and trying to optimize its utility function, where the utility function represents the cognitive user's performance and controls the outcomes of the game. There are many advantages of applying game theory to dynamic spectrum usage for cognitive radios:

1. *Only local information and distributive implementation*: The individual cognitive user observes the outcome of the game and adjusts only its parameters in response to optimize its own benefit. As a result, there is no need for collecting all the information and conducting optimization in a centralized way.
2. *More robust outcome*: For the centralized optimization, if the information for optimization is not quite accurately obtained, the optimized results can be far away from optimality. In contrast, the local information is always accurate, so the outcome of the distributed game approaches is robust.
3. *Combinatorial nature*: For traditional optimization technique such as programming, it is hard to handle the combinatorial problems. For game theory, it is natural to discuss the problem in a discrete form. In the problems such as spectrum access, to analyze the combinatorial problems by game theory is considerably convenient.
4. *Rich mathematics for optimization*: There are many mathematical tools available to analyze the outcome of the game. Specifically, if the (non-cooperative) game is played once, the static game can be studied. If the game is played multiple times, dynamic game theory is employed. If some contracts and mutual benefits can be obtained, cooperative game explains how to divide the profits. Auction theory studies the behaviors of both seller and bidder. We will study some of those techniques in the following sections.

Next, we define some basic game concepts and study two ways to present a game. Then we give some properties of the game, such as dominance, Nash equilibrium, Pareto optimality and mixed strategies. Further, we discuss the low efficiency of the outcome for non-cooperative static games. Finally, some methods are briefly discussed to improve the game outcomes.

A game can be roughly defined as each user adjusts its strategy to optimize its own utility to compete with others. Strategy and utility can be defined as:

Definition 9.1. *A strategy σ is a complete contingent plan, or a decision rule, that defines the action an agent will select in every distinguishable state Ω of the world.*

Definition 9.2. *In any game, utility (payoff) u represents the motivations of players. A utility function for a given player assigns a number for every possible outcome of the game with the property that a higher (or lower) number implies that the outcome is more preferred.*

One of the most common assumptions made in game theory is rationality. Generally speaking, rationality implies that all players are motivated by maximizing their own utilities. In a stricter sense, it implies that every player always maximizes its utility, thus being able to perfectly calculate the probabilistic result of every action. A game can be defined as follows.

Definition 9.3. *A game G in the strategic form has three elements: the set of players $i \in \mathcal{I}$, which is a finite set $\{1, 2, ..., K\}$; the strategy space Ω_i for each player i; and* *utility function u_i, which measures the outcome of the ith user for each strategy profile $\boldsymbol{\sigma} = (\sigma_1, \sigma_2, ..., \sigma_K)$. We define $\boldsymbol{\sigma}_{-i}$ as the strategies of player i's opponents, i.e., $\sigma_{-i} = (\sigma_1, ..., \sigma_{i-1}, \sigma_{i+1}, ..., \sigma_K)$. In static games, the interaction between users occurs only once, while in dynamic games the interaction occurs several times.*

One of the most simple games is the non-cooperative static game which can be presented by the strategic (normal) form.

Definition 9.4. *A non-cooperative game is one in which players are unable to make enforceable contracts outside of those specifically modeled in the game. Hence, it is not defined as games in which players do not cooperate, but as games in which any cooperation must be self-enforcing.*

Definition 9.5. *A static game is one in which all players make decisions (or select a strategy) simultaneously, without knowledge of the strategies that are being chosen by other players. Even though the decisions may be made at different points in time, the game is simultaneous because each player has no information about the decisions of others; thus, it is as if the decisions are made simultaneously.*

Definition 9.6. *The strategic (or normal) form is a matrix representation of a simultaneous game. For two players, one is the "row" player, and the other, the "column" player. Each row or column represents a strategy and each box represents the payoffs to each player for every combination of strategies.*

To analyze the outcome of the game, the Nash equilibrium is a well-known concept which states that in the equilibrium every agent will select a utility-maximizing strategy given the strategies of every other agent.

Definition 9.7. *Define a strategy vector $\sigma = [\sigma_1 \ldots \sigma_K]$ and define the strategy vector of the ith player's opponents as $\sigma_i^{-1} = [\sigma_1 \ldots \sigma_{i-1} \ \sigma_{i+1} \ldots \sigma_K]$, where K is the number of users and σ_i is the ith user's strategy. u_i is the ith user's utility. Nash equilibrium point σ^* is defined as:*

$$u_i(\sigma_i^*, \sigma_i^{-1}) \geq u_i(\tilde{\sigma}_i, \sigma_i^{-1}), \ \forall i, \ \forall \tilde{\sigma}_i \in \Omega, \ \sigma_i^{-1} \in \Omega^{K-1}, \tag{9.1}$$

i.e., given the other users' resource allocations, no user can increase its utility alone by changing its own resource allocation.

In other words, a Nash equilibrium, named after John Nash, is a set of strategies, one for each player, such that no player has incentive to unilaterally change its action. Players are in an equilibrium if a change in strategies by any one of them will lead that player to earn less than if it remains with its current strategy.

Until now, we have only discussed the strategy that is deterministic, or pure strategy. A pure strategy defines a specific move or action that a player will follow in every possible attainable situation in a game. Such moves may not be random, or drawn from a distribution, as in the case of mixed strategies.

Definition 9.8. *Mixed Strategy: A strategy consisting of possible moves and a probability distribution (collection of weights) which corresponds to how frequently each move is about to play. A player will only use a mixed strategy when it is indifferent about several pure strategies. Moreover, if the opponent can benefit from knowing the next move, the mixed strategy is preferred since keeping the opponent guessing is desirable.*

There might be an infinite number of Nash equilibriums. Among all these equilibriums, we need to select the optimal one. There are many criteria by which to judge if the equilibrium is optimal or not. Among these criteria, Pareto optimality is one of the most important definitions.

Definition 9.9. *Pareto optimal: Named after Vilfredo Pareto, Pareto optimality is a measure of efficiency. An outcome of a game is Pareto optimal if there is no other outcome that makes every player at least as well off and at least one player strictly better off. That is, a Pareto optimal outcome cannot be improved upon without hurting at least one player. Often, a Nash equilibrium is not Pareto optimal implying that the players' payoffs can all be increased.*

Since the individual user has no incentive to cooperate with the other users in the system and imposes harm to the other users, the outcome of the non-cooperative static game might not be optimal from the system point of view. To overcome this problem, *pricing (or taxation)* has been used as an effective tool both by economists and researchers in computer networks. The pricing technique is motivated by the following two objectives:

1. The revenue for the system is optimized.
2. The cooperation for resource usage is encouraged.

An efficient pricing mechanism can make the distributed decisions compatible with the system efficiency obtained by centralized control. A pricing policy is called *incentive compatible*, if pricing enforces a Nash equilibrium that achieves the system optimum. Specifically, the new utility with pricing is

$$u' = u - \alpha Q \tag{9.2}$$

where u is the original utility, α the price for user's resource Q and the price can be different for different users. It is known that the above utility function can achieve Pareto optimality, if the utility is quasi-convex or quasi-concave.

In cognitive radio literature, it is worth mentioning the following games that can be modeled for spectrum access technology. First, a game is considered a *potential game* if the incentive of all players to change their strategy can be expressed in one global function, the potential function. The potential function is a useful tool to analyze equilibrium properties of games, since the incentives of all players are mapped into one function, and the set of pure Nash equilibria can be found by simply locating the local optima of the potential function. In [10], a potential game was utilized for problems such as interference avoidance.

In [11], a *Cournot game* was used to model the spectrum sharing problem as an oligopoly market in which a few firms compete with each other in terms of amount of commodity supplied to the market to gain the maximum profit. In this case the secondary users are analogous to the firms who compete for the spectrum offered by the primary user and the cost of the spectrum is determined by using a pricing function. Both static and dynamic Cournot games were investigated.

In some cognitive scenarios, the primary users and secondary users can be formulated as the multiple level market game, so that both types of users can be satisfied with the shared spectrum size and the charge pricing. The available techniques are the demand-and-request functions [17], Stackleberge game [13] and so on. The multiple level game is non-cooperative game and Nash equilibria can be derived for spectrum usage.

9.3 Correlated Equilibrium Through Learning

One of the major design challenges for cognitive radios is to coordinate and cooperate in accessing the spectrum opportunistically among multiple distributive users with only local information. In this section, we study the behavior of an individual distributed secondary user to control its rate when the primary user is absent. Each secondary user seeks to maximize its rates over different channels. However, excessive transmissions can cause the collisions with the other secondary users. The collisions reduce not only the system throughput but also individual performances. We propose a new solution concept, the correlated equilibrium, which is better compared to the non-cooperative Nash equilibrium in terms of spectrum utilization efficiency and fairness among the distributive users. Using the correlated equilibrium concept, the distributive users adjust their transmission probabilities over the available channels, so that the collisions are avoided and the users' benefits are optimized. We exhibit the adaptive no-regret algorithm [24] to learn the correlated equilibrium in a distributed manner. We show that the proposed learning algorithm converges to a set of correlated equilibria with probability one.

For the system model, we consider the general models for dynamic opportunistic spectrum access for cognitive radios, in which there exist several primary users with a set of available channels and a large number of secondary users. The channel availability of secondary users inherently depends on the activities of the primary users. Moreover, the secondary users have to compete for the idle channels among the interfering secondary users. If collisions occur, there are some penalties in the forms of packet loss and power waste. This is the major focus here. We consider that there are N channels in the wireless network. Without loss of generality, each channel has a unit bandwidth. These channels are shared among M primary users and K secondary users seeking channel access opportunistically.

For adjacent secondary users, they can interference with each other. We use *interference matrix* $\mathbf{L}$ to depict the interference graph. The interference matrix has the dimension of K by K, and its elements are defined as

$$\mathbf{L}_{ij} = \begin{cases} 1, & \text{if } i \text{ and } j \text{ interfere with each other} \\ 0, & \text{otherwise.} \end{cases} \tag{9.3}$$

The interference matrix depends on the relative location of secondary users.

Next, we define *channel availability matrix* as a K by N matrix, $\mathbf{A}(t)$. Each user can transmit over a specific channel with a set of different rates. The elements of the matrix are defined as

$$\mathbf{A}_{in}(t) = \begin{cases} 1, & \text{if channel } n \text{ is available for secondary user } i \text{ at time } t \\ 0, & \text{otherwise.} \end{cases} \tag{9.4}$$

We note that the channel availability matrix $\mathbf{A}(t)$ varies over time. This matrix is the result of a sensing task done by secondary users and depends on the primary users' traffic, relative location between the secondary users and the primary users. Notice that each individual secondary user only knows its corresponding row of matrix $\mathbf{L}$ and $\mathbf{A}(t)$.

Define the set of secondary user i as I which is the finite set $\{1, 2, \ldots, K\}$. For each available channel, a secondary user can select $L+1$ discrete rates $\Upsilon = \{0, \upsilon_1, \ldots, \upsilon_L\}$. The strategy space Ω_i for secondary user i is on the available channels and can be denoted as $\Omega_i = \prod_{n=1}^{N} \Upsilon^{\mathbf{A}_{in}}$. The action of user is $r_i^n = \upsilon_l$ representing secondary user i occupies channel n by rate υ_l. We define the strategy profile $\mathbf{r}^n = (r_1^n, r_2^n, \ldots, r_K^n)'$, and we define $\mathbf{r}_{-i}^n$ as the strategies of user i's opponents (interference neighbors defined in $\mathbf{L}$) for channel n. We also define $\mathbf{r}_i = (r_i^1, \ldots, r_i^N)'$ as the action of secondary users over all channels, and $\mathbf{r}_{-i}$ as the secondary user i's opponents' actions.

The utility function U_i measures the outcome of secondary user i for each strategy profile $\mathbf{r}^1, \ldots, \mathbf{r}^N$ over different channels. We define the utility function as the maximum achievable rate for the secondary users over all the available channels as:

$$U_i = \sum_{n=1}^{N} \mathbf{A}_{in} R_i(r_i^n, \mathbf{r}_{-i}^n) \tag{9.5}$$

where $R_i(r_i^n, \mathbf{r}_{-i}^n)$ is the outcome of resource competition for user i and the other users. Notice that the utility function represents the maximum achievable rate. In practice, the secondary users need not occupy all the available channels.

We consider un-slotted One-persistent CSMA as the random multiple access protocol for the secondary users. Since the channel can be occupied by the primary user again in the near future, each secondary user transmits whenever the channel is idle. From [14], we have

$$R_i(r_i^n, \mathbf{r}_{-i}^n) = \begin{cases} \dfrac{r_i^n S^n}{\sum_i r_i^n}, & \text{if } G \le G_0 \\ 0, & \text{otherwise} \end{cases} \tag{9.6}$$

where

$$S^n = \frac{G^n[1 + G^n + \tau G^n(1 + G^n + \tau G^n/2)]e^{-G^n(1+2\tau)}}{G^n(1+2\tau) - (1 - e^{-\tau G^n}) + (1 + \tau G^n)e^{-G^n(1+\tau)}}, \tag{9.7}$$

$G^n = \sum_i r_i^n$, and τ is the propagation delay over packet transmission time. When the network payload increases, more collisions happen and consequently the average delay for each packet increases. For some types of payloads like multimedia services, the delayed packets can cause significant QoS loss. In [15], it has been shown that the average delay can be unbounded for a sufficiently large load. Moreover, for cognitive radios, since the primary users can reoccupy the channel in the near future, a certain delay can cause the second user to lose the opportunity for transmission entirely. So we define G_0 as the maximum network payload. Any network payload larger than G_0 will cause an unacceptable average delay. As a result, the utility function is zero.

In the following, we first propose a new solution concept, correlated equilibrium. Then, we investigate a linear programming method to calculate the optimal correlated equilibrium. Finally, we utilize a no-regret algorithm to learn the correlated equilibria in a distributed way.

To analyze the outcome of the game, Nash equilibrium is a well-known concept, which states that in the equilibrium every user will select a utility-maximizing strategy given the strategies of every other user. If a user follows an action in every possible attainable situation in a game, the action is called pure strategy, in which the probability of using action ν_l, $p(r_i^n = \nu_l)$, has only one non-zero value 1 for all l. In the case of mixed strategies, the user will follow a probability distribution over different possible action, i.e., different rate l.

In Table 9.1, we illustrate an example of two secondary users with different actions. In Table 9.1a, we list the utility function for two users taking action 0 and 1. We can see that when two users take action of 0, they have the best overall benefit. We can see this action as a cooperative action (in our case the users transmit less aggressively). But if any user plays more aggressively using action 1 while the other still plays action 0, the aggressive user has a better utility, but the other user has a lower utility and the overall benefit is reduced. In our case, the aggressive user can achieve a higher rate. However, if both users play aggressively using action 1, both users obtain very low utilities. This situation represents the congested network with low throughput of CSMA. In Table 9.1b, we show two Nash equilibria, where one of the users dominates the other. The dominating user has the utility of 6 and the dominated user has the utility of 3, which is unfair. In Table 9.1c, we show the mixed Nash equilibrium where two users have the probability 0.75 for action 0 and 0.25 for action 1, respectively. The utility for each user is 4.5.

Table 9.1. Example of two secondary users game (a) reward table (left most); (b) Nash equilibrium (middle left); (c) mixed Nash equilibrium (middle right); (d) correlated equilibrium (right most).

	0	1		0	1		0	1		0	1
0	(5,5)	(6,3)	0	0	(0 or 1)	0	9/16	3/16	0	0.6	0.2
1	(3,6)	(0,0)	1	(1 or 0)	0	1	3/16	1/16	1	0.2	0

Next, we study a new concept of correlated equilibrium which is more general than Nash equilibrium and was first proposed by Nobel Prize winner, Robert J. Aumann [16], in 1974. The idea is that a strategy profile is chosen randomly according to a certain distribution. Given the recommended strategy, it is to the players' best interests to conform with this strategy. The distribution is called the correlated equilibrium.

We assume $N = 1$ and we omit the notation n. Define a finite K-user game in strategic form as $\mathcal{G} = \{K, (\Omega_i)_{i \in K}, (U_i)_{i \in K}\}$, where Ω_i is the strategy space for user i and U_i is the utility function for user i. Define Ω_{-i} as the strategy space for user i's opponents. Denote the action for user i and its opponents as $\mathbf{r}_i$ and $\mathbf{r}_{-i}$, respectively. Then, the correlated equilibrium is defined as:

Definition 9.10. *A probability distribution p is a correlated strategy of game $\mathcal{G}$, if and only if, for all $i \in K$, $\mathbf{r}_i \in \Omega_i$, and $\mathbf{r}_{-i} \in \Omega_{-i}$,*

$$\sum_{\mathbf{r}_{-i} \in \Omega_{-i}} p(\mathbf{r}_i, \mathbf{r}_{-i})[U_i(\mathbf{r}'_i, \mathbf{r}_{-i}) - U_i(\mathbf{r}_i, \mathbf{r}_{-i})] \leq 0, \forall \mathbf{r}'_i \in \Omega_i. \tag{9.8}$$

By dividing inequality in (9.8) with $p(\mathbf{r}_i) = \sum_{\mathbf{r}_{-i} \in \Omega_{-i}} p(\mathbf{r}_i, \mathbf{r}_{-i})$, we have

$$\sum_{\mathbf{r}_{-i} \in \Omega_{-i}} p(\mathbf{r}_{-i}|\mathbf{r}_i)[U_i(\mathbf{r}'_i, \mathbf{r}_{-i}) - U_i(\mathbf{r}_i, \mathbf{r}_{-i})] \leq 0, \forall \mathbf{r}'_i \in \Omega_i. \tag{9.9}$$

The inequality in (9.9) means that when the recommendation to user i is to choose action $\mathbf{r}_i$, then choosing action $\mathbf{r}'_i$ instead of $\mathbf{r}_i$ cannot obtain a higher expected payoff to i.

We note that the set of correlated equilibria is non-empty, closed and convex in every finite game. Moreover, it may include the distribution that is not in the convex hull of the Nash equilibrium distributions. In fact, every Nash equilibrium is a correlated equilibrium and Nash equilibria correspond to the special case where $p(\mathbf{r}_i, \mathbf{r}_{-i})$ is a product of each individual user's probability for different actions, i.e., the play of the different players is independent [16–18]. In Table 9.1b and c, the Nash equilibria and mixed Nash equilibria are all within the set of correlated equilibria. In Table 9.1d, we show an example where the correlated equilibrium is outside the convex hull of the Nash equilibrium. Notice that the joint distribution is not the product of two users' probability distributions, i.e., the two users' actions are not independent. Moreover, the utility for each user is 4.8, which is higher than that of the mixed strategy.

The characterization of the correlated equilibria set illustrates that there are solutions of correlated equilibria that achieve strictly better performance compared to the Nash equilibria in terms of the spectrum utilization efficiency and fairness. However, the correlated equilibrium defines a set of solutions which is better than Nash equilibrium, but it does not tell any more information regarding which correlated equilibrium is most suitable in practice. We propose two refinements. The first one is the maximum sum correlated equilibrium that maximizes the sum of utilities of the secondary users. The second is the maxmin fair correlated equilibrium that seeks to improve the worst-case situation. The problem can be formulated as a linear programming problem as:

$$\max_p \sum_{i \in K} E_p(U_i) \text{ or } \max_p \min_i E_p(U_i) \tag{9.10}$$

$$\text{s.t.} \begin{cases} p(\mathbf{r}_i, \mathbf{r}_{-i})[U_i(\mathbf{r}_i', \mathbf{r}_{-i}) - U_i(\mathbf{r}_i, \mathbf{r}_{-i})] \le 0 \\ \forall \mathbf{r}_i, \mathbf{r}_i' \in \Omega_i, \forall i \in K \end{cases}$$

where $E_p(\cdot)$ is the expectation over p. The constraints guarantee the solution is within the correlated equilibrium set.

Next, we will exhibit a class of algorithm called regret-matching algorithm [18]. In particular, for any two distinct actions $\mathbf{r}_i \neq \mathbf{r}_i'$ in Ω_i and at every time T, the regret of user i at time T for not playing $\mathbf{r}_i'$ is

$$\mathcal{R}_i^T(\mathbf{r}_i, \mathbf{r}_i') := \max\{D_i^T(\mathbf{r}_i, \mathbf{r}_i'), 0\} \tag{9.11}$$

where

$$D_i^T(\mathbf{r}_i, \mathbf{r}_i') = \frac{1}{T} \sum_{t \le T} (U_i^t(\mathbf{r}_i', \mathbf{r}_{-i}) - U_i^t(\mathbf{r}_i, \mathbf{r}_{-i})). \tag{9.12}$$

$D_i^T(\mathbf{r}_i, \mathbf{r}_i')$ has the interpretation of average payoff that user i would have obtained, if it had played action $\mathbf{r}_i'$ every time in the past instead of choosing $\mathbf{r}_i$. The expression $\mathcal{R}_i^T(\mathbf{r}_i, \mathbf{r}_i')$ can be viewed as a measure of the average regret. The probability $p_i(\mathbf{r}_i)$ for user i to take action $\mathbf{r}_i$ is a linear function of the regret. The algorithm was named regret-matching (no-regret) algorithm, because the stationary solution of the learning algorithm exhibits no regret and the play probabilities are proportional to the "regrets" for not having played other actions. The detail regret-matching algorithm is shown in Table 9.2. The complexity of the algorithm is $O(L)$.

For every period T, let us define the relative frequency of users' action $\mathbf{r}$ played till T periods of time as follows

$$z_T(\mathbf{r}) = \frac{1}{T} \#\{t \le T : \mathbf{r}_t = \mathbf{r}\} \tag{9.13}$$

where $\#(\cdot)$ denotes the number of times the event inside the bracket happens and $\mathbf{r}_t$ is all users' action at time t. The following theorem guarantees that the adaptive learning algorithm shown in Table 9.2 has the property that z_T converges almost surely to a set of the correlated equilibria.

Theorem 9.1. [18] *If every player plays according to adaptive learning algorithm in Table 9.2, then the empirical distributions of play z_T converge almost surely to the set of correlated equilibrium distributions of the game G, as $T \to \infty$.*

In the simulations, we employ the maximal sum utility function as the objective. In Fig. 9.1a, we show the different equilibria as a function of G_0 for the three-user game. We show the results of the gain obtained by the greedy user in the Nash equilibrium point (NEP), the gain obtained by the victim of the greedy user in NEP,

Table 9.2. The regret-matching learning algorithm © *2007 IEEE. Reprinted, with permission, from [4]*.

Initialize arbitrarily probability for taking action of user i, $p_i^1(\mathbf{r}_i), \forall i \in K$
for t = 1,2,3,...
1. Find $D_i^T(\mathbf{r}_i, \mathbf{r}_i')$ as in (9.12)
2. Find average regret $\mathcal{R}_i^T(\mathbf{r}_i, \mathbf{r}_i')$ as in (9.11)
3. Let $\mathbf{r}_i \in \Omega_i$ be the strategy last chosen by user i,
i.e., $\mathbf{r}_i^t = \mathbf{r}_i$. Then probability distribution action for
next period, p_i^{t+1} is defined as
$p_i^{t+1}(\mathbf{r}_i') = \frac{1}{\mu}\mathcal{R}_i^T(\mathbf{r}_i, \mathbf{r}_i') \quad \forall \mathbf{r}_i' \neq \mathbf{r}_i$
$p_i^{t+1}(\mathbf{r}_i) = 1 - \sum_{\mathbf{r}_i' \neq \mathbf{r}_i} p_i^{t+1}(\mathbf{r}_i'),$
where μ is a certain constant that is sufficiently large.

the learning result and the optimal correlated equilibrium calculated by linear programming. Here the action space is $[0.1, 0.2, \ldots, 1.5]$. When G_0 is large, there is less penalty for greedy behaviors. So all users tend to transmit as aggressively as possible. This results in the prisoners' dilemma [19], where all users suffer. When G_0 is less than 2.8, the greedy user can have better performance (NEP best) than that (NEP worst) of the cooperative user. Due to the less significant penalty if all users transmit aggressively, the game will not degrade to the prison dilemma. However, the performances are quite unfair for the greedy users with best NEP and the cooperative users with worst NEP. All users have the same utility in the correlated equilibrium and learning result. So fairness is better than that of the NEP. When G_0 is from 2.2 to 2.8, the correlated equilibrium has a better performance even than that of the greedy user (NEP best). When G_0 is from 1.4 to 2.8, the optimal correlated equilibrium has a better performance than that of the learning result. When G_0 is sufficiently small, most of the uncooperative strategies are eliminated by significant penalty. Consequently, the learning result has the same performance as that of the optimal correlated equilibrium.

In Fig. 9.1b, we show the network performance of the proposed algorithm. For simplicity, we assume the hidden terminal problem [14] has been solved. We show the average user utility per channel as a function of the network density. When the network density is small, the average utility increases since there is an increasing number of users occupying the channel. When the user density is sufficiently large, the utility begins to decrease due to the collisions. The best NEP and worst NEP are different while the correlated equilibrium and learning result achieve almost the same performance as the best NEP and 5–15% better than the worst NEP.

There are many other works for learning based on finite-state Markov decision process (MDP), such as the decentralized cognitive medium access based on partial observable Markov decision process (POMDP), which is presented in [20]. Some other learning schemes include reinforcement learning, Q-learning and so on. All these techniques can be utilized for the spectrum access for cognitive radios.

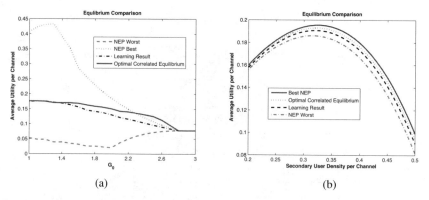

(a)   (b)

Fig. 9.1. (**a**) Utility function versus. G_0 (for three users) and (**b**) network performances [© 2007 IEEE. Reprinted, with permission, from [4]].

9.4 Referee for Mediation

In this section, a concept of virtual referee [5,21] is introduced to improve wireless resource usage of cognitive radios. We use this referee approach for an example application to conduct channel assignment, adaptive modulation and power control for multi-cluster cognitive networks. The goal is to minimize the overall transmitted power under the constraints that each cognitive user has the desired throughput and each cognitive user's power is bounded. Each cognitive user in the different clusters minimizes its own utility function, e.g., transmitted power, in a distributed and non-cooperative game by employing water-filling scheme [22]. We define the channel set that the ith cognitive user can allocate to its throughput R_i as transmission channel set S_i. Each channel can be occupied by more than one cognitive user but not necessarily by all users. Within the transmission channel set, the user would allocate the throughput to different channels by the algorithms, such as water-filling [22], so that the utility such as the power can be optimized. When the interferences are severe, the channel will be over crowded with users and consequently, the radio resource cannot be efficiently utilized. Under this condition, a virtual referee will be introduced to mediate the resource usage, so that the game outcome can be improved. This virtual referee can be the base station, access point or cluster head. This approach can significantly improve the network performance without adding much hardware to cognitive networks.

The K co-channel clusters are taken into consideration. Each cluster consists of one cognitive radio link. The total number of channels is L. The ith user's signal to interference noise ratio (SINR) at channel l can be expressed as:

$$\Gamma_i^l = \frac{P_i^l G_{ii}^l}{\sum_{k \neq i} P_k^l G_{ki}^l + N_0} \tag{9.14}$$

where P_k^l and G_{ki}^l is the transmitted power and propagation loss from the kth cognitive source to the ith cognitive destination in the lth channel, respectively, and N_0 is the thermal noise level.

Rate adaptation such as adaptive modulation provides each channel with the ability to match the effective bit rates, according to the interference and channel conditions. MQAM is a modulation method with high spectrum efficiency. In [23], for a desired rate r_i^l of MQAM, the BER of the lth channel of the ith user can be approximated as a function of the received SINR Γ_i^l by:

$$\text{BER}_i^l \approx c_1 e^{-c_2\left(\Gamma_i^l/2^{r_i^l}-1\right)} \tag{9.15}$$

where $c_1 \approx 0.2$ and $c_2 \approx 1.5$ with small BER_i^l. Rearranging (9.15), for a specific desired BER_i^l, the ith user's transmission rate of the lth channel for the SINR Γ_i^l and the desired BER_i^l can be expressed as:

$$r_i^l = W \log_2(1 + c_3^i \Gamma_i^l). \tag{9.16}$$

In order to compare the Nash equilibriums (NEP) and the optimal solution for power minimization, a simple two-user two-channel example is illustrated as follows. The simulation setup is: $\text{BER} = 10^{-3}$, $N_0 = 10^{-3}$, the maximal power for each user over different channels is $P_{\max} = 10^4$ and channel gain matrices are

$$G^1 = \begin{bmatrix} 0.0631 & 0.0100 \\ 0.0026 & 0.2120 \end{bmatrix}, G^2 = \begin{bmatrix} 0.4984 & 0.0067 \\ 0.0029 & 0.9580 \end{bmatrix}.$$

Figure 9.a shows the overall power contour as a function of two users' rate allocations, where each user's minimal rate requirement $R_1 = R_2 = 6$. The two curves show the minimal locations for the two users' own power when the interference from the other user is fixed, respectively. Each user tries to minimize its power by adjusting its rate allocation so that the operating point is more close to the curve. Consequently, the cross is a Nash equilibrium, where no user can reduce its power alone. We can see that the Nash equilibrium under this setup is unique and optimal for the overall power. (It is worth mentioning that the feasible domain is not convex at all.) Figure 9.b shows the situation when $R_1 = R_2 = 8$. Because the rate is increased, the co-channel interferences are increased and the NEP is no longer the optimum. There exists more than one local optimum, and the global optimum occurs when user does not occupy the channel 1. Figure 9.c shows the situation when $R_1 = R_2 = 8.5$. The contour graph is not connected. There are two NEPs and two local optima. Under the above two conditions, we need to remove users from using the channels. If we further increase $R_1 = R_2 = 10$, there exists no feasible area, i.e., neither user can have a resource allocation that satisfies both power and rate constraints. In this case, the minimal rate requirement should be reduced.

From the above observations, we can see that the behaviors of the optimal power minimization solution and NEP depend on how severe interferences are. In order to let NEP converge to the desired solution, we need to find a criterion to decide whether the users can make a good use of the channels like the situation in Figure 9.a.

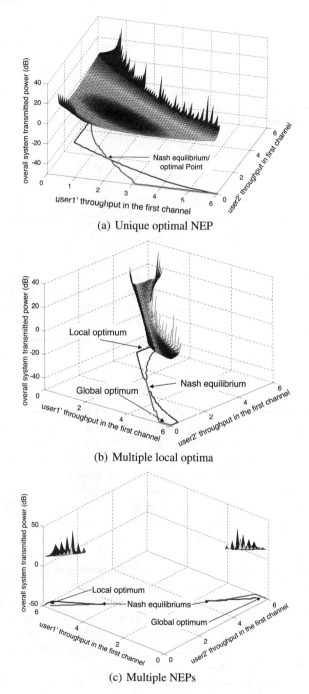

(a) Unique optimal NEP

(b) Multiple local optima

(c) Multiple NEPs

Fig. 9.2. Two-user example [© *2007 IEEE. Reprinted, with permission, from [5]*].

If not, we should decide which user should be kicked out of using specific channels. The criterion is to check whether the KKT condition [24] is satisfied. Specifically, if the co-channel interferences are too severe, the constraints of throughput and maximal transmitted power are not satisfied. As a result, NEP is not a local optimum.

Before developing the proposed algorithm, we analyze two extreme cases. In the first case, the groups of channels are assigned to different clusters without overlapping such that there are no co-channel interferences among clusters. We call it the fixed channel assignment scheme. However, this extreme method has the disadvantage of low spectrum efficiency because of the low frequency re-usage. In the second extreme case, all cognitive users share all the channels. We call it pure water-filling scheme. From Fig. 9.b and Fig. 9.c, we can see that the system can be balanced at the undesired point, because of the severe inter-cluster co-channel interferences. So the facts motivate us to believe that the optimal resource allocation is between these two extreme cases, i.e., each channel can be shared by only a group of selected users for transmission.

In Fig. 9.3, we show the block diagram of the proposed algorithm from system point of view. We initially set S_i to have all channels. Then the non-cooperative competition for radio resources is employed. After the system is iteratively balanced by the water-filling among cognitive radio users, if the system is balanced in a desired solution, the water-filling is continuously employed. Otherwise, some users must remove some channels from the transmission group S_i. If the removal can make all users balanced in the desired NEP, the algorithm continues in the water-filling step. Otherwise, the user removal step is continued, until no user can be removed or the desired NEP is achieved. If no user can be removed and the desired NEP is still not achieved, the desired throughput requirement R_i has to be reduced.

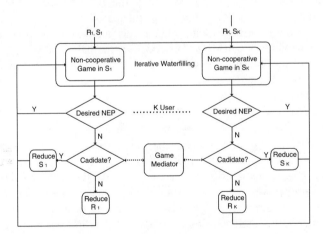

Fig. 9.3. Proposed distributed referee approach [© *2007 IEEE. Reprinted, with permission, from [5]*].

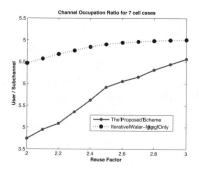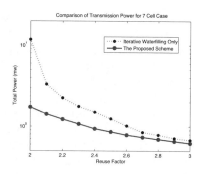

Fig. 9.4. (**a**) User per channel and (**b**) power saving [© *2007 IEEE. Reprinted, with permission, from [5]*].

The complexity of the proposed referee-based scheme is $O(N \log N)$, where N is the number of channels. The convergence speed of the non-cooperative competition is similar to that of closed-loop power control proposed in [25,26]. The overhead for the proposed scheme occurs only when the system cannot be balanced in a good Nash equilibrium. Under this condition, a referee needs to collect information from all the co-channel interfered clusters. The frequency for this overhead is much lower than that of the non-cooperative competition. The collected information includes power value, channel gain value and noise-plus-interference variance value over all channels. Since all these values are consistently obtained by all the distributed users at any time, there is no need for extra measurement. The amount of this information is also small and can be exchanged among the cells with few packets. So the overhead is negligible. In summary, this referee-based scheme imposes little burden on wireless sensor network implementation.

We consider the simulations with 32 channels and seven cognitive radio links. The overall bandwidth is 6.4 MHz. The receiver thermal noise is -70 dBm. The required BER of the transmitted symbols is 10^{-3} for every subchannel and user. We define the reuse factor R_u as the distance between two base stations D over the cell radius r which is set as 100 m, which is one of the main factors to affect the severeness of co-channel interference. The rate constraint is set as 10 Mbits for each user. In Fig. 9.4a, we show the average number of users per channel. In Fig. 9.4b, we show the overall transmitted power versus reuse distance R_u for the pure water-filling algorithm and the proposed algorithm. The smaller the reuse distance R_u is, the higher the co-channel interference. We can see that the proposed algorithm can reduce the overall power about 90% when the co-channel interferences are severe ($R_u = 2$), because more users are kicked out in this case. When R_u increases, the co-channel interferences reduce. Consequently, water-filling and proposed schemes yield the same overall transmitted power.

The referee-based approach creates a virtual referee to mediate the network performances. If the autonomous cognitive users cannot share the network resources efficiently, the referee will make some mandatory changes for resource usage so as to improve the system performances and game outcomes. There is no need to add additional hardware, while the performances can be greatly improved.

9.5 Threat and Punishment Using Repeated Interactions

In some types of the autonomous and distributed wireless cognitive radio networks, tasks need to be performed cooperatively while greediness might lead to the performance breakdown. The individual user may act cooperatively such that the overall system performance is high, or they may act non-cooperatively where everybody suffers low efficiency. However, if only one user deviates from the cooperative agreement, it can get benefits. In order to prevent users from greediness, repeated interaction, such as repeated game, is proposed to enforce cooperation among cognitive users. The basic rationale is to punish the user that deviates by playing non-cooperatively in the near future, such that the benefits obtained in a short-term deviation will be eliminated by a long-term punishment. In this section, we outline the punishment approach and give two examples.

The basic idea of the threat and punishment using repeated game comes from the concept of Cartel in the economics literature [7]. Cartel means the combination of independent commercial or industrial enterprises designed to limit competition. The soul of Cartel maintenance is to construct contracts among independent individuals for cooperative benefits and non-cooperative punishment, so as to limit inefficient competition. Next, we combine the idea with the repeated game theory, so that the new approach will punish anyone who deviates from cooperation.

To analyze the outcome of a game, the *Nash equilibrium* is a well-known concept, which states that in the equilibrium every agent selects a utility-maximizing strategy given the strategies of other agents. However, one problem with an NEP is that it is not necessarily very efficient in performances. If the users can play cooperatively, the performances can be greatly improved. Thus, the question arises as to how to enforce the greedy users to cooperate with each other. The repeated game provides us possible mechanisms to enforce the users to cooperate by considering long-term scenarios. In the repeated games, the players face the same static game in every period, and the player's overall payoff is a weighted average of the payoffs in each stage over time. In the repeated game, the players can observe some information reflecting their opponents' past play. Hence, they are able to condition their future plays on the observed information in history to obtain better equilibriums.

Definition 9.11. *Let G be a static game and β be a discount factor. The T-period repeated game, denoted as $G(T, \beta)$, consists of game G repeated T times. The payoff for such a game is given by*

$$V_i = \sum_{t=1}^{T} \beta^{t-1} u_i^t \tag{9.17}$$

where u_i^t denotes the payoff to player i in period t. If T goes to infinity, then $G(\infty, \beta)$ is referred as the infinitely repeated game. In the following, we use infinitely repeated game.

Now the question is whether cooperation among users can be enforced by the repeated games to generate better performances. The Folk's theorem [19] for infinitely repeated games asserts that if the player's discount factor β approaches 1, any feasible, individually rational payoff can be enforced by an equilibrium. This equilibrium can yield better performances than those of static game NEP. We need to further develop the game rule for enforcing cooperation among users to achieve this better equilibrium.

The basic idea for the proposed Cartel maintenance repeated game framework is to provide enough threat to greedy users so as to prevent them from deviating from cooperation. First the cooperative point is obtained so that all users have better performances than those of non-cooperative NEP. However, if any user deviates from cooperation while others still play cooperatively, this deviating user has a better utility, while others have relatively worse utilities. If no rule is employed, the cooperative users will also have incentives to deviate. Consequently, the network deteriorates to non-cooperation with inefficient performances. The proposed framework provides a mechanism so that the current defecting gains of the selfish user will be outweighed by future punishment strategies from other users. For any rational user, this threat of punishment prevents them from deviation. So cooperation is enforced.

To implement the mechanism, we propose a trigger strategy to introduce punishment on the defecting users. In the trigger strategy, the players start with cooperation. Assume that each user can observe the public information (e.g., the outcome of the game), P_t at time t. Examples of this public information can be the successful transmission rate, network throughput, etc. Notice that such public information is mostly imperfect or simply partial information about the users' strategies. Here we assume a larger P_t stands for a higher cooperative level, resulting in higher performances for all users. Let the cooperative strategies be $\bar{\lambda} = [\lambda_1, \lambda_2, ..., \lambda_K]^T$ and the non-cooperative strategies be $\bar{s} = [s_1, s_2, ..., s_K]^T$, respectively. The trigger-punishment game rule is characterized by three parameters: the optimal punishment time T, trigger threshold P^* and the cooperative strategy $\bar{\lambda}$. Trigger punishment strategy $(\bar{\lambda}, P^*, T)$ for distributed user i is given as follows:

(a) User i plays the strategy of the cooperative phase, $\bar{\lambda}$, in period 0.
(b) If the cooperative phase is played in period t and $P_t > P^*$, user i plays the cooperative phase in period $t + 1$.
(c) If the cooperative phase is played in period t and $P_t < P^*$, user i switches to a punishment phase for $T - 1$ periods, in which the players play a static Nash equilibrium $\bar{s}$ regardless of the realized outcomes. At the Tth period, play returns to the cooperative phase.

Note that $\bar{s}$ generates the non-cooperative outcome, which is much worse than that generated by the cooperative strategy $\bar{\lambda}$. Therefore, the selfish users that deviate will have much lower utilities in the punishment phase. Moreover, the punishment time T is designed to be long enough to let all cheating gains of the selfish users be outweighed by the punishment. So the users have no incentive to deviate from cooperation, since the users aim to maximize the long-run payoffs over time.

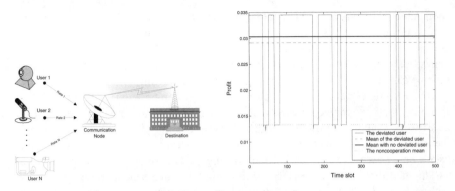

Fig. 9.5. (a) Wireless network block diagram, (b) Punishment for deviation [© *2004 IEEE. Reprinted, with permission, from [7]*].

Next, we study two examples using the proposed framework. The first one is for cognitive radio multiple access networks. The second example further investigates the learning schemes if the cognitive radio users do not know how to cooperate.

In the first example, we employ the proposed framework to a multi-user network shown in Fig. 9.5a. There are many distributed users and one communication node (e.g., cluster head). Each user can transmit its data packets to the communication node by using the multiple access protocols such as Aloha, CSMA, etc. The communication node has the ability to transmit the data packets to the remote destination via a wireless link. We assume that there is a reliable feedback channel. So, the system can be described as multiple users sharing a communication link. Each user can control its transmission rate. The users need to compete with each other for the communication link which is fluctuating due to the wireless channel conditions. Thus one user's rate can affect the performances of other users and the whole system. So it is necessary to find a rate control algorithm such that the system can operate at the optimal point. Moreover, it is hard to have communication channels among cognitive users. Therefore, a distributed algorithm is required for rate control.

For distributed users in the network, there are costs to transmit their packets and benefits if their packets are successfully transmitted. Each user's profit is defined as the benefits minus the cost. The users are able to adapt their packet transmission rates for the cooperation or punishment play. They can observe their successful packet transmission probability, and correspondingly play cooperation or non-cooperation. Based on the proposed framework, we derive the optimal parameters of the packet transmission rate, punishment time and trigger threshold for the distributed greedy users. In Fig. 9.5 b, we show how the scheme punishes the cheating user. We assume one user deviates from the cooperative rate λ^* and transmit at the higher rate s, while others transmit at λ^*. We show that the profit of this deviating user fluctuates over time. For comparison, we also show the average profits (as the straight lines) when the user transmits at optimal rate from overall system point of view, cooperative rate λ^* and non-cooperative rate. We can see that at first the user does get more profit than the mean without the deviated user by diverging from λ^*. However, this deviation is soon detected by others' and the punishment phase is performed by other users. The non-cooperation mean is much lower than that during cooperation. The mean of this deviated user is lower than the mean without the deviated user, because the deviation gain is eliminated by others over time. This shows the reason why the proposed scheme can enforce cooperation among users by threatening punishment.

In the second example, we further investigate the combination of learning schemes. In some ad hoc cognitive networks, cognitive users need to forward others' packet so as to communicate with each other. Forwarding the others' packets consumes the user's own limited battery resource. Therefore, it may not be of the autonomous user's best interest to forward all the arriving packets. In fact, it is reasonable to assume that the users are selfishly maximizing their own benefits by dropping others' packets. However, not forwarding others' packets will severely affect the network connectivity and the proper functionality of the network, which in turn impairs the users' own benefits as well. The non-cooperation usually causes very low system and users' performances. Therefore, it is very crucial to design a mechanism to enforce cooperation among greedy users. Moreover, even though the users would like to cooperate, they might not know how to cooperate. So it is important to develop self-learning algorithms so that the cooperative points can be studied distributively in the autonomous users.

We try to propose a distributed self-learning repeated game framework to enforce cooperation in performing packet-forwarding tasks as shown in Fig. 9.6a. The framework has two major schemes: first, an adaptive repeated game scheme ensures cooperation among ad hoc cognitive users, which maintains the current cooperative packet-forwarding probabilities. The repeated game scheme provides the users with a mechanism that any deviating user would be punished enough by others in the future, so that no user has incentive to deviate. Second, a self-learning scheme tries to find the better cooperative probabilities that are feasible and benefit all users. Starting from non-cooperation, the above two proposed schemes are employed iteratively. Better cooperation is discovered and maintained over iterations, until convergence to some close optimal solution.

In Fig. 9.6b, we show the simulation results of the proposed framework for utility and packet forwarding probability over time. Initially, packet-forwarding probability $\alpha = 0$, because of the non-cooperative transmission. Then the system tries to find a better packet transmission rate. When it finds a better solution, all users adapt its α to the value. However, since the punishment period T is not adjusted to an optimal value, the deviation can have benefits. So there exists a period that the utility and α switch from cooperation to non-cooperation. In this period, T is increased until every user realizes that there is no benefit for deviation because of the long period of punishment. If the system is stable for a period of time, a new α is determined to see whether the performance can be improved. If so, the new value is adopted, otherwise the original value is restored. So the packet-forwarding probability is adjusted until the optimal solution is found, and the learned utility function is a non-decreasing function.

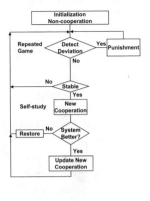
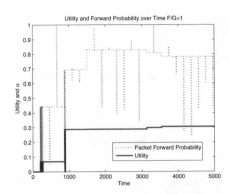

Fig. 9.6. (**a**) Proposed self-learning repeated game and (**b**) self-learning curve [© *2005 IEEE. Reprinted, with permission, from [6]*].

9.6 Spectrum Auction

In this section, we first discuss the basics of auction theory. Then we investigate the mechanism design for auctions. Finally, we use an example to explain how to utilize the auction theory for spectrum usage in cognitive radios.

Auction theory is important for practical, empirical and theoretical reasons. First, a large amount of wireless networking and resource allocation problems can be formulated as auction theory. For example, the routing problem for self-interested users is studied in [27]. Second, the auction theory has a simple game setup, and many theoretical results are available for analysis. The definition of auction is as follows.

Definition 9.12. *A market mechanism in which an object, service or set of objects, is exchanged on the basis of bids submitted by participants. Auction provides a specific set of rules that will govern the sale or purchase (procurement auction) of an object to the submitter of the most favorable bid.*

The interactions and outcome of an auction are determined by the *rules*, which include four components:

- *Information*: what the auctioneer and bidders know before the auction starts.
- *Bids*: what the bidders submit to the auctioneer to express their interests in the good.
- *Allocation*: how the good is allocated among the bidders as a function of the bids.
- *Payments*: how the bidders pay the auctioneer as functions of the bids and allocation.

To implement auction theory in wireless networking and resource allocation, the credit-based system is usually proposed. The individual user can select to pay for some kind of services such as a route. The payment can be implemented via a certain central "bank" system. However, this requires more control than the other game theory approaches, such as non-cooperative games. Moreover, in order to achieve different design goals such as the network total benefit, the auction method shall be designed according to different available information. Mechanism design is the tool for game and auction design.

Mechanism design is the subfield of microeconomics and game theory that considers how to implement good system-wide solutions to problems that involve multiple self-interested agents, each with private information about their preferences. The goal is to achieve a social choice function implemented in distributed systems with private information and rational agents. The design criteria can be different as follows:

1. *Efficiency*: select the outcome that maximizes total utility.
2. *Fairness*: select the outcome that minimizes the variance in utility.
3. *Revenue maximization*: select the outcome that maximizes revenue to a seller (or more generally, utility to one of the agents).
4. *Budget-balance*: implement outcomes that have balanced transfers across agents.
5. Pareto optimality.

One well-known auction mechanism that achieves the efficient allocation is the Vickery–Clarke–Groves (VCG) auction [28]. In a VCG auction, the bidders are asked to reveal their bids simultaneously, from which the auctioneer determines the efficient allocation. The auctioneer then asks each bidder i to pay for the "performance loss" of other bidders due to bidder i's participation in the auction, which involves solving one additional optimization problem for each bidder. It is well known that it is a (weakly) dominant strategy for the bidders to bid truths in the VCG auction, i.e., revealing their true rate increase functions. As a result, the VCG auction achieves the efficient allocation.

The limitations of the VCG auction for cognitive radio users are as follows. First, the users (bidders) need to submit the complete information to the central control unit serving as the auctioneer, which involves revealing users' complete private information. This might be overheard by other users and so can lead to security problems. Also, accurately specifying the information requires much signaling overhead and communication bandwidth, which may significantly reduce the network performance. Furthermore, it is usually computationally expensive for solving the optimization problems.

Due to these concerns, in [29], two simpler share auctions are proposed for cognitive radios. First, we discuss the system model. Suppose K user-CDMA is utilized with processing gain B. The received SINR is given by

$$\Gamma_i = \frac{P_i G_{ii}}{N_0 + \frac{1}{B}\left(\sum_{j \neq i} P_j G_{ji}\right)} \tag{9.18}$$

where P_i is the transmit power, G_{ij} is the channel gain and N_0 is the thermal noise power. User i receives a strictly concave increasing utility as $U_i(\Gamma_i, \theta_i)$, where θ_i is user-dependent priority parameter.

Next, we discuss the two share auctions, namely the *SNR auction* and the *power auction*. The main advantages of the two auctions are the simplicities of bids and allocation. The rules of the two auctions are described below, with the only difference being in payment determination.

9.6.1 Share Auction

- *Information*: The auctioneer (which can be a cluster head) announces a positive *reserve bid* $\zeta > 0$ and a unit *price* $\pi > 0$ to all users before the auction starts. Here ζ ensures unique outcome, and π is for unit SINR or received power.
- *Bids*: User i submits $b_i \geq 0$ to the auctioneer.
- *Allocation*: The auctioneer allocates transmit power according to

$$P_i G_{ii} = \frac{b_i}{\sum_{j=1}^{N} b_j + \zeta} P \qquad (9.19)$$

where P is the overall allowable power.
- *Payments*: In an SNR auction, cognitive user i pays the auctioneer

$$C_i = \pi \triangle \text{SNR}_i. \qquad (9.20)$$

In a power auction, source i pays the relay

$$C_i = \pi P_i G_{ii}. \qquad (9.21)$$

A bidding profile is defined as the vector containing the users' bids, $\mathbf{b} = (b_1, ..., b_K)$. The bidding profile of user i's opponents is defined as $b_{-i} = (b_1, ..., b_{i-1}, b_{i+1}, ..., b_K)$, so that $\mathbf{b} = (b_i; b_{-i})$. User i chooses b_i to maximize its payoff $U_i(b_i; b_{-i}, \pi)$. The desirable outcome of an auction is called a *Nash equilibrium* (NE), which is a bidding profile $\mathbf{b}^*$ such that no user wants to deviate unilaterally, i.e.,

$$U_i\left(b_i^*; b_{-i}^*, \pi\right) \geq U_i\left(b_i; b_{-i}^*, \pi\right), \forall i \in 1, \ldots, K, \forall b_i \geq 0. \qquad (9.22)$$

Define user i's *best response* (for fixed b_{-i} and price π) as

$$\mathcal{B}_i\left(b_{-i}, \pi\right) = \{b_i | b_i = \arg\max_{\tilde{b}_i \geq 0} U_i(\tilde{b}_i; b_{-i}, \pi)\} \qquad (9.23)$$

which in general could be a set. An NE is also a fixed point solution of all users' best responses. In [29], the following four questions for both auctions are answered. First, an NE does exist, and in some mild conditions, the NE is unique. This NE can be converged by using a distributed iterative algorithm with some partial information that is private and local. The SNR auction with log utility can achieve weighted max–min fairness, while the power auction can achieve social optimum for large bandwidth.

9.7 Mutual Benefits Through Bargaining

In order for the distributed cognitive users to cooperate with each other, one method is to give the individuals mutual benefits for cooperative behavior. In most existing literature, the benefit incentive approach is performed in a framework of "pricing anarchy," where a price is announced by the system, so that the distributed users have to pay high price for non-cooperation and the cooperative behaviors will be rewarded. However, there are many potential design challenges for the pricing technique to be employed in cognitive networks. First, the price itself may not represent the true benefits of the cognitive users. Instead, the price might be artificial so that autonomous users may just ignore it. Furthermore, pricing technique needs a lot of computation power and signaling to calculate the optimal price. This is especially hard to implement in cognitive networks. In addition, if the utility of each user is not convex, there might be many local optima for the pricing methods. Finally, for the heterogeneous networks and for the resource allocation with integer/combinatorial optimization, the pricing techniques are hard to be effective. Because of the above reasons, we need to have novel perspective and find new approaches to give users mutual benefits to cooperate.

In daily life, a market is served as a central gathering point, where people can exchange goods and negotiate transactions, so that people will be satisfied through bargaining. Similarly, in wireless cognitive networks, there exist some nodes, like cluster heads, that can serve as a function of the market. The distributed cognitive users can negotiate via these nodes to cooperate in making the decisions on the resource usage, such that each of them will operate at its optimum and joint agreements are made about their operating points. Such a fact motivates us to employ the cooperative game theory [3,30,31], which can achieve the crucial notion of fairness and maximize the overall system performances. The idea is to negotiate among users so that the mutual benefits can be obtained, which enlightens us with the new perspective on how to provide incentive for cooperation. In the following, we list one possible problem formulation, a basic illustration of the proposed approaches and some simulation results.

We have proposed the cooperative game theory approaches for resource allocation in multiple-user multiple-channel scenario within a cluster of cognitive networks. The problem can be formulated in the following example. There are K users and a total of N channels. Each channel can be occupied by only one user so as to avoid severe co-channel interferences and maintain the basic link quality. Since a channel condition for a specific channel may be good for more than one user, there is a competition among users for their transmissions over these good channels. So this is where the game concept comes in. Moreover there are some other practical constraints. For example, the maximal transmitted power for each user is bounded by the maximal transmitted power $P_{\max}$, and each user has a minimal rate requirement $R^i_{\min}$. To formulate the problem, we define $a_{ij} = [\mathbf{A}]_{ij} = 1$, if the ith user occupies the jth channel; $a_{ij} = 0$, otherwise, and $[\mathbf{P}]_{ij}$ as the corresponding power. One example of the optimization goal is to determine different users' channel assignment matrix $\mathbf{A}$ and power matrix $\mathbf{P}$ such that the network objective function U will be maximized, i.e.,

$$\max_{\mathbf{A},\mathbf{P}} U \tag{9.24}$$

$$\text{subject to} \begin{cases} \text{Assignment: } \sum_{i=1}^{K} a_{ij} = 1, \forall j \\ \text{Minimal rate: } R_i \geq R_{\min}^i, \forall i \\ \text{Maximal power: } \sum_{j=1}^{N} P_{ij} \leq P_{\max}, \forall i \end{cases}$$

where R_i is the ith user's rate and U can have different definitions for network objectives such as:

- Maximal rate: $U = \sum_{i=1}^{N} R_i$.
- Max–min fairness: $U = \min R_i$.
- Nash bargaining solutions: $U = \prod_{i=1}^{K} \left(R_i - R_{\min}^i \right)$.

The first two network objectives are widely studied in the literature. In [3], we proposed the concept of Nash bargaining solution (NBS), because of the following two reasons: first, it can be shown that this network objective will ensure NBS fairness of allocation in the sense that this NBS fairness is a generalized proportional fairness. From the simulation results, this NBS fairness ensures that users' allocated resource are not affected by other users' situations. Second, cooperative game theories p that there exists a unique and efficient solution under the six axioms shown The intelligent merit of this NBS solution is that it can provide a special new tr. between the fairness and efficiency, which is widely researched recently in acade. and industry.

The difficulty to solve (9.24) by traditional methods lies in the fact that the problem itself is a constrained combinatorial problem and the constraints are non-linear. Thus the complexities of the traditional schemes are high especially with a large number of users. Moreover, distributed algorithms are desired for cognitive networks, while centralized schemes are dominant in the literature. To develop algorithms that can be easily deployed in distributed cognitive networks, we outline the ideas of the proposed approaches as follows:

Bargaining for two-user case: Due to the facts that in social life most negotiations are taken between two parties, we first consider the case in which the number of users $K = 2$ and we will develop a fast two-user bargaining solution. Since different users might have different gains over the same channel, the intuitive idea is to allow two users to negotiate and exchange their occupied channels such that mutual benefits will be obtained. The difficulty is to determine how to optimally exchange channels, which is a complex integer programming problem. An interesting low complexity algorithm was given in [22]. The idea is to sort the order of channels first and then to use a simple two-band partition for the channel assignment. When signal to noise ratio (SNR) is high, the two-band partition for two-user channel assignment can be near optimal for the optimization goal. The possible solution has the complexity of $O(N^2)$ and can be further improved by using a binary search algorithm with a complexity of only $O(N \log N)$.

Multiple users using coalitions: For the case in which the number of users is larger than two, the computational complexity is very high with respect to the number of channels. Here, we propose a two-step iterative scheme: first, users are

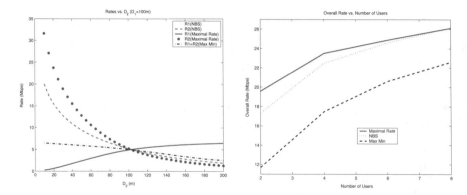

Fig. 9.7. (**a**) Each user's rate versuss. D_2 and (**b**) overall rate versus. number of users [©] 2005 IEEE. Reprinted, with permission, from [3]].

grouped into pairs, which are called *coalitions*. Then with each coalition, the two-user solution is employed for two users to negotiate and improve their performances by exchanging channel sets. Further, the users are regrouped and then renegotiated again. The above regrouping-and-negotiation iteration is repeated until convergence. By using this scheme, the computational cost can be greatly reduced. Some algorithms such as Hungarian method [32] can be utilized to find the optimal coalition pairs in each round. These minimal optimization efforts can be performed in the central point, such as the cluster head, while lower implementation costs are imposed on the distributed less-sophisticated users. Moreover, the above-mentioned approach can also be generalized to other formulated problems dealing with multi-user communications with different optimization goals and constraints.

To demonstrate the effectiveness of the proposed scheme, a simulation is conducted for a multiple-cognitive-user cluster with 32 channels. In Fig. 9.7a, a two-user case is studied. The rates of both users for the NBS, maximal rate and max–min schemes are shown versus the second user's distance from base station D_2. Here the first user's location is fixed at 100 m ($D_1 = 100$). For the maximal rate scheme, the user closer to the base station has a higher rate, and the rate difference is very large when D_1 and D_2 are different. For the max–min scheme, both users have the same rate which is reduced when D_2 increases. This is because the system has to accommodate the user with the worst channel condition. While for the NBS scheme, user 1's rate is almost the same regardless of D_2 and user 2's rate is reduced when D_2 increases. This shows that the NBS solution is fair in the sense that the user's rate is determined only by its channel condition and not by other interfering users' conditions.

In Fig. 9.7b, we show the sum of all users' rates versus the number of users in the system for three schemes. We can see that all three schemes have better performances when the number of users increases. This is because of multi-user diversity, provided by the independent varying channels across the different users. The performance improvement saturates gradually. The NBS scheme has a similar performance to that of the maximal rate scheme and has a much better performance than that of the max–min scheme. The performance gap between the maximal rate scheme and the NBS scheme reduces when the number of users is large. This is because more bargain pair choices are available to increase the system performance. The simulation results show that the proposed NBS scheme achieves a good tradeoff between fairness and efficiency.

We propose the idea of mutual benefits using bargaining for ensuring cooperation of resource allocation. By using cooperative game theory such as Nash bargaining solution and coalition, users' performances can be improved by locally exchanging the resources. The wireless cognitive network performance can be significantly improved and fairness among distributed users can be ensured in a self-organized way. Many other works are based on the proposed idea. In [33], a dynamic spectrum access scheme was proposed for ad hoc networks using the bargaining scheme. In [34], the idea was extended to cognitive radios. In [35], mesh networks were investigated, and in [36], multimedia source coding was also considered.

9.8 Contract Using Cooperative Game

Until now, we have discussed how to play the cooperative game and obtain mutual benefits by bargaining. To further analyze the benefits and rewards, we investigate a game coalition that describes how much collective payoff a set of nodes can gain and how to divide the payoff. The associated analysis concepts include core, Shapley function and nucleolus. In the following, we will explain these concepts and explain how to use them in the cognitive radio networks.

Definition 9.13. *A coalition S is defined to be a subset of the total set of player K, $S \in K$. The users in a coalition try to cooperate with each other. The coalition form of a game is given by the pair (K, v), where v is a real value function, called characteristic function. $v(S)$ is the value of the cooperation for coalition S with the following properties:*

1. *$v(\emptyset) = 0$*
2. *(Superadditivity) if S and T are disjoint coalitions ($S \cap T = \emptyset$), then $v(S) + v(T) \leq v(S \cup T)$.*

The coalition states the benefit obtained via cooperation agreement. But we still need to study how to divide the benefit to the cooperative users. One of the possible properties of an agreement is mutual benefit. The agreement is stable since no coalition shall have the incentive and power to upset the cooperative agreement. The set of such division of v is called the core defined in the following definitions.

Definition 9.14. *A payoff vector $x = (x_1, \ldots, x_K)$ is said to be group rational or efficient if $\sum_{i=1}^{K} x_i = v(K)$. A payoff vector x is said to be individually rational if the user can obtain the benefit no less than acting alone, i.e., $x_i \geq v(\{i\})$, $\forall i$. An imputation is a payoff vector satisfying the above two conditions.*

Definition 9.15. *An imputation* x *is said to be unstable through a coalition* S *if* $v(S) > \sum_{i \in S} x_i$, *i.e., the users have incentive for coalition* S *and upset the proposed* x. *The set* C *of a stable imputation is called the core, i.e.,*

$$C = \{x : \sum_{i \in K} x_i = v(K) \text{ and } \sum_{i \in S} x_i \geq v(S), \forall S \subset K\}. \quad (9.25)$$

Core gives a reasonable set of possible shares. A combination of shares is in a core if there exists no subcoalition in which its members may gain a higher total outcome than the share of concern. If the share is not in a core, some members may be frustrated and may think of leaving the whole group with some other members and form a smaller group.

To illustrate the idea of core, we give the following example. Suppose the game with the following characteristic functions:

$$v(\emptyset) = 0, v(\{1\}) = 1, v(\{2\}) = 0, v(\{3\}) = 1, \quad (9.26)$$
$$v(\{1,2\}) = 4, v(\{1,3\}) = 3, v(\{2,3\}) = 5, v(\{1,2,3\}) = 8.$$

By using $v(\{2,3\}) = 5$, we can eliminate the payoff vector (such as $(4,3,1)$), since user 2 and user 3 can achieve better payoff by forming coalition themselves. Using the same analysis, the final core of the game is (3,4,1), (3,3,2), (3,2,3), (3,1,4), (2,5,1), (2,4,2), (2,3,3), (2,2,4), (1,5,2), (1,4,3) and (1,3,4).

Core concept defines the stability of an allocation of payoff. However, it does not define how to allocate the utility. Next, we study each individual player's power in the coalition by defining a value called Shapley function.

Definition 9.16. *A Shapley function* ϕ *is a function that assigns to each possible characteristic function* v *a real number, i.e.,*

$$\phi(v) = (\phi_1(v), \phi_2(v), \ldots, \phi_K(v)) \quad (9.27)$$

where $\phi_i(v)$ *represents the worth or value of player* i *in the game. The Shapley axioms for* $\phi(v)$ *is*

1. *Efficiency:* $\sum_{i \in K} \phi_i(v) = v(K)$.
2. *Symmetry: If* i *and* j *are such that* $v(S \cup \{i\}) = v(S \cup \{j\})$ *for every coalition* S *not containing* i *and* j, *then* $\phi_i(v) = \phi_j(v)$.
3. *Dummy Axiom: If* i *is such that* $v(S) = v(S \cup \{i\})$ *for every coalition* S *not containing* i, *then* $\phi_i(v) = 0$.
4. *Additivity: If* u *and* v *are characteristic functions, then* $\phi(u + v) = \phi(v + u) = \phi(u) + \phi(v)$.

It can be proved that there exists a unique function ϕ satisfying the Shapley axioms. To calculate the Shapley function, suppose we form the grand coalition by entering the players into this coalition one at a time. As each player enters the coalition, he receives the amount by which his entry increases the value of the coalition he enters. The amount a player receives by this scheme depends on the order in which the players are entered. The Shapley value is just the average payoff to the players if the players are entered in completely random order, i.e.,

$$\phi_i(v) = \sum_{S \subset K, i \in S} \frac{(|S| - 1)!(K - |S|)!}{K!} [v(S) - v(S - \{i\})]. \tag{9.28}$$

For the example in (9.26), it can be shown that the Shapley value is $\phi = (14/6, 17/6, 17/6)$.

Another concept for multiple cooperative games is nucleolus. For a fixed characteristic function, an imputation x is found such that the worst inequity is minimized, i.e., for each coalition S and its associated dissatisfaction, an optimal imputation is calculated to minimize the maximum dissatisfaction. First we define the concept of excess which measures the dissatisfactions.

Definition 9.17. *The measure of the inequity of an imputation x for a coalition S is defined as the excess:*

$$e(x, S) = v(S) - \sum_{j \in S} x_j. \tag{9.29}$$

Obviously, any imputation x is in the core, if and only if all its excesses are negative or zero.

Among all allocation, kernel is a fair allocation, defined as in the following

Definition 9.18. *A kernel of v is the set of all allocations x such that*

$$\max_{S \subseteq K - j, i \in S} e(x, S) = \max_{T \subseteq K - i, j \in T} e(x, T). \tag{9.30}$$

If players i and j are in the same coalition, then the highest excess that i can make in a coalition without j is equal to the highest excess that j can make in a coalition without i.

Finally, we define nucleolus as follows.

Definition 9.19. *Nucleolus is the allocation x which minimizes the maximum excess.*

$$x = \arg \min_x (\max e(x, S), \forall S). \tag{9.31}$$

The nucleolus has the following property: the nucleolus of a game in coalitional form exists and is unique. The nucleolus is group rational, individually rational and satisfies the symmetry axiom and the dummy axiom. If the core is not empty, the nucleolus is in the core and kernel. In other words, the nucleolus is the best allocation with the min–max criteria.

To utilize the cooperative game in dynamic spectrum allocation for cognitive networks, the cognitive users sign a contract for spectrum usage before accessing the spectrum. This contract ensures that the benefits of cooperation are greater than those of the individual actions. The core concepts can test whether or not the cooperation is stable. Then if the average fairness is considered, Shapley values can allocate different cognitive users their share of cooperation benefits. On the other hand, if the max–min fairness is considered, the concepts of excess, kernel and nucleolus define the allocation. Overall, the cognitive users seek the contracts for resource usage that can benefit all.

9.9 Centralized Optimization

In this section, we discuss how to formulate the centralized optimization for resource allocation of cognitive radios. Specifically, we study what the resources are, what the parameters are, what the practical constraints are and what the optimized performances across the different layers are. In addition, we address how to perform resource allocation in multi-user scenarios. The tradeoffs between the different optimization goals and different users' interests are also investigated. This centralized optimization can serve as the performance upper bound for the other approaches and can also provide insights for the design of other schemes.

Many resource allocation problems for cognitive radios can be formulated as constrained optimization problems, which can be optimized from the network point of view or from the individual point of view. The general formulation can be written as:

$$\min_{\mathbf{x} \in \Omega} f(\mathbf{x}) \tag{9.32}$$

$$\text{s.t.} \begin{cases} g_i(\mathbf{x}) \leq 0, \text{ for } i = 1, \ldots, m \\ h_j(\mathbf{x}) = 0, \text{ for } j = 1, \ldots, l \end{cases}$$

where $\mathbf{x}$ is the parameter vector for optimizing the resource allocation, Ω is the feasible range for the parameter vector and $f(\mathbf{x})$ is the optimization goal matrix, objective goal or utility function that represents the performance or cost. Here, $g_i(\mathbf{x})$ and $h_j(\mathbf{x})$ are the inequality and equality constraints, respectively, for the parameter vector. The optimization process finds the solution $\bar{\mathbf{x}}$ that satisfies all the inequality and equality constraints. For the optimal solution, $f(\bar{\mathbf{x}}) \leq f(\mathbf{x}), \forall \mathbf{x} \in \Omega$.

If the optimization goal, the inequality constraints, and the equality constraints are all linear functions of the parameter vector $\mathbf{x}$, then the problem in (9.32) is called a *linear program*. One important characteristic of a linear program problem is that there is a global optimal point that is very easy to obtain by linear programming. But on the other hand, one major drawback of linear program is that most of the practical problems in wireless networking and resource allocation are non-linear. Therefore, it is hard to model these practical problems as linear programs. If either the optimization goal or the constraint functions are non-linear, the problem in (9.32) is a *non-linear program*. In general, there are multiple local optima in a non-linear program, and to find the global optimum is not an easy task. Furthermore, if the feasible set Ω contains some integer sets, the problem in (9.32) is an *integer program*. Most integer programs are NP-hard problems which cannot be solved by polynomial time.

One special kind of non-linear program is a convex optimization problem in which the feasible set Ω is a convex set, and the optimization goal and the constraints are convex/concave/linear functions. A convex set is defined as follows.

Definition 9.20. *A set Ω is a convex set if for any $x_1, x_2 \in \Omega$ and any θ with $0 \leq \theta \leq 1$, we have $\theta x_1 + (1 - \theta)x_2 \in \Omega$.*

A convex function f is defined as follows.

Definition 9.21. *A function f is a convex function in* x, *if the feasible range* Ω *of parameter vector* x *is a convex set, and if for all* $x_1, x_2 \in \Omega$ *and* $0 \leq \theta \leq 1$,
$$f(\theta x_1 + (1 - \theta)x_2) \leq \theta f(x_1) + (1 - \theta)f(x_2).$$
A function f is strictly convex if the strict inequality holds whenever $x_1 \neq x_2$ *and* $0 < \theta < 1$. *A function f is called concave if* $-f$ *is convex.*

If function f is differentiable, and if either the following two conditions hold, then f is a convex function.

First order condition: $f(\mathbf{x}_2) \geq f(\mathbf{x}_1) + \nabla f(\mathbf{x}_1)^T (\mathbf{x}_2 - \mathbf{x}_1)$.

Second order condition: $\nabla^2 f(\mathbf{x}) \succeq 0$.

One important application of the convex function is Jensen's inequality. Suppose function f is convex and the parameter $\mathbf{x}$ has any arbitrary random distribution over Ω then the following equality holds
$$f(E(\mathbf{x})) \leq E(f(\mathbf{x}))$$
where E denotes expectation.

The advantages of convex optimization for wireless-networking-and-resource-allocation problems are shown as follows:

- There are a variety of applications such as automatic control systems, estimation and signal processing, communications and networks, electronic circuit design, data analysis and modeling and statistics.
- Computation time is usually quadrature. Problems can then be solved, very reliably and efficiently, using interior-point methods or other special methods for convex optimization.
- Solution methods are reliable enough to be embedded in a computer-aided design or analysis tool, or even a real-time reactive or automatic control system.
- There are also theoretical or conceptual advantages of formulating a problem as a convex optimization problem.

The challenges of the convex optimization are to recognize and model the problem as a convex optimization. Moreover, there are many tricks for transforming problems into convex forms.

We have discussed the basics for constrained optimization problems. Next we will see how the problem can be formulated. In resource allocation for cognitive networks, the parameters, functions and constraints in (9.32) can have the following physical meaning:

- Parameters
    1. *Physical layer*: transmitted power, modulation level, channel coding rate, channel/code selection and others.
    2. *MAC layer*: transmission time/frequency, service rate, priorities for transmission and others.
    3. *Network layer*: route selection, routing cost and others.
    4. *Application layer*: source-coding rate, buffer priority, packet arrival rate and others.
- Optimization Goals

1. *Physical layer*: minimal overall power, maximal throughput, maximal rate per joule, minimal bit error rate, and others.
2. *MAC layer*: maximal overall throughput, minimal buffer overflow probability, minimal delay and others.
3. *Network layer*: minimal cost, maximal profit and others.
4. *Application layer*: minimal distortion, minimal delay and others.

- Constraints
  1. *Primary user*: channel occupancy, interference level and others.
  2. *Physical layer*: maximal mobile transmitted power, available modulation constellation, available channel coding rate, limited energy and others.
  3. *MAC layer*: contentions, limited time/frequency slot, limited information about other mobiles and others.
  4. *Network layer*: maximal hops, security concerns and others.
  5. *Application layer*: the base layer transmission, limited source rate, strict delay requirement, security and others.

After formulating the constrained optimization problem for resource allocation over cognitive networks, we need to find solutions. In general for centralized optimization, we classify the different approaches as the following categories.

- *Closed-form solution*: One of the most important methods used to find a closed form solution for constrained optimization is the Lagrangian method, which has the following steps
  1. Rewrite (9.32) as a Lagrangian multiplier function J as

$$J = f(\mathbf{x}) + \sum_{i=1}^{m} \lambda_i g_i(\mathbf{x}) + \sum_{j=1}^{l} \mu_j h_j(\mathbf{x}) \tag{9.33}$$

where λ_i and μ_j are Lagrangian multipliers.
  2. Differentiate J over $\mathbf{x}$ and set to zero as

$$\frac{\partial J}{\partial \mathbf{x}} = 0. \tag{9.34}$$

  3. From (9.34), solve λ_i and μ_j.
  4. Replace λ_i and μ_j in the constraints to get optimal $\mathbf{x}$.

Notice that the difficulty in the Lagrangian method is Step (3) and Step (4), where the closed form solution is obtained for the Lagrangian multipliers. Some approximations and mathematical tricks are necessary to obtain the closed form solutions.

- *Mathematical programming*: If the optimization problem is used to find the best objective function within a constrained feasible region, such a formulation is sometimes called a mathematical program. Many real-world and theoretical problems can be modeled in this general framework. There are the four major subfields of the mathematical programming:
  1. Linear programming studies the case in which the objective function is linear and the feasible set is specified using only linear equalities and inequalities.

2. Convex programming studies the case where the constraints and the optimization goals are all convex or linear.

3. Non-linear programming studies the general case in which the objective function or the constraints or both contain non-linear parts.

4. Dynamic programming studies the case in which the optimization strategy is based on splitting the problem into smaller subproblems, or considers the optimization problems over time.

- *Integer/combinatorial optimization*: The discrete optimization is the problem in which the decision variables assume discrete values from a specified set. The combinatorial optimization problems, on the other hand, are problems of choosing the best combination out of all possible combinations. Most combinatorial problems can be formulated as integer programs. In cognitive radio resource allocation, many variables have only integer values such as the modulation rate, and other variables such as the channel allocation have a combinatorial nature. Integer optimization is the process of finding one or more best (optimal) solutions in a well-defined discrete problem space. The major difficulty with these problems is that we do not have any optimality conditions to check if a given (feasible) solution is optimal or not. There are several possible solutions such as relaxation and decomposition, enumeration, cutting planes and the knapsack problem.

Overall, the centralized scheme has the best performance but needs considerable signaling and overheard. The centralized scheme can fit the network scenarios where the topology is simple, or can be served as a performance upper bound to compare with other more practical schemes.

9.10 Degrees of Cooperation

In this section, we conclude this chapter by discussing the degrees of cooperation for the different approaches. As we have mentioned previously, the non-cooperation among cognitive radios can significantly reduce the network performances and in turn the users' own benefits. The cooperation can bring mutual benefits to cognitive radio users. However, these benefits do not come for free. Some network infrastructure is needed to build up these mutual benefits, which cause some design issues. In the previous sections, we have already discussed the different approaches. Next, we concentrate on the design issues such as signaling and complexity. Then the pros and cons are investigated. We further study the best network scenarios under which a certain approach fits best. By understanding the above issues, finally, we compare the different approaches.

In cognitive networks, in order to obtain the information such as channel conditions, signaling is performed so that resource allocation can be conducted in an optimal way. However, signaling incur considerable communication overhead. Most of the current wireless networks have more than 20% of overhead. Reducing the overhead can greatly enhance the spectrum utilization, increase the number of users and improve the network performance. One of the possible ways to reduce overhead is to conduct resource optimization using only local information. This is very important especially if the system topology is distributed as in cognitive radio networks.

Since the cognitive radios are usually equipped with simple transceivers, the complexity issue has to be considered. There are two concerns for optimization complexity. First one is how complex the optimization algorithm is, and the second one is where the optimization is performed. A large number of optimization problems especially for those with integer nature are NP-hard. To solve the problem, some suboptimal simple solutions should be developed. Currently, the common hardware and software can solve the problems with the complexity up to ON^2 where N is the bottleneck parameter. The complexity for the distributed cognitive radio users should be even lower.

The next important problem for design of wireless cognitive networks is mobility. Due to the topology changes and channel variation, the optimization needs to be performed in a timely fashion. This requirement casts a significant challenge for the iterative solutions and demand for the information without delay. The convergence speed for the iterative algorithms should be at least as fast as the variation caused by mobility. For example, in a 3G UMTS system, the closed-loop power control signal is performed 1500 times per second. This fast update for the iteration can improve the convergence speed, but on the other hand cause additional overhead for the signaling. For a non-iterative algorithm, the information must be accurate without delay. Otherwise, the optimization results will become obsolete and generate inferior performances.

Different approaches have their own pros and cons, and there is no one "elixir" that can handle all design problems for all types of networks. In addition, there are some other design issues that need to be paid attention to. So we need to understand the strength and weakness for different schemes, so that we can select the one that fits the network scenario best. In the sequel, we discuss and compare all types of schemes discussed in this chapter. Table 9.3 summarizes some of the discussions.

- *Non-cooperative competition*: The cognitive radio users have their own autonomy and they access the spectrum in a fully distributed way. The cognitive users utilize only local information for resource allocation, and no signaling or overheard is necessary. The complexity of the non-cooperative competition algorithms is usually low, due to the commonly used convex (or concave) utility function. This type of approaches can fully adapt to the user mobility, since the users can simply change their strategies for better payoff if the situations change. All the above factors are the advantages of non-cooperative competition. However, the significant problem for such approaches is the possible low performance, due to the severe non-cooperation. Even though the problem can be improved using techniques such as pricing, the solutions do not come for free. For example, calculating the optimal price is a difficult problem and might need considerable signaling, which counteracts the advantages for such approaches. So the best network scenarios for non-cooperative competition are those where the Nash equilibria have similar performances to those of the optima. Specifically for the interference avoidance, if the clusters are located sufficiently far away, the

Table 9.3. Degrees of cooperation.

Types	Cooperation	Signaling	Pros	Cons
Non-cooperative competition	Nash equilibria	None	No overhead, simple	Less efficient
Correlated equilibrium and learning	Outside convex hull of Nash equilibria	None	No overhead, better performance	Convergence slow, little mobility
Referee mediation	Only good Nash equilibria	Some for referee	Better performance	Bad stability, low mobility
Repeated interaction	Any feasible solution better than Nash equilibria	Perfectly observed public information	Local information, better performance	No mobility, need mutual dependency, false war
Spectrum auction	Nash equilibria	Some for auctioneer	Simple, fair	No mobility, signaling
Bargaining	Fair Pareto optimum	Only to partners	Simple, mobility	Monopoly
Cooperative game	Fair mutual benefits	Global information before participation	Stable, fair, autonomous	Signaling before contract, no mobility
Centralized optimization	Global optimum	Global information	Optimum, mobility	Overhead, estimator errors

non-cooperative competition has good performance due to less co-channel interference.

- *Learning for better equilibria*: Nash equilibria might not be the best equilibria for distributed cognitive radio users. Learning scheme can achieve the better equilibria using only the past history and without requiring more signaling and overhead. The complexity of learning algorithms can be relatively high. Moreover, there is a tradeoff between the convergence speed and complexity. To achieve the fast convergence speed, the complexity of the learning algorithms can be high. Some simple learning algorithms have been proved to converge to the optimal solution with sufficiently long learning time. However, the long learning time causes a problem for mobility. If the users move frequently, before the learning algorithms converge, the situations such as network topologies and channel conditions may change. This is similar to slope overload distortion in ADPCM or delta modulation. Moreover, if the non-cooperative competition is too severe, the learning algorithms might converge too slowly, fluctuate or become very sensitive to randomness. So the learning schemes fit the situation in which the non-cooperative competition is not so severe; there is an achievable gap between Nash equilibria and the optimal solutions; and the network mobility is sufficiently low.

- *Referee mediation*: To overcome the challenges for the learning schemes, a virtual referee can improve the outcome of non-cooperative competition by intervening in the game rules. The virtual referee needs to collect the information so as to improve the equilibria. However, this information exchange burden is not severe since it is only necessary when the networks are balanced to undesired equilibria. The complexity for this type of approaches is relatively low. Mobility is not an issue, if the network changes can be handled mostly by the non-cooperative competitions and the frequency for virtual referee's mediation is not too high. However, the referee mediation approaches require the assumptions that all the cognitive radios are able to follow the instructions to change the game rules. So the cognitive radio users are not fully autonomous. Moreover, too much intervention by the virtual referee can cause a network stability problem. This type of approach fits the similar scenarios like the learning schemes except that the cognitive users can have a certain extent of mobility.

- *Threat and punishment from repeated interactions*: If the cognitive radio users belong to different authorities, they will not listen to the virtual referee. Under this condition, threat and punishment from repeated interactions can be utilized to enforce user cooperation. There is public information that needs to be received accurately by all cognitive users who use this information to determine if any other user deviates from cooperation. Because of this reason, if this public information is not accurate, some "false war" can happen among distributed cognitive users. To a certain extent, the network can deteriorate to total non-cooperation. The complexity of such an approach is not high, since only detection, cooperation and non-cooperation need to be performed. This approach can hardly handle the mobility, since the deviating users can move from cluster to cluster to escape future punishment, or equivalently saying that mobile users might not care too much about future punishment so that they would rather behave non-cooperatively now. In addition, if some cognitive users have less dependency on other users, the other users can arbitrarily play non-cooperatively with these users without worrying about revenge. So this type of approach fits the network scenarios where the cognitive users have less mobility, have mutual dependency, and can access public information accurately.

- *Spectrum auction*: Similar to an auction in real life, a spectrum auction requires an auctioneer who can handle the bidding and resulting resource allocations. The information exchange requires the signaling of bidding and allocation results, which can be relatively trivial. The complexity of auction algorithms can be very high, for example the VCG auction. But the computation burden is for the auctioneer only. The spectrum auction cannot handle mobility. If the mobile users move, a new auction needs to be implemented. Similar to the referee case, the cognitive users are required to follow the instructions for resource usage from the auctioneer. The spectrum auction fits the network scenario without mobility, and there should be some semi-centralized nodes, such as cluster heads, that can serve as auctioneers.

- *Mutual benefits via bargaining*: The bargaining approach can provide the local mutual benefits to the adjacent cognitive radios. The cognitive users can exchange the information locally to bargain on spectrum usage. The overhead is limited to local users only and the complexity of algorithms is usually low. This type of approach can handle the mobility, since the bargaining can take place whenever new mutual benefits appear. Moveover, this bargaining process fits the situations with integer and combinatorial optimization well. However, if one user occupies most of the spectrum, it is less efficient for the other cognitive users to negotiate with this monopolist.

- *Contract using cooperative game*: This approach is similar to a spectrum auction, except that there is no need for an auctioneer. Instead, all participant users "put their cards on the table" and figure out the best strategies for coalitions. The resulting mutual benefits are divided to cognitive radios according to different fairness criteria. A lot of information signaling is necessary before the contract is agreed by all users, but no signaling is needed after that. The complexity of coalition formation can be high. The mobility is required to be limited, otherwise the contract becomes obsolete too quickly. The cooperative game fits better if the users are located densely, so that the information exchange can be easy.

- *Centralized scheme*: The cognitive radios are the slave type, which means the users fully cooperate and follow the instructions from the centralized node. The optimization requires the accurate channel information without delay. For the scenario of multiple cognitive radio users talking to one common destination such as a base station, centralized control can be utilized since the channel information is constantly collected by the destination to maintain the links. On the other hand, for the network scenarios like the ad hoc case or the multiple cluster case, it is very difficult for the channel information to be exchanged over different destinations. In this situation, the centralized control can hardly be implemented but can serve as a performance upper bound for the other distributed schemes. The complexity of the centralized schemes are usually high, due to the non-linear, non-convex and probably integer or dynamic nature of the optimization problem. However, the optimization is usually performed in the destination where the computation ability is relatively high. For mobility, if the channel information is prompt, the centralized scheme is robust with the channel variation. However, if the channel information needs to be fed back or sent via signaling, the delay can significantly degrade the performance of the centralized scheme.

References

1. S. Haykin, "Cognitive radio: Brain-empowered wireless communications," *IEEE J. Select. Areas Commun.*, vol. 23, no. 2, pp. 201–220, Feb. 2005.
2. I. F. Akyildiz, W. Y. Lee, M. C. Vuran, and S. Mohanty, "Next generation/dynamic spectrum access/cognitive radio wireless networks: A survey," *Comput. Netw. Int. J. Comput. Telecommun. Netw.*, vol. 50, no. 13, pp. 2127–2159, Sept. 2006.
3. Z. Han, Z. Ji, and K. J. Ray Liu, "Fair multiuser channel allocation for OFDMA networks using Nash bargaining solutions and coalitions," *IEEE Trans. Commun.*, vol. 53, no. 8, pp. 1366–1376, 2005.

4. Z. Han, C. Pandana, and K. J. Ray Liu, "Distributive opportunistic spectrum access for cognitive radio using correlated equilibrium and no-regret learning," in *Proc. of IEEE Wireless Communications and Networking Conference* (Hong Kong, China), Mar. 2007.

5. Z. Han, Z. Ji, and K. J. Ray Liu, "A referee-based distributed scheme of resource competition game in multi-cell multi-user OFDMA networks," *IEEE J. Select. Areas Commun.*, Special Issue on Non-cooperative Behavior in Networking, 2nd Quarter, 2007.

6. Z. Han, C. Pandana, and K. J. R. Liu, "A self-learning repeated game framework for optimizating packet forwarding networks," in *Proc. of IEEE Wireless Communications and Networking Conference* (New Orleans, LA), pp. 2131–2136, Mar. 2005.

7. Z. Han, Z. Ji, and K. J. Ray Liu, "Dynamic distributed rate control for wireless networks by optimal cartel maintenance strategy," in *Proc. of IEEE Global Telecommunications Conference*, vol. 6 (Dallas, TX), pp. 3454–3458, Nov. 2004.

8. X. Liu and W. Wang, "On the characteristics of spectrum-agile communication networks," in *Proc. IEEE International Symposium on New Frontiers in Dynamic Spectrum Access Networks, 2005 (DySPAN 2005)* (Baltimore, MD), pp. 214–223, Nov. 2005.

9. Z. Han and K. J. Ray Liu, "Non-cooperative power control game and throughput game over wireless networks," *IEEE Trans. Commun.*, vol. 53, no. 10, pp. 1625–1629 Oct. 2005.

10. A. B. MacKenzie and L. A. DaSilva, *Game theory for wireless engineers.* Morgan and Claypool Publishers, 2006.

11. D. Niyato and E. Hossain, "A game-theoretic approach to competitive spectrum sharing in cognitive radio networks," in *Proc. of IEEE Wireless Communication and Networking Conference* (Hong Kong), March 2007.

12. D. Niyato and E. Hossain, "Hierarchical spectrum sharing in cognitive radio: A microeconomic approach," in *Proc. of IEEE Wireless Communication and Networking Conference* (Hong Kong), March 2007.

13. B. Wang, Z. Han, and K. J. Ray Liu, "Distributed relay selection and power control for multiuser cooperative communication networks using buyer/seller game," in *Proc. of Annual IEEE Conference on Computer Communications, INFOCOM* (Anchorage, Alaska), May 2007.

14. D. P. Bertsekas and R. G. Gallager, *Data networks*, 2nd ed. Prentice-Hall, 1992.

15. Y. Yang and T. S. P. Yum, "Delay distributions of slotted ALOHA and CSMA," *IEEE Trans. Commun.*, vol. 51, no. 11, pp. 1846–1857, Nov. 2003.

16. R. J. Aumann, "Subjectivity and correlation in randomized strategy," *J. Math. Econ.*, vol. 1, no. 1, pp. 67–96, 1974.

17. R. J. Aumann, "Correlated equilibrium as an expression of Bayesian rationality," *Econometrica*, vol. 55, no. 1, pp. 1–18, Jan. 1987.

18. S. Hart and A. Mas-Colell, "A simple adaptive procedure leading to correlated equilibrium," *Econometrica*, vol. 68, no. 5, pp. 1127–1150, Sept. 2000.

19. R. B. Myerson, *Game theory: Analysis of conflict,* 5th ed. Harvard University Press 2002.

20. Q. Zhao, L. Tong, and A. Swami, "Decentralized cognitive MAC for dynamic spectrum access," in *Proc. of IEEE International Symposium on New Frontiers in Dynamic Spectrum Access Networks, 2005 (DySPAN 2005)* (Baltimore, MD), pp. 224–232, Nov. 2005.

21. Z. Han, Z. Ji, and K. J. Ray Liu, "Power minimization for multi-cell OFDM networks using distributed non-cooperative game approach," in *Proc. of IEEE Global Telecommunications Conference* (Dallas, TX), Nov. 2004.

22. S. T. Chung, S. J. Kim, and J. M. Cioffi, "A game-theoretic approach to power allocation in frequency-selective Gaussian interference channels," in *Proc. of IEEE International Symposium on Information Theory* (Pacifico Yokohama, Kanagawa, Japan), June 2003.

23. S. T. Chung and A. J. Goldsmith, "Degrees of freedom in adaptive modulation: A unified view," *IEEE Trans. Commun.*, vol. 49, no. 9, pp. 1561–1571, Sept. 2001.
24. D. P. Bertsekas, *Nonlinear programming*, 2nd ed. Athena Scienific, 1999.
25. R. Yates, "A framework for uplink power control in cellular radio systems," *IEEE J. Select. Areas Commun.*, vol. 13, no. 7, pp. 1341–1348, Sept. 1995.
26. G. J. Foschini and Z. Miljanic, "A simple distributed autonomous power control algorithms and its convergence," *IEEE Trans. Veh. Technol.*, vol. 40, no. 4, pp. 641–646, Nov. 1993.
27. L. Anderegg and S. Eidenbenz, "Ad hoc-VCG: A truthful and cost-efficient routing protocol for mobile ad hoc networks with selfish agents," in *Proc. of ACM Ninth Annual International Conference on Mobile Computing and Networking (MobiCom)* (San Diego, CA), Sept. 2003.
28. V. Krishna, *Auction theory*. Academic Press, 2002.
29. J. Huang, R. Berry, and M. L. Honig, "Auction-based spectrum sharing," *ACM/Springer Mobile Netw. Appl. J. (MONET)*, vol. 11, no. 3, pp. 405–418, June 2006.
30. H. Yaiche, R. R. Mazumdar, and C. Rosenberg, "A game theoretic framework for bandwidth allocation and pricing in broadband networks," *IEEE/ACM Trans. Netw.*, vol. 8, no. 5, pp. 667–678, Oct. 2000.
31. D. Grosu, A. T. Chronopoulos, M. Y. Leung, "Load balancing in distributed systems: An approach using cooperative games," in *Proc. of IPDPS 2002*, pp. 52–61, 2002.
32. H. W. Kuhn, "The Hungarian method for the assignment problem," *Naval Res. Log.*, Quarterly 2, 1955.
33. J. E. Suris, L. DaSilva, Z. Han, and A. MacKenzie, "Cooperative game theory approach for distributed spectrum sharing," in *Proc. of IEEE International Conference on Communications* (Glasgow, Scotland), June 2007.
34. H. Zheng and L. Cao, "Device-centric spectrum management," in *Proc. of IEEE International Symposium on New Frontiers in Dynamic Spectrum Access Networks, 2005 (DySPAN 2005)* (Baltimore, MD), pp. 56–65, Nov. 2005.
35. K.-D. Lee and V. C. M. Leung, "Fair allocation of subcarrier and power in an OFDMA wireless mesh network," *IEEE J. Select. Areas Commun.*, vol. 24, no. 11, pp. 2051–2060, Nov. 2006.
36. F. Fu, A. Fattahi, and M. van der Schaar, "Game-theoretic paradigm for resource management in spectrum agile wireless networks," in *Proc. of IEEE International Conference on Multimedia and Expo* (Toronto, Canada), July 2006.

Additional Reading

1. X. Liu and S. Shankar, "Sensing-based opportunistic channel access," *ACM MONET*, vol. 11, no. 4, pp. 577–591, Aug. 2006.
2. J. Huang, R. A. Berry, and M. L. Honig, "Spectrum sharing with distributed interference compensation," in *Proc. of IEEE International Symposium on New Frontiers in Dynamic Spectrum Access Networks, 2005 (DySPAN 2005)* (Baltimore, MD), pp. 88–93, Nov. 2005.
3. V. Srivastava, J. Neel, A. MacKenzie, R. Menon, L. A. DaSilva, J. Hicks, J. H. Reed, and R. Gilles, "Using game theory to analyze wireless ad hoc networks," *IEEE Commun. Surv. Tutor.*, vol. 7, no. 4, pp. 46–56, 4th quarter 2005.
4. R. Etkin, A. Parekh, and D. Tse, "Spectrum sharing for unlicenced bands," in *Proc. of IEEE International Symposium on New Frontiers in Dynamic Spectrum Access Networks, 2005 (DySPAN 2005)* (Baltimore, MD), pp. 251–258, Nov. 2005.

5. N. Nie and C. Comaniciu, "Adaptive channel allocation spectrum etiquette for cognitive radio networks," *ACM MONET (Mobile Networks and Applications)*, Special Issue on Reconfigurable Radio Technologies in Support of Ubiquitous Seamless Computing, 2006.
6. J. Neel, J. Reed, and R. Gilles, "Game models for cognitive radio algorithm analysis," in *Proc. of SDR Forum Technical Conference* (Phoenix, Arizona), pp. 15–18, Nov. 2004.
7. C. Saraydar, N. Mandayam, and D. Goodman, "Efficient power control via pricing in wireless data networks," *IEEE Trans. Commun.*, vol. 50, no. 2, pp. 291–303, Feb. 2002.
8. C. Pandana and K. J. Ray Liu, "Near-optimal reinforcement learning framework for energy-aware sensor communications," *IEEE J. Select. Areas Commun.*, vol. 23, pp. 259–268, Apr. 2005.

Cognitive MAC Protocols for Dynamic Spectrum Access

Qing Zhao[1], Yunxia Chen[1], and Ananthram Swami[2]

[1] Department of Electrical and Computer Engineering
    University of California at Davis[†], USA
    {qzhao,yxchen}@ece.ucdavis.edu
[2] Army Research Laboratory, USA
    aswami@arl.army.mil

10.1 Introduction

The paradox between the overly crowded spectrum and the pervasiveness of idle frequency bands in both time and space indicates that spectrum shortage results from the current static spectrum management policy rather than the physical scarcity of usable radio frequencies [1]. To improve spectrum efficiency, researchers in the engineering, economics, and regulation communities have been actively searching for better spectrum management strategies. Under the general term of dynamic spectrum access, various spectrum reform ideas have been proposed. We provide below a taxonomy to illustrate the relationship among these diverse ideas.

10.1.1 Dynamic Spectrum Access

The term "dynamic spectrum access" has broad connotations that encompass various approaches to spectrum reform, and should be contrasted with the current static spectrum management policy. As illustrated in Fig. 10.1, dynamic spectrum access strategies can be generally categorized under three models.

1. Dynamic exclusive use model: This model maintains the basic structure of the current spectrum regulation policy: spectrum bands are licensed to services for exclusive use. The main idea is to introduce flexibility to improve spectrum efficiency. Two approaches have been proposed under this model: *spectrum property rights* [2, 3] and *dynamic spectrum allocation* [4]. The former approach allows licensees to sell and trade spectrum and to freely choose technology. Economy

---

[†] This work was supported in part by the Army Research Laboratory CTA on Communication and Networks under Grant DAAD19-01-2-0011 and by the National Science Foundation under Grants CNS-0627090 and ECS-0622200.

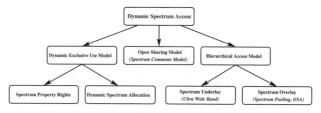

Fig. 10.1. A taxonomy of dynamic spectrum access.

and market will thus play a more important role in driving toward the most profitable use of this limited resource. Note that even though licensees have the right to lease or share the spectrum for profit, such sharing is not mandated by the regulation policy.

The second approach, dynamic spectrum allocation, was brought forth by the European DRiVE project [4]. It aims to improve spectrum efficiency through dynamic spectrum assignment by exploiting the spatial and temporal traffic statistics of different services. Similar to the current static spectrum allotment policy, such strategies allocate, at a given time and region, a portion of the spectrum to a radio access network for its exclusive use. This allocation, however, varies at a much faster scale.

Based on an exclusive-use model, these approaches cannot eliminate white space in spectrum resulting from the bursty nature of wireless traffic.

2. Open sharing model: Also referred to as spectrum commons [5,6], this model employs open sharing among peer users as the basis for managing a spectral region. Advocates of this model draw support from the phenomenal success of wireless services operating in the unlicensed ISM band (e.g., WiFi). Centralized [7,8] and distributed [9–11] spectrum sharing strategies have been initially investigated to address technological challenges under this spectrum management model.

3. Hierarchical access model: Built upon a hierarchical access structure with primary and secondary users, this model can be considered as a hybrid of the above two. The basic idea is to open licensed spectrum to secondary users and limit the interference perceived by primary users (licensees). Two approaches to spectrum sharing between primary and secondary users have been considered: *spectrum underlay* and *spectrum overlay*.

The underlay approach imposes severe constraints on the transmission power of secondary users so that they operate below the noise floor of primary users. By spreading transmitted signals over a wide frequency band (UWB), secondary users can potentially achieve short-range high data rate with extremely low transmission power. Based on a worst-case assumption that primary users transmit all the time, this approach does not exploit spectrum white space.

Spectrum overlay was first envisioned by Mitola [12] under the term "spectrum pooling" and then investigated by the DARPA XG program [13] under the term "opportunistic spectrum access (OSA)". Differing from spectrum underlay, this approach does not necessarily impose severe restrictions on the transmission power of secondary users, but rather on when and where they may transmit. It directly targets at spatial and temporal spectrum white space by allowing secondary users to identify and exploit local and instantaneous spectrum availability in a non-intrusive manner.

Compared to the dynamic exclusive use and open sharing models, this hierarchical model is perhaps the most compatible with the current spectrum management policy and legacy wireless systems. Furthermore, the underlay and overlay approaches can be employed simultaneously to further improve spectrum efficiency.

We point out that the hierarchical access model is sometimes categorized under the open sharing model (see, e.g., [6]). Spectrum sharing between primary and secondary users is, however, fundamentally different from spectrum sharing among peer users in both technical and regulatory aspects. We have thus separated the hierarchical access model from the open sharing model in the above taxonomy.

10.1.2 Cognitive Radio

Cognitive radio is often used as a synonym for dynamic spectrum access. We provide below a brief introduction to software-defined radio and cognitive radio.

The terms "software-defined radio" and "cognitive radio" were coined by Mitola in 1991 and 1998, respectively. Software-defined radio, sometimes shortened to software radio, is generally a multi-band radio that supports multiple air interfaces and protocols and is reconfigurable through software run on DSP or general-purpose microprocessors [14]. Cognitive radio, built upon a software radio platform, is a context-aware intelligent radio capable of autonomous reconfiguration by learning and adapting to the communication environment [15]. While dynamic spectrum access is certainly an important application of cognitive radio, cognitive radio represents a much broader paradigm where many aspects of communication systems can be improved via cognition.

10.2 Cognitive MAC for Opportunistic Spectrum Access

In this chapter, we focus on the overlay approach under the hierarchical access model (see Fig. 10.1). The term opportunistic spectrum access (OSA) will be adopted throughout. Our emphasis is on the design of cognitive medium access control (MAC) protocols for secondary users in OSA networks.

10.2.1 Basic Components of Cognitive MAC

Basic design components of cognitive MAC for OSA include (1) a sensing policy for real-time decisions about whether to sense and where in the spectrum to sense and (2) an access policy that determines whether to access based on the sensing outcomes.

The purpose of the sensing policy is twofold: to identify a spectrum opportunity for immediate access and to obtain statistical information on spectrum occupancy for improved future decisions. A balance must be reached between these two often conflicting objectives, and the trade-off should adapt to the bursty traffic and energy constraint of the secondary user. For example, when there are energy costs associated with sensing, a secondary user may decide to skip sensing when its current estimate of spectrum occupancy indicates that no channels are likely to be idle. Clearly, such decisions should balance the reward in energy savings with the cost in lost spectrum information and potentially missed spectrum opportunities.

The objective of the access policy, on the other hand, is to minimize the chance of overlooking an opportunity without violating the constraint of being non-intrusive. Whether the secondary user should adopt an aggressive or a conservative access policy depends on the operating characteristics (probability of false alarm vs. probability of miss detection, and permissible level of interference) of the spectrum sensor. A joint design of MAC protocols and spectrum sensors at the physical layer is thus necessary to achieve optimality. Energy constraints will further complicate the design of access policies. For energy-constrained OSA in fading environments, the secondary user may avoid transmission when the sensed channel is in a deep fade. Even the residual energy level will play an important role in decision-making. When the battery is depleting, should the user wait for increasingly better channel conditions for transmission or should it lower the requirement on channel conditions given that sensing also costs energy? How is such a decision affected by the accuracy and energy consumption characteristics of the spectrum sensor? And how sensitive are such policies to incomplete models and inaccurate model parameter estimates?

The above discussion highlights some of the complexities in the design of a cognitive MAC for OSA in a dynamic network environment with fading, sensing errors, and energy constraints. It demonstrates that the optimal design of cognitive MAC for OSA calls for a cross-layer approach that integrates signal processing with networking.

In this chapter, we aim to illuminate the interactions between the physical and the MAC layers in OSA networks. We focus, in particular, on the impact of sensing errors and channel fading conditions at the physical layer on the optimal sensing and access policies at the MAC layer. In particular, we present a decision-theoretic framework first developed in [16–19]. Based on the theory of partially observable markov decision process (POMDP), this framework integrates the basic components of OSA, leading to an optimal joint design of signal processing algorithms for opportunity identification and MAC protocols for opportunity exploitation.

10.2.2 Related Work

A majority of the existing work focuses on spatial spectrum opportunities that are static or slowly varying in time. Example applications include the reuse of certain TV-bands that are not used for TV broadcast in a particular region. Due to the slow temporal variation of spectrum occupancy, real-time opportunity identification is not as critical a component in this class of applications, and the prevailing approach to OSA tackles network design in two separate steps: (1) opportunity identification assuming continuous full-spectrum sensing; (2) opportunity allocation among secondary users assuming full knowledge of spectrum opportunities. Opportunity identification in the presence of fading and noise uncertainty has been studied in [20–24]. Spatial opportunity allocation among secondary users can be found in [25–28] and references therein. Differing from these works, we focus on the exploitation of temporal spectrum opportunities resulting from the bursty traffic of primary users, For an overview of challenges and recent development in OSA, readers are referred to [29, 30].

10.3 The Network and Protocol Model

10.3.1 The Network Model

Consider a spectrum consisting of N channels,[1] each with bandwidth B_n ($n = 1, \cdots, N$). These N channels are licensed to a primary network whose users communicate according to a synchronous slot structure. The traffic statistics of the primary network are such that the occupancy of these N channels follows a discrete-time Markov process with 2^N states. Specifically, the network state in slot t is given by $S(t) \triangleq [S_1(t), \cdots, S_N(t)]$ where $S_n(t) \in \{0 \text{ (occupied)}, 1 \text{ (idle)}\}$ is the occupancy state of channel n. The state diagram for $N = 3$ and a sample path of the state evolution are illustrated in Figs. 10.2 and 10.3, respectively. We assume that the spectrum usage statistics of the primary network remain unchanged for T slots. We further assume that the state transition probabilities of the underlying Markov model are known: $P_{s,s'} \triangleq \Pr\{S(t+1) = s' \mid S(t) = s\}$, for every $s, s' \in \{0,1\}^N$. In Section 10.4.3, we discuss OSA with unknown or mismatched Markov model.

We consider a secondary network that seeks spectrum opportunities in these N channels (see Fig. 10.3). We focus on an ad hoc network where secondary users join/exit the network and sense/access the spectrum independently without exchanging local information. In each slot, a secondary user chooses a set of channels to sense and a set of channels to access. Limited by its hardware constraints and energy supply, a secondary user can sense no more than L_1 ($L_1 \leq N$) and access no more than L_2 ($L_2 \leq L_1$) channels in each slot.[2] For the ease of presentation, we assume $L_1 = L_2 = 1$. Results presented in this chapter can be extended to general cases as discussed in [17, 18, 31].

Our goal is to develop cognitive MAC protocols for the secondary network. For an ad hoc OSA network without a central coordinator or a dedicated communication channel, it is desirable to have a decentralized MAC protocol where each secondary user independently searches for spectrum opportunities, aiming at optimizing

---

[1] Here we use the term channel broadly. A channel can be a frequency band with specified bandwidth, a collection of spreading codes in DS-CDMA network, a set of hopping codes in FH-SS, or a set of subcarriers in an OFDM system.

[2] In principle, we can let $L_2 = N$, i.e., access decisions need not be confined to the currently sensed set of channels.

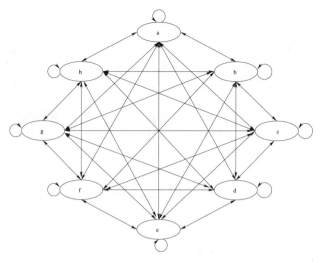

Fig. 10.2. The underlying Markov process for $N = 3$.

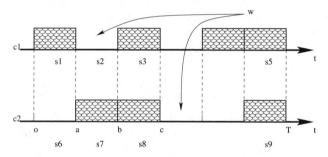

Fig. 10.3. A sample path of spectrum occupancy.

its own performance. Such decentralized protocols do not rely on cooperation among secondary users.

10.3.2 The Basic Protocol Structure

Without delving into protocol details (which are given in Sect. 10.6), we present here the basic protocol structure. At the beginning of each slot,[3] a secondary user with data to transmit chooses a channel to sense and decides whether to access based on the sensing outcome. When the secondary user decides to transmit, it generates a random backoff time, and transmits when this timer expires and no other secondary user has already accessed that channel during the backoff time. At the end of the slot, the receiver acknowledges a successful data transmission. The basic slot structure is illustrated in Fig. 10.4.

---

[3] Secondary users can synchronize to a slot structure broadcasted by the primary network.

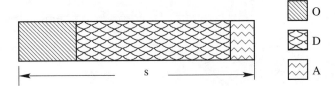

Fig. 10.4. The slot structure.

10.4 The Impact of Sensing Errors on Non-intrusive Cognitive MAC

We study the impact of sensing errors at the physical layer on the design of cognitive MAC protocols. We formulate the joint PHY-MAC design of OSA networks as a constrained partially observable Markov decision process (POMDP). Involved in the design are three basic components: a spectrum sensor at the physical layer; a sensing policy, and an access policy, both at the MAC layer.

10.4.1 Problem Formulation

10.4.1.1 Spectrum Sensor

The spectrum sensor of a secondary user detects, at the beginning of each slot, the availability of the chosen channel. It essentially performs a binary hypotheses test: $\mathcal{H}_0$ (null hypothesis indicating that the sensed channel is idle) vs. $\mathcal{H}_1$ (alternative indicating a busy channel). Let Θ_a be the sensing outcome (the result of the hypotheses test): $\Theta_a = 1$ (idle) and $\Theta_a = 0$ (busy).

If the sensor mistakes $\mathcal{H}_0$ for $\mathcal{H}_1$, a false alarm occurs, and a spectrum opportunity is overlooked by the sensor. On the other hand, when the sensor mistakes $\mathcal{H}_1$ for $\mathcal{H}_0$, we have a miss detection. Let $\epsilon \overset{\Delta}{=} \Pr\{\Theta_a = 0 \,|\, S_a = 1\}$ and $\delta \overset{\Delta}{=} \Pr\{\Theta_a = 1 \,|\, S_a = 0\}$ denote, respectively, the probabilities of false alarm and miss detection. The performance of a sensor is specified by the receiver operating characteristic (ROC) curve which gives the probability of detection $1 - \delta$ as a function of ϵ (see Fig. 10.5). We point out that analyzing the ROC curve of the spectrum sensor in a wireless network environment can be complex. We assume here that the ROC curve of the spectrum sensor has already been obtained, and we focus on the tradeoff between false alarm and miss detection. Specifically, we seek to answer the following question: which point δ on the given ROC curve should the spectrum sensor operate at?

If the secondary user completely trusts the sensing outcome in decision-making, false alarms result in wasted spectrum opportunities whereas miss detections lead to collisions with primary users. To optimize the performance of the secondary user while limiting its interference to the primary network, we should carefully choose the sensor operating point. Meanwhile, the spectrum access decisions should be made by taking into account the sensor operating characteristics. A joint design of the

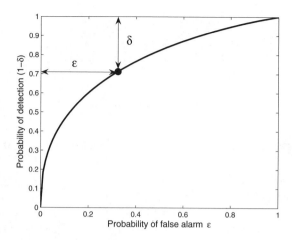

Fig. 10.5. The ROC curve of a spectrum sensor.

spectrum sensor at the physical layer and the access policy at the MAC layer is thus necessary to achieve optimality.

10.4.1.2 Sensing and Access Policies

The sensing policy specifies, in each slot, which channel to sense, and the access policy determines whether to transmit based on the sensing outcome. At the beginning of a slot, a secondary user with data to transmit chooses a channel $a \in \{1, \ldots, N\}$ to sense. Based on the sensing outcome Θ_a, the secondary user decides whether to transmit over the sensed channel: $\Phi_a \in \{0 \text{ (no access), } 1 \text{ (access)}\}$. At the end of the slot, the receiver acknowledges a successful data transmission: $K_a \in \{0 \text{ (unsuccessful), } 1 \text{ (successful)}\}$. Note that an acknowledgement $K_a = 1$ is obtained if and only if the secondary user chooses to access $\Phi_a = 1$ and the channel is idle $S_a = 1$, i.e.,

$$K_a = 1_{[S_a=1, \Phi_a=1]}. \tag{10.1}$$

A reward $R_{K_a}^{(a, \Phi_a)}$ is accrued depending on K_a. Assuming that the number of information bits that can be transmitted is proportional to the channel bandwidth, we define the reward $R_{K_a}^{(a, \Phi_a)}$ obtained by choosing sensing and access action (a, Φ_a) as

$$R_{K_a}^{(a, \Phi_a)} = K_a B_a. \tag{10.2}$$

Due to partial spectrum monitoring and sensing errors, the secondary user and the receiver cannot directly observe the current state of the spectrum occupancy. We thus have a POMDP.

It has been shown in [32] that the knowledge of the current spectrum occupancy state based on all past decisions (i.e., sensing and access actions) and observations (i.e., acknowledgements) can be summarized by a belief state

$\lambda(t) \triangleq \{\lambda_s(t)\}_{s \in \{0,1\}^N}$, where $\sum_s \lambda_s(t) = 1$. Each element $\lambda_s(t)$ of the belief state $\lambda(t)$ is the conditional probability (given the decision and observation history) that the current spectrum occupancy state is given by $s \in \{0,1\}^N$ prior to the state transition in slot t. Hence, a sensing policy π_s is given by a sequence of functions: $\pi_s = [\mu_1, \ldots, \mu_T]$ where $\mu_t : [0,1]^{2^N} \rightarrow \{1, \ldots, N\}$ maps the belief state $\lambda(t) \in [0,1]^{2^N}$ at the beginning of slot t to a channel $a \in \{1, \ldots, N\}$ to be sensed. An access policy π_c is given by a sequence of functions: $\pi_c = [\nu_1, \ldots, \nu_T]$ where $\nu_t : [0,1]^{2^N} \times \{0,1\} \rightarrow \{0,1\}$ maps the belief state $\lambda(t) \in [0,1]^{2^N}$ and the sensing outcome $\Theta_a \in \{0,1\}$ of the chosen channel a to an access action $\Phi_a \in \{0,1\}$.

10.4.1.3 Design Objective

We want to determine the optimal sensor operating point δ and the optimal sensing and access policies $\{\pi_s, \pi_c\}$. The objective is to maximize the total expected reward (equivalently the throughput of the secondary user) in T slots under the collision constraint:

$$\{\delta^*, \pi_s^*, \pi_c^*\} = \arg \max_{\delta, \pi_s, \pi_c} \mathbb{E}_{\{\delta, \pi_s, \pi_c\}} \left[\sum_{t=1}^T R_{K_a}^{(a, \Phi_a)}(t) \,\middle|\, \lambda(1) \right]$$

$$\text{s.t. } P_a(t) = \Pr\{\Phi_a(t) = 1 \mid S_a(t) = 0, \lambda(t)\} \leq \zeta \text{ holds}$$

$$\text{for any } a \text{ and } t \text{ such that } \Pr\{S_a(t) = 0 \mid \lambda(t)\} > 0 \qquad (10.3)$$

where $\mathbb{E}_{\{\delta, \pi_s, \pi_c\}}$ is the expectation given that sensing and access policies $\{\pi_s, \pi_c\}$ are employed and sensor operates at point δ, $\lambda(1)$ is the initial belief state which is usually given by the stationary distribution of the spectrum occupancy states. Note that when $\Pr\{S_a(t) = 0 \mid \lambda(t)\} = 0$, i.e., channel a is available with probability 1 in slot t, the constraint in (10.3) becomes irrelevant and the secondary user's access decision is simply $\Phi_a(t) = 1$. In the rest of this section, we consider the non-trivial case where $\Pr\{S_a(t) = 0 \mid \lambda(t)\} > 0$ in any channel a and slot t.

10.4.2 Separation Principle for Optimal Joint Design

The design objective given in (10.3) is a constrained POMDP, which usually requires randomized policies to achieve optimality. In this case, a sensing policy determines the mapping from the current belief state to the probability of choosing each channel and an access policy the mapping from the current belief state to the transmission probabilities under different sensing outcomes. Since there exist uncountably many probability distributions, randomized policies are computationally prohibitive. In this section, we establish a separation principle for the optimal joint design. This separation principle reveals the existence of deterministic optimal sensing and access policies, leading to significant complexity reduction. It also enables us to obtain, in closed-form, the optimal sensor operating point and the optimal access policy.

10.4.2.1 The Impact of Sensor Operating Point on Access Policy

Let $f_a^\theta(\lambda(t), t)$ be the probability of transmitting over chosen channel a given sensing outcome $\Theta_a = \theta$ and belief state $\lambda(t)$ at the beginning of slot t. In Theorem 10.1, we provide closed-form optimal transmission probabilities $(f_a^1(\lambda(t), t), f_a^0(\lambda(t), t))$ for different sensor operating points δ.

Theorem 10.1. The optimal access policy is time-invariant and belief-independent. Specifically, the optimal transmission probabilities are solely determined by the sensor operating point δ and the maximum allowed probability of collision ζ, i.e., for any chosen channel a, belief state $\lambda(t)$, and slot t, we have

$$(f_a^1(\lambda(t), t), f_a^0(\lambda(t), t)) = \begin{cases} (1, \frac{\zeta - \delta}{1 - \delta}), & \delta < \zeta \\ (1, 0), & \delta = \zeta \\ (\frac{\zeta}{\delta}, 0), & \delta > \zeta. \end{cases} \qquad (10.4)$$

Proof. See [33] for details.

Theorem 10.1 enables us to study the impact of sensor operating characteristics on the optimal access policy. As illustrated in Fig. 10.6, the ROC curve can be partitioned into two regions: the "conservative" region ($\delta > \zeta$) and the "aggressive"

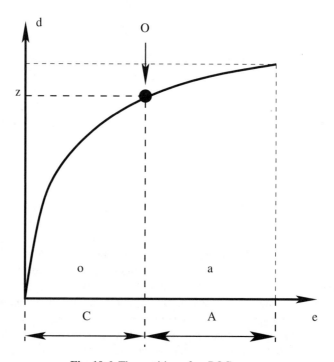

Fig. 10.6. The partition of an ROC curve.

region ($\delta < \zeta$). When $\delta > \zeta$, the spectrum sensor is more likely to misidentify an opportunity (i.e., a busy channel is sensed to be idle). Hence, the access policy should be conservative to ensure that the probability of collision is bounded below ζ. Specifically, even when the sensing outcome $\Theta_a = 1$ indicates that the channel is available, the user should only transmit with probability $\frac{\zeta}{\delta} < 1$. When the channel is sensed to be busy: $\Theta_a = 0$, the user should trust the sensing outcome and refrain from transmission. On the other hand, when $\delta < \zeta$, the spectrum sensor is more likely to overlook an opportunity (i.e., an idle channel is sensed to be busy). Hence, the user should adopt an aggressive access policy: always transmit when the channel is sensed to be available and transmit with probability $\frac{\zeta-\delta}{1-\delta} > 0$ even when the channel is sensed to be busy. When $\delta = \zeta$, the optimal access policy is deterministic: always trust the sensing outcome.

10.4.2.2 The Separation Principle

Given belief state $\lambda(t)$ at the beginning of slot t, we rewrite the design constraint in (10.3) as

$$P_a(t) = \sum_{\theta=0}^{1} \Pr\{\Phi_a = 1 \mid \Theta_a = \theta\} \Pr\{\Theta_a = \theta \mid S_a(t) = 0\}$$
$$= \delta f_a^1(\lambda(t), t) + (1 - \delta) f_a^0(\lambda(t), t). \tag{10.5}$$

Careful inspection of (10.4) and (10.5) reveals that the constraint given in (10.3) is satisfied regardless of the chosen channel. We thus have a separation principle (Theorem 10.2) for the optimal joint OSA design, which decouples the design of spectrum sensor and access policy from that of sensing policy. Following this separation principle, we obtain closed-form optimal sensor operating point δ^* and access policy π_c^* in Theorem 10.3.

Theorem 10.2. Separation Principle The joint design of OSA formulated in (10.3) can be obtained in two steps without losing optimality. First, choose sensor operating point δ and access policy π_c according to (10.4) to maximize the expected immediate reward. Second, choose sensing policy π_s to maximize the expected total reward.

Proof. See [33] for details.

Theorem 10.3. The optimal sensor operating point is $\delta^* = \zeta$. The optimal access policy π_c^* is given by $\Phi_a^* = \Theta_a$.

Proof. See [33] for details.

Theorem 10.3 reveals the existence of deterministic optimal access policy for the constrained POMDP given in (10.3). Specifically, the optimal access policy π_c^* is to simply trust the sensing outcome: $\Phi_a^* = \Theta_a$, i.e., access if and only if the channel is detected to be available.

10.4.2.3 The Optimal Sensing Policy

In Theorem 10.3, we have obtained the optimal sensor operating point δ^* and the optimal access policy π_c^*. Since δ^* and π_c^* have been chosen to ensure the constraint regardless of the chosen channel, we are free to search for the optimal sensing policy π_s^* over the whole design space. The design of the sensing policy thus becomes an unconstrained POMDP, where optimality can be achieved by deterministic policies.

Let $V_t(\boldsymbol{\lambda}(t))$ denote the maximum total expected reward obtained from slot t, $1 \leq t \leq T$, given the belief state $\boldsymbol{\lambda}(t)$ at the beginning of slot t. Given sensor operating point δ^* and access policy π_c^*, we obtain $V_t(\boldsymbol{\lambda}(t))$ recursively by

$$
\begin{aligned}
V_t(\boldsymbol{\lambda}(t)) &= \max_a \sum_{s \in \{0,1\}^N} \sum_{s' \in \{0,1\}^N} \lambda_{s'}(t) P_{s',s} \sum_{k_a=0}^{1} Q_s(k_a) \\
&\quad \times [k_a B_a + V_{t+1}(\mathcal{T}(\boldsymbol{\lambda}(t)\,|\,a, k_a))], \quad 1 \leq t < T \\
V_T(\boldsymbol{\lambda}(T)) &= \max_a \sum_{s \in \{0,1\}^N} \sum_{s' \in \{0,1\}^N} \lambda_{s'}(t) P_{s',s} Q_s(1) B_a
\end{aligned}
\tag{10.6}
$$

where $Q_s(0) = 1 - Q_s(1)$, $Q_s(1) \triangleq \Pr\{K_a = 1 \,|\, S(t) = s\} = 1_{[s_a=1]}(1 - \epsilon^*)$ is the probability of successful transmission when the current spectrum occupancy $S(t)$ is in state $s = [s_1, \ldots, s_N]$. Note that $1_{[s_a=1]}$ indicates whether channel a is idle given $S(t) = s$ and ϵ^* is the probability of false alarm that can be achieved when the spectrum sensor operates at δ^*. The updated belief state $\boldsymbol{\lambda}(t+1) = \mathcal{T}(\boldsymbol{\lambda}(t)\,|\,a, k_a)$ can be obtained via Bayes rule as

$$
\lambda_s(t+1) = \frac{\sum_{s' \in \{0,1\}^N} \lambda_{s'}(t) P_{s',s} Q_s(k_a)}{\sum_{s \in \{0,1\}^N} \sum_{s' \in \{0,1\}^N} \lambda_{s'}(t) P_{s',s} Q_s(k_a)}.
\tag{10.7}
$$

The optimal sensing policy π_s^* can be obtained by solving the optimality equation given in (10.6). It is shown in [32] that $V_t(\boldsymbol{\lambda}(t))$ is piecewise linear and convex, leading to a linear programming procedure for calculating π_s^*.

Suboptimal sensing policies with reduced complexity are developed in [16,17,34].

10.4.3 Simulation Examples

In this section, we provide simulation examples to study the cognitive nature of the MAC protocols developed within the POMDP framework and the impacts of sensor operating point δ and mismatched Markov model on the performance of the optimal OSA.

10.4.3.1 Simulation Setup

We consider $N = 3$ independently evolving channels with the same bandwidth $B_n = 1$. As illustrated in Fig. 10.7, the state transition of spectrum occupancy can be

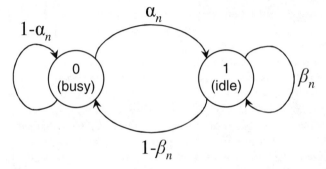

Fig. 10.7. The Markov model of independently evolving channels.

characterized by $\boldsymbol{\alpha} \triangleq [\alpha_1, \alpha_2, \alpha_3]$ and $\boldsymbol{\beta} \triangleq [\beta_1, \beta_2, \beta_3]$, where α_n denotes the probability that channel n transits from state 0 (busy) to state 1 (idle) and β_n denotes the probability that it stays in state 1. We assume that the spectrum occupancy dynamics remain unchanged over $T = 10$ slots. The throughput of the secondary user is measured by the expected total reward per slot, i.e., $V_1(\boldsymbol{\lambda}(1))/T$, where $\boldsymbol{\lambda}(1)$ is given by the stationary distribution of the underlying Markov process.

At the beginning of each slot, the spectrum sensor takes M measurements $\{Y_i\}_{i=1}^M$ of the chosen channel. We assume that both the channel noise and the signal of primary users can be modeled as white Gaussian processes $\mathcal{N}$. Then, the spectrum sensor performs the following hypotheses test:

$$\begin{cases} \mathcal{H}_0 \text{ (idle channel)} : & Y_i \sim \mathcal{N}(0, \sigma_0^2), \; i = 1, \cdots, M \\ \mathcal{H}_1 \text{ (busy channel)} : & Y_i \sim \mathcal{N}(0, \sigma_1^2), \; i = 1, \cdots, M \end{cases}$$

where σ_0^2 is the noise power and σ_1^2 is the primary signal power. The energy detector is optimal under Neyman–Pearson (NP) criterion [35, sect. 2.6.2]:

$$\sum_{i=1}^{M} Y_i^2 \underset{\mathcal{H}_0}{\overset{\mathcal{H}_1}{\gtrless}} \eta \tag{10.8}$$

where the threshold η determines the false alarm and miss detection rates of the detector. The ROC curve of the energy detector is given by [35, Sect. 2.6.2]

$$1 - \delta = 1 - \gamma\left(\frac{M}{2}, \eta \frac{\sigma_0^2}{\sigma_1^2}\right), \quad \epsilon = 1 - \gamma\left(\frac{M}{2}, \eta\right) \tag{10.9}$$

where $(\sigma_1^2 - \sigma_0^2)/\sigma_0^2$ is the SNR and $\gamma(n, a) = \frac{1}{\Gamma(n)} \int_0^a t^{n-1} e^{-t} \, dt$ is the incomplete gamma function. In all the figures, we assume $M = 10$ and SNR = 5 dB.

10.4.3.2 The Cognitive Nature of POMDP Modeling

As discussed in Sect. 10.2.1, a fundamental tradeoff in the design of sensing policies is between obtaining immediate spectrum access and gaining spectrum statistical information for future use. To illustrate this, we consider a simple static sensing strategy that chooses the channel most likely to be available (weighted by its bandwidth) based on the stationary distribution of the underlying Markov process. In this case, the secondary user simply waits on a particular channel predetermined by the spectrum occupancy statistics and the channel bandwidths. Such an approach ignores information, about the underlying state of the Markov process, that can be obtained from the sensing outcomes. Missing in this approach is that every sensing outcome provides information on the state of the underlying Markov process. Channel selection should be based on the *a posterior* distribution of channel availability that exploits the whole history of sensing outcomes, i.e., the belief state. As demonstrated in this section, the optimal sensing strategy is one of sequential decision making that achieves the best trade-off between gaining immediate access in the current slot and gaining system state information for future use. We illustrate in Fig. 10.8 the potential gain of optimally using the observation history assuming perfect sensing. Plotted in Fig. 10.8 is the throughput of the secondary user as a function of time. We see from this figure that the performance of the optimal approach improves over time, which results from the increasingly accurate information on the system state obtained by accumulating observations. Approximately 40% improvement is achieved over the static approach.

10.4.3.3 Impact of Sensor Operating Point on MAC Performance

Figure 10.9 illustrates the impact of sensor operating point δ on the throughput and the optimal access policy of the secondary user. The upper figure plots the maximum throughput of the secondary user for each given sensor operating point δ. The optimal access policy is specified by the transmission probabilities (f_a^0, f_a^1), which are shown in the middle and the lower figures, respectively. We can see that the maximum throughput is achieved at $\delta^* = \zeta = 0.05$ and the transmission probabilities change with δ as given by Theorem 10.1. Interestingly, the throughput curve

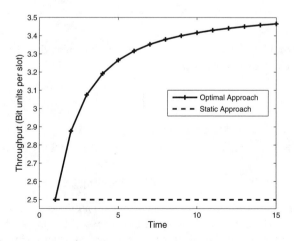

Fig. 10.8. The cognitive nature of POMDP modeling.

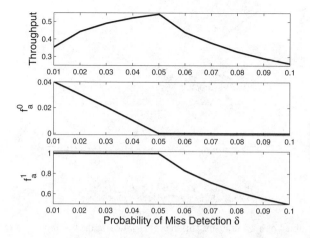

Fig. 10.9. The impact of sensor operating point on the throughput in normalized units. $\alpha = [0.2, 0.4, 0.6]$, $\beta = [0.8, 0.6, 0.4]$, $\zeta = 0.05$.

is concave with respect to δ in the "aggressive" region ($\delta < \zeta$) and convex in the "conservative" region ($\delta > \zeta$). The performance thus degrades at a faster rate when the sensor operating point drifts toward the "conservative" region. This suggests that miss detections (which lead to collisions) are more harmful to the performance of OSA than false alarms (which represent missed opportunities).

10.4.3.4 OSA with Unknown or Mismatched Model

If the transition probabilities of the Markov model are unknown, formulations and algorithms for POMDP with an unknown model exist in the literature [36] and can be applied to the problem of OSA design. Here we study the impact of mismatched Markov model on the performance of the optimal OSA.

We assume that the spectrum occupancy evolves according to the transition probabilities given by α and β while the secondary user employs the optimal OSA policy based on inaccurate transition probabilities α' and β'. In the upper plot of Fig. 10.10, we plot the relative throughput loss of the secondary user as a function of the relative error ψ in transition probabilities which is given by $\psi = \frac{\alpha'_n - \alpha_n}{\alpha_n} \times 100\% = \frac{\beta'_n - \beta_n}{\beta_n} \times 100\%$. Clearly, the maximum throughput is achieved when the relative error is zero (i.e., the secondary user has accurate information on transition probabilities). Inaccurate transition probabilities can cause performance loss. We find that the relative performance loss is below 4% even when the absolute relative error is up to 20%. In the lower figure, we examine the probability of collision perceived by

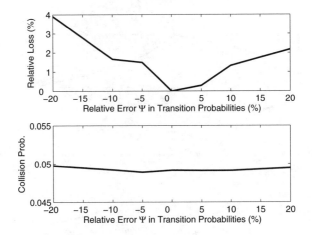

Fig. 10.10. The impact of inaccurate transition probabilities on the throughput of the secondary user. $\alpha = [0.2, 0.4, 0.6]$, $\beta = [0.8, 0.6, 0.4]$, $\zeta = 0.05$.

the primary network. We find that the probability of collision is not affected by mismatched transition probabilities. The reason behind this observation is the separation principle: the optimal sensor operating point and the optimal access policy, which determine the probability of collision, are independent of the spectrum occupancy dynamics.

10.5 The Impact of Fading on Energy-Constrained Cognitive MAC

In this section, we study the impact of channel fading conditions at the physical layer on the design of cognitive MAC protocols under energy constraints. We show that the problem can again be formulated within the framework of POMDP. Optimal and suboptimal sensing and access policies with reduced complexity are obtained for energy-constrained OSA networks in fading environments. To isolate the effect of energy constraint on the design of cognitive MAC, we assume that sensing errors are negligible.

10.5.1 Energy and Fading Model

The network model is the same as that given in Sect. 10.3. We present below the energy and channel fading model.

We assume that channels between the secondary user and its destination follow a block fading model. That is, the channel gain in a slot is a random variable (RV) identically and independently distributed (i.i.d.) across slots but not necessarily i.i.d. across channels.

Let $E_s(n)$ and $E_{tx}(n)$ denote, respectively, the energy consumed in sensing and accessing channel n in a slot. For simplicity, we assume that sensing energy consumption $E_s(n)$ is identical for all channels: $E_s(n) = e_s$ for every n. Note that the transmission energy consumption $E_{tx}(n)$ is a RV depending on the current fading condition of channel n. In general, the better the channel condition, the lower the required transmission energy. Let L be the number of power levels at which the secondary user can transmit and ε_k the energy consumed in transmitting at the kth power level in a slot. The transmission energy consumption $E_{tx}(n)$ thus has realizations restricted to a finite set $\mathcal{E}_{tx}$ given by

$$E_{tx}(n) \in \mathcal{E}_{tx} \triangleq \{\varepsilon_k\}_{k=0}^L \qquad (10.10)$$

where $0 < \varepsilon_1 < \cdots < \varepsilon_L < \infty$ and $\varepsilon_0 = 0$ indicates that the secondary user does not transmit. We also consider the energy e_p consumed in the sleeping mode of the secondary user.

Let E denote the residual energy level of a secondary user at the beginning of a slot. Note that E is an RV determined by the channel conditions and the sensing and access decisions in all previous slots. Thus, E belongs to the finite set $\mathcal{E}_r$ given by

$$E \in \mathcal{E}_r \triangleq \{e : e = \mathcal{E}_0 - \sum_{k=0}^L c_k(e_s + \varepsilon_k) - ce_p, e \geq 0, c, c_k \geq 0, c, c_k \in \mathbb{Z}\} \cup \{0\} \qquad (10.11)$$

where c_k is the number of slots when the secondary user chooses to sense a channel and then transmit over it at the kth power level and c is the number of slots when the secondary user turns to sleeping mode. Section 10.6 discuses how the secondary user can obtain knowledge of the required power level.

10.5.2 Optimal Energy-Constrained OSA

The energy-constrained OSA can be formulated as a constrained POMDP, which is usually more difficult to solve than an unconstrained one. By absorbing the residual energy level of the secondary user into the state space, we reduce a constrained POMDP to an unconstrained one. Based on the theory of POMDP, we obtain the optimal sensing and access policies.

10.5.2.1 An Unconstrained POMDP Formulation

State Space. In each slot, the network state is characterized by the current spectrum occupancy $S \in \{0,1\}^N$ and the residual energy level $E \in \mathcal{E}_r$ of the secondary user at the beginning of this slot. The state space $\mathcal{S}$ can be defined as

$$(S, E) \in \mathcal{S} \triangleq \{(s, e) : s \in \{0,1\}^N, e \in \mathcal{E}_r\}. \qquad (10.12)$$

Action Space. After the state transition of spectrum occupancy at the beginning of each slot, the secondary user can either choose a channel $a \in \{1, \ldots, N\}$ to sense or go to sleep ($a = 0$). If the secondary user chooses channel a to sense, then it will obtain a sensing outcome $\Theta_a \in \{0, 1, \ldots, L\}$ which reflects the occupancy state and the fading condition of the chosen channel: $\Theta_a = 0$ indicates that channel a is busy (i.e., $S_a = 0$) and $\Theta_a = k$ ($k = 1, \ldots, L$) indicates that channel a is idle (i.e., $S_a = 1$) and the fading condition requires the secondary user to transmit at the k-th power level (i.e., $E_{\text{tx}}(a) = \varepsilon_k$). Given sensing outcome Θ_a, the secondary user decides whether to transmit over the chosen channel. Let $\Phi_a(k) \in \{0 \text{ (no access)}, 1 \text{ (access)}\}$ ($k = 0, \ldots, L$) denote the access decision under sensing outcome $\Theta_a = k$. Since we have assumed perfect spectrum sensing, the access decision under $\Theta_a = 0$ (busy) is simple: $\Phi_a(0) = 0$ (no access). In this case, secondary users will not collide with primary users.

The action space $\mathcal{A}$ consists of all sensing decisions a and access decisions $\bar{\Phi}_a \triangleq [\Phi_a(1), \ldots, \Phi_a(L)]$:

$$(a, \bar{\Phi}_a) \in \mathcal{A} \triangleq \{(0, [0, \ldots, 0])\} \cup \{(a, \phi) : a \in \{1, \ldots, N\}$$

$$\phi \triangleq [\phi(1), \ldots, \phi(L)] \in \{0, 1\}^L\}. \tag{10.13}$$

Note that the access decision $\bar{\Phi}_0$ associated with sensing action $a = 0$ (sleeping mode) is determined by $\Phi_0(k) = 0$ for all $1 \le k \le L$.

Network State Transition.   Recall that the network state consists of two parts: the spectrum occupancy S and the residual energy E of the secondary user. At the beginning of each slot, the spectrum occupancy S transits independently of the residual energy E according to transition probabilities $\{P_{s,s'}\}$. As stated in Section 10.3, we assume that the spectrum occupancy dynamics $\{P_{s,s'}\}$ are known and remain unchanged during the battery lifetime of the secondary user.

If the secondary user decides to sense channel $a \in \{1, \ldots, N\}$ in this slot, then it will consume e_s in sensing and $\Phi_a(\Theta_a)\varepsilon_{\Theta_a}$ in transmitting. Thus, at the end of this slot, the residual energy of the secondary user reduces to $E' = \mathcal{T}_E(E \mid a, \Theta_a, \Phi_a(\Theta_a))$:

$$\mathcal{T}_E(E \mid a, \Theta_a, \Phi_a(\Theta_a)) = \begin{cases} E - e_p, & a = 0 \\ \max\{E - e_s - \Phi_a(\Theta_a)\varepsilon_{\Theta_a}, 0\}, & a \ne 0 \end{cases} \tag{10.14}$$

where e_p is the energy consumed in the sleeping mode.

Observations.   Due to partial spectrum sensing, the secondary user does not have full knowledge of the spectrum occupancy state in each slot. It, however, can obtain the occupancy state of the chosen channel $a \in \{1, \ldots, N\}$ from sensing outcome (i.e., observation) $\Theta_a \in \{0, 1, \ldots, L\}$. Let $q_s^{(a)}(k)$ be the probability that the secondary user observes $\Theta_a = k$ in the chosen channel a given current spectrum occupancy state $S = s$. Under perfect spectrum sensing, we have that

$$q_s^{(a)}(k) = \Pr\{\Theta_a = k \mid S = s\} = \begin{cases} 1_{[k \ne 0]} p_a(k), & \text{if } a \ne 0, s_a = 1 \\ 1_{[k = 0]}, & \text{if } a \ne 0, s_a = 0 \end{cases} \tag{10.15}$$

where $p_a(k) \triangleq \Pr\{E_{\text{tx}}(a) = \varepsilon_k\}$ is the probability that the fading condition of channel n requires the secondary user to transmit at the k-th power level, and $1_{[x]}$ is the indicator function: $1_{[x]} = 1$ if x is true and 0 otherwise. Note that $\{p_a(k)\}_{k=1}^{L}$ are determined by the fading statistics of channel a and are independent of the spectrum occupancy state. From (10.15), we can see that $\sum_{k=0}^{L} q_s^{(a)}(k) = 1$ for any spectrum occupancy state $s \in \mathcal{S}$ and any chosen channel $a \in \{1, \dots, N\}$.

Note that if the secondary user turns to sleep, then it will not have any sensing outcome. We can define $\{q_s^{(0)}(k)\}$ as arbitrary values that satisfy $\sum_{k=0}^{L} q_s^{(0)}(k) = 1$. For simplicity, we define $q_s^{(0)}(k) = 1_{[k=0]}$.

Reward Structure. At the end of each slot, the secondary user obtains a non-negative reward $R_{E,\Theta_a}^{(a,\Phi_a(\Theta_a))}$ depending on its residual energy E at the beginning of this slot, the sensing outcome Θ_a, and the sensing and access decisions $(a, \Phi_a(\Theta_a))$. Assuming that the number of information bits that can be transmitted over a channel in one slot is proportional to the channel bandwidth, we define immediate reward $R_{E,\Theta_a}^{(a,\Phi_a(\Theta_a))}$ as

$$R_{E,\Theta_a}^{(a,\Phi_a(\Theta_a))} \triangleq \begin{cases} 0, & a = 0 \\ \Phi_a(\Theta_a) B_a 1_{[E-e_s-\varepsilon_{\Theta_a} \geq 0]}, & a \neq 0. \end{cases} \quad (10.16)$$

That is, a reward is obtained if and only if the secondary chooses to sense and access (i.e., $a \neq 0$, $\Phi_a(\Theta_a) = 1$) an idle channel (i.e., $\Theta_a \neq 0$) and its residual energy is enough to cope with the channel fade in the selected channel (i.e., $E - e_s - \varepsilon_{\Theta_a} \geq 0$). Note that no reward will be accumulated once the battery energy level drops below $e_s + \varepsilon_1$, where ε_1 is the least required transmission energy. Hence, the total expected accumulated reward represents the total expected number of information bits that can be delivered by the secondary user during its battery lifetime.

Belief State At the beginning of a slot, the secondary user has the information of its own residual energy E but not the current spectrum occupancy state S. As stated in Section 10.4, its knowledge of S based on all past decisions and observations can be summarized by a belief state $\boldsymbol{\lambda} = \{\lambda_s\}_{s \in \{0,1\}^N}$ [32], where λ_s is the conditional probability (given the decision and observation history) that the spectrum occupancy is in state s at the beginning of this slot prior to the state transition.

At the end of a slot, the secondary user can update the belief state $\boldsymbol{\lambda}$ for future use based on sensing action a and sensing outcome Θ_a in this slot. Specifically, let $\boldsymbol{\lambda}' \triangleq \mathcal{T}_{\lambda}(\boldsymbol{\lambda} \mid a, k)$ denote the updated belief state whose element λ_s' denotes the probability that the current spectrum occupancy state is $S = s$ given belief state $\boldsymbol{\lambda}$ at the beginning of this slot and the observation $\Theta_a = k$ of the chosen channel a in the current slot. Applying Bayes rule, we obtain λ_s' as

$$\lambda_s' = \Pr\{S = s \mid \boldsymbol{\lambda}, a, k\}$$
$$= \begin{cases} \sum_{s'} \lambda_{s'} P_{s',s}, & a = 0 \\ \dfrac{\sum_{s'} \lambda_{s'} P_{s',s} 1_{[s_a=1_{[k \neq 0]}]}}{\sum_{s} \sum_{s'} \lambda_{s'} P_{s',s} 1_{[s_a=1_{[k \neq 0]}]}}, & a \neq 0 \end{cases} \quad (10.17)$$

where the summations are taken over the space $\{0, 1\}^N$ of spectrum occupancy state S. Note that when the secondary user turns to sleeping mode ($a = 0$), no observation is made and the belief state is updated according to the spectrum occupancy dynamics $\{P_{s,s'}\}$.

Unconstrained POMDP Formulation. We have formulated the energy-constrained OSA as a POMDP problem. A policy π of this POMDP is defined as a sequence of functions:

$$\pi \overset{\triangle}{=} [\mu_1, \mu_2, \ldots], \quad \mu_t : [0, 1]^{2^N} \times \mathcal{E}_r \to \mathcal{A}$$

where $\{a, \bar{\Phi}_a\} = \mu_t(\lambda, E)$ maps every information state (λ, E), which consists of belief state $\lambda \in [0, 1]^{2^N}$ and residual energy $E \in \mathcal{E}_r$, at the beginning of slot t to a sensing decision $a \in \{0, 1, \ldots, N\}$ and a set of access decisions $\bar{\Phi}_a = [\Phi_a(1), \ldots, \Phi_a(L)] \in \{0, 1\}^L$.

The design objective is to find the optimal policy π^* that maximizes the total expected reward:

$$\pi^* = \arg \max_{\pi} \mathbb{E}_{\pi} \left[\sum_{t=1}^{\infty} R_{E,\Theta_a}^{(a,\Phi_a(\Theta_a))}(t) \,\middle|\, \lambda_0 \right] \tag{10.18}$$

where λ_0 is the initial belief state given by the stationary distribution of spectrum occupancy. We thus have an unconstrained POMDP.

10.5.2.2 Optimal Policy

Let $V(\lambda, E)$ be the value function, which denotes the maximum expected remaining reward that can be accrued when the current information state is (λ, E). We notice from (10.16) that the value function is given by $V(\lambda, E) = 0$ for any information state (λ, E) with residual energy $E < e_s + \varepsilon_1$. For any other information state, its value function $V(\lambda, E)$ is the unique solution to the following equation:

$$V(\lambda, E) = \max_{(a,\Phi) \in \mathcal{A}} \sum_{k=0}^{L} u_k^{(a)} [R_{E,k}^{(a,\phi(k))} + V(\mathcal{T}_\lambda(\lambda \,|\, a, k), \mathcal{T}_E(E \,|\, a, k, \phi(k)))] \tag{10.19}$$

where $\mathcal{T}_\lambda(\lambda \,|\, a, k)$ is the updated belief state given in (10.17), $\mathcal{T}_E(E \,|\, a, k, \phi(k))$ is the reduced battery energy given in (10.14), and $u_k^{(a)} \overset{\triangle}{=} \Pr\{\Theta_a = k \,|\, \lambda\}$ is the probability of observing $\Theta_a = k$ given belief state λ, which is determined by the spectrum occupancy dynamics and the channel fading statistics:

$$u_k^{(a)} = \sum_{s' \in \{0,1\}^N} \lambda_{s'} \sum_{s \in \{0,1\}^N} P_{s',s} \, q_s^{(a)}(k). \tag{10.20}$$

In principle, by solving (10.19), we can obtain the optimal sensing and access actions (a^*, Φ_a^*) that achieve the maximum expected reward $V(\lambda, E)$ for each possible information state (λ, E). We can also obtain the maximum expected number of information bits V_{opt} that can be delivered by a secondary user during its battery lifetime as $V_{\text{opt}} = V(\lambda_0, \mathcal{E}_0)$, where λ_0 is the initial belief state.

10.5.3 Optimal Policy with Reduced Complexity

Although the value function given in (10.19) can be solved iteratively, it is computationally expensive. In this section, we first identify the sources of high complexity of the optimal policy and then reduce the complexity accordingly.

10.5.3.1 Complexity of the Optimal Policy

We measure the computational complexity of a policy as the number of multiplications required to obtain all sensing and access actions during the secondary user's battery lifetime T when initial belief state and battery energy are given.

We notice from (10.19) that the optimal sensing and access action in the first slot depends on the value functions of all possible information states during the battery lifetime T. Hence, the computational complexity of the optimal policy is determined by the number of multiplications required to calculate the value functions of all possible information states.

Following the complexity analysis in [34], we can calculate the number of all possible information states $(\boldsymbol{\lambda}, E)$ during the secondary user's battery lifetime. Specifically, noting from (10.17) that the updated belief state is the same under all non-zero sensing outcomes ($k \neq 0$), we can see that each information state $(\boldsymbol{\lambda}, E)$ can transit to at most $L + 1$ different information states under sensing action $a \neq 0$ but only one under sensing action $a = 0$. Hence, for fixed initial information state $(\boldsymbol{\lambda}_0, \mathcal{E}_0)$, the number of all possible information states is on the order of $O((N(L + 1))^{T-1})$, which is exponential in the battery lifetime T and polynomial in the number N of channels. Moreover, from (10.19) and (10.20), we can see that it requires $O(3|\mathcal{A}|2^N 2^N (L + 1))$ multiplications to calculate each value function, where $|\mathcal{A}|$ is the size of the action space, 2^N is the dimension of the belief state, and $L+1$ is the number of possible observations. Therefore, the computational complexity of the optimal policy is on the order of $O(3|\mathcal{A}|2^N 2^N (L + 1)(N(L + 1))^{T-1})$. We can see that the complexity is mainly caused by the following three factors: (1) the number $O((N(L+1))^{T-1})$ of possible information states; (2) the size $|\mathcal{A}|$ of the action space, and (3) the dimension 2^N of the belief state. We will address the first factor in Section 10.5.4. In this section, we focus on the other two factors.

10.5.3.2 Reduction of Action Space Size

Careful inspection of (10.14), (10.16) and (10.19) reveals that the quantity $R_{E,k}^{(a,\phi(k))} + V(\mathcal{T}_{\lambda}(\boldsymbol{\lambda} \,|\, a, k), \mathcal{T}_E(E \,|\, a, k, \phi(k)))$ inside the square parenthesis of (10.19) only depends on the k-th entry $\phi(k)$ of the access decision $\bar{\phi}$ and is independent of $\phi(i)$ ($i \neq k$). We can thus simplify (10.19) as

$$V(\boldsymbol{\lambda}, E) = \max_{a \in \{0,1,\ldots,N\}} \{\sum_{k=0}^{L} u_k^{(a)} \max_{\phi(k) \in \{0,1\}} [R_{E,k}^{(a,\phi(k))}$$
$$+ V(\mathcal{T}_{\lambda}(\boldsymbol{\lambda} \,|\, a, k), \mathcal{T}_E(E \,|\, a, k, \phi(k)))]\}. \tag{10.21}$$

The maximization in (10.21) is taken over a space with size $O(2NL)$, increasing linearly with the number L of power levels, while that in (10.19) is taken over the action space $\mathcal{A}$ whose size $O(N2^L)$ increases exponentially with L.

Proposition 10.1 states that the optimal access decision Φ_a^* is a threshold policy.

Proposition 10.1. Given the belief state λ and the residual energy level E of the secondary user at the beginning of a slot, there exists a threshold k_a^* associated with sensing action $a \in \{1, \ldots, N\}$ such that the optimal access decision $\Phi_a^* = [\phi_a^*(1), \ldots, \phi_a^*(L)]$ is given by

$$\phi_a^*(k) = \begin{cases} 1, & \text{if } k \leq k_a^* \\ 0, & \text{if } k > k_a^*. \end{cases} \tag{10.22}$$

Proof. See [18]. □

Proposition 10.1 can help us avoid the search for optimal access decisions in some scenarios, resulting in further complexity reduction. Specifically, for each sensing action $a \neq 0$, we can calculate the optimal access decisions $\phi_a^*(k)$ in a decreasing order of sensing outcome k. Once we have $\phi_a^*(k^*) = 1$ for a certain value of k^*, we can determine the optimal access decisions for all remaining sensing outcomes $k < k^*$ without further computation.

10.5.3.3 Reduction of Belief State Dimension

Assume that the spectrum occupancy evolves independently across channels. It has been shown in [16] that $\omega \overset{\Delta}{=} [\omega_1, \ldots, \omega_N]$, where ω_n denotes the probability (conditioned on all previous decisions and observations) that channel n is available at the beginning of a slot prior to the state transition, is a sufficient statistic for belief state λ. Note that the dimension of ω increases linearly $O(N)$ with the number N of channels while that of λ increases exponentially $O(2^N)$.

Using the belief state ω, we can simplify the value function given in (10.21). Specifically, let $\alpha_n = \Pr\{S_n' = 1 \mid S_n = 0\}$ denote the probability that channel n transits from 0 (busy) to 1 (idle) and $\beta_n = \Pr\{S_n' = 1 \mid S_n = 1\}$ the probability that channel n remains idle. Then, (10.21) reduces to

$$\hat{V}(\omega, E) = \max_{a \in \{0,1,\ldots,N\}} \{(1 - \omega_a')\hat{V}(\hat{T}_\lambda(\omega \mid a, 0), \mathcal{T}_E(E \mid a, 0, 0))$$

$$+ \omega_a' \sum_{k=1}^{L} p_a(k) \max_{\phi(k) \in \{0,1\}} [R_{E,k}^{(a,\phi(k))} + \hat{V}(\hat{T}_\lambda(\omega \mid a, k), \mathcal{T}_E(E \mid a, k, \phi(k)))]\} \tag{10.23}$$

where $\omega_0' \overset{\Delta}{=} 0$, $\omega_a' = \omega_a \beta_a + (1 - \omega_a)\alpha_a$ ($a \in \{1, \ldots, L\}$) is the probability that channel a is available in the current slot given ω, $\mathcal{T}_E(E \mid a, k, \phi_a(k))$ is the reduced battery energy given in (10.14), and the updated belief state $\hat{\omega} \overset{\Delta}{=} [\omega_1, \ldots, \omega_N] = \hat{T}_\lambda(\omega \mid a, k)$ is given by

$$\hat{\omega}_n = \begin{cases} 0, & \text{if } a \neq 0, n = a, k = 0 \\ 1, & \text{if } a \neq 0, n = a, k \neq 0 \\ \omega'_n, & \text{otherwise.} \end{cases} \qquad (10.24)$$

10.5.4 Suboptimal Cognitive MAC with Reduced Complexity

We notice from (10.19) that the optimal sensing and access decisions in a slot rely on the value functions of all possible information states in the remaining slots, which significantly increases the computational complexity of the optimal policy. In this section, we provide a suboptimal solution to energy-constrained OSA, which reduces the number of value functions used in decision-making. We show that the computational complexity of this suboptimal strategy can be very favorably traded off with its performance.

10.5.4.1 The Greedy-w Approach

Referred to as greedy-w approach, the proposed strategy maximizes the total expected reward in a time window of w slots. Let $Y_w^{(a)}(\boldsymbol{\lambda}, E)$ denote the maximum reward that can be accumulated in a window of w slots given information state $(\boldsymbol{\lambda}, E)$ and sensing action a. We can calculate $Y_w^{(a)}(\boldsymbol{\lambda}, E)$ recursively by

$$Y_0^{(a)}(\boldsymbol{\lambda}, E) = 0$$

$$Y_w^{(a)}(\boldsymbol{\lambda}, E) = \sum_{k=0}^{L} u_k^{(a)} \max_{\phi(k) \in \{0,1\}} [R_{E,k}^{(a,\phi(k))} \qquad (10.25)$$
$$+ \max_{b \in \{0,1,\ldots,N\}} Y_{w-1}^{(b)}(\mathcal{T}_\lambda(\boldsymbol{\lambda} \,|\, a, k), \mathcal{T}_E(E \,|\, a, k, \phi(k)))]$$

where $u_k^{(a)}$, $\mathcal{T}_\lambda(\boldsymbol{\lambda} \,|\, a, k)$, and $\mathcal{T}_E(E \,|\, a, k, \phi(k))$ are given in (10.20), (10.17), and (10.14), respectively. From (10.25), we can see that for any w, $Y_w^{(a)}(\boldsymbol{\lambda}, E) = 0$ if $E < e_s + \varepsilon_1$.

Given belief state $\boldsymbol{\lambda}$ and residual energy E of the secondary user at the beginning of a slot, the greedy-w approach chooses channel a_w that maximizes the reward obtained in the next w slots to sense, i.e.,

$$a_w = \arg \max_{a \in \{0,1,\ldots,N\}} Y_w^{(a)}(\boldsymbol{\lambda}, E). \qquad (10.26)$$

Given sensing outcome $k \in \{1, \ldots, L\}$, the access decision $\phi_{a_w}(k)$ of the greedy-w approach is given by

$$\phi_{a_w}(k) = \arg \max_{\phi \in \{0,1\}} \{R_{E,k}^{(a_w,\phi)} \qquad (10.27)$$
$$+ \max_{b \in \{1,\ldots,N\}} Y_{w-1}^{(b)}(\mathcal{T}_\lambda(\boldsymbol{\lambda} \,|\, a_w, k), \mathcal{T}_E(E \,|\, a_w, k, \phi))]\}.$$

Next, we consider two extreme cases of the greedy-w strategy.

Case 1: When $w = 1$, the greedy-1 approach focuses solely on maximizing the immediate reward. Specifically, the secondary user employing greedy-1 approach chooses the channel with the maximum expected immediate reward and transmits whenever the channel is sensed to be available:

$$a_1 = \arg \max_{a \in \{1, \ldots, N\}} \sum_{k=1}^{L} u_k^{(a)} R_{E,k}^{(a,1)}$$

$$\phi_{a_1}(k) = 1_{[k \neq 0]}.$$

(10.28)

The greedy-1 approach has the lowest computational complexity.

Case 2: Consider the case when window size w exceeds the maximum battery lifetime of the secondary user. In this case, the network reaches a terminating state in less than w slots regardless of the sensing and access strategies. Since no reward is accumulated after the network reaches a terminating state, the greedy-w approach is equivalent to the optimal strategy.

10.5.4.2 Complexity Vs. Performance

We can see from (10.26) and (10.27) that the sensing and access decisions made by the greedy-w approach in a slot only depend on the value functions of all possible information states in the next w slots. Hence, the total number of value functions required to determine the sensing and access decisions during battery lifetime T is on the order of $O((N(L+1))^{w-1}T)$, which is linear in T. Clearly, the computational complexity of greedy-w approach increases with w.

Next, we compare the performance of the greedy-w approach with the optimal performance $V(\lambda_0, \mathcal{E}_0)$. In Fig. 10.11, we plot the total expected number of information bits that can be delivered by the secondary user during its battery lifetime. We

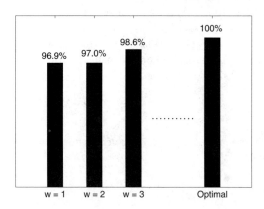

Fig. 10.11. Throughput comparison of the greedy-w and the optimal approaches.

consider $N = 2$ independently evolving channels with different occupancy dynamics. As the window size w increases, the performance of the greedy-w approach improves. It quickly approaches the optimal performance as w increases.

The above observations show that the computational complexity of the greedy-w approach increases while its performance loss as compared to the optimal performance decreases as the window size w increases. Hence, by choosing a suitable w, usually small, the greedy-w approach can achieve a desired tradeoff between complexity and performance.

10.5.5 Numerical Examples

Equation (10.19) indicates that a sensing and access action $(a, \phi) \in \mathcal{A}$ affects the total expected reward in three ways: (1) it yields an immediate reward $R_{E,k}^{(a,\phi(k))}$ in this slot; (2) it transforms the current belief state $\boldsymbol{\lambda}$ to $\mathcal{T}_\lambda(\boldsymbol{\lambda}, a, k)$ which summarizes the spectrum occupancy information up to this slot; (3) it causes a reduction in battery energy from E to $\mathcal{T}_E(E, a, k, \phi(k))$, decreasing the remaining battery lifetime. Hence, to maximize the total expected reward during battery lifetime, the optimal sensing and access policy should achieve a tradeoff among gaining instantaneous reward, gaining information for future use, and conserving energy. In this section, we study the impact of spectrum occupancy dynamics, channel fading statistics, and energy consumption characteristics on the optimal sensing and access actions.

10.5.5.1 To Sense or Not to Sense?

The secondary user may choose to sense in order to gain immediate reward and spectrum occupancy information, but not to sense in order to conserve energy. Hence, the optimal decision on whether to sense should strike a balance between gaining reward/information and conserving energy. In Table 10.5.5.1, we study the optimal sensing decision $1_{[a^* \neq 0]}$ in a particular slot under different spectrum occupancy dynamics and belief states.

We consider $N = 2$ independently evolving channels with identical spectrum occupancy dynamics $\alpha_1 = \alpha_2 = \alpha$ and $\beta_1 = \beta_2 = \beta$. We assume that $\beta = 1 - \alpha$. Hence, the stationary distribution of spectrum occupancy state S is given by $\boldsymbol{\omega}_1 = [0.5, 0.5]$. Consider another belief state $\boldsymbol{\omega}_2 = [0, 0]$ with which the secondary user has full information on the spectrum occupancy prior to the state transition in this

Table 10.1. The impact of spectrum occupancy dynamics α and belief states $\boldsymbol{\omega}$ on the optimal sensing decision $1_{[a^* \neq 0]}$. $N = 2$, $[B_1, B_2] = [1, 1]$, $\mathcal{E}_0 = 4$, $e_s = 0.6$, $e_p = 0.1$, $L = 2$, $\mathcal{E}_{\text{tx}} = \{1, 2\}$, $p_n(1) = p_n(2) = 0.5$ for $n = 1, 2$.

α	0.05		0.1		0.4		0.8	
$\boldsymbol{\omega}$	[0.5,0.5]	[0,0]	[0.5,0.5]	[0,0]	[0.5,0.5]	[0,0]	[0.5,0.5]	[0,0]
Sense	X		X		X	X	X	X
Do not sense		X		X				

slot. Conditioned on the belief states at the beginning of this slot, the conditional probability that channel n is available can be calculated as $\Pr\{S_n = 1 \,|\, \boldsymbol{\omega}_1\} = 0.5$ and $\Pr\{S_n = 1 \,|\, \boldsymbol{\omega}_2\} = \alpha$ for $n = 1, 2$. From Table 10.5.5.1, we find that the secondary user chooses not to sense only when the conditional probability $\Pr\{S_n = 1 \,|\, \boldsymbol{\omega}\}$ that the channel is available is very small. We also find that the secondary user always chooses to sense if the belief state is given by the stationary distribution $\boldsymbol{\omega}_1$ of the spectrum occupancy states. The reason behind this is the monotonicity of the value function $\hat{V}(\boldsymbol{\omega}, E)$ in terms of battery energy E. Specifically, if the secondary user chooses not to sense, then its belief state at the beginning of next slot will remain $\boldsymbol{\omega}_1$ but its battery energy will be reduced by e_{p} due to energy consumption in the sleeping mode. The maximum total expected reward that can be obtained is thus given by $\hat{V}(\boldsymbol{\omega}_1, E - e_{\mathrm{p}})$. Since $\hat{V}(\boldsymbol{\omega}, E)$ increases with the battery energy E for every fixed $\boldsymbol{\omega}$, we have $\hat{V}(\boldsymbol{\omega}_1, E) \geq \hat{V}(\boldsymbol{\omega}_1, E - e_{\mathrm{p}})$ and hence the secondary user should choose to sense whenever it has a stationary belief state.

10.5.5.2 To Access or Not to Access?

Without an energy constraint, the secondary user should always access the channel that is sensed to be available. However, under the energy constraint, the access decision should take into account both the energy consumption characteristics and the channel fading statistics. For example, when the sensed channel is available but has poor fading condition, should the secondary user access this channel to gain immediate reward or wait for better channel realizations to conserve energy? In Table 10.2, we study the impact of sensing energy consumptions e_s and channel fading statistics $\{p_n(k)\}_{k=1}^{L}$ on the optimal access decision $\phi^*(k)$ under different observations k. We find that when sensing energy consumption e_s is negligible, the secondary user should refrain from transmission under poor channel conditions and wait for the best channel realization. However, when e_s is large, it should always grab the instantaneous opportunity regardless of the fading condition because the sensing energy consumed in waiting for the best channel realization may exceed the extra energy consumed in combating the poor channel fading.

The access decision should also take into account the channel fading statistics. Comparing the optimal access decisions in the two cases of Table 10.2 when sens-

Table 10.2. The impact of sensing energy consumptions e_s and channel fading statistics on the optimal access decision $\phi^*(k)$ under different observations k. $N = 2$, $[B_1, B_2] = [1, 1]$, $\mathcal{E}_0 = 8$, $e_{\mathrm{p}} = 0.1$, $L = 3$, $\mathcal{E}_{\mathrm{tx}} = \{1, 2, 3\}$. Case 1: $p_n(1) = 0.5, p_n(2) = 0.3, p_n(3) = 0.2$ for $n = 1, 2, 3$. Case 2: $p_n(1) = 0.3, p_n(2) = 0.3, p_n(3) = 0.4$.

Sensing energy e_s		0			0.7			0.8			1.0		
Observation k		1	2	3	1	2	3	1	2	3	1	2	3
Case 1	Access	X			X	X		X	X		X	X	X
	Do not access		X	X			X			X			
Case 2	Access	X			X	X		X	X	X	X	X	X
	Do not access		X	X			X						

ing energy is $e_s = 0.8$. We find that if the probability that the channel experiences deep fading is small (case 1), the secondary user should avoid transmitting under poor channel realizations because the waiting time for a better channel realization is short and hence the energy wasted in waiting can still be lower than the extra energy needed to combat the poor channel condition. On the other hand, if the channel tends to have poor fading conditions (case 2), the secondary user should focus on gaining immediate reward because of the long waiting time for better channel realizations.

10.6 Protocol Specifics of Decentralized Cognitive MAC

In this section, we present protocol specifics of the cognitive MAC strategies presented in Sects. 10.4 and 10.5.

10.6.1 Transceiver Synchronization

Without a dedicated communication or control channel, transceiver synchronization is a key issue in distributed cognitive MAC for OSA networks [16, 17, 37]. Specifically, a secondary user and its intended receiver need to hop to the same channel at the beginning of each slot in order to carry out the communication. The synchronization problem can be separated into two phases: the initial handshake between the transmitter and the receiver and the synchronous hopping in the spectrum after the initial establishment of communication.

There are a number of standard implementations to facilitate the initial handshake. As given in [16, 17, 37], we can borrow the idea of receiver-oriented code assignment in CDMA ad hoc networks [38]. Specifically, each secondary user is assigned a set of channels (not necessarily unique) which it monitors regularly to check whether it is an intended receiver. A user with a message for, say, user A will transmit a handshake signal over one of the channels assigned to user A. Once the initial communication is established, the transmitter and the receiver will implement the same spectrum sensing and access strategy which governs channel selection in each slot. As detailed in [17, 18], the sensing and access strategies presented in Sects. 10.4 and 10.5 ensure synchronous hopping between the transmitter and the receiver in the presence of collisions, sensing errors, and fading.

Specifically, the structure of the cognitive MAC protocols developed within the POMDP framework ensures that both the transmitter and the receiver have the same information on the occupancy state and the fading condition of the sensed channel in each slot. Hence, at the end of each slot, the transmitter and the receiver will reach the same updated belief state λ. Since the channel selection is determined by the information state λ, the transmitter and the receiver will hop to the same channel in the next slot, i.e., transceiver synchronization is maintained.

10.6.2 Identification of Spectrum Opportunity and Fading Condition

When every secondary user is affected by the same set of primary users, the state of a channel is the same at both the transmitter and the receiver. Detection of spectrum opportunity can thus be carried out at the transmitter alone. When secondary users are affected by different sets of primary users, however, the state of spectrum occupancy is location dependent; a channel that is idle at a transmitter may not be idle at the corresponding receiver. In this case, spectrum opportunities need to be identified jointly by the transmitter and the receiver[4] [17, 34]

To achieve joint opportunity identification at both transmitter and receiver, a scheme based on RTS-CTS exchange is proposed in [17, 37]. We briefly comment on this scheme using the energy-constrained cognitive MAC given in Sect. 10.5 as an example, where we show that this scheme also facilitates the estimation of channel fading conditions.

At the beginning of a slot, the transmitter and the receiver hop to the same channel. If the channel is sensed to be available, the transmitter generates a random backoff time. If the channel remains idle when its backoff time expires, it transmits a short request-to-send (RTS) message to the receiver, indicating that the channel is available at the transmitter. Upon receiving the RTS, the receiver estimates the channel fading condition using the RTS, and then replies with a clear-to-send (CTS) message if the channel is also available at the receiver. The receiver also informs the transmitter of the current fading condition by piggybacking the estimated channel state to the CTS. After a successful exchange of RTS-CTS, the transmitter and the receiver can communicate over this channel. At the end of this slot, the receiver acknowledges every successful data transmission. Note that at the beginning of each slot, the transmitter and the receiver can also choose not to hop to any channel and turn to sleep mode until the beginning of next slot.

We point out that the RTS-CTS exchange has multiple functions. Besides facilitating opportunity identification and channel fading estimation, it also mitigates the hidden and exposed terminal problem as in a conventional communication network [39]. Other collision avoidance schemes such as busy tone and dual busy tone may be incorporated to further reduce the occurrence of collision among secondary users.

Conclusion

In this chapter, we have discussed some of the technical challenges of cognitive MAC for OSA and made an initial attempt to establish a theoretical framework within which these challenges can be systematically and collectively addressed. In particular, the framework of POMDP makes the MAC cognitive; an opportunistic user makes optimal decisions for sensing and access based on the belief state that summarizes the knowledge of the network state based on all past decisions and observations.

---

[4] In this case, $S_n(t) = 1$ if channel n is available at both the transmitter and the receiver. Otherwise, $S_n(t) = 0$. Strictly speaking, the availability of a channel at the secondary transmitter is determined by primary receivers rather than primary transmitters in its neighborhood [29]. The detection of primary receivers can be transformed to the detection of primary transmitters. A detailed presentation can be found in [29].

This decision-theoretic framework also allows the integration of sensing errors, hardware limitations, and energy constraints into the modeling of cognitive MAC design.

References

1. "FCC spectrum policy task force: Report of the spectrum efficiency working group," Nov. 2002.
2. R. Coase, "The federal communications commission," J. Law Econ., pp. 1–40, 1959.
3. D. Hatfield and P. Weiser, "Property rights in spectrum: Taking the next step," in *Proc. of the first IEEE Symp. New Frontiers Dynamic Spectr. Access Networks*, pp. 43–55, Nov. 2005.
4. L. Xu, R. Tonjes, T. Paila, W. Hansmann, M. Frank, and M. Albrecht, "DRiVE-ing to the Internet: Dynamic radio for IP services in vehicular environments," in *Proc. of 25th Annual IEEE Conference on Local Computer Networks*, pp. 281–289, Nov. 2000.
5. Y. Benkler, "Overcoming agoraphobia: Building the commons of the digitally networked environment," *Harvard J. Law Technol.*, Winter 1997–1998.
6. W. Lehr and J. Crowcroft, "Managing shared access to a spectrum commons," in *Proc. of the first IEEE Symp. New Frontiers Dynamic Spectr. Access Networks*, pp. 420–444, Nov. 2005.
7. C. Raman, R. Yates, and N. Mandayam, "Scheduling variable rate links via a spectrum server," in *Proc. of the first IEEE Symposium on New Frontiers in Dynamic Spectrum Access Networks*, pp. 110–118, Nov. 2005.
8. O. Ileri, D. Samardzija, and N. Mandayam, "Demand responsive pricing and competitive spectrum allocation via a spectrum server," in *Proc. of the first IEEE Symposium on New Frontiers in Dynamic Spectrum Access Networks*, pp. 194–202, Nov. 2005.
9. S. Chung, S. Kim, J. Lee, and J. Cioffi, "A game-theoretic approach to power allocation in frequency-selective Gaussian interference channels," in *Proc. of IEEE International Symposium on Information Theory*, pp. 316–316, June 2003.
10. R. Etkin, A. Parekh, and D. Tse, "Spectrum sharing for unlicensed bands," in *Proc. of the first IEEE Symposium on New Frontiers in Dynamic Spectrum Access Networks*, pp. 251–258, Nov. 2005.
11. J. Huang, R. Berry, and M. Honig, "Spectrum sharing with distributed interference compensation," in *Proc. of the first IEEE Symposium on New Frontiers in Dynamic Spectrum Access Networks*, pp. 88–93, Nov. 2005.
12. J. Mitola, "Cognitive radio for flexible mobile multimedia communications," in *Proc. of IEEE International Workshop on Mobile Multimedia Communications*, pp. 3–10, Nov. 1999.
13. "DARPA: The Next Generation (XG) Program." http://www.darpa.mil/ato/programs/xg/index.htm.
14. J. Mitola, *Software Radios: Wireless Architecture for the 21st Century*. Wiley, 2000.
15. J. Mitola, "Cognitive radio." Licentiate proposal, KTH, Stockholm, Sweden.
16. Q. Zhao, L. Tong, and A. Swami, "Decentralized cognitive MAC for dynamic spectrum access," in *Proc. of the first IEEE Symposium on New Frontiers in Dynamic Spectrum Access Networks*, pp. 224–232, Nov. 2005.
17. Q. Zhao, L. Tong, A. Swami, and Y. Chen, "Decentralized cognitive MAC for opportunistic spectrum access in ad hoc networks: A POMDP framework," in *IEEE Journal on Selected Areas in Communications: Special Issue on Adaptive, Spectrum Agile and Cognitive Wireless Networks*, Apr., 2007.

18. Y. Chen, Q. Zhao, and A. Swami, "Distributed cognitive MAC for energy-constrained opportunistic spectrum access," in *Proc. of IEEE Military Communication Conference*, Oct. 2006.
19. Y. Chen, Q. Zhao, and A. Swami, "Joint design and separation principle for opportunistic spectrum access," in *Proc. of IEEE Asilomar Conference on Signals, Systems, and Computers*, Oct. 2006.
20. A. Sahai and N. Hoven and R. Tandra, "Some fundamental limits on cognitive radio," in *Proc. of Allerton Conference on Communication, Control, and Computing*, Oct. 2004.
21. D. Cabric, S. M. Mishra, and R. W. Brodersen, "Implementation issues in spectrum sensing for cognitive radios," in *Proc. of IEEE Asilomar Conference on Signals, Systems, and Computers*, pp. 772–776, Oct. 2004.
22. K. Challapali, S. Mangold, and Z. Zhong, "Spectrum agile radio: Detecting spectrum opportunities," in *Proc. of International Symposium on Advanced Radio Technologies*, 2004.
23. B. Wild and K. Ramchandran, "Detecting primary receivers for cognitive radio applications," in *Proc. of the first IEEE Symposium on New Frontiers in Dynamic Spectrum Access Networks*, pp. 124–130, Nov. 2005.
24. A. Ghasemi and E. Sousa, "Collaborative spectrum sensing for opportunistic access in fading environments," in *Proc. of the first IEEE Symposium on New Frontiers in Dynamic Spectrum Access Networks*, pp. 131–136, Nov. 2005.
25. H. Zheng and C. Peng, "Collaboration and fairness in opportunistic spectrum access," in *Proc. of IEEE International Conference on Communications*, vol. 5, pp. 3132–3136, May 2005.
26. W. Wang and X. Liu, "List-coloring based channel allocation for open-spectrum wireless networks," in *Proc. of IEEE Vehicular Technology Conference*, vol. 1, pp. 690–694, Sept. 2005.
27. S. Sankaranarayanan, P. Papadimitratos, A. Mishra, and S. Hershey, "A bandwidth sharing approach to improve licensed spectrum utilization," in *Proc. of the first IEEE Symposium on New Frontiers in Dynamic Spectrum Access Networks*, pp. 279–288, Nov. 2005.
28. M. Steenstrup, "Opportunistic use of radio-frequency spectrum: A network perspective," in *Proc. of the first IEEE Symposium on New Frontiers in Dynamic Spectrum Access Networks*, pp. 638–641, Nov. 2005.
29. Q. Zhao and B. Sadler, "A survey of dynamic spectrum access: Signal processing, networking, and regulatory policy," in *IEEE Signal Process. Magazine*, May, 2007.
30. Q. Zhao and A. Swami, "A decision-theoretic framework for dynamic spectrum access", to appear in *IEEE Wireless Communications Magazine*, 2007.
31. Y. Chen, Q. Zhao, and A. Swami, "Joint PHY-MAC design for opportunistic spectrum access in the presence of sensing errors," submitted to *IEEE Trans. Signal Process.*, Jan. 2007.
32. R. Smallwood and E. Sondik, "The optimal control of partially observable Markov processes over a finite horizon," *Operat. Res.*, vol. 21, pp. 1071–1088, 1973.
33. Y. Chen, Q. Zhao, and A. Swami, "Proof of the separation principle for opportunistic spectrum access," Technical report, University of California at Davis, 2006. http://www.ece.ucdavis.edu/_qzhao/Report.html.
34. D. Djonin, Q. Zhao, and V. Krishnamurthy, "Optimality and complexity of opportunistic spectrum access: A truncated Markov decision process formulation," in *Proc. of IEEE International Conference on Communications (ICC)*, 2007.
35. H. L. V. Trees, Detection, Estimation, and Modulation Theory, Part I. Wiley-Interscience, 2001.

36. D. Aberdeen, "A survey of approximate methods for solving partially observable markov decision processe," Technical report, National ICT Australia, Dec. 2003. http://users.rsise.anu.edu.au/~daa/papers.html.
37. Q. Zhao, L. Tong, and A. Swami, "A cross-layer approach to cognitive mac for spectrum agility," in *Proc. of IEEE Asilomar Conference on Signals, Systems, and Computers*, pp. 200–204, Oct.–Nov. 2005.
38. M. B. Pursley, "The role of spread spectrum in packet radio networks," *Proc. IEEE*, vol. 75, pp. 116–134, Jan. 1987.
39. A. S. Tanenbaum, Computer Networks, 3rd ed. Upper Saddle River, NJ: Prentice-Hall PTR, 1996.

11

Game Theoretic Learning and Pricing for Dynamic Spectrum Access in Cognitive Radio

Michael Maskery[1], Vikram Krishnamurthy[1], and Qing Zhao[2]

[1] University of British Columbia, Canada
 {mikem,vikramk}@ece.ubc.ca
[2] University of California at Davis, USA
 qzhao@ece.ucdavis.edu

11.1 Introduction

This chapter deals with game theoretic methods for dynamic spectrum access in cognitive radio systems. Cognitive radio systems need to employ dynamic spectrum access methods to efficiently share radio spectrum with other cognitive radios while avoiding interference with legacy systems. Due to the inherent decentralized nature of cognitive radio, dynamic spectrum access strategies need to be decentralized. To address this, we formulate a model in which cognitive radios are players competing for spectrum resources in a game theoretic setting. The players need to access channels in a dynamic and uncertain environment to satisfy demand while respecting system-imposed sharing incentives.

The reader is undoubtedly familiar with the term *Nash equilibrium* in non-cooperative games. In this paper we use a more general equilibrium concept called *correlated equilibrium*. The concept of correlated equilibria in game theory was introduced by Aumann [1,2].[3] Correlated equilibria are easier to characterize and more natural to decentralized adaptive algorithms such as those considered here.

The problem of non-cooperative radio resource allocation is addressed elsewhere in [2–4] from a non-game theoretic perspective, and in [5,6] from a game theoretic one. Of these, [7] is auction-based and does not fit in our framework. Reference [5] is very similar to our approach, even employing similar learning-based ideas, but for a fundamentally different scenario.

Before presenting our main results, including our game theoretic dynamic spectrum access model and adaptive learning algorithm, we begin by reviewing the main ideas in dynamic spectrum access and game theory.

---

[3] Aumann was awarded the 2005 Nobel Prize in Economics. The Nobel Prize press release in October 2005 reads: "Aumann also introduced a new equilibrium concept, correlated equilibrium, which is weaker than Nash equilibrium, the solution concept developed by John Nash, an Economics Laureate in 1994. Correlated equilibrium can explain why it may be advantageous for negotiating parties to allow an impartial mediator to speak to the parties either jointly or separately, and in some instances give them different information".

11.1.1 Brief Overview of Dynamic Spectrum Access

The proliferation of a wide range of wireless devices and their applications has resulted in an overly crowded radio spectrum; almost all usable frequencies have already been assigned. This makes one pessimistic about the feasibility of integrating emerging wireless services such as large-scale sensor networks into the existing communication infrastructure.

In contrast to the apparent spectrum scarcity is the pervasiveness of spectrum opportunity. Extensive measurements indicate that, at any given time and location, a large portion of licensed spectrum lies unused. For example, over 62% white space exists in the spectrum under 3 GHz [8]. This paradox between the overly crowded spectrum and the pervasiveness of idle frequency bands in both time and space indicates that spectrum shortage results from the spectrum management policy rather than the physical scarcity of usable frequencies.

The underutilization of spectrum has stimulated a flurry of exciting activities in search for dynamic spectrum access strategies for improved efficiency. Approaches envisioned for dynamic spectrum access fall under three general models: dynamic exclusive use, open sharing and hierarchical access.

The dynamic exclusive use model aims to introduce flexibility to the current command-and-control spectrum regulation policy while maintaining the spectrum licensees' right of exclusive use. Specific approaches include spectrum property rights [9] and dynamic spectrum allotment brought forth by the European DRiVE project [10]. The open sharing model, also referred to as the spectrum commons model [11], draws support from the phenomenal success of wireless services operating in the unlicensed ISM band. It employs open sharing among peer users as the basis for spectrum management. The hierarchical access model can be considered as a hybrid of the above two. The basic idea is to open licensed spectrum to secondary users and limit the interference perceived by primary users (licensees). One approach to spectrum sharing between primary and secondary users is spectrum overlay, which was first envisioned by Mitola [12] under the term "spectrum pooling" and then investigated by the DARPA XG program [13] under the term "opportunistic spectrum access". Another approach is spectrum underlay enabled by the technology of ultra wide band. A more detailed taxonomy of dynamic spectrum access can be found in Chapter 10.

In this chapter, we focus on the overlay approach to dynamic spectrum access. This approach directly targets at idle frequency bands in both time and space by allowing secondary users to identify and exploit instantaneous and local spectrum availability without causing unacceptable interference to primary users.

While conceptually simple, spectrum overlay presents technical challenges across the entire networking protocol stack. Basic components of spectrum overlay include spectrum opportunity identification and spectrum opportunity exploitation. The opportunity identification module is responsible for accurately identifying and intelligently tracking idle frequency bands that are dynamic in both time and space. The opportunity exploitation module takes input from the opportunity identification module and decides whether and how a transmission should take place. The overall design objective of OSA is to provide sufficient benefit to secondary users while protecting spectrum licensees from interference. We present below a brief overview of major technical issues and recent development in each module. A more detailed survey of technical and regulatory issues in spectrum overlay can be found in [14].

11.1.1.1 Spectrum Opportunity Identification

As shown in [15], in a general network setting with spatially varying primary user activity, spectrum opportunity detection needs to be performed jointly by a secondary transmitter and its intended receiver. Specifically, a channel is an opportunity when no primary users in the neighborhood of the secondary transmitter are *receiving* over this channel and no primary users in the neighborhood of the secondary receiver are *transmitting* over this channel. Spectrum opportunity detection thus has both signal processing and networking aspects. The problem can, however, be reduced to a classic signal processing problem: detecting the presence of primary users' signals [15]. Based on the secondary user's knowledge of the signal characteristics of primary users, three traditional signal detection techniques can be employed: matched filter, energy detector (radiometer) and cyclostationary feature detector [16]. A matched filter performs coherent detection. It requires the least number of samples to achieve a given detection power but relies on synchronization and a priori knowledge of primary users' signaling. On the other hand, the non-coherent energy detector requires only basic information of primary users' signal characteristics but suffers from long detection time. Cyclostationary feature detector can improve the performance over an energy detector by exploiting an inherent periodicity in the primary users' signal. Details of this type of detectors can be found in [17]. While classic signal detection techniques exist in the literature, detecting primary transmitters in a dynamic wireless environment with noise uncertainty, shadowing, and fading is a challenging problem that has attracted much research attention [18].

Due to hardware limitation and energy cost associated with spectrum monitoring, a secondary user may not be able to sense all channels in the spectrum simultaneously. In this case, the secondary user needs a sensing strategy for intelligent channel selection to track the time varying spectrum opportunities. The purpose of the sensing strategy is twofold: catch a spectrum opportunity for immediate access and obtain statistical information on spectrum occupancy so that more rewarding sensing decisions can be made in the future. A tradeoff has to be reached between these two often conflicting objectives. Within the framework of partially observable Markov decision processes, optimal opportunity tracking strategies have been studied in [3, 19] and reviewed in Chapter 10.

11.1.1.2 Spectrum Opportunity Exploitation

Once spectrum opportunities are detected, secondary users need to decide whether and how to exploit them. Specific issues include whether to transmit given that opportunity detectors may make mistakes, what modulation and transmission power to use and how to share opportunities among secondary users to achieve a network-level objective.

The optimal design of spectrum access strategies in the presence of spectrum sensing errors has been addressed in [20,21]. Specifically, the interaction between the spectrum access protocols at the MAC layer and the operating characteristics of the spectrum opportunity detector at the physical layer is quantitatively characterized, and the optimal joint design of opportunity detectors, access strategies and opportunity tracking strategies is obtained. A review of these results is given in Chapter 10.

Modulation and power control in spectrum overlay networks also present unique challenges not encountered in the conventional wired or wireless networks. Since secondary users often need to transmit over non-contiguous frequency bands, orthogonal frequency division multiplexing (OFDM) has been considered as an attractive candidate for modulation in spectrum overlay networks [21–23]. Power control for secondary users needs to take into account the detection range of the opportunity detector, the maximum allowable interference level and the transmission power of primary users [15]. This complex networking issue remains largely open.

Spectrum opportunity sharing among secondary users has been addressed in the context of exploiting locally unused TV broadcast bands (see [1,2,24,25] and references therein). For this type of applications, spectrum opportunities are considered static or slowly varying in time. Real-time opportunity identification is not as critical a component as in applications that exploit temporal spectrum opportunities. It is often assumed that spectrum opportunities at any location over the entire spectrum are known.

In this chapter, we focus on distributed sharing of slowly varying spectrum opportunities among competing secondary users. Differing from the graph coloring approach considered in [1,24], game theory is employed to capture the distributed interaction among selfish secondary users with individual resource demands.

11.1.2 Organization of Chapter

The rest of this chapter is organized as follows. In Sect. 11.2 we introduce the game theoretic equilibrium and learning concepts that are needed to analyze our decentralized spectrum access model. In Sect. 11.3 we present the spectrum access model itself, along with algorithms for estimating channel competition, simultaneous adaptive learning of distributed resource allocation policies and centralized optimization of system-level spectral efficiency. The chapter concludes with a brief summary and discussion.

11.2 Review of Nash and Correlated Equilibrium in Games

Because our dynamic spectrum access model relies on a decentralized decision approach among secondary users, we rely on game theory to provide operational algorithms and performance analysis in this chapter. Thus, in this section, we present a brief discussion of game theoretic concepts which are to be used, such as Nash and correlated equilibria, as well as an overview of game theoretic learning algorithms by which cognitive radios can adaptively discover how to allocate resources in a competitively optimal fashion.

11.2.1 Equilibrium Definitions

For a game with L players, the problem of each player $l = 1, 2, \ldots, L$ is to devise a rule for selecting their own action X^l from a set S^l (with size S^l), in order to maximize the expected value of a given utility function $u^l(X^1, X^2, \ldots, X^L)$. Since each player only controls one of L variables, the problem requires careful consideration of the actions of other players, which are unknown in advance.

The central concept in non-cooperative game theory is an equilibrium, which identifies stable operating points of the system under certain conditions, such as common knowledge of rationality. The most common such equilibrium is due to Nash [24], defined as follows:

Definition 11.1. *For each player l, who takes random action X^l, define a strategy π^l to be a probability distribution on S^l, so that $\pi^l(x^l) = Pr(X^l = x^l)$ for all $x^l \in S^l$. Label the joint (random) action of all players by X, and define the strategy profile π to be the product of all individual strategies, so that $\pi(x) = Pr(X = x) = \prod_{k=1}^{L} \pi^k(x^k)$. ($X$ resides on the space $\mathbb{S} = S^1 \times S^2 \times \ldots \times S^L$.) We may write any strategy profile π as (π^l, π^{-l}) for any l, where π^{-l} is the strategy profile of all players but l. The expected utility to l resulting from π is*

$$u^l(\pi) = \sum_{x \in \mathbb{S}} u^l(x)\pi(x). \tag{11.1}$$

Now, π is a Nash equilibrium if each π^l is an optimal response to the collection π^{-l} of strategies of other players. That is,

$$u^l(\pi^l, \pi^{-l}) \geq u^l(\sigma^l, \pi^{-l}) \tag{11.2}$$

for all $l = 1, 2, \ldots, L$ and all possible alternative strategies σ^l.

The notation (σ^l, π^{-l}) means that l uses strategy σ^l instead of π^l.

In this chapter, we find it useful to focus on an important generalization of the Nash equilibrium, which was proposed in [1,2] and is known as the *correlated equilibrium*. This is defined as follows:

Definition 11.2. *Define a joint strategy π to be a probability distribution on the product space $\mathbb{S} = S^1 \times S^2 \times \ldots \times S^L$. That is, $\pi(x) = Pr(X = x)$ for joint actions $X, x \in \mathbb{S}$. (The expected utility to l resulting from π is again as in (11.1).) We may decompose any strategy π into marginals (π^l, π^{-l}) for any l, where π^l is the marginal action distribution (strategy) of l, and π^{-l} is the marginal strategy of all players but l. Now, π is a correlated equilibrium if*

$$u^l(\pi^l, \pi^{-l}) \geq u^l(\sigma^l, \pi^{-l}), \tag{11.3}$$

for all $l = 1, 2, \ldots, L$ *and all possible alternative marginal strategies* σ^l *that are a function of* π^l.

In the correlated equilibrium, strategy π provides each player l with an action "recommendation" a^l. Based on this, and knowing π, a player could calculate an a posteriori probability distribution for the actions of other players, and hence an expected utility for each action. The equilibrium condition states that there is no deviation rule (represented by a function σ^l of π^l) that would award l a better expected utility. Combining (11.1) and (11.3), we obtain the equivalent condition:

$$\sum_{x^{-l} \in \boldsymbol{S}^{-l}} \pi(j, x^{-l})[u^l(k, x^{-l}) - u^l(j, x^{-l})] \leq 0 \tag{11.4}$$

for all $l = 1, 2, \ldots, L$, and $j, k \in \boldsymbol{S}^l$. That is, for any recommendation j to l, there is no profitable deviation k. The correlated equilibria comprise a convex set, given by:

$$\mathcal{CE} = \{\pi \in \Delta(\mathbb{S}) : \pi \text{ satisfies } (11.4) \; \forall \; l, j, k\}. \tag{11.5}$$

The correlated equilibrium concept permits coordination between players, and can lead to improved performance over a Nash equilibrium [1]. If a correlated equilibrium distribution $\pi(s)$ can be written as a product of independent marginals $\pi(s) = \prod_{k=1}^{L} \pi^k(s^k)$, then it also satisfies the definition of a Nash equilibrium. The set (11.5) is also structurally simpler than the set of Nash equilibria; it is a convex set, whereas the Nash equilibria are isolated points at the extrema of this set [25]. Since the set of correlated equilibria is convex, fairness between players can also be addressed in this domain. Finally, decentralized, online adaptive procedures (see below) naturally converge to (11.5), whereas the same is not true for Nash equilibria (the so-called law of conservation of coordination [26]).

11.2.2 Adaptive Learning of Equilibria

A particularly interesting application of game theory is its usefulness in developing adaptive procedures in multiagent environments. Such procedures enable components of a system to learn a satisfactory (in game theory, equilibrium) policy for action through repeated interaction with their common environment. Moreover, these procedures are completely decentralized; each component interacts with others only through the effects of the environment, so explicit coordination is not necessary.

We outline the most well-known adaptive game theoretic learning schemes. In what follows, let $n = 0, 1, 2, \ldots$ be discrete time, let $\boldsymbol{X}_n^l$ denote the action of player l at time n, and let $\boldsymbol{X}_n^{-l}$ denote the joint actions of all players but l at time n.

1. *Best response:* If the common interaction between players is ignored, each player will simply attempt to maximize its performance, assuming the environment will remain the same. In the best response scheme, each player simply takes action

$$X^l_{n+1} = \text{argmax}_{x \in S^l} \{u^l(x, X^{-l}_n)\}.$$

That is, each player l acts optimally, assuming the other players will repeat their previous actions. Although it fails to account for simultaneous adaptation from multiple players, this approach can be shown to converge in some special cases, such as two-player zero-sum games, supermodular games, potential games and certain types of submodular games.

2. *Fictitious play:* The most well-known procedure, fictitious play was introduced in [27] and has been extensively studied since, see [28]. In this scheme, each player calculates a best response assuming the historical distribution of play is a good predictor of future actions. That is,

$$X^l_{n+1} = \text{argmax}_{x \in S^l} \{u^l(x, \bar{z}^{-l}_n)\}$$

where $\bar{z}^{-l}_n$ is the empirical joint distribution of play up to time n. Fictitious play enjoys good convergence properties in practice, although convergence to Nash equilibrium is known to be false in general. One drawback is the need to explicitly observe and model the behavior of all opponents, which may not be appropriate for cognitive radios with limited awareness.

3. *Regret-based algorithms:* More recently, a general class of algorithms has been proposed in the form of regret-based learning [28–31]. Regret-based algorithms are, in a sense, a generalization fictitious play, which replace explicit opponent modeling with an implicit "regret matrix," θ^l_n. This tracks, for every pair of actions $j, k \in S^l$, the difference in utility if l had taken action k in the past everywhere it took action j. Given $X^l_n = j$, the probability of $X^l_{n+1} = k$ is proportional to $\theta^l_{n,jk}$, the regret from j to k. Learning proceeds by exploring and switching to actions that are perceived as "better" according to this regret measure.

In this paper we focus on regret-based procedures, as they are simple to implement and have well-understood convergence properties. Maintenance of θ^l_n requires minimal computation and no explicit awareness of other players. The main disadvantage is that players are required to know $u^l(k, X^{-l}_n)$ for all possible $k \in S^l$ at each n. This requirement is removed in modified regret matching [29], which is presented (modified for our purposes) in Algorithm 11.1.

11.3 Decentralized Dynamic Spectrum Access Through Adaptive Reinforcement Learning

We consider a system of L cognitive radios, competing for access to C wireless communication channels which may be occupied at any time by primary users, who have priority in access. At successive time intervals of length Λ, each radio determines which of the C channels are unoccupied by primary users, and of these, the transmission rate (quality) sustainable by each channel. The objective of each radio $l = 1, 2, \ldots, L$ is to select a subset of unoccupied channels for use, in order to satisfy its current demand level. However, since there is competition, there is no guarantee that the selected channels will be captured for exclusive use by l. Instead, we consider a simple slotted CSMA scheme for sharing each channel among users who select it, and propose a decentralized reinforcement learning scheme that allows each radio to find a satisfactory channel allocation through repeated channel selections and performance measurements.

Once each radio selects a subset of channels for use in a particular time interval, repeated competition takes place for each selected channel, as follows. Divide time into K subintervals of length λ/K. In each subinterval $k = 1, 2, \ldots, K$, all radios l active on channel i generate a backoff time $\tau_k^l(i)$; the smallest backoff time captures the channel for transmission for the remainder of the subinterval, as in a typical CSMA MAC protocol.

The history of successes and failures over these K channel capture attempts is used to give performance feedback to each user, i.e., as a sample of how much data it can expect to transmit over each selected channel. However, we can get more information out of these attempts. Specifically, we show how to couple the success/failure history with the history of backoff times used to estimate the number of users competing for these channels in Sect. 11.3.2. This extra information allows us to increase the level of cooperation in the cognitive radio problem; instead of merely trying to satisfy their own demand, users can attempt to minimize their interference with each other by explicitly favoring uncrowded channels over crowded ones. The complete radio utility function to accomplish this is formulated in Sect. 11.3.1.

We assume that the environment of the cognitive radio users varies slowly in time relative to the decision interval length λ. The variation we consider here is in terms of the channel occupancy of primary users, and the traffic demand level of individual cognitive radios. Furthermore, the cognitive radio utility function may be periodically updated by a central base station (see below). These slow variations in parameters motivate us to consider an *adaptive* reinforcement learning strategy, which allows radios to respond to changes in their environment without discarding everything they have learned to date. This adaptive strategy is based on the decentralized, game theoretic learning procedure of modified regret matching [29], and is outlined in Sect. 11.3.3.

Finally, even when radios act in the decentralized fashion described above, we may be able to improve performance by occasionally adjusting the behavior of each radio from a central controller. Since each radio acts to maximize a utility in our framework, we propose a scheme which parameterizes the radio utility, and periodically broadcasts parameter updates from a central controller, or base station, so as to improve global system performance.

We formulate this parameter adjustment scheme as an optimal "pricing" problem for the system, which we approach through stochastic optimization techniques. Suppose that a parameter (price) ϕ can be periodically broadcast (on a slow time scale) to each radio. Upon receiving the price update, each radio adjusts its utility as a function of ϕ and continues its usual behavior under the new utility. The aim of the central controller is to discover that ϕ which maximizes a global utility $G(\pi(\phi))$, where $\pi(\phi)$ is the long-run (equilibrium) behavior of the radios under the utility priced by ϕ.

Since $\pi(\phi)$ and hence $G(\pi(\phi))$ is difficult or impossible to calculate a priori, a stochastic approximation approach is necessary for the discovery of the optimal ϕ. We propose to investigate Robbins–Monro type algorithms for this purpose [32]. By estimating the derivative $g(\phi) \approx dG(\pi(\phi))/mathrmd\phi$, we can use for example the steepest ascent method

$$\widehat{\phi}_{k+1} = \widehat{\phi}_k + \alpha_k \widehat{g}(\widehat{\phi}_k) \qquad (11.6)$$

to successively approach an optimal ϕ, where $\alpha_k > 0$.

We propose to use spectral efficiency for our performance measure $G(\pi(\phi))$, which measures the time average proportion of available channels actually used by cognitive radios during a given period. Since radio decisions are decentralized, we do not expect the spectral efficiency to be 100%, but we hope to make incremental improvements through our pricing procedure.

A block diagram of our system is given in Fig. 11.1. In total, there are four time scales in our problem formulation. At the slowest time scale, the base station sets pricing parameters. Next is the time scale of variation of primary user activity and demand levels. Third, and much faster, is the decision time scale (intervals of length λ) of the cognitive radios themselves, and fourth, the fastest time scale (intervals of length λ/K), are the CSMA channel access attempts. For definiteness, we assume that pricing changes are on the order of hours, while primary user and demand variations are on the order of seconds. We take λ to be approximately 1 ms, and $K \approx 10$ CSMA attempts per channel allocation decision.

11.3.1 Decentralized Dynamic Spectrum Access Model and Radio Utility

In this section we present a mathematical outline of the decentralized dynamic spectrum access problem, which is used to formulate a utility function which each cognitive radio user attempts to maximize.

As above, we divide time into equal slots of length λ, and label each slot by $n = 1, 2, \ldots$. At the beginning of the nth time slot, we assume each cognitive radio $l = 1, 2, \ldots, L$ has the following information:

1. $\mathcal{C}$, the number of channels available for transmission use in the radio system.
2. $C \in \mathbb{R}^{\mathcal{C}}$, a vector giving the quality (bits transmissible per time slot) of each available channel.
3. $Y_n \in \Psi = \{x \in \mathbb{R}^{\mathcal{C}} : x(i) \in \{0,1\} \text{ for all } i = 1, 2, \ldots, \mathcal{C}\}$, a vector showing the current channel usage pattern of primary users; channel i is in use if $Y_n(i) = 1$.
4. $d_n^l \in \mathbb{R}$, the current demand level of the cognitive radio user l (in bits per time slot).
5. A pricing parameter $\phi(i)_n$ for each channel $i = 1, 2, \ldots, \mathcal{C}$, obtained from the base station.

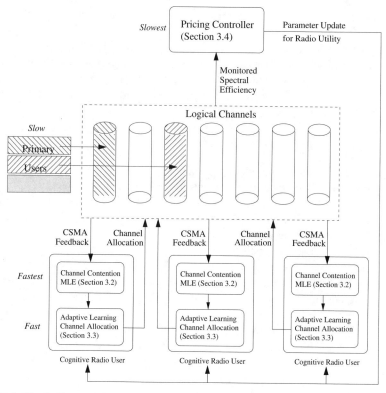

Fig. 11.1. Block diagram of decentralized learning system for cognitive radio dynamic spectrum access.

All these quantities are static or vary slowly in time, hence each radio knows their value before a channel allocation decision is made. For example, we will suppose that the primary user activity Y_n evolves according to a Markov chain with transition matrix $I + \varepsilon Q$, where $0 < \varepsilon \ll 1$ and Q is a generator matrix with each row summing to zero.

Next, each radio l chooses a channel allocation action $\boldsymbol{X}_n^l$, according to the learning scheme outlined in Sect. 11.3.3. For any $y \in \Psi$, define

$$\Psi_\perp(y) = \{x \in \Psi : x \cdot y = 0\} \tag{11.7}$$

to be the set of vectors in Ψ orthogonal to y. Action $\boldsymbol{X}_n^l$ then belongs to the slowly varying space $\boldsymbol{S}_n^l = \Psi_\perp(Y_n)$. That is, each player can select any collection of unused channels. For notational convenience, we adopt the following definition:

Definition 11.3. *For any index $i = 1, 2, \ldots, \mathcal{C}$ and any vector $x \in \Psi$, we say that $i \in x$ if and only if $x(i) = 1$.*

We also denote the joint action of all l decision makers by $\boldsymbol{X}_n$.

The joint channel allocation action $\boldsymbol{X}_n$ is then fixed for K successive CSMA transmission slots $n_1, n_2, \ldots, n_K$, each of length Λ/K. In each transmission slot n_k, each radio l generates a backoff time $\tau_{n_k}^l(i)$ for each selected channel $i \in \boldsymbol{X}_n^l$. Backoff times are generated according to a uniform distribution on the interval $(0, \tau_{\max})$ for some fixed parameter $\tau_{\max}$. Each radio waits until its backoff time expires then transmits data in the remainder of the slot only if the channel is sensed clear. If the smallest backoff time is sufficiently smaller than the next smallest backoff time (allowing time to sense the channel clear and switch from receive to transmit mode), then the radio with the smallest backoff time transmits successfully. Otherwise, there is a collision since two radios will have sensed the channel to be clear and transmitted data. Thus, each transmission slot is used at most by one radio. For each n, $i \in \boldsymbol{X}_n^l$, and $k = 1, 2, \ldots, K$, define

$$\gamma_{n_k}^l(i) = I\{ \text{ channel } i \text{ captured by } l \text{ in slot } n_k\} \tag{11.8}$$

where $I\{\cdot\}$ is the usual indicator function.

At the end of the decision time slot n (of length Λ), each radio l has collected the following information on its CSMA attempts:

$$\gamma_n^l = \left\{ \gamma_{n_k}^l(i) : i \in \boldsymbol{X}_n^l, k = 1, 2, \ldots, K \right\} \tag{11.9}$$

$$\tau_n^l = \left\{ \tau_{n_k}^l(i) : i \in \boldsymbol{X}_n^l, k = 1, 2, \ldots, K \right\}. \tag{11.10}$$

This information is used for performance feedback. For each $i \in \boldsymbol{X}_n^l$, we calculate the proportional throughput achieved:

$$R_n^l(i) = \frac{1}{K} \sum_{k=1}^{K} \gamma_{n_k}^l(i). \tag{11.11}$$

Since the CSMA MAC is random, (11.11) is a random function of the joint decision $\boldsymbol{X}_n$. We note that we can take $R_n^l(i) = 0$ for all $i \notin \boldsymbol{X}_n^l$, and that $E[R_n^l(i)]$ clearly decreases in the contention level $\sum_{l=1}^{L} \boldsymbol{X}_n^l(i)$.

Section 11.3.2 also shows how to use (γ_n^l, τ_n^l) to obtain an estimate $\widehat{N}_n^l(i)$ for the number of users contending for channel i in decision time slot n. We show there that the maximum likelihood estimate for the contention level is given by $\widehat{N}_n^l(i) = 1 + \theta$, where θ solves

$$\sum_{k:\gamma_{n_k}^l(i)=0} \frac{a_k^\theta \log(a_k)}{1 - a_k^\theta} = \sum_{k:\gamma_{n_k}^l(i)=1} \log(a_k) \tag{11.12}$$

where $a_k = 1 - (\tau_k^l(i) + \delta)/\tau_{\max}$, for CSMA parameters $(\delta, \tau_{\max})$. We will also give an approximate solution to (11.12).

Given the information from (11.11) and (11.12), we propose a utility function to guide the reinforcement learning procedure. The utility for radio user l is given by:

$$\hat{u}^l(Y_n, d^l, X_n^l) = -\left(d^l - \sum_{i=1}^{c} C(i) R_n^l(i)\right)^2 - \sum_{i=1}^{c} \phi(i) \widehat{N}_n^l(i) R_n^l(i). \tag{11.13}$$

The following remarks on (11.13) are in order:

1. The utility is implicitly a function of X_n, the actions of all players, through $\widehat{N}_n^l(i)$ and $R_n^l(i)$.
2. It is negative; to maximize (11.13), a radio must match its resources to its demand (first term), and simultaneously avoid designated crowded channels (second term).
3. The objective of avoiding other users as directed by the base station, and not exceeding the demand level d^l, enables cooperation between cognitive radio users.

The observed utility (11.13) is used as feedback to guide future channel allocation decisions in a decentralized fashion. To accomplish this, each radio takes a sequence of actions $\{X_1^l, X_2^l, \ldots, X_n^l\}$ and observes corresponding rewards $\{u_1^l, u_2^l, \ldots, u_n^l\}$. This data is used to generate a new action X_{n+1}^l through a decentralized, adaptive, regret-based reinforcement learning procedure, as described in Sect. 11.3.3. This procedure is game theoretic in nature, that is, it converges even when other cognitive radio users are simultaneously adapting their behavior. This is a critical observation, since naive, single-agent reinforcement learning procedures rely heavily on a static environment for convergence, which is not present in a multiagent situation. Game theoretic algorithms such as the one studied here enables cognitive radio activity to converge to an equilibrium (specifically a correlated equilibrium), which implies that each radio adopts a channel allocation that maximizes its own utility in response to the actions of others. This allows the cognitive radio system to learn, in a completely decentralized manner, to equitably share the available radio channels.

11.3.2 Channel Contention Estimate

In this section we show how to use the information obtained from repeated CSMA attempts to estimate the number of cognitive radio users competing for a given channel. This estimate is required for computing the utility (11.13) for reinforcement learning, and is based solely upon the history of successes and failures of repeated CSMA channel access attempts over a fixed period, along with the associated backoff times used in each attempt. This information is given in (11.9) and (11.10).

Consider a fixed channel i and a specific active user l. We wish to estimate $\widehat{N}_n^l(i)$, the number of users competing for resource i during decision slot n, based on K CSMA channel access attempts within that slot.

First, consider a single, general CSMA channel access attempt on channel i, and suppose there are $\theta(i)$ other active users competing for that channel. Each of these users $m \neq l$ chooses a random backoff time $\tau^m(i)$ uniformly on $(0, \tau_{\max})$. If l chooses $\tau^l(i) = t$, it captures the channel if $t < \tau^m(i) + \delta$ for all $\theta(i)$ users $m \neq l$, where δ is the time required to sense the channel clear and switch from RX to Tx mode. The probability of this event is given according to the order statistic $\tau_{\theta(i)}^{(1)}$, by

$$\Pr(l \text{ captures channel}) = \Pr\left(\tau_{\theta(i)}^{(1)} > t + \delta\right). \tag{11.14}$$

Likewise, we have

$$\Pr(l \text{ fails to capture channel}) = \Pr\left(\tau_{\theta(i)}^{(1)} < t + \delta\right). \tag{11.15}$$

It is well known that the order statistics for uniform random variables are given by the beta distribution. For the first order statistic $\tau_\theta^{(1)}$ on the interval $(0, \tau_{\max})$, the distribution simplifies to:

$$\Pr\left(\tau_{\theta(i)}^{(1)} > t + \delta\right) = \begin{cases} \left(1 - \frac{t+\delta}{\tau_{\max}}\right)^{\theta(i)}, & t \le \tau_{\max} - \delta \\ 0, & t > \tau_{\max} - \delta. \end{cases} \tag{11.16}$$

A bit of reflection reveals that this indeed satisfies probabilistic intuition.

Suppose now that during decision interval n, l has recorded the success or failure of K CSMA attempts, along with the backoff time used in each attempt. These attempts are labeled $n_1, n_2, \ldots, n_K$. Note that $\theta(i)$ is held fixed over the K attempts by the decision structure. Then l can obtain a maximum likelihood estimate for the $\theta(i)$ by maximizing the quantity:

$$L(\theta(i)) = \prod_{k:\gamma_{n_k}^l = 1} \Pr\left(\tau_{\theta(i)}^{(1)} > \tau_{n_k}^l(i) + \delta\right) \cdot \prod_{k:\gamma_{n_k}^l = 0} \Pr\left(\tau_{\theta(i)}^{(1)} < \tau_{n_k}^l(i) + \delta\right) \tag{11.17}$$

$$= \prod_{k:\gamma_{n_k}^l = 1} \left(1 - \frac{\tau_{n_k}^l(i) + \delta}{\tau_{\max}}\right)^{\theta(i)} \cdot \prod_{k:\gamma_{n_k}^l = 0} \left(1 - \left(1 - \frac{\tau_{n_k}^l(i) + \delta}{\tau_{max}}\right)^{\theta(i)}\right) \tag{11.18}$$

where $\tau_{n_k}^l(i)$ is the backoff time of user l at time index k on channel i and $\gamma_{n_k}^l(i)$ denotes success or failure of the corresponding CSMA attempt, as in (11.8). The MLE is simply $\widehat{N}_n^l(i) = 1 + \arg\max_\theta L(\theta(i))$.

Differentiating the likelihood (or log likelihood) with respect to θ, we obtain that the maximizing θ must solve

$$\sum_{k:\gamma_{n_k}^l(i) = 0} \frac{a_k^\theta \log(a_k)}{1 - a_k^\theta} = \sum_{k:\gamma_{n_k}^l(i) = 1} \log(a_k) \tag{11.19}$$

where $a_k = 1 - (\tau_k^l(i) + \delta)/\tau_{\max}$.

Equation (11.19) is difficult to solve analytically. Numerically, we can state the following general properties:

1. $\widehat{N}_n^l(i)$ increases with the number of channel access failures.
2. $\widehat{N}_n^l(i)$ increases on average with the maximum successful backoff time.
3. $\widehat{N}_n^l(i)$ increases on average with the minimum unsuccessful backoff time.

Behaviour of Contention MLE

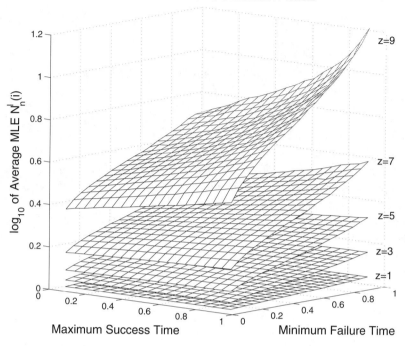

Fig. 11.2. Numerical plot of the (log) average of estimates $\widehat{N}_n^l(i)$. z denotes the number of failed CSMA attempts out of $K = 10$, and the maximum backoff time is $\tau_{\max} = 1$. Each data point represents an average over 5000 randomly generated observations satisfying the given limits for success and failure times.

A plot of $\widehat{N}_n^l(i)$ is given in Fig. 11.2 for $K = 10$ CSMA attempts and $\delta = 0$. For each data point, we specified the number of CSMA channel access failures $z = 1, 3, 5, 7, 9$ as well as the maximum and minimum backoff times of the successful and unsuccessful CSMA attempts, respectively ($\tau_{\max}$ is normalized to one). We then generated 5000 data samples corresponding to the specified limits and plotted an average of the results on a logarithmic scale, to emphasize the importance of the number of failures z on the estimate.

If we approximate a_k on the left-hand side by

$$\bar{a}_0 = \frac{1}{|I_0|} \sum_{k:\gamma_{n_k}^l(i)=0} a_k$$

where $|I_0| = \sum_{k=1}^K (1 - \gamma_{n_k}^l(i))$ is the number of terms in that summation (the number of channel access failures), (11.19) becomes:

$$|I_0| \frac{\bar{a}_0^\theta \log(\bar{a}_0)}{1 - \bar{a}_0^\theta} = \sum_{k:\gamma_{n_k}^l(i)=1} \log(a_k). \qquad (11.20)$$

The approximation in (11.20) corresponds to replacing the backoff times of failed channel access attempts by their average. From this, we can obtain an analytic solution:

$$\theta = -\log\left(1 + \frac{|I_0| \log(\bar{a}_0)}{\sum_{k \in I_1} \log(a_k)}\right) / \log(\bar{a}_0). \qquad (11.21)$$

Numerical studies show that the approximation (11.21) is quite accurate on average, but can have a large variance in unfavorable conditions. Experimentally, it can be shown that the approximation error in (11.21) is small when either the number of channel access failures $|I_0|$ is small, or when the successful backoff times are small. In other cases, it may be preferable to use (11.21) to generate an initial guess, which may be refined by the Newton–Raphson method.

11.3.3 Adaptive Learning for Channel Allocation

In this section we describe our decentralized learning approach to the cognitive radio dynamic spectrum assignment problem. Our approach is based on the modified regret matching procedure of [29], which is formulated here as a distributed stochastic approximation algorithm. This formulation allows us to specify an adaptive variant of the original procedure, called "modified regret tracking," which uses a constant stepsize to dynamically adapt to time varying conditions, thus allowing users to function in a changing environment.

As is usual in reinforcement learning, each user takes a sequence of actions $\{X_n^l \in S_n^l : n = 0, 1, 2, \ldots\}$ and observes a sequence of rewards $\{u_n^l \in \mathbb{R} : n = 0, 1, 2, \ldots\}$. The action at time $n + 1$ is a random function of this history of actions and rewards.

At each decision period n, users take joint action $X_n \in \mathbb{S}$, with user l taking action $X_n^l \in S^l$. To implement the algorithm, each user l uses the observed utilities associated with past joint actions $\{X_n : n = 1, 2, \ldots\}$ to derive regret values $\theta_{n,jk}^l$: $j, k \in S^l$, according to:

$$\theta_{n,jk}^l = \sum_{\tau \leq n: X_\tau^l = k} \varepsilon_{\tau-1} \left(\prod_{\sigma=\tau}^{n-1}(1 - \varepsilon_\sigma)\right) \frac{p_\tau^l(j)}{p_\tau^l(k)} u^l(k, X_\tau^{-l})$$

$$- \sum_{\tau \leq n: X_\tau^l = j} \varepsilon_{\tau-1} \left(\prod_{\sigma=\tau}^{n-1}(1 - \varepsilon_\sigma)\right) u^l(j, X_\tau^{-l}). \qquad (11.22)$$

If $\varepsilon_\tau = 1/(\tau + 1)$, this is simply the average:

$$\theta_{n,jk}^l = \frac{1}{n} \sum_{\tau \leq n: X_\tau^l = k} \frac{p_\tau^l(j)}{p_\tau^l(k)} u^l(k, X_\tau^{-l}) - \frac{1}{n} \sum_{\tau \leq n: X_\tau^l = j} u^l(j, X_\tau^{-l}). \qquad (11.23)$$

If $\varepsilon_\tau = \varepsilon$, it is the exponentially weighted moving average:

$$\theta^l_{n,jk} = \sum_{\tau \leq n: X^l_\tau = k} \varepsilon(1-\varepsilon)^{n-\tau} \frac{p^l_\tau(j)}{p^l_\tau(k)} u^l(k, X^{-l}_\tau)$$

$$- \sum_{\tau \leq n: X^l_\tau = j} \varepsilon(1-\varepsilon)^{n-\tau} u^l(j, X^{-l}_\tau). \qquad (11.24)$$

These values are computed recursively in the algorithm below.

To gain some intuition, we refer back to the original regret matching algorithm of [28]. Here, the regret value is taken as the simple time average

$$\theta^l_{n,jk} = \frac{1}{n} \sum_{\tau \leq n: X^l_\tau = j} \left(u^l(k, X^{-l}_\tau) - u^l(j, X^{-l}_\tau) \right). \qquad (11.25)$$

That is, the regret measures the average gain that l would have received had he played k in the past instead of j. If the gain is positive, then clearly l should be more likely to switch to action k in the future, and in fact regret matching does exactly this by switching to each action k at time $n+1$ with probability proportional to the positive component of $\theta^l_{n,jk}$. Note, however, that (11.25) requires that l knows what utility he would have received for each action, *even if that action was not taken*. To overcome this difficulty, [29] approximates the first term of the summation (11.25) by the first summation in (11.23).

The complete procedure, including the exact formulation of action probabilities, is summarized in Algorithm 11.1, which is carried out independently by each user.

Algorithm 11.3.1 Adaptive Learning for Channel Allocation: The regret-based algorithm for user activation has parameters $(u^l, \mu, \delta, \{\varepsilon_n : n = 1, 2, \ldots\}, \theta^l_0, X^l_0)$, where u^l are the user utilities, μ is a function of the utilities as in (11.29), δ is a small probability with which actions are chosen from a uniform distribution, $\{\varepsilon_n\}$ is a small stepsize, and θ^l_0, X^l_0 are arbitrary initial regrets and actions.

Define the $S^l \times S^l$ matrix with entries:

$$\boldsymbol{H}^l_{jk}(X_n) = I\{X^l_n = k\} \frac{p^l_n(j)}{p^l_n(k)} u^l(k, X^{-l}_n) - I\{X^l_n = j\} u^l(j, X^{-l}_n). \qquad (11.26)$$

The Procedure Is As Follows:

1. *Initialization:* Set $n = 0$ and take action X^l_0. Initialize regret $\theta^l_0 = \boldsymbol{H}^l(X_0)$.
   Repeat for $n = 0, 1, 2, \ldots$:
   Action update: Choose $X^l_{n+1} = k$ with probability

$$\Pr(X^l_{n+1} = k | X^l_n = j, \theta^l_n = \theta^l)$$

$$= \begin{cases} (1-\delta) \min\left(\max\{\theta^l_{jk}, 0\}/\mu, \frac{1}{S^l-1}\right) + \frac{\delta}{S^l}, & k \neq j, \\ 1 - \sum_{i \neq j} \left[(1-\delta) \min\left(\max\{\theta^l_{ji}, 0\}/\mu, \frac{1}{S^l-1}\right) + \frac{\delta}{S^l}\right] & k = j. \end{cases}$$

$$(11.27)$$

Regret value update: Calculate $\boldsymbol{H}^l(X_{n+1})$, and update θ_{n+1} using the stochastic approximation (SA):

$$\theta_{n+1}^l = \theta_n^l + \varepsilon_n(\boldsymbol{H}^l(X_{n+1}) - \theta_n^l). \tag{11.28}$$

In (11.27), μ is a normalization constant, which is chosen

$$\mu > (S^l - 1)(u_{\max}^l - u_{\min}^l) \tag{11.29}$$

over all $l = 1, 2, \ldots, L$, where $(u_{\max}^l, u_{\min}^l)$ are obtained from (11.13).

Note that θ_n^l is a moving average of the updates $\{H^l(X_k) : k = 1, 2, \ldots, n\}$. Because of this, Algorithm 11.1 can be viewed as a stochastic approximation with a constant stepsize $\varepsilon_n \equiv \varepsilon > 0$; actions are chosen with probability proportional to their (moving) average potential performance in the past. (This differs from best response, which would base action choices on the immediately previous result, essentially setting $\varepsilon = 1$.) For the original modified regret matching algorithm of [29], one would instead use $\varepsilon_n = 1/(n + 1)$.

Since the utility varies, a constant stepsize in Algorithm 11.3.1 is needed to keep users responsive to the changes.

11.3.3.1 Convergence of Regret-Based Learning

When a decreasing stepsize $\varepsilon_n = 1/(n+1)$ is used in Algorithm 11.1, it is proven in [29] that the global empirical distribution of play (defined below) converges almost surely to the set of ε-correlated equilibria. If, in addition, the "tremble" term δ in (11.27) is decreased sufficiently slowly, convergence is to the set of correlated equilibria proper (11.5).

It is therefore reasonable to expect similar convergence results of the constant stepsize version of Algorithm 11.3.1, with fixed small $\varepsilon_n = \varepsilon$ and a fixed small tremble δ. The general relation between decreasing and constant stepsize stochastic approximation (SA) algorithms is well known [32]. Essentially, when a decreasing stepsize SA converges almost surely, it can be shown that the constant stepsize version converges weakly, as the stepsize $\varepsilon \to 0$. Intuitively then, our adaptive version of Algorithm 11.3.1 should track the set of correlated equilibria, with the benefit that changes to the utility functions are handled smoothly by the constant stepsize.

We now describe in detail what is meant by this type of convergence. First, convergence is stated in terms of the empirical distribution of play, which can be viewed as a diagnostic that monitors the performance of the entire cognitive radio network. This is defined as follows:

Definition 11.4. *The empirical distribution of play up to time n is:*

$$\bar{z}_n = \sum_{\tau \leq n} \varepsilon_{\tau-1} \Big(\prod_{\sigma=\tau}^{n-1}(1 - \varepsilon_\sigma)\Big)e_{X_\tau} \tag{11.30}$$

where $e_x = [0, 0, \ldots, 1, 0, \ldots, 0]$ with the one in the xth position.

Here ε_n is a weighting factor. If $\varepsilon_n = 1/(n+1)$, the empirical distribution is simply

$$\bar{z}_n = 1/n \sum_\tau e_{X_\tau}. \tag{11.31}$$

If $\varepsilon_n \equiv \varepsilon > 0$ is constant, it is the exponentially weighted moving average

$$\bar{z}_n = \varepsilon \sum_\tau (1-\varepsilon)^{n-\tau} e_{X_\tau}. \tag{11.32}$$

Note that in both cases $\bar{z}_n$ is an empirical frequency, since $\sum_i \bar{z}_n(i) = 1$. We point out here that $\bar{z}$ satisfies the following recursion:

$$\bar{z}_{n+1} = \bar{z}_n + \varepsilon_n(e_{X_{n+1}} - \bar{z}_n) \tag{11.33}$$

where X_{n+1} is constructed according to Step (2a) of Algorithm 11.3.1. When a decreasing stepsize $\varepsilon_n = 1/(n+1)$ is used, (11.33) directly yields (11.31). With a constant stepsize $\varepsilon_n = \varepsilon$, (11.33) directly yields (11.32).

Second, in contrast to most convergence results, convergence of the empirical distribution of play for Algorithm 11.3.1 is not to a specific point, but to the *set* of correlated equilibria ($\mathcal{CE}$). This property is as follows:

Definition 11.5. *$\bar{z}_n$ converges to the set $\mathcal{CE}$ if for any $\varepsilon > 0$ there exists $N_0(\varepsilon)$ such that for all $n > N_0$ we can find $\psi \in \mathcal{CE}$ at a distance less than ε from $\bar{z}_n$.*

The actual proof of weak convergence for the adaptive modified regret tracking algorithm can be approached in two ways. First, one can attempt to adapt the original proof in [29] for a constant stepsize. Second, one can take a differential inclusion approach, similar to that found in [31,33]. The first approach appears plausible, but technically difficult. The proof in [29] is based on the idea of Blackwell approachability [34], to which the existence of a decreasing stepsize is central. One would therefore be forced to begin by modifying Blackwell's 1956 result, then proceed to carry the modifications through the proof in [29]. The differential inclusion approach therefore appears more promising. Although [31,33] still assumes here a decreasing stepsize, it treats the convergence of the original (non-modified) regret matching algorithm of [28] in such a way that the constant stepsize result can easily be obtained through the methods of [32]. Since the modified procedure (used here) of [29] is obtained from [28], it should not be too difficult to use similar methods here.

11.3.4 Stochastic Optimization of Spectral Efficiency via Centralized Pricing

In this section we describe a simple stochastic optimization approach for improving spectral efficiency in the decentralized channel access learning environment. The approach relies on a base station, which monitors only the outcome of cognitive radio activities. That is, the base station is not aware of the actions of individual cognitive radios, but only of how often free channels are used by the group for data transmission. It attempts to influence the behavior by periodically broadcasting a common C-*valued* parameter vector to each radio, which is used by the cognitive radios to update their utility function. The results of this section are not integral to the decentralized learning scheme; the pricing scheme describes a way to improve the equilibrium behavior obtained in the previous sections from a global perspective, but is not necessary to the operation of the cognitive radio system as already described.

Recall the utility function (11.13) of cognitive radio users, which is parameterized by pricing vector ϕ. For each channel i, $\phi(i)$ represents a unit interference penalty; user l essentially pays a cost $\phi(i)$ for each portion of channel i it uses and each user present on channel i. This is meant as a simple disincentive so that a base station, seeing that channel i is too crowded, can encourage users to move to other channels by imposing a high cost $\phi(i)$. Conversely, users may be attracted to low-cost channels in order to balance load across the spectrum.

Although it is possible to devise much more sophisticated incentive rules, possibly through mechanism design theory, than the one presented here, we feel that at least elementary control can be imposed through our formulation, and that it provides a sufficient proof of concept of stochastic optimization-based pricing for tuning cognitive radio networks. Moreover, the basic stochastic optimization approach, which we outline here, will remain the same regardless of the particular pricing parametrization used.

The objective of the base station is to discover, through experimentation, a pricing parameter ϕ which maximizes the spectral efficiency of the cognitive radio system. The spectral efficiency is defined as the average proportion of available radio channels that are used by the cognitive radios, which may be sampled over T decision intervals as:

$$\widehat{\text{SE}}(\phi) = \frac{\sum_{n=1}^{T} \sum_{l=1}^{L} \min\left\{\sum_{i=1}^{C} C(i) R_n^l(i), d^l\right\}}{\sum_{n=1}^{T} \sum_{i=1}^{C} [C(i)(1 - Y_n(i))]}. \tag{11.34}$$

Note that (11.34) is based only on the observable channel usage.

To incrementally improve the spectral efficiency, we propose the following algorithm:

Algorithm 11.3.2 Stochastic Optimization-Based Pricing: For large T, set pricing interval length $T\Lambda$. A decision time n of the cognitive radios (on time scale Λ) is said to belong to pricing interval m if $mT + 1 \le n \le (m + 1)T$. For pricing intervals $m = 0, 1, 2, \ldots$, repeat the following:

1. Monitor the decisions in pricing interval m. That is, collect data $Y_n(i)$ and $R_n^l(i)$ for $i = 1, 2, \ldots, C$, $l = 1, 2, \ldots, L$, and n in pricing interval m.
2. At the end of interval m, calculate the spectral efficiency according to (11.34) using the data gathered. Estimate the derivative $\widehat{\text{dSE}}/\text{d}\phi$.
3. Broadcast a new pricing vector for pricing interval $m + 1$, according to

$$\widehat{\phi}_{m+1} = \widehat{\phi}_m + \alpha_m \frac{\widehat{\text{dSE}}}{\text{d}\phi}(\widehat{\phi}_m). \tag{11.35}$$

The derivative may be estimated using standard approximation methods, for example the finite difference or simultaneous perturbation methods [36].

Conclusion

In this chapter we have presented an iterative, decentralized method for discovering efficient dynamic spectrum access policies for cognitive radio. Under the spectrum overlay model, we have shown how the spectrum access problem can be treated as a game theoretic problem and given algorithms that allow cognitive radios to independently assess and adapt to their environment in real time.

The key advantage of our approach is complete decentralization, that is, the lack of requirement for any collaboration or communication between cognitive radios. We do require, in the centralized pricing scheme of Sect.11.3.4, the ability to receive occasional updates from a central base station, but this feature is meant only as an optional enhancement to the decentralized system. We are able to obtain effective performance from the decentralized scheme essentially for two reasons. First, since radios are aware of the presence of competitors, they are able to estimate and adapt to channel competition by leveraging game theoretic algorithms specifically designed to converge in a multiuser setting. Second, we have built cooperative tendencies into the utility function (11.13) itself; radios are penalized for obtaining more resources than they require and are bound to obey direction from the base station through the pricing function ϕ. While this second consideration might be negated by selfish design, the structural results would not change; the equilibrium obtained would simply be less efficient than that obtained through cooperative design. Let us emphasize: cooperative design in essentially decentralized systems allows us to achieve many of the benefits of a completely integrated architecture without the same costly infrastructure.

The constant step size learning algorithm presented in this chapter converges weakly to the set of correlated equilibria of a non-cooperative game. Moreover, the algorithm can be used to track a slowly time varying correlated equilibrium set caused due to changing activity of primary users, with the limiting behavior of the algorithm captured by a differential inclusion. Suppose we were to assume that primary user activity evolves according to a slow Markov chain with transition probability matrix $I + \epsilon Q$ (where $\epsilon > 0$ is a small parameter and Q is a generator matrix with each row summing to zero). With this assumption, how can one analyze the tracking performance of the learning algorithm with step size ϵ? Note that the adaptation speed (step size ϵ) of the algorithm matches the speed at which the correlated equilibrium set changes (transition matrix $(I + \epsilon Q)$). In our recent work [36,37], we have shown that the limiting behavior of the stochastic approximation algorithm for tracking a parameter evolving according to a Markov chain is captured by a Markovian switched ordinary differential equation. This result was somewhat remarkable, since typically the limiting process of a stochastic approximation algorithm is a deterministic ordinary differential equation. We conjecture that the limiting behavior of Algorithm 11.3.1 is captured by a Markovian switched differential inclusion (see [38]). This analysis requires use of yet another extremely powerful tool in stochastic analysis namely, the so-called "martingale problem" of Strook and Varadhan, see [41,42] for comprehensive treatments of this area.

There are many other interesting avenues for continuation of this research. Aside from improving and validating the algorithms presented here, one can modify the problem to consider the case of partial channel observation. This is especially important when the number of channels becomes too large for simultaneous monitoring. Moreover, for this situation, it is important to identify initial methods for eliminating a large number of channels from consideration, in order to improve the convergence rate and memory requirements of the adaptive learning approach considered here. Finally, we can expand our scope from static games to stochastic games, in which the player actions not only determine their immediate utility, but also give a probability distribution over new games to be played in future rounds. A stochastic game approach for similar sensor-based systems has been carried out in [41,42].

References

1. R. Aumann, "Correlated equilibrium as an expression of Bayesian rationality," *Econometrica*, vol. 55, no. 1, pp. 1–18, 1987.
2. R. Aumann, "Subjectivity and correlation in randomized strategies," *J. Math. Econ.*, vol. 1, pp. 67–96, 1974.
3. Q. Zhao, L. Tong, and A. Swami, "Decentralized cognitive MAC for dynamic spectrum access," in *Proc. IEEE DySPAN 2005*, pp. 224–232, 2005.
4. H. Zheng and L. Cao, "Device-centric spectrum management," in *Proc. IEEE DySPAN 2005*, pp. 56–65, 2005.
5. N. Nie and C. Comaniciu, "Adaptive channel allocation spectrum etiquette for cognitive radio networks," in *Proc. IEEE DySPAN 2005*, pp. 269–278, 2005.
6. J. Robinson, "An iterative method of solving a game," *Ann. Math.*, vol. 54, pp. 298–301, 1951.
7. J. Huang, R. Berry, and M. Honig, "Auction-based spectrum sharing," *Springer, Mobile Netw. Appl.*, vol. 11, no. 3, pp. 405–418, 2006.
8. M. McHenry, "Spectrum white space measurements," June 2003. Presented to New America Foundation Broadband Forum; Measurements by Shared Spectrum Company, Available at http://www.newamerica.net/Download Docs/pdfs/Doc File 185 1.pdf.
9. D. Hatfield and P. Weiser, "Property rights in spectrum: Taking the next step," in *Proc. First IEEE Symposium on New Frontiers in Dynamic Spectrum Access Networks*, Nov. 2005.
10. L. Xu, R. Tonjes, T. Paila,W. Hansmann, M. Frank, and M. Albrecht, "DRiVE-ing to the Internet: Dynamic radio for IP services in vehicular environments," in *Proc. 25th Annual IEEE Conference on Local Computer Networks*, pp. 281–289, Nov. 2000.
11. Y. Benkler, "Overcoming agoraphobia: Building the commons of the digitally networked environment," *Harvard J. Law Technol.*, Winter 1997–1998.
12. J. Mitola, "Cognitive radio for flexible mobile multimedia communications," in *Proc. IEEE International Workshop on Mobile Multimedia Communications*, pp. 3–10, 1999.
13. "DARPA: the next generation (XG) program." http://www.darpa.mil/ato/programs/xg/index.htm.

14. Q. Zhao and B. M. Sadler, "A survey of dynamic spectrum access: signal processing, networking, and regulatory policy," *IEEE Signal Processing Magazine*, vol. 55, no. 5, pp. 2294–2309, May, 2007.
15. Q. Zhao, "Spectrum opportunity and interference constraint in opportunistic spectrum access," in *Proc. IEEE International Conference on Acoustics, Speech, and Signal Processing (ICASSP)*, Apr. 2007.
16. D. Cabric, S. M. Mishra, and R. W. Brodersen, "Implementation issues in spectrum sensing for cognitive radios," in *Proc. 38th. Asilomar Conference on Signals, Systems, and Computers*, pp. 772–776, 2004.
17. W. Gardner, "Signal interception: A unifying theoretical framework for feature detection," *IEEE Trans. Commun.*, vol. 36, pp. 897–906, Aug. 1988.
18. A. Sahai, N. Hoven, and R. Tandra, "Some fundamental limits on cognitive radio," in *Proc. Allerton Conference on Communication, Control, and Computing*, Oct. 2004.
19. Q. Zhao, L. Tong, A. Swami, and Y. Chen, "Decentralized cognitive MAC for opportunistic spectrum access in ad hoc networks: A POMDP framework," *IEEE J. Select. Areas Commun.*, Special Issue on Adaptive, Spectrum Agile and Cognitive Wireless Networks, Apr. 2007.
20. Y. Chen, Q. Zhao, and A. Swami, "Joint PHY/MAC design of opportunistic spectrum access in the presence of sensing errors," submitted to *IEEE Trans. Signal Process.* in Jan. 2007.
21. T. Weiss and F. Jondral, "Spectrum pooling: An innovative strategy for enhancement of spectrum efficiency," *IEEE Commun. Mag.*, vol. 42, pp. 8–14, Mar. 2004.
22. U. Berthold and F. K. Jondral, "Guidelines for designing OFDM overlay systems," in *Proc. First IEEE Symposium on New Frontiers in Dynamic Spectrum Access Networks*, Nov. 2005.
23. H. Tang, "Some physical layer issues of wide-band cognitive radio systems," in *Proc. First IEEE Symposium on New Frontiers in Dynamic Spectrum Access Networks*, Nov. 2005.
24. J. Nash, "Non-cooperative games," *Ann. Math.*, vol. 54, no. 2, pp. 286–295, 1951.
25. R. Nau, S. Canovas, and P. Hansen, "On the geometry of Nash equilibria and correlated equilibria," *Int. J. Game Theory*, vol. 32, no. 4, pp. 443–453, 2004.
26. S. Hart and A. Mas-Colell, "Uncoupled dynamics do not lead to Nash equilibrium," *Am. Econ. Rev.*, vol. 93, no. 5, pp. 1830–1836, Dec. 2003.
27. D. Fudenberg and D. Levine, *The theory of learning in games*. MIT Press, 1999.
28. S. Hart and A. Mas-Colell, "A simple adaptive procedure leading to correlated equilibrium," *Econometrica*, vol. 68, no. 5, pp. 1127–1150, 2000.
29. S. Hart and A. Mas-Colell, "A reinforcement procedure leading to correlated equilibrium," *Economic Essays*, Springer, 2001, pp. 181–200.
30. A. Cahn, "General procedures leading to correlated equilibria," *Int. J. Game Theory*, vol. 33, no. 1, pp. 21–40, 2004.
31. M. Benaim, J. Hofbauer, and S. Sorin, "Stochastic approximations and differential inclusions ii: Applications," *UCLA Department of Economics, Levine's Bibliography*, May 2005.
32. H. Kushner and G. Yin, *Stochastic approximation and recursive algorithms and applications*, 2nd ed. New York, NY: Springer-Verlag, 2003.
33. M. Benaim, J. Hofbauer, and S. Sorin, "Stochastic approximations and differential inclusions," *SIAM J. Control Optim.*, vol. 44, no. 1, pp. 328–348, 2005.
34. D. Blackwell, "An analog of the minimax theorem for vector payoffs," *Pacific J. Math.*, vol. 6, pp. 1–8, 1956.

35. J. Spall, *Introduction to stochastic search and optimization: estimation, simulation, and control*. Wiley Press, 2003.
36. G. Yin and V. Krishnamurthy "Least mean square algorithms with Markov regime switching limit," *IEEE Trans. Autom. Control,* vol. 50, no. 5, pp. 577–593, 2005.
37. G. Yin, V. Krishnamurthy, and C. Ion, "Regime switching stochastic approximation algorithms with application to adaptive discrete stochastic optimization," *SIAM J. Optim.*, vol. 14, no. 4, pp. 1187–1215, 2004.
38. A. Benveniste, M. Metivier, and P. Priouret "Adaptive Algorithms and Stochastic Approximations," in *Applications of Mathematics*, vol. 22, Springer-Verlag, 1990.
39. S. Ethier and T. Kurtz, *Markov processes–characterization and convergence*. Wiley, 1986.
40. H. Kushner, *Approximation and weak convergence methods for random processes, with applications to stochastic systems theory*. Cambridge, MA: MIT Press, 1984.
41. M. Maskery and V. Krishnamurthy, "Decentralized algorithms for netcentric force protection against anti-ship missiles," (preprint) *IEEE Trans. Aerospace Electr. Syst.*, 2007.
42. M. Maskery and V. Krishnamurthy, "Network enabled missile deflection: Games and correlated equilibrium," (preprint), *IEEE Trans. Aerospace Electr. Syst.*, 2007.

Additional Reading

1. S. Sankaranarayanan, P. Papadimitratos, A. Mishra, and S. Hershey, "A bandwidth sharing approach to improve licensed spectrum utilization," in *Proc. First IEEE Symposium on New Frontiers in Dynamic Spectrum Access Networks (DySPAN)*, 2005.
2. Y. Chen, Q. Zhao, and A. Swami, "Joint design and separation principle for opportunistic spectrum access," in *IEEE Asilomar Conference on Signals, Systems, and Computers*, 2006.
3. H. Zheng and C. Peng, "Collaboration and fairness in opportunistic spectrum access," in *Proc. IEEE International Conference on Communications (ICC)*, 2005.
4. W. Wang and X. Liu, "List-coloring based channel allocation for open-spectrum wireless networks," in *Proc. IEEE VTC*, 2005.
5. M. Steenstrup, "Opportunistic use of radio-frequency spectrum: A network perspective," in *Proc. First IEEE Symposium on New Frontiers in Dynamic Spectrum Access Networks*, 2005.
6. M. Maskery and V. Krishnamurthy, "Decentralized activation in a ZigBee-enabled unattended ground sensor network: A correlated equilibrium game theoretic analysis," submitted to *IEEE/ACM Trans. Netw.*, 2006.
7. V. Krishnamurthy, G. Yin, and M. Maskery "Stochastic approximation based tracking of correlated equilibria for game-theoretic reconfigurable sensor network deployment," in *Proc. IEEE Conference on Decision and Control*, 2006.
8. V. Krishnamurthy, M. Maskery, and M. Hanh Ngo, "Scalable sensor activation and transmission scheduling in sensor networks over Markovian fading channels," in *Wireless sensor networks. Signal processing and communications perspectives*, Wiley Press, 2007.
9. M. Maskery and V. Krishnamurthy, "Decentralized activation in a ZigBee-enabled unattended ground sensor network: A correlated equilibrium game theoretic analysis," in *Proc. IEEE International Conference on Communications*, 2007.
10. M. Maskery and V. Krishnamurthy, "Decentralized management of sensors in a multiattribute environment under weak network congestion," in *Proc. IEEE International Conference on Acoustics, Speech, and Signal Processing*, 2006.

12

Decentralized Spectrum Management Through User Coordination *

Haitao Zheng and Lili Cao

Department of Computer Science
University of California, Santa Barbara, CA 93106 USA
{htzheng,lilicao}@cs.ucsb.edu

12.1 Introduction

Wireless devices are becoming ubiquitous – they are a vital component in our daily life. Unfortunately, the deployment and expansion of new wireless technologies is being slowed down or even blocked by the inefficient access of radio spectrum. Historical (and current) spectrum allocation policies assign a fixed spectrum band to each wireless technology. Over the time, this static assignment results in an artificial "spectrum scarcity," including over-allocation and under-utilization of licensed bands, and an increasingly crowded unlicensed band [1].

To address spectrum scarcity and realize the potential of radio spectrum, we need new mechanisms to dynamically distribute spectrum among competing wireless devices according to their demand and usage. Dynamic spectrum allocation is feasible since new generation of wireless devices can quickly adjust their radio transmission frequencies with a wide range of spectrum, enabled by recent advance in Cognitive Radio hardware [2, 3]. While maximizing spectrum utilization is the primary goal of dynamic spectrum systems, a good allocation scheme needs to provide fairness across users.

Dynamic spectrum management is challenging, particularly in large-scale wireless networks. Due to the phenomenon of *radio interference* [4], allocation of spectrum exhibits a form of externality. A user seizing spectrum without coordinating with others can cause harmful interference with its surrounding neighbors, and thus reducing spectrum utilization and degrading others' performance. Therefore, spectrum allocation needs to address the constraint of radio interference, which makes the problem NP-hard [4, 5].

There are multiple complimentary ways to address the problem of spectrum allocation, using approximation algorithms. The widely used approach is central-

---

* Portions reprinted, with permission, from "Distributed Spectrum Allocation via Local Bargaining" by Lili Cao and Haitao Zheng, published at the Proceedings of IEEE Communications Society Conference on Sensor, Mesh and Ad Hoc Communications and Networks (SECON), 2005. ©[2005] IEEE.

ized approximations. Given a fixed topology, prior work [5, 6] reduces the allocation problem into a conventional graph coloring problem or its variants. A central manager obtains the conflict topology that specifies interference constraints among users [4], and performs coloring algorithms to derive a conflict-free spectrum assignment that intends to maximize a *system utility*. While operating based on global network knowledge, good centralized algorithms like graph-coloring face high complexity cost, and hence are not efficient in dynamic or large-scale networks. In particular, a topology-optimized allocation algorithm begins with no prior information, and assigns each user a close-to-optimal assignment. When network topology or spectrum availability change, the network needs to completely recompute spectrum assignments for all users after each change, resulting in high computational and communication overhead. This costly operation needs to be repeated frequently to maintain spectrum utilization and fairness. In addition, centralized algorithms require the existence of a centralized server.

In this chapter, we consider a decentralized approach to spectrum allocation where instead of relying on any central servers, users perform local coordinations to modify their spectrum usage to approach a new conflict free spectrum assignment that maximizes system utility. In addition to being low-cost, this approach provides quick adaptation to topology variations. When network dynamics occur, our approach starts from the previous spectrum allocation, and performs a limited number of computations to arrive at a new solution to the new topology and spectrum availability.

The rest of the chapter is organized as follows. We begin in Sect. 12.2 by defining the spectrum allocation problem and the system utility functions. Next, we propose a local coordination framework in Sect. 12.3 and develop specific strategies to improve system utilization and fairness in Sect. 12.4. We then in Sect. 12.5 provide a set of theoretical analysis to evaluate system utility and algorithm complexity. Next in Sect. 12.6, we conduct experiments to evaluate the performance of bargaining strategy and validate the theoretical lower bound. We summarize related work in Sect. 12.7, discuss implications and future directions before we conclude the chapter.

12.2 The Problem of Spectrum Allocation and Its Solution via Centralized Approximation

As background, we describe in this section the problem of spectrum allocation, and the solutions via centralized approximation algorithms. We start with the theoretical model used to represent the general allocation problem, and two utility functions that maximize spectrum utilization and fairness. We describe how to reduce the optimal allocation problem to a variant of a graph multi-coloring problem and describe the previous solutions that optimize the spectrum allocation for a given topology.

12.2.1 Problem Model and Utility Functions

We consider the case where the collection of available spectrum ranges forms a spectrum pool, divided into non-overlapping orthogonal channels. We assume a network of N users indexed from 0 to N–1 competing for M spectrum channels indexed 0 to M–1. Each user can be a transmission link or a broadcast access point. We consider an Open Spectrum based system where users can only access a spectrum channel if the usage will not produce interference to any primary users.[1] Users select communication channels and adjust transmit power accordingly to avoid interfering with primaries. The channel availability and throughput for each user can be calculated based on the location and channel usage of nearby primaries. The spectrum access problem becomes a channel allocation problem, i.e. to obtain a conflict free channel assignment for each user that maximizes system utility. The key components of our model are:

Channel availability $L(n)$. $\Gamma = \{l_{m,n} | l_{m,n} \in \{0,1\}\}_{M \times N}$ is an M by N binary matrix representing the channel availability: $l_{m,n} = 1$ if and only if channel m is available at user n. In general, $l_{m,n} = 0$ when channel m is occupied by a primary user who conflicts with user n, so that the transmissions of n on this channel will interfere with the primary's activity if they use channel m concurrently. Let $L(n) = \{0 \leq m \leq M - 1 | l_{m,n} = 1\}$ be the set of channels available at n.

Interference constraint C. Let $C = \{c_{n,k} | c_{n,k} \in \{0,1\}\}_{N \times N}$, an $N \times N$ matrix, represents the interference constraints among users. If $c_{n,k} = 1$, users n and k would interfere with each other if they use the same channel. The interference constraint depends on the signal strength of transmissions and the distance between users. A simple model of interference constraint is the binary geometry metric, i.e. two transmissions conflict if they are within π distance from each other. This provides an approximation to the effects of interference in real wireless systems.

It should be noted that the interference constraint could also depend on the frequency location of the channel (i.e. m), since power and transmission regulations vary significantly across frequencies. The work in [5, 7] considers the channel-dependency and uses an $M \times N \times N$ interference matrix C. In this chapter, for simplicity, we consider a channel-independent interference constraint, assuming channels have similar power and transmission regulations. It is straightforward to extend the proposed approaches to account for channel-dependent or other interference conditions.

User dependent channel throughput B. Let $B = \{b_{m,n} > 0\}_{M \times N}$ describe the reward that a user gets by successfully acquiring a spectrum band, i.e. $b_{m,n}$ represents the maximum bandwidth/throughput that user n can acquire through using spectrum band m (assuming no interference from other neighbors). Let $b_{m,n} = 0$ if $l_{m,n} = 0$. So that B represents the bandwidth weighted user available spectrum.

---

[1] In Open Spectrum Systems [8,9], primaries have the highest priority to access spectrum.

Conflict free assignment A. $A = \{a_{m,n} | a_{m,n} \in \{0,1\}\}_{M \times N}$ where $a_{m,n} = 1$ denotes that spectrum band m is assigned to user n, otherwise 0. A satisfies all the constraints defined by C, that is,

$$a_{m,n} + a_{m,k} \leq 1, \text{ if } c_{n,k} = 1, \forall \, n, k < N, m < M.$$

Let $\Lambda_{N,M}$ denote the set of conflict free spectrum assignments for a given set of N users and M spectrum bands.

User throughput of a conflict free assignment. Let $T_A(n)$ represent the throughput that user n gets under assignment A, i.e. $T_A(n) = \sum_{m=0}^{M-1} a_{m,n} \cdot b_{m,n}$.

Given this model, the goal of spectrum allocation is to maximize network utilization, defined by U. We can define the spectrum assignment problem by the following optimization function:

$$A^* = \max_{A \in \Lambda_{N,M}} \arg \max U(A).$$

System Utility Function U. We can obtain utility functions for specific application types using sophisticated subjective surveys. An alternative is to design utility functions based on traffic patterns and fairness inside the network. In this chapter, we consider and address fairness based system utility. Consistent with prior work [10–12], we address fairness for single-hop flows since they are the simplest format in wireless transmissions. We postpone the discussion of routing related utility functions to the future work. Similar to [11], we define fairness in terms of maximizing total logarithmic user throughput, refereed to as proportional fairness. The utility can be expressed as

$$U(A) = \sum_{n=0}^{N-1} \log T_A(n) = \sum_{n=0}^{N-1} \log \sum_{m=0}^{M-1} a_{m,n} \cdot b_{m,n}. \tag{12.1}$$

As a reference, another utility function is the total spectrum utilization in terms of total user throughput

$$U(A) = \sum_{n=0}^{N-1} T_A(n). \tag{12.2}$$

Maximizing utilization does not consider fairness, and the resulting channel assignment is in general unbalanced. Note that the two key goals of a spectrum allocation algorithm are spectrum utilization and fairness. Combinations of these two goals form specific utility functions that can be customized for different types of network applications.

12.2.2 Color-Sensitive Graph Coloring

The work in [5, 7] shows that by mapping each channel into a color, the spectrum assignment problem can be reduced to a graph multi-coloring (GMC) problem.

Definition 12.1. *Given the channel assignment problem in above, the system can be represented by a* Conflict Graph $G = (V, E, B)$ *where V is a set of vertices denoting the users that share the spectrum, B represents the bandwidth weighted available spectrum, mapping to the color list at each vertex, and E is a set of undirected edges between vertices representing interference constraint between two vertices defined by C. For any two distinct vertices $u,v \in V$, an edge between u and v, is in E if and only if $c_{u,v} = 1$.*

Figure 12.1 illustrates an example of GMC graph. There are five colors available. The numbers outside the brackets attached to each node denote the colors assigned to that node, while the numbers inside the brackets denote the available color list of each node.

A GMC problem is to color each vertex using a number of colors from its color list, and find the color assignment that maximizes system utility. The coloring is constrained by that if an edge exists between any two distinct vertices, they cannot be colored with the same color. Most importantly, the objective of coloring is to maximize system utility. This is different from traditional graph color solutions that assign one color per vertex. Notice that the solution to this graph coloring problem is to maximize system utility for a given graph, i.e. a given topology and channel availability. This characterizes the optimal solution for a static environment.

The optimal coloring problem is known to be NP-hard [14]. Efficient algorithms to optimize spectrum allocation for a given network topology exist. In [7], the authors presented a set of sequential heuristic based approaches that produce good coloring solutions. The algorithm starts from an empty color assignment and iteratively assign

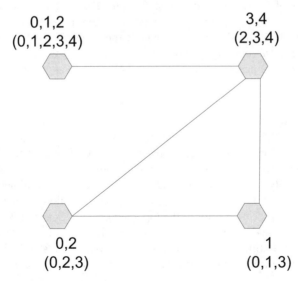

Fig. 12.1. An example of GMC graph [© *2005 IEEE. Reprinted, with permission, from [13]*].

colors to vertices to approximate the optimal assignment. In each stage, the algorithm labels all the vertices with a non-empty color list according to some policy-defined labeling. The algorithm picks the vertex with the highest valued label and assigns the color associated with the label to the vertex. The algorithm then deletes the color from the vertex's color list, and from the color lists of the constrained neighbors. The color list and the interference constraint of a vertex keep on changing as other vertices are processed, and the labels of the colored vertex and its neighbor vertices are modified according to the new graph. The algorithm can be implemented using a centralized controller who observes global topology and makes decisions, or through a distributed algorithm where each vertex performs a distributed voting process. Results in [5, 7] show that the heuristic based algorithms perform similarly to the global optimum (derived off-line for simple topologies), and the centralized and distributed algorithms perform similarly.

12.3 Decentralized Coordination Framework

The approach described in Sect. 12.2 attempts to globally optimize spectrum allocation for a given topology. In a mobile network model, node movements lead to constant changes in network topology. Using the existing approach, we can reapply the spectrum allocation algorithm after each change in the conflict graph. This approach assumes no prior allocation information, and incurs high computation and communication overheads. Finally, to efficiently perform the graph-coloring, the system requires a central controller who has access to the global information and performs the coloring algorithm.

To reduce these overheads and avoid the dependence on the central server, we propose the use of an adaptive and robust distributed algorithm that takes prior allocation into account in new spectrum assignments. More importantly, users consider both needs of neighboring devices and available spectrum when determining its spectrum usage. When sub-optimal spectrum usage is detected, users trigger coordination events among nearby peers and apply local adjustments to approach a globally optimized spectrum assignment.

12.3.1 Overview

An efficient dynamic allocation algorithm can run every time user movement causes a change in the corresponding network conflict graph. Therefore, an adaptive algorithm needs to only compensate for small changes affecting a local network region. The algorithm starts from a non-optimal spectrum allocation, which can be constructed from the allocation prior to the topology change. Consider a conflict graph with N nodes (indexed from 0 to $N-1$) and M channels (indexed from 0 to $M-1$), where the optimized assignment is $A_{M \times N}$. When a new node (indexed as N) joins the network, the assignment after introduction of the node becomes $A'_{M \times (N+1)}$ where

$$A'_{m,n} = \begin{cases} A_{m,n} & : \quad 0 \le n \le N - 1, \ 0 \le m < M \\ 0 & : \quad n = N, \ 0 \le m < M. \end{cases}$$

Or if a primary user i enters the network and wants to use channel m_0, the nodes within impact of primary i (denoted by $R(i)$) need to stop using channel m_0 within a given time. Hence, the assignment becomes $A'_{M \times N}$ such that

$$A'_{m,n} = \begin{cases} 0 & : \quad m = m_0 \text{ and } n \in R(i) \\ A_{m,n} & : \quad \text{otherwise.} \end{cases}$$

Assuming the spectrum allocation was near optimal before the topology change, local coordination between affected vertices can quickly optimize allocations for utilization and fairness. During local coordination, sets of neighboring vertices, each of which form a connected component of the conflict graph, self-organize into coordination groups. Each group modifies spectrum assignment within the group to improve system utility while ensuring that the change in spectrum assignment does not require any change at other nodes outside the group (due to interference constraints). Note that a node can represent a transmission link or an access point. Coordination related to a transmission link is carried out by the transmitter or receiver while coordination related to an access point is carried out by the access point.

There are two types of coordination: an *explicit bargaining-based* approach where devices negotiate spectrum usage through message exchange [13], and an *implicit rule-based* approach where devices observe behavior of neighbors and independently adjust usage following predefined rules [15].

The main challenge in designing a good decentralized framework is how to use recursive local improvements to approach the global optimal, particularly how to avoid ping-pong effects and provide fast convergence. Next, we show that by imposing reasonable constraints, designing an efficient coordination protocol, our proposed approach is a low-complexity alternative to the centralized approximations. Note that the detailed coordination procedure depends on the format of system utility function, hence we provide a design example in Sect. 12.4 using proportional fairness as the utility function.

12.3.2 Explicit Coordination via Bargaining

In explicit coordination, users self-organize in local bargaining groups defined by physical proximity, and adjust spectrum allocation within each group to maximize local system utility metrics such as proportional fairness. Groups form on-demand when a non-optimal band allocation is discovered, and dissolve following negotiations for spectrum. By participating in different groups at different times, a device can repeatedly negotiate with multiple neighbors to improve system performance.

12.3.2.1 Bargaining Constraints

To perform bargaining/coordination, we must first determine the size and membership of local coordination groups. Large groups increase the complexity of coordination due to high synchronization and communication costs. In addition, interactions might occur between groups if they share neighboring users. To facilitate bargaining, we propose two constraints to regulate the procedure and simplify the process.

Constraint 1: Limited Neighbor Bargaining
While pair-wise bargaining is already a hard convex optimization problem, bargaining within a large group implies even higher computation and communication overheads. Coordinating around a central leader per group can greatly simplify spectrum assignment. In this chapter, we propose a simple group formation where a node that wants to improve its spectrum assignment broadcasts a bargaining request to its k-hop neighbors, where k is the ratio of interference range to transmission range. These neighbors are connected to the node in the corresponding conflict graph. Neighbors willing to participate reply to the sender and form a bargaining group. Note that it is possible that two connected neighbors in a conflict graph might not be able to communicate directly with each other, i.e. when $k > 1$. The bargaining information can be relayed by the nodes in between. These relay nodes do not necessarily participate in the bargaining group. In this following, we will use node to present any vertex in a conflict graph.

For each bargaining group, the requester becomes the group coordinator and performs the bargaining computation. The bargaining strategies can be divided into the following formats:

(1a) *One-to-One Bargaining*: Node n_1 who initiates the bargaining can choose to coordinate with only one neighboring node n_2 at a time. They exchange some channels to improve system utility while complying with the conflict constraints from the other neighbors. This is the simplest bargaining process and the requester only needs approval from one of his neighbors to perform the bargaining. When multiple neighbors, e.g. n_2 and n_3 acknowledge the bargaining request, n_1 can sequentially compute assignment assuming bargaining with n_2 first, and then with n_3. n_1 broadcasts the assignment to both n_2 and n_3. This expands the bargaining group to (n_1, n_2, n_3) without adding extra signaling overhead. However, this also requires that n_1 chooses a sequential bargaining order and gets approval from all the group members on the order before conducting bargaining. If one of neighbors disapproves the request, n_1 needs to perform another request. Hence, for simplicity, we restrict this format to only one-to-one bargaining. Figure 12.2 illustrates an example of one-to-one bargaining.

(1b) *One-Buyer-Multi-Seller Bargaining*: A buyer node n_1 purchases a set of channels M_0, from its neighbors who are currently using any channel in M_0, such that to improve system utility. In this case, the bargaining requires concurrent approval from multiple neighbors. As we will show later, this type of bargaining is necessary to eliminate user starvation. Figure 12.2 illustrates an example with one buyer and four sellers.

Constraint 2: Self-contained Group Bargaining
Once the bargaining groups are organized, the bargaining inside each group should not disturb the spectrum assignment at nodes outside the group. That is, after the bargaining, the modified channel assignment should not lead to any conflict with

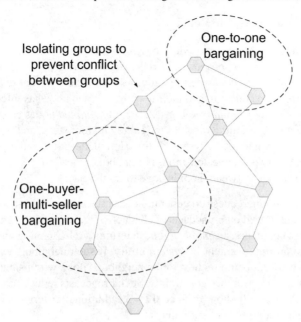

Fig. 12.2. An example of bargaining groups [© *2005 IEEE. Reprinted, with permission, from [13]].*

nodes outside the group. This helps to maintain system stability, so that a bargaining may not invoke a series of reactions due to violations in interference constraints. More importantly, this guarantees that if a bargaining improves the utility in a local area, it also improves the system utility. Or in other words, a local improvement will lead to a system improvement. This constraint has two components.

(2a) *Restricted negotiable channels*: This restricts the set of channels that are exchangeable between nodes inside each bargaining group, such that when a node gets one channel from its neighbor, the assignment does not conflict with its neighbors outside the bargaining group.

(2b) *Isolated bargaining group*: This not only restricts each node to participate in at most one bargaining group at any time, but also requires that the members of any two bargaining groups cannot be directly connected. Having nodes between groups regulate spectrum adjustment and prevents conflict between groups. The necessity of this requirement can be explained by the following example. Assume there are two neighboring nodes A and B (two nodes in the conflict graph connected with an edge) who are the members of two different bargaining groups. Before assignment, A and B are not using channel 0. After the bargaining, both A and B are granted with channel 0 from their bargaining neighbors. However, as A and B conflict with each other if using the same channel, the bargaining produces interference conflict among nodes. The detailed procedure to form isolated bargaining groups will be introduced in Sect. 12.3.2.2. An example of isolation between bargaining groups is shown in Fig. 12.2.

12.3.2.2 Bargaining Steps in Detail

We design a local bargaining procedure based on the above constraints, assuming a distributed architecture. We propose a *distributed, iterative grouping and bargaining process*. We assume that nodes periodically broadcast their current channel assignment and interference constraints to their neighbors. Each node has three states: *bargaining*, *disabled* and *enabled* (see Fig. 12.3). Only *enabled* nodes can perform bargaining. The actual bargaining involves the following four steps, and repeats until no further bargaining can improve system utility. Here nodes refer to the vertices in the conflict graph, and two "connected" nodes might be physically k ($k > 1$) hops away. Information exchange between them is done through relay.

1. Initialize bargaining request: In general, nodes affected by mobility events initialize bargaining. Based on broadcasts of channel assignments and interference constraints from neighbors, an *enabled* node determines if bargaining with a neighbor will lead to an improvement in system utility. If such neighbors exist, the node broadcasts a coordination request to the neighbors along with its current channel assignment and interference constraints. Such broadcasts reduces communication overhead. As we will show in Sect. 12.4.3, additional criterion exists to guide nodes on generating bargaining requests.

2. Acknowledge bargaining request: Neighbors who are *enabled* and willing to coordinate reply an *ACK* message with its current channel assignment and interference constraints. We assume that nodes are willing to collaborate to improve system utility, and accept requests that improve system utility even if it might degrade their individual channel assignments. Incentive systems to encourage such collaboration will be investigated in a future work. If a node receives multiple concurrent requests from its neighbors, it acknowledges the request that leads to the highest bargaining gain calculated from information embedded in the request.

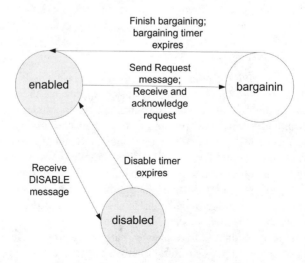

Fig. 12.3. Node state and transitions [© *2005 IEEE. Reprinted, with permission, from [13]*].

3. Bargaining group formation: When the requester receives the replies, it selects the members of the bargaining group, and broadcasts this information along with the proposed modification of the channel assignment to neighbors. Once the bargaining group is set, its members enter *bargaining* state. They broadcast a *DISABLE* message with a timer equal to the estimated duration of the coordination process to neighbors not in the bargaining group. Note that the *DISABLE* message can be embedded in the *ACK* messages to reduce overhead. Nodes receiving the message enter *disabled* state for the duration of the timer. This procedure prevents nodes who are neighbors of existing bargaining group to participate in any future bargaining before the timer expires. Following this, all bargaining groups are isolated.

4. Negotiation: Once all members acknowledge the changes to the channel assignment, each member updates its local channel assignment. This is straightforward for one-to-one bargaining. For one-buyer-multiple-seller bargaining, interactions among members can be coordinated by the bargaining requestor. After bargaining, each member enters *enabled* state. Figures 12.3 and 12.4 illustrate the node state transition and messages during bargaining.

12.3.3 Implicit Coordination via Rules Based Adjustment

For energy-constrained devices such as sensors and mobile ad hoc devices, frequent communication between devices is undesirable. In addition, future networks are likely to have heterogeneous devices and access technologies [16], making the implementation of a common coordination protocol a significant challenge. For these types of networks, we propose a device-centric management scheme, where users

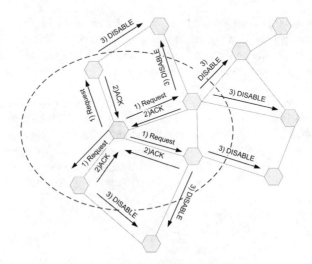

Fig. 12.4. Messages exchanged during bargaining [© *2005 IEEE. Reprinted, with permission, from [13]*].

observe local conditions and neighbors' actions, and independently adapt their spectrum usage using a set of rules predefined by spectrum regulators. In contrast to an explicit bargaining-based approach, users tend to prioritize their own performance with minimal regard to system utility. Their compliance with the rules promotes efficient and fair spectrum sharing. This implicit approach greatly simplifies implementation and significantly reduces coordination traffic. In this approach, users rely on the rules to determine the appropriate channels to use. The spectrum rules are the key to making these tradeoffs, and are highly dependent on the format for system utility. The work in [15] proposed five rules to address different network scenarios. Interesting readers should refer to [15] for the detailed rule design.

12.3.4 Coordination to Improve Utilization

Once the local bargaining procedure is set, the specific bargaining strategy may be customized for different utility functions. It is easy to show that for utilization based utility (total user throughput), the optimization can be reduced to solving M optimization problems for each color respectively. On each color, the corresponding optimization problem is exactly a Weighted Independent Set (WIS) problem [17]. WIS problem is a special case of Weighted Set Packing problem, and can be approximated by a local improvement heuristic algorithm, generally called $t - improvement$ [17, 18]. It is straightforward to convert this algorithm to local bargaining among neighbors. We omit the bargaining procedure due to space constraints, and next focus on the local bargaining strategy for the fairness-based utility.

12.4 Decentralized Coordination to Improve Fairness

In this section, we focus on the coordination strategy optimizing for fairness.[2] Based on its definition in (12.1), the optimization aims to maximize the total logarithmic user throughput, i.e. the product of user throughput. Therefore, the global fairness utility increases if nodes with many assigned channels "give" some channels to nodes with few assigned channels.

We start by describing basic one-to-one bargaining where two unbalanced nodes exchange channels to improve the local throughput product. We show that such bargaining is limited by the number of negotiable channels and thus not effective against the node starvation problem. We then develop a special case of one-buyer-multi-seller bargaining, referred to as *Feed Poverty* to eliminate node starvation. We also derive a theoretical lower bound of user throughput using local bargaining under a simplified network configuration.

We first define the following notations.

n: a node n in the conflict graph ($0 \leq n \leq N - 1$);

---

[2] In this section, we focus on the procedure for explicit bargaining-based coordinations. The detailed rule design for implicit coordination is omitted due to space limit. Readers can refer to [15] for detailed description.

$R(n) = \{v \in V | (n, v) \in E\}$: neighbors of n;

$R(X) = \bigcup_{n \in X} R(n) \setminus X$: neighbors of node set X;

$f_A(n) = \{0 \leq m \leq M - 1 | a_{m,n} = 1\}$: the set of channels assigned to node n under current assignment A.

12.4.1 One-to-One Fairness Bargaining

As we described, one-to-one bargaining allows two neighboring nodes n_1 and n_2 to exchange channels to improve system utility while complying with conflict constraints from the other neighbors. For n_1 and n_2 to negotiate, they need to first obtain the channels that are negotiable to avoid disturbing other neighbors, referred to as $C_b(n_1, n_2)$:

$$C_b(n_1, n_2) = L(n_1) \cap L(n_2) \cap \{\{0..M - 1\} \setminus \bigcup_{n \in R(n_1, n_2)} f_A(n)\}.$$

Given $C_b(n_1, n_2)$, we can define the one-to-one bargaining regarding fairness as follows:

Definition 12.2. *For an assignment $A_{M \times N}$, an* One-to-One Fairness Bargaining *finds nodes n_1 and n_2, and their bargaining channel set $C_b(n_1, n_2)$, and modifies $A_{M \times N}$ to $A'_{M \times N}$ related to n_1, n_2 and channels $C_b(n_1, n_2)$, such that*

$$T_{A'}(n_1) \cdot T_{A'}(n_2) > T_A(n_1) \cdot T_A(n_2).$$

The One-to-One Fairness Bargaining increases the product of the bargaining users while other nodes' throughput values remain unaffected. Hence, the system fairness increases with each bargaining. The improvement between each pair of nodes (n_1, n_2) can be calculated as $G(n_1, n_2) = \frac{T_{A'}(n_1) T_{A'}(n_2)}{T_A(n_1) T_A(n_2)} - 1$. This is used in the bargaining process (in Sect. 12.3) to determine whether a bargaining can improve system utility.

Given (n_1, n_2), assigning channels to n_1 and n_2 to maximize their throughput product is a difficult task. This is because node throughput depends on all channels (including non-negotiable ones) assigned to a node, and the available bandwidth on a channel differs between nodes. The problem is shown to belong to the class of convex programming problems [19]. When the number of negotiable channels ($|C_b(n_1, n_2)|$) is small (e.g. $|C_b(n_1, n_2)| < 10$), exhaustive search may be feasible. Otherwise we need to use approximations based on heuristics such as the one given in [19]: first sort channels in $C_b(n_1, n_2)$ (by channel bandwidth), then use a two-band partition to determine the allocation.

The effectiveness of One-to-One Fairness Bargaining is constrained by the size of $C_b(n_1, n_2)$. In general, due to heavy interference constraints among neighboring nodes, $C_b(.)$ could be very small. Figure 12.5 illustrates an example where the conflict graph is a chain topology consisting of three nodes A, B, and C. Node B is not assigned with any channel and the system utility is zero. We refer to this as

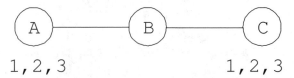

Fig. 12.5. An example of starvation [© *2005 IEEE. Reprinted, with permission, from [13]*].

user *starvation*. Node A and B cannot negotiate due to the constraint from C (i.e. $C_b(A, B) = \emptyset$), while node b and c also cannot negotiate due to the constraint from a (i.e. $C_b(B, C) = \emptyset$). Hence, the *Fairness Bargaining* is not effective to eliminate user *starvation*.

12.4.2 Feed Poverty Bargaining

We observe that user *starvation* in most cases is a result of the lack of flexibility in bargaining. As for the example in Figure 12.5, by allowing A and C to give up channel 1 at the same time and feed it to B, we can remove the starvation at B. This is an example of one-buyer-multi-seller bargaining. In this chapter, we propose a special one-buyer-multi-seller bargaining, called *Feed Poverty* where if a node (buyer) has very poor channel assignment, the neighboring nodes can collaborate together to feed it with some channels.

Definition 12.3. *For an assignment* $A_{M \times N}$, *a feed poverty bargaining is to find some node* n_0 *and channel* m_0, *modify* $A_{M \times N}$ *to* $A'_{M \times N}$, *such that*

$$
A'_{m,n} = \begin{cases} 1 & : & m = m_0 \text{ and } n = n_0 \\ 0 & : & m = m_0 \text{ and } n \in R(n_0) \\ A_{m,n} & : & \text{otherwise} \end{cases}
$$

(intuitively, the assignment lets some of n_0's *neighbors give up channel* m_0 *and feed it to* n_0*) and*

$$
G_{FP}(n_0) = (T_{A'}(n_0)) \cdot \prod_{n \in R(n_0) \wedge A_{m_0,n}=1} (T_{A'}(n))
$$
$$
- (T_A(n_0)) \cdot \prod_{n \in R(n_0) \wedge A_{m_0,n}=1} (T_A(n))
$$
$$
> 0.
$$

This means the product-throughput of the users involved in the bargaining is locally increasing, while the other users' throughput are not affected. So generally the bargaining improves system utility, except that, in case of starvation of other users, the system utility remains $-\infty$. A special case of Feed Poverty is when $A_{m_0,n} = 0$ for all $n \in R(n_0)$. This means none of n_0's neighbors are using channel m_0, and n_0 simply seizes it.

When there is no feasible *One-to-One Fairness Bargaining*, i.e. $|C_b| = \emptyset$, the requestor initializes a *Feed Poverty Bargaining* on all neighbors who acknowledge the request. The requestor sequentially selects multiple channels to maximize group utility.

12.4.3 BF-Optimal Assignment

We propose to combine *One-to-One Fairness Bargaining* and *Feed Poverty Bargaining* into a *Fairness Bargaining with Feed Poverty* (BF). Each node who wants to improve its spectrum usage starts with negotiating *One-to-One Fairness Bargaining* with its neighbors to improve system utility. If there is no negotiable channel between it and any of its neighbors, a *poor* node can broadcast a Feed-Poverty request to its neighbors to initialize *Feed Poverty Bargaining*. Overall, a channel assignment A is said to be *BF-optimal* if no further *Fairness Bargaining with Feed Poverty* can be performed.

12.5 Theoretical Analysis

In this section, we perform theoretical analysis on the explicit negotiation approach. We examine system utility, user throughput and algorithm complexity for BF-optimal assignments. The analytical results provide insights to fairness provided and system efficiency.

12.5.1 Lower Bound on User Throughput

We first examine the user throughput under any BF-optimal assignment, i.e. the user performance when system stabilizes. With the goal of maximizing system proportional fairness, we show that each user's throughput is lower-bounded.

Theorem 12.1. *Under a BF-optimal assignment A, for each vertex n in the conflict graph G, $0 \leq n \leq N - 1$, its throughput $T(n)$ has a lower bound, i.e.*

$$T(n) > \frac{B(n)}{d(n) + 1} - \mathrm{MB}(n)$$

$$B(n) \triangleq \sum_{m=0}^{M-1} b_{m,n}$$

$$\mathrm{MB}(n) \triangleq \max_{m=0}^{M-1} b_{m,n} \tag{12.3}$$

where $d(n)$ represents the number of conflicting neighbors of n or the degree of n in the conflict graph, $B(n)$ represents n's total available bandwidth, and $\mathrm{MB}(n)$ represents n's maximum channel bandwidth.

The proof is included in Appendix.

Theorem 12.1 shows that the lower bound of each user's throughput depends on its interference condition. First, the bound of user n scales inversely with the number of competing users in n's neighborhood, i.e. $d(n) + 1$. A user should obtain less throughput in crowded areas (with more conflicting neighbors) than in sparse areas. This scaling, also the spirit of centralized greedy allocation algorithms [5, 7], provides an immediate intuition of fairness.

Second, the bound scales linearly with the node's total available throughput $B(n)$. This trend demonstrates that the proposed strategy provides similar level of fairness to users regardless of their channel conditions. If a user improves its channel bandwidth using sophisticated physical layer techniques, without changing its interference condition, the throughput bound increases linearly, but the number of assigned channels remains the same. This also shows that our proposed strategy will not favor users in good channel conditions and starve users in bad channel conditions.

Under certain circumstances, channel quality fluctuates due to fading, shadowing and environmental factors, making it impractical to collect channel quality of neighbors in real time. Hence, a reasonable approach is to assume all channels are identical (i.e. with bandwidth 1 if it is available, and with bandwidth 0 if it is not). Next, we show that Theorem 12.1 can be reduced to the following:

Theorem 12.2. *Under a BF-optimal assignment A, for each vertex n in the conflict graph G, $0 \le n \le N - 1$ with degree $d(n)$ and channel availability list $L(n)$, its spectrum usage $T(n)$ has a lower bound,* i.e.

$$T(n) \ge \left\lfloor \frac{|L(n)|}{d(n) + 1} \right\rfloor \triangleq \Gamma(n).$$

The proof is straightforward using Theorem 12.1, included in Appendix. Theorem 12.2 shows that the proposed *Fairness Negotiation with Feed Poverty* guarantees a *poverty line* $\Gamma(n)$ to each vertex n. The *poverty line* provides a guideline in negotiation in real systems where a vertex is entitled to request negotiation if its current throughput is below its poverty line. We refer to this as the *Poverty guided negotiation*.

It is easy to show that if channels are fully available at each vertex, i.e. $|L(n)| = M$, a BF-optimal assignment can eliminate user starvation if the number of channels $M \ge \Delta + 1$, where Δ is the maximum degree in the graph $\Delta = \max_{0 \le n < N} d(n)$. This matches to the well-known conclusion in graph coloring where the chromatic number of a graph is at most $\Delta + 1$ [20]. It can also be shown that the bound in Theorem 12.2 is tight, for several types of network topology: clique, ring, star, and straight line. We omit the proof due to space constraints. For randomly generated topologies, we obtain statistical results on the tightness of the bound using experiments in Sect. 12.6.2.

12.5.2 Bound on Complexity of *Poverty Guided Negotiation*

Using *Poverty Guided Negotiation (PGB)*, only the node below its *poverty line* is qualified to initiate a negotiation. The requestor n can sequentially select the best (i.e.

providing the highest improvement to system utility) channels, and negotiate with the neighboring nodes to take over these channels. Following the detailed proof for Theorem 12.2, we can show that after each negotiation iteration, one node will reach its poverty line. When the system reaches an equilibrium, every node is guaranteed with its poverty line. This also allows us to derive an upper bound on the number of negotiation iterations.

Theorem 12.3. *In a system with N nodes, with uniform channel bandwidth, the Poverty guided negotiation will reach an equilibrium after an expected number of at most $O(N^2)$ iterations. By optimizing the order of negotiation, the system can reach an equilibrium in at most N iterations.*

The proof is in Appendix.

12.5.3 Theoretical Distance to Social Optimal

We also compare the performance of the proposed negotiation strategy, to the social optimal assignment. A social optimal assignment maximizes the global system utility. We focus on the *price of anarchy (POA)* [21], defined as the ratio of the system utility of social optimal assignment over the worse case of the proposed assignment strategies. However, without restriction on channel bandwidth or number of nodes, the POA is unbounded in general. Therefore, we restrict ourselves to the case with uniform channel bandwidth, i.e. $b_{m,n} = 1$ for all m, n, and a system of N users. Next we derive the upper bound of the POA.

Theorem 12.4. *For a topology with uniform channel availability and channel bandwidth, i.e. $b_{m,n} = 1$, $l_{m,n} = 1$, and $M > \Delta,$[3] the price of anarchy for a BF-optimal channel assignment is at most*

$$\frac{M^N \cdot \prod_{\langle u,v \rangle} \left(\frac{d_v}{d_u + d_v} \right)^{\frac{1}{d_u}} \cdot \left(\frac{d_u}{d_u + d_v} \right)^{\frac{1}{d_v}}}{\prod_{n=0}^{N-1} \left\lfloor \frac{M}{d_n + 1} \right\rfloor}. \tag{12.4}$$

The proof can be found in [13].

12.6 Experimental Results

We conduct experimental simulations to quantify the performance of bargaining-based spectrum allocation. We also validate the proposed local bargaining algorithms against the theoretical lower bounds. For simulations, we assume a noiseless, mobile radio network. We simulate an ad-hoc network by randomly placing a set of nodes on a 100×100 area. We assume that each active node broadcasts data packets to some of its neighbors. We further abstract the network into a *Conflict Graph* where

---

[3] Otherwise there may exist starvation and the ratio may be meaningless.

each vertex represents a transmitting node. Any two nodes interfere with each other (i.e. connected in the conflict graph) if they are within distance of 20. The actual distance threshold depends on the choice of transmission power and radio hardware. We simply use 20 as an illustrative example. For default scenarios, we assume that channels are equally weighted and all the channels are available for each node, i.e. $l_{n,m} = 1$, $b_{n,m} = 1$. In terms of traffic demands, all transmitting nodes are assumed backlogged. We focus on maximizing fairness, because bargaining under utility based on spectrum utilization can be reduced to the classical local search of weighted independent set problem, and has been investigated extensively.

Under the simulation settings, for *One-to-One Fairness Bargaining*, the optimal assignment of channels between two nodes (n_1, n_2) can be derived easily to maximize the product of the number of channels assigned to n_1 and n_2. For *Feed Poverty Bargaining*, we select channel m_0, i.e. that produces the minimum disturbance to the neighbors,

$$m_0^* = \arg\min_{m_0} \prod_{n \in Nbr(n_0) \wedge A_{m_0,n}=1} \frac{T_A(n)}{T_{A'}(n)}.$$

Topology dynamics are modeled by having nodes randomly moving to new locations. We divide time into slots, and in each time slot, $p\%$ of nodes move to a new randomly selected location. The model captures the way mobility is manifested in ad hoc networks without delving into complex protocols. A moving node takes the original channel assignment but disables the channels that conflict with its new neighbors. In each time slot, after the topology change, nodes adjust their channel usage.

We use two metrics to evaluate the performance.

- System utility: We consider fairness defined in (12.1). Note that if there exists a user with no channel assigned, the utility becomes $-\infty$. For better representation, we modify the utility to $U(A) = \sqrt[N]{\prod_{n=0}^{N-1} T_A(n)}$ and $U(A) = 0$ if there is any $T_A(n) = 0$.
- Communication overhead: We quantify algorithm complexity as the communication overhead, i.e. total number of messages exchanged among nodes, since transmission and handling of messages will likely dominate computations for channel assignment. In both local bargaining and graph coloring approaches, each iteration of spectrum assignment or bargaining involves a four-way handshake between neighbors, i.e. (request, acknowledgement, action, acknowledgement).

We first compare the performance of local bargaining to the graph coloring approaches that approximate to the solution that maximizes system utility for a given conflict graph [5, 7]. We also validate the impact on system performance using poverty line guided bargaining. We also examine the effectiveness of using local bargaining to optimize spectrum assignment for fixed topologies and algorithm scalability.

12.6.1 Comparison with Centralized Graph Coloring Approach

We compare the proposed local bargaining to the graph-coloring approach. We refer to these two as BARGAINING and GREEDY, respectively. We randomly deploy 40 links with 30 channels in a given area and produce the corresponding conflict graph. We use the graph coloring approach to derive an initial spectrum assignment for the given conflict graph. We simulate mobility events in the next 100 time slots, one event per time slot where up to six nodes move to new locations. After each event, we apply both local bargaining and graph-coloring approaches to derive the new spectrum allocation.

Figure 12.6a illustrates the sorted fairness utility using both approaches, and local bargaining performs nearly as good as the graph coloring approach. The graph col-

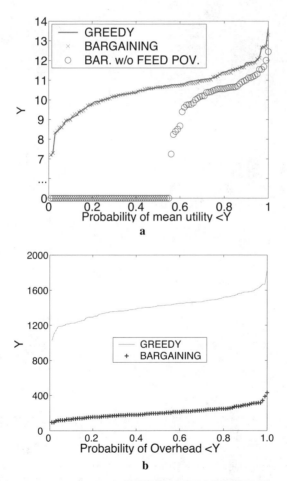

Fig. 12.6. Performance comparison of GREEDY and BARGAINING. **a.** System utility. **b.** Algorithm complexity. [© *2005 IEEE. Reprinted, with permission, from [13]*].

oring approach makes decisions with the knowledge of global topology, while using local bargaining, each user makes decisions based on only neighbor information. Figure 12.6b compares the communication overhead in each time slot. We observe that local bargaining achieves similar performance while incurring much lower complexity in terms of messages exchanged. This significant overhead reduction allows quick adaptation to network dynamics. In Fig. 12.6a, we also examine the performance of local bargaining using only *one-to-one fairness bargaining*, without *Feed Poverty Bargaining*. The results confirm that *Feed Poverty Bargaining* is required to eliminate *starvation*.

Next, we extend the simulation to allow $p\%$ of vertices to move to new randomly selected locations. In general, larger p implies more disturbance to the conflict graph

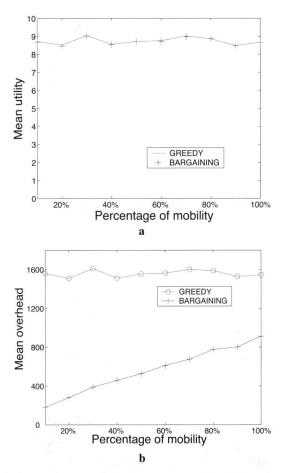

Fig. 12.7. Performance comparison of GREEDY and BARGAINING under different level of network dynamics. **a** System utility. *b* Algorithm complexity. [© *2005 IEEE. Reprinted, with permission, from [13]*].

and thus more vertices will perform bargaining to adapt their spectrum usage to the new topology. Figure 12.7 illustrates the system utility and algorithm overhead for local bargaining and graph coloring for increasing values of p. The utility is geometrically averaged over 100 time slots, and overhead is averaged over 100 time slots. As before, local bargaining performs similarly to graph coloring approach in system utility. The overhead of graph coloring is not sensitive to the value of p as it mainly depends on the number of vertices and channels. The overhead complexity of local bargaining increases with p as more vertices need to perform bargaining. We observe that even under 100% mobility, local bargaining reduces overhead by half compared to the graph coloring approach. Therefore, local bargaining appears to be an attractive alternative to graph coloring for spectrum allocation on a given network topology.

In Fig. 12.8, we examine the impact of the number of vertices at $p = 20\%$ mobility and 40 channels. Increasing vertex density results in additional interference and

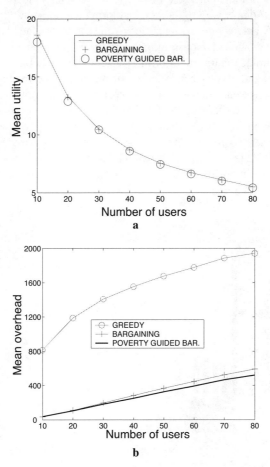

Fig. 12.8. Performance comparison of GREEDY and BARGAINING under different vertex density **a**System utility. **b**Algorithm complexity. [© *2005 IEEE. Reprinted, with permission, from [13]*].

increases average vertex degree in the conflict graph. Therefore, system utility scales inversely with the number of vertices while the algorithm complexity increases. As before, results show that local bargaining compares favorably to graph coloring in quality of allocation while incurring significantly less overhead.

12.6.2 Tightness of the Poverty Bound

We now examine the appropriateness of the user poverty bound derived in Theorem 12.2. Figure 12.9 illustrates the histogram of the ratio of the actual user throughput and the poverty bound assuming 40 vertices and 100 time slots. Results show that the theoretical bound is valid and fairly tight. As we described, nodes can use the poverty line to decide whether further local bargaining is necessary. A vertex with an assignment below the poverty line should bargain with additional neighbors to acquire additional channels. In Fig. 12.8, we also compare the performance of *Poverty guided bargaining* to the graph coloring and bargaining approaches. We show that *Poverty guided bargaining* performs close to that of the bargaining but with 10% less overhead, again demonstrating the tightness of the poverty line bound.

Next, we examine the impact of primary user deployment on the bound. We randomly deploy 10 and 30 primary users to our previous experiments. Primary users have the same interference range as secondary users, and occupy one randomly chosen channel. Figure 12.10 shows the CDF of the ratio between actual node throughput and lower bound. We observe that the bound becomes slightly looser as the number of primary users increases. This is mainly due to the mismatch between $L(n)$ and $d(n)$. We calculate $d(n)$ to include all the interfering neighbors who has at least one channel in common with n. Therefore, $d(n)$ is the upper bound of the conflicting neighbors on each channel.

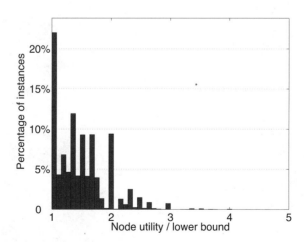

Fig. 12.9. Histogram of the ratio between the actual user throughput and the poverty bound [© 2005 IEEE. Reprinted, with permission, from [13]].

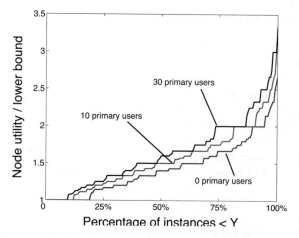

Fig. 12.10. Tightness of the bound when there are primary users [© *2005 IEEE. Reprinted, with permission, from [13]*].

When channel bandwidth is non-uniform, we compute the bound following Theorem 12.1. Figure 12.11 compares the tightness of the bound when channel bandwidth varies between 1–3 and 1–5. The result is relatively smoother by eliminating the truncation effect in Theorem 12.2. We see that the bound is looser compared to that of the uniformed bandwidth. This is mainly due to the fact that the lower bound is only a sub-bound, i.e. a node's throughput is always *strictly* larger than the bound. The figure also shows that the bound is sensitive to the variance in bandwidth across channels.

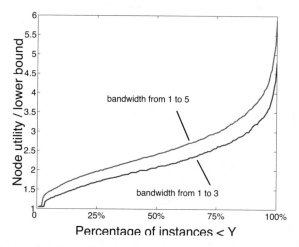

Fig. 12.11. Tightness of the bound in theorem 12.1 when channel bandwidth is not uniform[© *2005 IEEE. Reprinted, with permission, from [13]*].

12.6.3 Use Local Bargaining to Optimize For a Given Topology

As stated before, it is possible to use local bargaining to approximate the graph coloring approach and derive the spectrum allocation for a given topology. Local bargaining starts from a random allocation and gradually improve the system utility. Figure 12.12 compares the system utility and algorithm complexity using graph coloring (GREEDY), local bargaining (BARGAINING) and random assignment (RANDOM). User starvation is common when using random assignment, resulting in many zero values for system utility. Local bargaining can effectively eliminate user starvation and performs only slightly worse compared to graph coloring approach. Figure 12.12b shows that local bargaining can significantly reduce communication overhead.

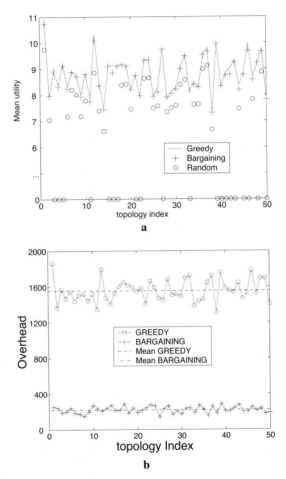

Fig. 12.12. Performance comparison of graph coloring, local bargaining and random spectrum assignment for static environments **a**System utility. **b**Algorithm complexity [© *2005 IEEE. Reprinted, with permission, from [13]*].

12.6.4 Scalability of Local Bargaining

We also evaluate the complexity of bargaining in large scale networks. We measure the total system overhead, i.e. the total number of message exchanges for a system to reach an equilibrium. We keep the user density constant and vary the system scale from 200 to 1000, and assume 20% of node mobility. Results in Fig. 12.13 together with those in Fig. 12.8b indicates that the system overhead scales linearly with the number of users. Hence, the average overhead per user is roughly constant, eight messages or two bargaining iterations under the above system configurations. This result demonstrates the efficiency of the proposed approach.

12.6.5 Comparison of Explicit and Implicit Coordination Approaches

Figure 12.14 compares all three approaches, graph coloring (CA (graph coloring)), explicit (CA (bargaining)) and implicit coordination (rule B, C, D) approaches. There is a noticeable performance gap between rule and collaboration based approaches. This confirms the effectiveness of using explicit collaboration when power and complexity are not considered. Compared to the explicit bargaining approach approach, rule based approach leads to a graceful 8% degradation in utilization and 25% (on log-scale) in fairness.

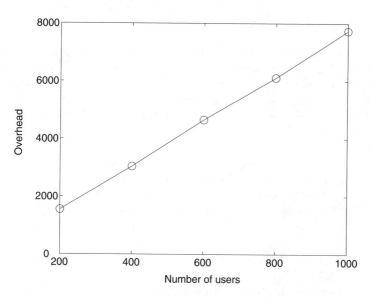

Fig. 12.13. System overhead in large scale networks [© *2005 IEEE. Reprinted, with permission, from [13]*].

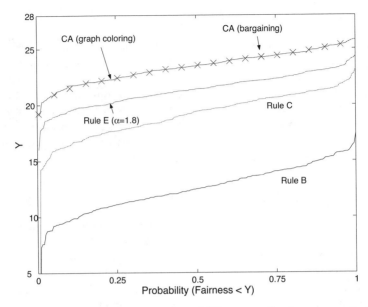

Fig. 12.14. Utilization and fairness comparison of different coordination schemes for 20 channels and 40 users [© *2005 IEEE. Reprinted, with permission, from [15]*].

12.7 Related Work

Optimal conflict-free channel assignment satisfying a global optimal objective is often NP-hard, even when global topology information is available [22]. Centralized approximations are widely used in single hop wireless networks such as cellular networks. This can be easily extended to multi-hop wireless networks by flooding connectivity and traffic requirements across the network, and requiring all users to run a variant of the centralized algorithm. However, this approach clearly does not scale as networks become larger and more dynamic.

An alternative decentralized allocation, where users act based on locally available information is much more attractive. Both analytical framework and practical strategies have been proposed. Analytical frameworks in [10, 11] address fairness for single-hop flows, and derive an estimate of the rate at each flow to achieve max-min fairness. However, there is no guarantee that a feasible scheme exists to achieve the rate.

Practical strategies have been proposed for sharing a single channel. Contention based schemes invoke a random access protocol like ALOHA and CSMA, where users contend in time to share a common channel [10, 11, 22]. While this scheme provides fairness and utilization on a single channel system probabilistically, its application to a multi-channel system requires each user to know how many and which channel(s) to access. Another approach, conflict free time slot scheduling, provides guaranteed channel usage by reserving time slots for each flow. Solutions in [6, 23, 24] assign exactly one time slot to each flow. This approach can be used in multi-channel systems if each user uses only one channel. Another solution [12] allows users to use multiple slots/channels to achieve Max–Min–fair, but does not consider interference from neighbor transmissions.

Multi-channel assignment strategies were developed mostly for cellular networks. The work in [25] provides solutions to assign frequency bands among base stations to minimize call blocking probability for voice traffic. There is no notion of fairness as the traffic determines the number of channels each base station should use. In [26], the authors proposed a graph-theoretic model and discussed the price of anarchy under various topology conditions such as different channel numbers and bargaining strategies. The main difference between [26] and the proposed work is that the proposed model allows multi-coloring of a vertex, while in [26] each vertex can only be assigned with at most one color.

In [5,7], the authors presented a generalized spectrum allocation problem where interference constraint C is channel dependent. The authors developed a set of greedy coloring approach to optimize spectrum allocation for a given conflict graph. We use the proposed approaches in [5,7] as the reference algorithm, i.e. GREEDY in this chapter.

Cooperative/non-cooperative bargaining is also used in previous research to optimize channel allocation for cellular networks. In [27] and [28], the authors proposed a set of bargaining strategies for OFDMA based network, focusing on one-to-one bargaining. In these cases, nodes are mutually interfered, i.e. the corresponding conflict graph is fully connected. Forming bargaining group is to find any two users network and let them exchange certain channels to improve system performance. The main difference between these work and the proposed work is that the proposed work provides solutions for general conflict graphs where group setup needs to consider local topology (i.e. isolated group and self-contained channel adjustment). We propose a feed-poverty bargaining to eliminate user starvation which can not be addressed by one-to-one bargaining. In addition, the proposed work derives a lower bound for each node's channel assignment that does not limit to fully connected topology.

Conclusion

In this chapter, we present an adaptive and distributed approach to spectrum allocation in wireless networks. Instead of relying on any central servers, users perform local coordinations to modify their spectrum usage to approach a new conflict free spectrum assignment that maximizes system utility. Both theoretical and experimental results show that the proposed approaches perform similarly as the topology-optimized approach but with much less complexity.

Appendix

Proof of Theorem 12.1: We prove the theorem by contradiction: Assume that the bound (12.3) does not hold for node n under a BF-optimal assignment, i.e.

$$T(n) \leq \frac{B(n)}{d(n)+1} - MB(n). \tag{12.5}$$

In the following we will show that n can request a Feed Poverty Bargaining to improve the system utility. This contradicts with the assumption that assignment A is BF-optimal.

We divide the proof into two steps. First, we demonstrate how to find the proper channel to feed n. Next, we prove that system utility increases from this channel feed.

Step 1: channel selection. We assume that user n has $d(n)$ neighbors indexed $\{0, 1, \cdots, d(n) - 1\}$. We assume that the channels assigned to user n, $f_A(n)$ are indexed by $\{M - 1, M - 2, \cdots, M - |f_A(n)|\}$. We define $t \triangleq M - |f_A(n)|$, where t is the number of channels not assigned to n. For each user n, we can organize the assignment matrix A using the following table, where an element is 1 if the channel indexed by the row number is assigned to the user indexed by the column number.

Index	0	$\cdots$	d(n)-1	n	$\cdots$
0	$\cdots$	$\cdots$	$\cdots$	0	$\cdots$
$\vdots$	$\vdots$	$\ddots$	$\vdots$	$\cdots$	$\vdots$
$t-1$	$\cdots$	$\cdots$	$\cdots$	0	$\cdots$
t	0	$\cdots$	0	1	$\cdots$
$\vdots$	$\vdots$	$\ddots$	$\vdots$	$\vdots$	$\vdots$
M-1	0	$\cdots$	0	1	$\cdots$

We focus on the first t rows and first $d(n)+1$ columns (i.e. column $0, 1, \cdots, d(n) -1, n$) of matrix A since they represent the channel assignment at user n's neighbors. We also examine $d(n)$ neighbors to exclude any starved user k, *i.e.* $T(k) = 0$. Let d represent the number of non-starved neighbors of n, $d \leq d(n)$, we re-index the neighbors of n to $(0, \cdots, d - 1)$. In addition, from (12.5) we have

$$T(n) \leq \frac{B(n)}{d(n)+1} - MB(n) \leq \frac{B(n)}{d+1} - MB(n). \tag{12.6}$$

Integrating A with channel bandwidth matrix B, we construct an auxiliary matrix R with t rows and $d+1$ columns. The element of R can be represented as

$$R_{m,i} = \begin{cases} \frac{A_{m,i} \cdot b_{m,i}}{T(i)} & : \quad 0 \leq i \leq d-1, \ 0 \leq m \leq t-1 \\ \frac{b_{m,i}}{B(i)/(d+1)} & : \quad i = n, \ 0 \leq m \leq t-1. \end{cases}$$

It is straightforward to show that for each column $0 \leq i \leq d-1$,

$$\sum_{m=0}^{t-1} R_{m,i} = \sum_{m=0}^{t-1} \frac{A_{m,i} \cdot b_{m,i}}{T(i)} = 1. \tag{12.7}$$

and

$$\sum_{m=0}^{t-1} \sum_{i=0}^{d-1} R_{m,i} = \sum_{i=0}^{d-1} \sum_{m=0}^{t-1} R_{m,i} = d. \tag{12.8}$$

We can also show that for column n,

$$\sum_{m=0}^{t-1} R_{m,n} = \sum_{m=0}^{t-1} \frac{b_{m,n}}{B(n)/(d+1)}$$

$$= \frac{\sum_{m=0}^{t-1} b_{m,n}}{B(n)/(d+1)}$$

$$= \frac{B(n) - T(n)}{B(n)/(d+1)}$$

$$(by(12.6)) \geq \frac{B(n) - (\frac{B(n)}{d+1} - MB(n))}{B(n)/(d+1)}$$

$$> d. \tag{12.9}$$

Combining (12.8) and (12.9), we prove that

$$\sum_{m=0}^{t-1} R_{m,n} > \sum_{m=0}^{t-1} \sum_{i=0}^{d-1} R_{m,i}.$$

Therefore, there must exist a channel m_0, *s.t.*

$$R_{m_0,n} > \sum_{i=0}^{d-1} R_{m_0,i}. \tag{12.10}$$

Next we show that we can improve system utility by letting user 0 to $d-1$ give up channel m_0 and feed user n.

Step 2. Verify system utility. First we show that

$$R_{m_0,n} \leq 1. \tag{12.11}$$

From (12.5),

$$\frac{B(n)}{d+1} - MB(n) \geq T(n) \geq 0$$

$$\implies \frac{B(n)}{d+1} \geq MB(n) \geq b_{m_0,n}$$

$$\implies R_{m_0,n} = \frac{b_{m_0,n}}{B(n)/(d+1)} \leq 1.$$

Combining (12.11) with (12.10), we have

$$\sum_{i=0}^{d-1} R_{m_0,i} < 1,$$

$$R_{m_0,i} < 1, \quad 0 \leq i \leq d-1. \tag{12.12}$$

This shows that for each neighbor i, m_0 is not the only channel assigned. Therefore, after feeding user n with channel m_0, each neighbor has at least one channel assigned. The system utility after the feeding, depends on the value of $T(n)$ before the feeding.

Case 1) $T(n) = 0$. From (12.12), the feeding will increase $T(n)$ to $b_{m0,n}$ without starving any neighbor (0 to $d - 1$). Clearly, the system utility is improved from $-\infty$ if n is the only starved user in the network before the feeding. However, when there are multiple starved users, the system utility remains $-\infty$. In this case, we define increasing system utility by reducing the number of starved users or improving the fairness utility of the non-starved users. Through a similar process, we can continue to remove other starved users to improve system utility.

Case 2) $T(n) > 0$. The ratio of system utility after the feeding to that before the feeding can be derived as

$$
\begin{aligned}
G_{\text{FP}} &= \frac{(T(n) + b_{m_0,n}) \cdot \prod_{i=0}^{d-1}(T(i) - A_{m_0,i} \cdot b_{m_0,i})}{(T(n)) \cdot \prod_{i=0}^{d-1}(T(i))} \\
&= \frac{T(n) + b_{m_0,n}}{T(n)} \cdot \left(\prod_{i=0}^{d-1} \frac{T(i) - A_{m_0,i} \cdot b_{m_0,i}}{T(i)} \right) \\
&= \left(1 + \frac{b_{m_0,n}}{T(n)}\right) \cdot \prod_{i=0}^{d-1}\left(1 - \frac{A_{m_0,i} \cdot b_{m_0,i}}{T(i)}\right) \\
&= \left(1 + \frac{b_{m_0,n}}{T(n)}\right) \cdot \prod_{i=0}^{d-1}(1 - R_{m_0,i}).
\end{aligned}
\tag{12.13}
$$

We can show that

$$
1 + \frac{b_{m_0,n}}{T(n)}
$$

$$
(\text{by}(12.5)) \geq 1 + \frac{B_{m_0,n}}{\frac{B(n)}{d(n)+1} - MB(n)} \geq 1 + \frac{B_{m_0,n}}{\frac{B(n)}{d(n)+1} - B_{m_0,n}}
$$

$$
= \frac{\frac{B(n)}{d(n)+1}}{\frac{B(n)}{d(n)+1} - B_{m_0,n}} = \frac{1}{1 - R_{m_0,n}}.
\tag{12.14}
$$

In addition, using (12.12) and Lemma 12.1, we have

$$
\prod_{i=0}^{d-1}(1 - R_{m_0,i}) \geq 1 - \sum_{i=0}^{d-1} R_{m_0,i}
$$

$$
(\text{by}(12.10)) > 1 - R_{m_0,n}.
\tag{12.15}
$$

By combining (12.14) and (12.15) into (12.13), we get

$$G_{\text{FP}} > \frac{1}{1 - R_{m_0,n}} \cdot (1 - R_{m_0,n}) = 1. \tag{12.16}$$

From the above, the system utility increases after the feeding. This contradicts with the assumption that assignment A is BF-optimal. $\qquad\square$

Lemma 12.1. *Suppose* $a_1, a_2, \cdots, a_n \geq 0$, *and* $a_1 + a_2 + \cdots + a_n = p < 1$. *Let* $f(a_1, a_2, \cdots, a_n) = (1 - a_1)(1 - a_2) \cdots (1 - a_n)$. *Then* $f(a_1, a_2, \cdots, a_n) \geq 1 - p$.

Proof. f is minimized when there is only one non-zero number, i.e. $a_i = p$ and $a_j = 0$ for $j \neq i$. To prove this, let us assume that there are two non-zero numbers, i.e. $a_i \neq 0$ and $a_j \neq 0$. We modify $a_i' = a_i + a_j$ and $a_j' = 0$, so $a_i + a_j = a_i' + a_j'$. However,

$$(1 - a_i')(1 - a_j') = 1 - a_i - a_j < 1 - a_i - a_j + a_i a_j = (1 - a_i)(1 - a_j). \tag{12.17}$$

Thus, when there are more than one non-zero numbers in a_i, $1 \leq i \leq n$, we can modify the a_is to reduce f. This shows that f is minimized by having only one positive a_i, and $f = 1 - p$. $\qquad\square$

Proof of Theorem 12.2: By Theorem 12.1, when all channels have bandwidth 1, we have

$$T(n) > \frac{|L(n)|}{d(n) + 1} - 1.$$

Since $T(n)$ is an integer, this is equivalent to

$$T(n) \geq \left\lfloor \frac{|L(n)|}{d(n) + 1} \right\rfloor. \square$$

Proof of Theorem 12.3: We denote a user as "satisfied" if it is on or above its *poverty line*, otherwise "unsatisfied". First we prove a key property of the PGB:

Property 12.1. With a bargaining iteration in which channels are fed to n, if a neighbor n_1 is converted from "satisfied" to "unsatisfied", then $\Gamma(n) < \Gamma(n_1)$.

Proof. First we assume node n is below its *poverty line* $\Gamma(n)$. From the proof of Theorem 12.1, by combining (12.13), (12.14) and (12.16), we show that for channel m_0,

$$\frac{1}{1 - R_{m_0,n}} \cdot \prod_{i=0}^{d-1} (1 - R_{m_0,i}) > 1.$$

Under uniform channel bandwidth, this shows that before the bargaining,

$$\left(\frac{1}{1 - \frac{d+1}{|L(n)|}} \right) \cdot \prod_{i=0}^{d-1} \left(1 - \frac{1}{T(i)} \right) > 1.$$

Since $(1 - \frac{1}{T(i)}) < 1$ and $\frac{d+1}{|L(n)|} \leq \frac{1}{\Gamma(n)}$, this implies that for any neighbor n_1, $0 \leq n_1 \leq d - 1$ involved in the bargaining,

$$\frac{\Gamma(n)}{\Gamma(n) - 1} \cdot \frac{T(n_1) - 1}{T(n_1)} > 1$$

and therefore, $T(n_1) > \Gamma(n)$. Hence, after feeding channel m_0 to n, no neighbor's throughput is reduced to below $\Gamma(n)$. Hence, if the throughput of n_1 is reduced to below $\Gamma(n_1)$, then it must be the case where $\Gamma(n) < \Gamma(n_1)$. ☐

Proof (Proof for Theorem 12.3). From Property 12.1, if the PGB can be conducted such that the user with lower *poverty line* among its (one-hop) neighborhood wins the priority to be fed first, then each "satisfied" user will never become an "unsatisfied" user after it is fed by the PGB. Hence, the number of iteration is at most N. If the user to be fed is selected randomly among requestors, then a user with lower *poverty line* among its neighborhood will wait for an expected number of $O(N)$ interations to win the opportunity, because at each iteration the number of users who request a PGB is bounded by N. The total number of iterations until the system reaches an equilibrium is $O(N^2)$. ☐

Bibliography

1. M. McHenry, "Spectrum white space measurements," *New America Foundation Broadband Forum*, June 2003.
2. R. W. Brodersen, A. Wolisz, D. Cabric, S. M. Mishra, and D. Willkomm, "Corvus: A cognitive radio approach for usage of virtual unlicensed spectrum." Whitepaper, July 2004.
3. S. Mangold, Z. Zhong, K. Challapali, and C. T. Chou, "Spectrum agile radio: Radio resource measurements for opportunistic spectrum usage," in *Proc. of IEEE Globecom*, Nov.–Dec. 2004.
4. K. Jain, J. Padhye, V. N. Padmanabha, and L. Qiu, "Impact of interference on multi-hop wireless network performance," in *Proc. of ACM MobiCom*, Sept. 2003.
5. C. Peng, H. Zheng, and B. Y. Zhao, "Utilization and fairness in spectrum assignment for opportunistic spectrum access," *Mobile Networks Appl. (MONET)*, vol. 11, pp. 555–576, May 2006.
6. S. Ramanathan, "A unified framework and algorithm for channel assignment in wireless networks," *Wireless Networks*, vol. 5, pp. 81–94, Mar. 1999.
7. H. Zheng and C. Peng, "Collaboration and fairness in opportunistic spectrum access," in *Proc. of IEEE ICC*, June 2005.
8. R. J. Berger, "Open spectrum: A path to ubiquitous connectivity," *ACM Queue*, vol. 1, May 2003.
9. J. M. Peha, "Approaches to spectrum sharing," *IEEE Commun. Mag.*, vol. 43, pp. 10–12, Feb. 2005.
10. X. Huang and B. Bensaou, "On max-min fairness and scheduling in wireless ad-hoc networks: analytical framework and implementation," in *Proc. of ACM MobiHoc*, 2001.
11. T. Nandagopal, T. Kim, X.Gao, and V. Bharghavan, "Achieving MAC layer fairness in wireless packet networks," in *Proc. of ACM MobiCom*, Aug. 2000.

12. T. Salonidis and L. Tassiulas, "Distributed on-line schedule adaptation for balanced slot allocation in wireless ad hoc networks," in *Proc. of IWQoS*, June 2004.

13. L. Cao and H. Zheng, "Spectrum allocation in ad hoc networks via local bargaining," in *Proc. of SECON*, Sept. 2005.

14. M. R. Garey and D. S. Johnson, *Computers and Intractability: A Guide to the Theory of NP-Completeness*. W. H. Freeman, 1990.

15. H. Zheng and L. Cao, "Device-centric spectrum management," in *Proc. of IEEE DySPAN*, Nov. 2005.

16. L. Berlemann, B. Walke, and S. Mangold, "Behavior based strategies in radio resource sharing games," in *Proc. of IEEE PIMRC*, Nov. 2004.

17. E. M. Arkin and R. Hassin, "On local search for weighted k-set packing.," in *Proc. of ESA, LNCS 1284*, pp. 13–22, 1997.

18. M. M. Halldórsson, "Approximations of independent sets in graphs," in *Proc. of the International Workshop on Approximation Algorithms for Combinatorial Optimization (APPROX), LNCS 1444*, pp. 1–13, New York, NY: Springer-Verlag, 1998.

19. W. Yu and J. Cioffi, "FDMA capacity of gaussian multi-access channels with ISI," *IEEE Trans. Commun.*, vol. 50, pp. 102–111, Jan. 2002.

20. G. Chartrand, *A Scheduling Problem: An Introduction to Chromatic Numbers*, New York: Dover, pp. 202–209, 1985.

21. E. Koutsoupias and C. Papadimitriou, "Worst-case equilibria," in *Proc. 16th Annual Conference Theoretical Aspects of Computer Science, vol. 1563 of Lecture Notes in Computer Science*, pp. 404–413, New York, NY: Springer-Verlag, 1999.

22. A. Ephremides and T. Truong, "Scheduling broadcasts in multihop radio networks," *IEEE Trans. on Commun.*, vol. 38, pp. 456–460, Apr. 1990.

23. H. Luo, S. Lu, and V. Bharghavan, "A new model for packet scheduling in multihop wireless networks," in *Proc. of ACM MobiCom*, Aug. 2000.

24. L. Bao and J. J. Garcia-Luna-Aceves, "Hybrid channel access scheduling in ad hoc networks," in *Proc. of ICNP*, Oct. 2002.

25. S. Ramanathan and E. Lloyd, "Scheduling algorithms for multihop radio networks," *IEEE/ACM Trans. Network.*, vol. 1, pp. 166–177, Apr. 1993.

26. I. Katzela and M. Naghshineh, "Channel assignment schemes for cellular mobile telecommunication systems," *IEEE Personal Commun.*, vol. 3, pp. 10–31, June 1996.

27. M. M. Halldórsson, J. Y. Halpern, L. E. Li, and V. S. Mirrokni, "On spectrum sharing games," in *Proc. of the Twenty-Third Annual ACM Symposium on Principles of Distributed Computing (PODC)*, pp. 107–114, ACM Press, 2004.

28. Z. Han, Z. Ji, and K. R. Liu, "Low-complexity OFDMA channel allocation with Nash bargaining solution fairness," in *Proc. of IEEE Globecom*, Nov.–Dec. 2004.

29. Z. Han, Z. Ji, and K. R. Liu, "Power minimization for multi-cell OFDM networks using distributed non-cooperative game approach," in *Proc. of IEEE Globecom*, Nov.–Dec. 2004.

30. G. Zussman and A. Segall, "Energy efficient routing in ad hoc disaster recovery networks," in *Proc. of IEEE INFOCOM*, Mar. 2003.

31. E. Setton et al., "Cross-layer design of ad hoc networks for real-time video streaming," *IEEE Wireless Communications*, vol. 12, pp. 59–65, Aug. 2005.

32. A. Adya, P. Bahl, J. Padhye, A. Wolman, and L. Zhou, "A multi-radio unification protocol for IEEE 802.11 wireless networks," Technical report, Microsoft Research, 2003.

33. N. Ahmed and R. Baraniuk, "Throughput Measures for Delay-constrained Communications in Fading Channels," in *Allerton Conference on Communication, Control and Computing*, (Allerton, IL), Oct. 2003.

34. N. Abramson and F. Kuo, editors, Computer Networks, ch. The ALOHA System, pp. 501–518. Englewood NJ: Prentice-Hall, 1973.

35. C. Perkins and E. Royer, "Ad hoc on-demand distance vector routing," in *Proc. of ACM WMCSA*, Feb. 1999.

36. L. Bao and J. Garcia-Luna-Aceves, "Collision-free topology-dependent channel access scheduling," in *Proc. of IEEE Milcom*, 2000.

37. M. Alicherry, R. Bhatia, and L. Li, "Joint channel assignment and routing for throughput optimization in mmulti-radio wireless mesh networks," in *Proc. of ACM MobiCom*, Aug. 2005.

38. V. Brik, A. Mishra, and S. Banerjee, "Eliminating handoff latencies in 802.11 WLANs using multiple radios: Applications, experience, and evaluation," in *Proc. IMC*, Oct. 2005.

39. R. A. Brualdi, *Introductory Combinatorics*, 3rd ed. Englewood Cliffs, NJ: Prentice Hall, 1999.

40. S. Buchegger and J.-Y. L. Boudec, "A robust reputation system for p2p and mobile ad-hoc networks," in *Proc. of the Second Workshop on the Economics of Peer-to-Peer Systems (P2PECON)*, 2004.

41. R. Ramanathan, "Challenges: A radically new architecture for next generation mobile ad hoc networks," in *Proc. of ACM MobiCom*, Aug. 2005.

42. A. Akella, G. Judd, S. Seshan, and P. Steenkiste, "Self-management in chaotic wireless deployments," in *Proc. of ACM MobiCom*, Aug. 2005.

43. P. Kyasanur, J. Padhye, and P. Bahl, "On the efficacy of separating control and data into different frequency bands," in *Proc. of Broadnets*, Oct. 2005.

44. J. Li, C. Blake, D. S. D. Couto, H. I. Lee, and R. Morris, "Capacity of ad hoc wireless networks," in *Proc. of ACM MobiCom*, July 2001.

45. D. S. J. D. Couto, D. Aguayo, J. Bicket, and R. Morris, "A high-throughput path metric for multi-hop wireless routing," in *Proc. of ACM MobiCom*, Sept. 2003.

46. T. Henderson, D. Kotz, and I. Abyzov, "The changing usage of a mature campus-wide wireless network," in *Proc. of ACM MobiCom*, Sept. 2004.

47. M. Neufeld, J. Fifield, C. Doerr, A. Sheth, and D. Grunwald, "SoftMAC—flexible wireless research platform," in *Proc. of HotNets*, Nov. 2005.

48. R. Draves, J. Padhye, and B. Zill, "Routing in multi-radio,multi-hop wireless mesh networks," in *Proc. of ACM MobiCom*, Sept. 2004.

49. R. Draves, J. Padhye, and B. Zill, "Comparison of routing metrics for static multi-hop wireless networks.," in *Proc. of ACM SIGCOMM*, Aug. 2004.

50. M. Heusse, F. Rousseau, G. Berger-Sabbatel, and A. Duda, "Performance anomaly of 802.11b," in *Proc. of IEEE INFOCOM*, Mar. 2003.

51. M. Heusse, F. Rousseau, R. Guillier, and A. Duda, "Idle sense: An optimal access method for high throughput and fairness in rate diverse wireless LANs," in *Proc. of ACM SIGCOMM*, Aug. 2005.

52. J. Elson, L. Girod, and D. Estrin, "Fine-grained network time synchronization using reference broadcasts," in *Proc. of OSDI*, Dec. 2002.

53. S. Biswas and R. Morris, "ExOR: Opportunistic multi-hop routing for wireless networks," in *Proc. of ACM SIGCOMM*, Aug. 2005.

54. FCC, "Notice of inquiry: Additional spectrum for unlicensed devices below 900 MHz and in the 3 GHz band," ET Docket No. 02-380, Dec. 2002

55. J. Cox, "FCC adds spectrum, channels for 5.4 GHz band." NetworkWorld.com Article, Feb. 2006. http://www.networkworld.com/weblogs/wireless/011178.html.

56. "FCC spectrum policy task force." http://www.fcc.gov/sptf.

57. G. Foschini and Z. Miljanic, "A simple distributed autonomous power control algorithm and its convergence," *IEEE Trans. Veh. Technol.*, vol. 42, pp. 641–646, Nov. 1994.

58. A. Goldsmith and P. Varaiya, "Capacity of fading channels with channel side information," *IEEE Trans. Inf. Theory*, vol. 43, Nov. 1997.

59. "Wireless networks: State-of-the-art survey," 2002. http://www.ceid.upatras.gr/crescco/archive.htm.

60. G. Hardin, "The tragedy of the commons," *Science*, vol. 162, pp. 1243–1248, Dec. 1968.

61. A. M. F. P. Kelly, and D. Tan, "Rate control in communication networks: Shadow prices, proportional fairness and stability," *J. Operat. Res. Soc.*, vol. 49, pp. 237–252, 1998.

62. M. S. Kodialam and T. Nandagopal, "Characterizing the achievable rates in multihop wireless networks," in *Proc. of ACM MobiCom*, Aug. 2003.

63. M. S. Kodialam and T. Nandagopal, "Characterizing the capacity region in multi-radio multi-channel wireless mesh networks," in *Proc. of ACM MobiCom*, Aug. 2005.

64. F. Xue and P. Kumar, "The number of neighbors needed for connectivity of wireless networks," *Wireless Networks*, vol. 10, pp. 169–181, Mar. 2004.

65. I. F. Akyildiz, et al., "Medium access control protocols for multimedia traffic in wireless networks," *IEEE Network*, vol. 13, pp. 39–47, July/Aug. 1999.

66. "Madwifi FAQ." http://www.madwifi.net.

67. J. So and N. Vaidya, "Multi-channel MAC for ad hoc networks: Handling multi-channel hidden terminals using a single transceiver," in *Proc. of ACM MobiHoc*, May 2004.

68. P. D. Vries and A. Hassan, "Spectrum sharing rules for new unlicensed bands." Draft proposal to FCC, Nov. 2003.

69. R. Nelson and L. Kleinrock, "Spatial TDMA: A collision-free multihop channel access protocol," *IEEE Trans. Commun.*, vol. 33, pp. 934–944, Sept. 1985.

70. P. Kyasanur and N. H. Vaidya, "Routing in multi-channel multi-interface ad hoc wireless networks," Technical report, UIUC, Dec. 2004.

71. J. So and N. Vaidya, "A routing protocol for utilizing multiple channels in multi-hop wireless networks with a single transceiver," Technical report, UIUC, Oct. 2004.

72. P. Kyasanur and N. H. Vaidya, "Routing and link-layer protocols for multi-channel multiinterface ad hoc wireless networks," Technical report, UIUC, May 2005.

73. A. Rao and I. Stoica, "An overlay MAC layer for 802.11 networks," in *Proc. of ACM MobiSys*, June 2005.

74. A. Adya, P. Bahl, J. Padhye, A. Wolman, and L. Zhou, "A multi-radio unification protocol for IEEE 802.11 wireless networks," in *Proc. of Broadnets*, Oct. 2004.

75. J. Padhye, S. Agarwal, V. N. Padmanabhan, L. Qiu, and A. Rao, "Estimation of link interference in static multi-hop wireless networks," in *Proc. of IMC*, Oct. 2005.

76. D. T. P. Viswanath and R. Laroia, "Opportunistic beamforming using dumb antennas," *IEEE Trans. Inf. Theory*, vol. 48, pp. 1277–1294, June 2002.

77. T. Harks, "Utility proportional fair bandwidth allocation: An optimization oriented approach," in *QoS-IP*, pp. 61–74, 2005.

78. S. Borst and P. Whiting, "Dynamic rate control algorithms for HDR throughput maximization," in *Proc. of IEEE INFOCOM*, pp. 976–985, Apr. 2001.

79. M. K. Powell, "Broadband migration III: New directions in wireless policy." Remarks at the Silicon Flatirons Telecommunications Program, Oct. 2002.

80. P. Kyasanur and N. H. Vaidya, "Routing and interface assignment in multi-channel multiinterface wireless networks," in *Proc. of IEEE WCNC*, Mar. 2005.

81. X. Zeng, R. Bagrodia, and M. Gerla, "GloMoSim: A library for parallel simulation of large-scale wireless networks," in *Proc. of PADS*, May 1998.

82. R. Ramanathan, "The performance of ad hoc networks with beamforming antennas," in *Proc. of ACM MobiHoc*, Oct. 2001.

83. R. Choudhury, X. Yang, R. Ramanathan, and N. H. Vaidya, "Using directional antennas for medium access control in ad hoc networks," in *Proc. of ACM MobiCom*, Sept. 2002.

84. C. McDiarmid, "Random channel assignment in the plane," *Random Struct. Algorithms*, vol. 22, no. 2, pp. 187–212, 2003.

85. A. Raniwala and T. cker Chiueh, "Architecture and algorithms for an IEEE 802.11-based multi-channel wireless mesh network," in *Proc. of IEEE INFOCOM*, Mar. 2005.

86. A. Raniwala, K. Gopalan, and T. cker Chiueh, "Centralized channel assignment and routing algorithms for multi-channel wireless mesh networks," *ACM MC2R*, vol. 8, no. 2, 2004.

87. K. Römer, "Time synchronization in ad hoc networks," in *Proc. of ACM MobiHoc*, pp. 173–182, ACM, Oct. 2001.

88. S. Sakai, M. Togasaki, and K. Yamazaki, "A note on greedy algorithms for the maximum weighted independent set problem," *Discrete Appl. Math.*, vol. 126, no. 2–3, pp. 313–322, 2003.

89. FCC, "Facilitating opportunities for flexible, efficient and reliable spectrum use employing coginitive radio technologies." FCSS 03-322.

90. "Software defined radio forum." http://www.sdrforum.org.

91. J. Mitola III, "Wireless architectures for the 21st century." http://ourworld.compuserve.com/homepages/jmitola.

92. A. Sharma, M. Tiwari, and H. Zheng, "MadMAC: Building a reconfigurable radio testbed using commodity 802.11 hardware," in *IEEE SECON Workshop on Networking Technologies for Software Defined Radio Networks*, Sept. 2006.

93. K. Sundaresan and R. Sivakumar, "Routing in ad-hoc networks with MIMO links," in *Proc. of ICNP*, Nov. 2005.

94. A. Velayutham, K. Sundaresan, and R. Sivakumar, "Non-pipelined relay improves throughput performance of wireless ad-hoc networks," in *Proc. of IEEE INFOCOM*, 2005.

95. J. So and N. Vaidya, "Multi-channel MAC for ad hoc networks: Handling multi-channel hidden terminals using a single transceiver," in *Proc. of ACM MobiHoc*, pp. 222–233, ACM, May 2004.

96. J. So and N. Vaidya, "A routing protocol for utilizing multiple channels in multi-hop wireless networks with a single transceiver," in *Proc. of ACM MobiHoc*, May 2004.

97. W. K. Edwards et al., "Using speakeasy for ad hoc peer-to-peer collaboration," in *Proc. of CSCW*, Nov. 2002.

98. P. Bahl, R. Chandra, and J. Dunagan, "SSCH: Slotted seeded channel hopping for capacity improvement in IEEE 802.11 ad-hoc wireless networks," in *Proc. of ACM MobiCom*, Sept. 2004.

99. M. Sánchez, J. Zander, and T. Giles, "Combined routing & scheduling for spatial TDMA in multihop ad hoc networks," in *Proc. of Adhoc*, Mar. 2002.

100. J. R. Douceur, "The Sybil attack," in *Proc. of IPTPS*, Mar. 2002.

101. M. Grossglauser and D. Tse, "Mobility increases the capacity of ad-hoc wireless networks," in *Proc. of IEEE INFOCOM*, Apr. 2001.

102. M. Weiser, "The computer for the twenty-first century," *Sci. Am.*, vol. 265, pp. 94–104, Sept. 1991.

103. J. Bicket, D. Aguayo, S. Biswas, and R. Morris, "Architecture and evaluation of an unplanned 802.11b mesh network," in *Proc. of ACM MobiCom*, Aug. 2005.

104. N. Vaidya and S. Hameed, "Scheduling data broadcast in asymmetric communication environments," Technical report TR96-022, Texas A & M University, 18 1996.

105. U. Lee, E. Magistretti, B. Zhou, M. Gerla, P. Bellavista, and A. Corradi, "Efficient data harvesting in mobile sensor platforms," in *Proc. of IEEE PerSeNS*, Mar. 2006.

106. D. Raychaudhuri, "Adaptive wireless networks using cognitive radios as a building block," MobiCom 2004 Keynote Speech, Sept. 2004. Philadelphia, PA.

107. X. Jing and D. Raychaudhuri, "A spectrum etiquette protocol for efficient coordination of radio devices in unlicensed bands," in *Proc. of IEEE PIMRC*, Sept. 2003.

108. E. C. Efstathiou and G. C. Polyzos, "A peer-to-peer approach to wireless LAN roaming," in *Proc. of WMASH*, Sept. 2003.

109. S. L. Wu, C. Y. Lin, Y. C. Tseng, and J. P. Sheu, "A new multi-channel MAC protocol with on-demand channel assignment for multi-hop mobile ad hoc networks," in *Proc. of I-SPAN*, pp. 232–237, 2000.

110. "XG working group RFC, the XG vision and the XG architecture." http://www.darpa.mil/ato/programs/XG.

111. "XG working group RFC, the XG architecture." http://www.darpa.mil/ato/programs/XG.

112. X. Liu and H. Xiao, "Exploring opportunistic spectrum availability in wireless communication networks." submitted for publication.

113. W. Zhao, M. Ammar, and E. Zegura, "A message ferrying approach for data delivery in sparse mobile ad hoc networks," in *Proc. of ACM MobiHoc*, May 2004.

114. Q. Wang and H. Zheng, "Route and spectrum selection in dynamic spectrum networks," in *Proc. of IEEE CCNC*, Jan. 2006.

115. Q. Wang and H. Zheng, "Route and spectrum selection in dynamic spectrum networks," in *Proc. of IEEE CCNC*, Jan. 2006.

116. H. Zheng and H. Viswanathan, "Optimizing ARQ performance in high speed downlink systems with scheduling," *IEEE Trans. Wireless Commun.*, vol. 4, pp. 495–506, Mar. 2005.

117. J. Chen, T. Lv, and H. Zheng, "Joint cross-layer design for wireless QoS content delivery," *EURASIP Journal on Applied Signal Process.*, Special Issue on Cross Layer Interaction for Communications, pp. 167–182, Feb. 2005.

118. H. Zheng and K. J. R. Liu, "The subband modulation: A joint power and rate allocation framework for subband image and video transmission," *IEEE Trans. Circuits Syst. Video Technol.*, vol. 9, pp. 823–838, Aug. 1999.

119. J. Zhao, H. Zheng, and G. Yang, "Distributed coordination in dynamic spectrum allocation networks," in *Proc. of IEEE DySPAN*, Nov. 2005.

120. H. Zheng, Y. Zhu, C. Shen, and X. Wang, "On the effectiveness of cooperative diversity in ad hoc networks: A system level study," in *Proc. of IEEE ICASSP*, Mar. 2005.

121. G. Yang, H. Zheng, J. Zhao, and V. Li, "Adaptive channel selection through collaborative sensing," in *Proc. of IEEE ICC*, June 2006.

122. H. Zheng and K. J. R. Liu, "Power optimized space-time coding for image and video transmission over wireless channels," in *Proc. of ICIP*, Oct. 1999.

123. H. Zheng and K. J. R. Liu, "Space-time diversity for image over wireless channels," in *Proc. of IEEE International Symposium on Circuit and Systems (ISCAS)*, May 2000.

124. L. Cao and H. Zheng, "Spectrum access through local bargaining," under preparation.

125. H. Zheng and K. J. R. Liu, "Optimization approaches for delivering multimedia services over digital subscriber lines," *IEEE Signal Process. Mag.*, vol. 17, pp. 44–60, July 2000.

126. H. Zheng and K. J. R. Liu, "Robust image and video transmission over spectrally shaped channel using multicarrier modulation," *IEEE Trans. Multimedia*, vol. 1, pp. 88–103, Mar. 1999.

127. H. Zheng, A. Lozano, and M. Haleem, "Multiple ARQ processes for MIMO systems," *Special Issue on MIMO Systems, EURASIP Journal on Applied Signal Process.*, vol. 2004, pp. 772–782, May 2004.

128. S. Das, T. Klein, K. Leung, S. Mukherjee, G. Rittenhouse, L. Samuel, H. Viswanathan, and H. Zheng, "Distributed paging and registration in wireless networks," *IEEE Network*, vol. 19, pp. 19–25, Oct. 2005.

129. G. Song, Y. Li, J. L. J. Cimini, and H. Zheng, "A joint channel-aware and queue-aware data scheduling in multiple shared wireless channels," in *Proc. of IEEE WCNC*, vol. 5, pp. 1922–1927, Mar. 2004.

130. T. E. Klein, K. K. Leung, and H. Zheng, "Enhanced scheduling algorithms for improved TCP performance in wireless IP networks," in *Proc. of IEEE Globecom*, Nov. 2004.

131. H. Zheng and J. Boyce, "An improved UDP protocol for video transmission over internet-to-wireless networks," *IEEE Trans. Multimedia*, vol. 3, pp. 356–365, Sept. 2001.

132. H. Zheng, G. Rittenhouse, and M. Recchione, "The performance of voice over IP over UMTS downlink shared packet channel under different delay budgets," in *Proc. of IEEE VTC*, Oct. 2003.

133. H. Zheng and G. Rittenhouse, "Providing VoIP service in UMTS-HSDPA with frame aggregation," in *Proc. of IEEE ICASSP*, Mar. 2005.

134. S. Das, T. Klein, K. Leung, S. Mukherjee, G. Rittenhouse, L. Samuel, H. Viswanathan, and H. Zheng, "Distributed radio link protocol in an All-IP cellular network," in *Proc. of IEEE VTC*, Fall 2005.

Optimal Spectrum Sensing Decision
for Hardware-Constrained Cognitive Networks

Qian Zhang[1], Juncheng Jia[1], and Xuemin (Sherman) Shen[2]

[1] Department of Computer Science and Engineering,
Hong Kong University of Science and Technology, Hong Kong
{qianzh,jiajc}@cse.ust.hk
[2] Department of Electrical and Computer Engineering,
University of Waterloo, Canada
xshen@bbcr.uwaterloo.ca

13.1 Introduction

Wireless industry has been undergoing a rapid development in recent years. New technologies, new devices and new services related to wireless communication are emerging rapidly in various environments (home, office, public zone and so on). The exponential growth in wireless devices and services has resulted in an overly crowded spectrum. It is known that radio spectrum is a shared media and that interference will occur among different devices using the same spectrum frequency. To eliminate the interference between different wireless services, *command and control* policy is adopted nowadays to allocate a fixed portion of spectrum to each service. There is a common belief that we are running out of usable radio frequencies under such a spectrum sharing policy. The overly crowded US frequency allocation chart and the multi-billion-dollar price at European 3G spectrum auction have certainly strengthened this belief. However, actual spectrum usage measurements indicate the significantly unbalanced use of licensee in different spectrum band: with a small portion of spectrum (e.g., cellular band, unlicensed band) increasingly crowded, most of the rest allocated spectrum is underutilized [1,2]. This paradox indicates that spectrum shortage results from the inefficiency of spectrum management policy rather than the physical scarcity of usable frequencies.

With the underutilization of valuable spectrum resource and the increasing demand of spectrum for wireless communication services, efficient spectrum management schemes are needed. This drives the emergence of concept of open spectrum, the basic idea of which is to open licensed spectrum to unlicensed users (secondary users) and limit the interference perceived by licensees (primary users). The essential of open spectrum system is opportunistic spectrum access, which allows secondary users to identify available spectrum resources that not occupied by primary users and communicate in a manner that limits the level of interference perceived by the primary users. The open spectrum system has been studied in many projects, such as DARPA XG program [3,4], DIMSUMnet project [5], DRiVE/OverDRiVE project [6], etc. The open spectrum system uses a cognitive radio to recognize the status of radio spectrum environment and change its transmission parameters online, i.e., has the spectrum sensing ability and frequency agility. The ultimate goal is for the secondary network equipped with cognitive radio (call it as cognitive network) to efficiently utilize the available spectrum by intelligently sensing and accessing spectrum opportunity.

With such powerful cognitive radio technology, wireless devices will have the capability to recognize the radio environment around it. Physical layer implementation of cognitive radio is critical, but how to leverage the advanced features of cognitive radio (e.g., spectrum sensing and adaptation) in upper layers is also of particular importance. In this chapter, we investigate the media access control (MAC) protocol which is of significant importance in an ad hoc cognitive network, where there is no central controller for the entire system. The cognitive MAC should fully utilize the advanced capability provided by cognitive radio to improve the network performance. Specifically, cognitive MAC should make the appropriate sensing decision to explore spectrum opportunity, which is different from the physical layer issue of how to detect the existence of primary signal, and to utilize such opportunity to conduct data transmission. Since there is no centralized controller in an ad hoc cognitive network, such sensing and transmission operation must be coordinated among multiple secondary users, which brings further challenges.

Several cognitive radio MAC protocols have appeared in the literature to address various issues in cognitive network. In the IEEE 802.22 standard, distributed sensing and synchronization are coordinated by a base station [7]. Two-stage sensing, i.e., fast sensing using energy detection and fine sensing using feature detection, are used to balance the tradeoff between primary user detection accuracy and time overhead of spectrum sensing. Dynamic open spectrum sharing (DOSS) MAC allows nodes to adaptively select an arbitrary spectrum for the incipient communication subject to spectrum availability which makes good use of idle spectrum [8]; ad hoc secondary MAC (AS-MAC) was proposed for secondary users to coexist with GSM network where the issues of transmission status management of secondary users are addressed [9,10]. However, all these aforementioned protocols, with or without the aid of infrastructure, pay little attention to the hardware limitation of cognitive radio. They either use just a few fixed number of spectrum channels or assume the full-spectrum sensing ability for wide spectrum band. To the best of our knowledge, decentralized cognitive MAC (DC-MAC) is the first work that assumes the partial sensing ability of the cognitive radio in a spectrum management system and studies a joint sensing and transmission decision [11]. However, the influence of sensing overhead for the multiple channel opportunity is not fully considered. Besides the open spectrum context, there are many existing research works about multiple channel MAC design. The absence of primary users makes them fundamentally different from cognitive MAC design.

Different from the existing approaches, we observe the current spectrum sensing limitation for practical cognitive radios. We identify the hardware constraints in two main aspects: (1) *Sensing constraint*: for a given geometrical area, spectrum opportunity of interest may span a wide range of bandwidth. However, at any given period, accurate and fine sensing can only be conducted within a small portion of spectrum. (2) *Transmission constraint*: spectrum used by secondary users has the maximum bandwidth limits and spectrum fragmentation number limits which stems from the number of radios and orthogonal frequency-division multiplexing (OFDM) technology limitations [12]. These constraints bring new research challenges and also opportunities in cognitive MAC design. Here, we consider the common situations where secondary users are all equipped with a single cognitive radio. The cognitive radio cannot sense and transmit simultaneously, and discontinuous OFDM is used for spectrum aggregation but the maximum spread bandwidth and the number of fragments are limited. To protect primary users, a maximum detection time interval is used similar to that in IEEE 802.22, which represents the maximum time of interference from secondary signal a primary user can tolerate before it wants to use the spectrum.

These constraints and assumptions impose a limit on continuous transmission by secondary users and require the secondary users to sense spectrum before transmission. It is known that, only when a certain band of spectrum is sensed, the status of the band is known for secondary users. There is a tradeoff between spectrum opportunity and sensing overhead. For a single transmission pair, the more the spectrum is sensed, the more the spectrum opportunity can be explored, however, the larger the sensing overhead will be. A fundamental problem is how secondary users sense the spectrum intelligently (e.g., whether or not to sense further based on the current situation) and optimize the expected throughput. To solve this problem, we propose a new cognitive MAC, hardware-constrained multi-channel cognitive MAC (HC-MAC), which incorporates the sensing overhead and the transmission parameter limitations into the MAC design. We model the sensing process as an optimal stopping problem which can be solved by the principle of backward induction. However, the computation overhead of such optimal solution is quite large which is not suitable for real-time MAC protocols. We use k-stage look-ahead method to approximate the optimal solution with reduced overhead. In the practical protocol design, sender and receiver synchronization is an issue because of the spectrum heterogeneity. Moreover, the multiple user contention for available spectrum should be considered, such as the hidden terminal problem. In the proposed HC-MAC, we use a common channel for control messages and contention of secondary users; the sensing and transmission of single pair is reserved to prevent message collisions from neighboring nodes, and makes use of the block of sensing decision as a basic component. Further, the protocol does not require global time synchronization.

The remainder of the chapter is organized as follow. In Sect. 13.2, we introduce the preliminaries of open spectrum system and some related work. The system architecture under the hardware constraint will be given in Sect. 13.3. The optimal sensing decision is discussed in Sect. 13.4, with approximation algorithms and numerical results. Sect 13.5 gives the detailed protocol design for cognitive network. We use NS2 simulator to evaluate the performance of the HC-MAC protocol in Sect. 13.6. Then we conclude the chapter.

13.2 Preliminaries and Related Work

In this section, we give a brief description of the open spectrum system, and then some existing work related to cognitive MAC and multi-channel MAC design are discussed.

13.2.1 Preliminaries of Open Spectrum System

Current fixed allocation of radio spectrum results in significant underutilization of spectrum resources. The open spectrum system makes flexible use of the radio spectrum resources. In an open spectrum system, some spectrum bands can be shared by *primary users* and *secondary users*. Primary users possess the license of these spectrums which are granted by government. Normally, these primary users are legacy systems previously deployed in an area and the actuall utilization of their spectrum is quite low. Since little spectrum is available in this area, new spectrum-based communication systems cannot be deployed. However, with the help of open spectrum, secondary users are able to request the opportunistic usage of these spectrums from the primary users. Secondary users can only use the spectrum on a lease or non-interference basis. Therefore, both primary users and secondary users can benefit from open spectrum system: primary users may generate revenue from the leasing contract while secondary users can enable the communication which is not possible before.

Generally, channel availability from a secondary device perspective can be modeled in two ways: first, a binary model, in which the secondary device considers a channel to be occupied or available depending on whether it does or does not detect the presence of any primary device signal on that channel; and second, interference temperature model, in which the secondary device considers a channel to be unavailable for transmission if its transmission on that channel would result in increase of interference temperature beyond the predefined threshold within its interference range. Typically in both the models, the operations of secondary devices usually have two stages: sensing and transmission. In this chapter, we assume the sequential execution of these operations (as shown in Fig. 13.1). During sensing stages, PHY-layer sensing and MAC-layer sensing are used to detect the primary users and protect their service quality. PHY-layer sensing adapts modulation schemes and parameters to measure and detect the primary users' signals on different channels while MAC-layer sensing is used to determine when and which channel the secondary user should sense. Techniques such as cyclostationary signal processing, matched filters and radiometric detectors (or energy detectors) are generally used by secondary devices for PHY-sensing. Please refer to [13] for a comparative study of different PHY sensing techniques. Note that MAC-layer sensing decision is the main focus in this chapter. After the information of spectrum has been collected in the sensing stage, actual data transmission can be conducted on the channels underutilized by primary users during the transmission stage.

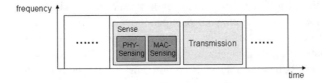

Fig. 13.1. Basic operations taken by a secondary device.

13.2.2 Related Work

Several MAC protocols have been developed for more flexible and efficient use of spectrum resource built on top of the cognitive radio. Some issues in the design of cognitive MAC also arise in general MAC protocol. In this section, we will briefly summarize these related works.

13.2.2.1 Cognitive MAC

There are several research efforts on cognitive MAC protocol design in both industry standardization and academic research projects. From the standardization point of view, the current IEEE 802.22 draft is the first worldwide standard related to cognitive radio. Its MAC employs the superframe structure [7]. Synchronized distributed sensing, fast sensing using energy detection and fine sensing using feature detection, are used. At the beginning of every superframe, the base station (BS) sends special preamble and SCH (superframe control header) through each and every TV channel (up to three contiguous) that can be used for communication and is guaranteed to meet the incumbent protection requirements. Because of the limited number of channels IEEE 802.22 adopts, the sensing overhead is not a major issue. In addition, IEEE 802.22 is operated in the point-to-multiple model, which is comparably easier than the cases without the control of the BS.

There are several ad hoc model MAC protocols for cognitive radio in academic research projects. Most of them do not consider the hardware constraints on spectrum sensing ability and assume full-spectrum sensing in a particular portion of spectrum. Dynamic open spectrum sharing (DOSS) MAC [8] protocol allows nodes to adaptively select an arbitrary spectrum for the incipient communication subject to spectrum availability. In this protocol, after the operation of detection of primary users' presence, three operational channels (a busy tone band, a control band and a data band) are set up. The biggest concern with this protocol is the need for multiple transceivers: one transceiver for each channel. Thus this protocol is not suitable for nodes with only one half-duplex radio. In [9], AS-MAC protocol was proposed to coexist with a GSM cellular system; one of the control channels in GSM band is used as the secondary common control channel. A common control channel facilitates many spectrum-sharing functionalities such as transmitter receiver handshake, communication with a central entity or sensing information exchange.

The sensing decision under hardware constraints of cognitive radio was first considered in [11]. It is not assumed that each secondary user has full knowledge of the availability of all channels, which implies continuous full-spectrum sensing by secondary users. With the channel occupancy by the primary network assumed to follow a discrete-time Markov process, at the beginning of each slot, a secondary user with data to transmit needs to choose a set of channels to sense and a set of channels to access based on the sensing outcome. Such spectrum sensing and access decisions are made to maximize the throughput of the secondary user while limiting the interference to the primary network by fully exploiting the sensing history and the spectrum occupancy statistics. Joint sensing and access decision was formulated as a partially observable Markov decision process (POMDP). However, the tradeoff between sensing overhead and transmission throughput gain was not considered.

13.2.2.2 General Multi-Channel MAC

Many MAC protocols have been proposed to exploit the multiple channels to increase the network capacity by using either multiple radios or just one single radio. A multi-radio multi-channel MAC in general assigns the radios of each node to different channels and enables more simultaneous transmissions so that multiple channels can be used simultaneously for each user. For single-radio multi-channel MAC, the idea is to let different users transmit in parallel on distinct channels, which also increases the throughput and reduces the delay.

For the dynamic channel assignment (DCA) algorithm [14], control messages (RTS/CTS) are exchanged over a control channel and data transfer takes place over a number of data channels. The dedicated radio at the control channel and the problem of control channel saturation are the main concerns. Slotted seeded channel hopping (SSCH) algorithm [15], where a number of channels are available for use and nodes exchange pseudo-random schedules for accessing the medium in a time-slotted manner. No dedicated control channel is needed so that the problem of control channel saturation is avoided. Multi-channel MAC (MMAC) [16] is proposed for single-radio ad hoc networks. Multi-channel hidden terminal problem is addressed within synchronized slotted frames. The assumption of global synchronization may incur great overhead for large systems. These works provide solutions for the problems in cognitive wireless network. However, note that, the presence of primary users makes a fundamental difference for MAC protocol design for cognitive wireless networks.

Optimal stopping rules were used by some of the existing works on MAC protocols. Multi-channel opportunistic auto-rate (MOAR) [16] explores opportunity to skip frequency channels in search for better quality channels. To balance the tradeoff between the time and resource cost of channel measurement/channel skipping and the throughput gain available via transmitting over a better channel, optimal stopping rule was devised to maximize the expected throughput. In our chapter, we focus on the gain from the simultaneous use of several channels with cognitive radio.

13.3 System Architecture of Hardware-Constrained Cognitive Network

In this section, we will first describe the practical hardware constraint of cognitive radio. Then we will present system architecture for a single cognitive radio MAC protocol and its key issue of sensing and access decision.

13.3.1 Hardware Constraints in Cognitive Radio

Current hardware development of cognitive radio is still at its infancy. One currently available cognitive radio product was developed by Adapt4 Inc., which can only work on the frequency band 217–220 MH and support 45 channels. Even in the future when cognitive radio is powerful enough to change its sensing and transmission parameter at its will, the cost to achieve this may still be quite high. Within the predictable near future, cognitive radio must have certain constraints as follows.

For wideband spectrum sensing, there exist certain limitations such as time consumption and energy constraints. Therefore, a common assumption is for a single cognitive radio, it can only sense a limited bandwidth of spectrum during a certain amount of time (call it *sensing constraint*). For different spectrum sensing approaches and different types of primary users, the time overhead varies.

Due to the dynamic behavior in the primary devices, secondary device can eventually find multiple discontinuous spectrum holes (fragmented spectrum) for potential transmission. Spectrum aggregation is a promising technology to leverage multiple available spectral fragments simultaneously to provide effective wide bandwidth communication services. However, for a single secondary device to utilize multiple fragmented spectrum for transmission, the hardware cost may be large [12]. Two potential hardware design options are available: a receiver chain per spectrum fragments provided that only a few fragments are to be aggregated or a single wide-band receiver for many fragments. The former is achievable using narrowband technologies but increasing component count may be a problem as the number of fragments increases. The latter, although more elegant, is more difficult due to technological limitations of wideband components, antenna sharing, and the challenge of managing inter-modulation products. Orthogonal frequency division multiplexing (OFDM) is very suitable to *aggregating* discontinuous spectrum due to the ability to *switch off* unwanted subcarriers, and hence produces a signal with a non-contiguous frequency spectrum which may be tailored to transmitting in available spectrum fragments, as shown in Fig. 13.2.

Because of the limited size and cost of the secondary device, in this chapter we consider the secondary networks consisting of devices equipped with single cognitive radio which can spread signal within a limited number of spectrum fragments in a spectrum band with limited bandwidth (call it *transmission constraint*). According to the recent report of Ofcom [12], using today's hardware technology, it is possible to aggregate fragments over a limited number of bands, each band being at most 50 MHz wide. The center frequencies of these bands can be anything from 100 MHz up to 1.5 GHz, and it is possible to have tunable bands in a single aggregating device.

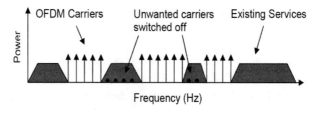

Fig. 13.2. Discontinuous OFDM for using fragmented spectrum.

The increased RF hardware costs for a two-fragment solution ranged between 70% and 600% depending on the technology and type of service involved.

13.3.2 System Architecture

The protection of privileges of primary users is of paramount importance in open spectrum system. In this chapter, we borrow some basic concepts, such as *channel detection time* from the MAC design of IEEE 802.22. Since the opportunistic spectrum access by secondary devices is completely transparent to primary users, the primary users may experience a short but measurable interference as soon as they start transmission on a channel that is currently used by a secondary device for its data transmission. In our system, we define a parameter as the maximum transmission time, which is the time during which an incumbent operation can withstand interference before the secondary device detects it. In other words, this parameter dictates how quickly and how well a secondary device must be able to detect incumbents.

The two limitations, i.e., sensing constraints and transmission constraints, raise the problem of how to optimize the sensing decision for each sensing slots. A simple example shown in Fig. 13.3 is used to illustrate the need for the sensing decision making. Each channel has the same bandwidth, B; the sensing time for a single channel is t and the maximum transmission time is T. Suppose that starting at the time t_0, a secondary user is about to take the next round of sensing and transmission. With the channel conditions unknown at that moment, it has to sense the spectrum. After two slots of sensing, the secondary user can just stop at time t_2 and use the available channels (one available channel) for transmission during the maximum transmission time of T with the achievable data rate $BT/(T+2t)$, which is depicted in *decision A* (Fig. 13.3a). Instead, it can aggressively continue to sense the next unknown channel as shown in *decision B*, which results in the data rate $2BT/(T+3t)$ if this channel is available as in Fig. 13.3b and $BT/(T+3t)$ if unavailable as in Fig. 13.3c. More spectrum band may be available for transmission, if more bands are sensed, but the sensing overhead is also increased. Moreover, the degree of availability of spectrum channels also influences the decision-making. The sensing decision made at each sensing slot, which is whether to stop sensing or continue sensing, then determines the achievable throughput. An alternative way is to simply fix the number of sensed channels. Although simples, it is suboptimal since the decision is not based on the information of spectrum availability.

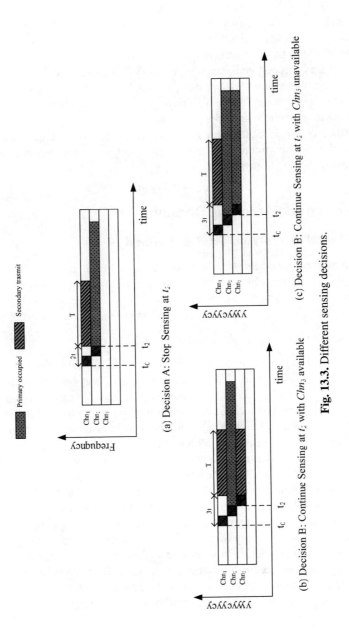

(a) Decision A: Stop Sensing at t_2

(b) Decision B: Continue Sensing at t_2 with Chn_3 available

(c) Decision B: Continue Sensing at t_2 with Chn_3 unavailable

Fig. 13.3. Different sensing decisions.

Besides the sensing overhead (t in the above example), we also know that the fragmented spectrum which can be utilized by a single secondary node for its transmission is also limited by hardware transmission constraint. Therefore, the sensing decision should take both hardware constraints (sensing and transmission) and spectrum opportunity into consideration.

As presented in the next section, we formulate the above decision problem with sensing and transmission constraints as an optimal stopping problem. In our MAC design, we use the simple design principle: sensing mechanism is used as a basic component in the protocol, where high throughput is achieved for a single transmission pair by using the efficient approximation algorithm; under the assumption that there exists a common available channel, contention-based random access in control channel is used by multiple secondary users to reserve the time interval for the following sensing and transmission.

13.4 Sensing and Accessing Decision

13.4.1 Channel Diversity and Sensing Overhead

There are multiple channels under consideration, and each channel is occupied by random primary traffic, which exposes itself as a spectrum opportunity with certain probability. According to the Shannon theory [17], for a single secondary user, the theoretical throughput upper bound is proportional to the bandwidth used:

$$R = W \log(1 + \text{SNR}) \tag{13.1}$$

where R is the data rate, W is the transmission bandwidth, and SNR is the received signal strength and noise rate. Therefore, if a secondary user can exploit more channels and utilize available channels, significant throughput increase can be achieved.

However, the idle channels at each node may be different because of the primary traffic variation and mobility. For the protection of primary users and for the exploitation of the spectrum opportunities, secondary users must sense channels with unknown condition before they can actually use them. Further negotiation between a sender and a receiver is also needed for exchanging their channel availability conditions. Only if a channel is available at both sides, it can be utilized for secondary use of that link. These operations consume the effective transmission time of the secondary users. Therefore, there is a tradeoff between exploring more idle channels and encountering more sensing overhead, which is of great importance in the design of a multiple channel cognitive MAC protocol. To express this issue more explicitly, suppose the maximum continuous transmission period for a secondary link is T, the sensing and negotiation overhead is t. Then the problem becomes how many channels a secondary user should explore so that the expected throughput is maximized.

13.4.2 The Theory of Optimal Stopping

The theory of optimal stopping is concerned with the problem of choosing a time to take an action based on sequentially observed random variables in order to maximize an expected payoff or to minimize an expected cost. Problems of this type are found in the area of statistics, where the action taken may be to test a hypothesis or to estimate a parameter, and in the area of operations research, where the action may be to replace a machine, hire a secretary or reorder stock, etc. The following is the definition of optimal stopping problems.

Stopping rule problems are defined by two objects:

1. A sequence of random variables, $X_1, X_2, \ldots$, whose joint distribution is assumed to be known,
2. A sequence of real-valued reward functions,

$$y_0, y_1(x_1), y_2(x_1, x_2), \ldots, y_\infty(x_1, x_2, \ldots).$$

Given these two objects, the associated stopping rule problem may be described as follows [18,19]. The sequence of $X_1, X_2, \ldots$ can be observed for as long as possible. For each $n = 1, 2, \ldots$, after observing $X_1 = x_1, X_2 = x_2, \ldots, X_n = x_n$, the decision may be to stop and receive the known reward $y_n(x_1, \ldots, x_n)$ or may be to continue and observe X_{n+1}. If the decision is not to take any observations, the received reward is a constant amount, y_0. If never stopping, the received reward is $y_\infty(x_1, x_2, \ldots)$.

The goal is to choose a time to stop such that the expected reward is maximized. It is allowed to use randomized decisions. That is, given that the process reaches stage n having observed $X_1 = x_1, \ldots, X_n = x_n$, it is to choose a probability of stopping that may depend on these observations. We denote this probability by $\phi_n(x_1, \ldots, x_n)$. A (randomized) stopping rule consists of the sequence of these functions,

$$\Phi = (\phi_0, \phi_1(x_1), \phi_2(x_1, x_2), \ldots) \tag{13.2}$$

where for all n and $x_1, \ldots, x_n$, $0 \leq \Phi_n(x_1, \ldots, x_n) \leq 1$. The stopping rule is said to be non-randomized if each $\phi_n(x_1, \ldots, x_n)$ is either 0 or 1. Thus, ϕ_0 represents the probability that no observations is taken at all. Given that the first observation is taken, and $X_1 = x_1$ is observed, $\phi_1(x_1)$ represents the probability to stop after the first observation, and so on. The stopping rule Φ and the sequence of observations $X = (X_1, X_2, \ldots)$, determine the random time N at which stopping occurs, $0 \leq N < \infty$, where $N = \infty$ if stopping never occurs.

A stopping rule problem has a finite horizon if there is a known upper bound on the number of stages at which one may stop. If stopping is required after observing $X_1, X_2, \ldots, X_T$, we say the problem has a horizon T. A finite horizon problem is a special case of the general stopping rule problem with $y_{T+1} = \ldots = y_\infty = -\infty$. Finite horizon stopping rule problems can be solved by the method of backward induction [18]. Since we must stop at stage T, we first find the optimal rule at stage $T - 1$. Thus, knowing the optimal rule at stage $T - 1$ we find the optimal stopping rule at stage $T - 2$ and so on, until back to the initial stage. In particular, we define

$$V_T^{(T)} = y_T(x_1, x_2, \ldots, x_T) \tag{13.3}$$

and then inductively for

$$V_j^{(T)} = \max\{y_j(x_1,\ldots,x_j), E(V_{j+1}^{(T)}(x_1,\ldots,x_j,X_{j+1})|X_1 = x_1,\ldots,X_j = x_j)\}.$$

$$(13.4)$$

13.4.3 Optimal Stopping of Spectrum Sensing

The spectrum sensing decision problem can be formulated as an optimal stopping problem. Let X_n denote the 0–1 (occupied–idle) state of the nth channel probed and the probability $\Pr(X_n = 1) = p$ is assumed to be equal. The expected value of X_n is $u = E(X_n)$. Let y_n denote the expected payoff of stopping probing and transmission after probing n channels. y_n is a function of the aggregated channel availability and depends on the radio technology. Here we generalize the constraints for the cognitive radio: the maximum number of adjacent channels a single secondary user can simultaneously use is W, the maximum number of spectrum fragments it can aggregate is F [12]. For a band of spectrum with adjacent channels $\{i, i + 1,\ldots, j\}$, we denote the number of fragments as $\mathrm{Frag}(i, j)$. Let b_n be the maximum number of idle channels within n adjacent channels (starting from 1), subject to the above constraints (W, F), namely

$$b_n(x_1,\ldots,x_n) = \max_{\substack{1 \le i \le j \le n \\ j-i+1 \le W \\ \mathrm{Frag}(i,j) \le F}} \sum_{k=i}^{j} x_k. \qquad (13.5)$$

The reward function y_n can be written as

$$y_n(x_1,\ldots,x_n) = \frac{T}{T + nt} b_n(x_1,\ldots,x_n) = \frac{c}{c+n} b_n(x_1,\ldots,x_n) \qquad (13.6)$$

where $c = T/t$. y_n is actually the effective data rate during the time interval T after make the stopping and transmission decision.

Assume the maximum number of channels a user can probe before make a stopping decision is at most $K(K \le N)$, which means this is a finite horizon problem, solvable by using the backward induction principle. Denote

$$V_K^{(K)}(x_1,\ldots,x_K) = y_K(x_1,\ldots,x_K) = \frac{c}{c+K} b_K(x_1,\ldots,x_K). \qquad (13.7)$$

Then

$$E(V_K^{(K)}(x_1,\ldots,x_{K-1},X_K)|X_1 = x_1,\ldots,X_{K-1} = x_{K-1})$$
$$= \frac{c}{c+K}[p \times b_K(x_1,\ldots,x_{K-1},1) + q \times b_K(x_1,\ldots,x_{K-1},0)] \qquad (13.8)$$

where p, q are the probabilities of $X_k = 1$ and $X_k = 0$, respectively; and inductively for $n = K - 1$ backward to $n = 2$,

$$V_n^{(K)}(x_1,\ldots,x_n)$$
$$= \max\{y_n(x_1,\ldots,x_n), E(V_{n+1}^{(K)}(x_1,\ldots,X_{n+1}|X_1=x_1,\ldots,X_n=x_n))\}$$
$$E(V_n^{(K)}(x_1,\ldots,x_{n-1},X_n)|X_1=x_1,\ldots,X_{n-1}=x_{n-1})$$
$$= p \times V_n^{(K)}(x_1,\ldots,x_{n-1},1) + q \times V_n^{(K)}(x_1,\ldots,x_{n-1},0). \tag{13.9}$$

Obviously, we should have a sensing at the beginning, with result x_1, since $E(V_2) \geq 0$. Then we compare y_1 with $E(V_2)$, make the decision and so on. At each stage, $\{E(V_n)\}$ defines the optimal stopping rule.

13.4.4 Complexity Reduction

Such a backward induction solution is a type of dynamic programming, which has the exponential complexity. For a small number of channels, direct computation is possible. However, with the increase in the number of channels, computation time grows exponentially. For a practical MAC protocol, we have to reduce the computational complexity to a reasonable level. In the following, we introduce the k-stage look-ahead rules to approximate the optimal stopping rule.

The k-stage look-ahead rules decide at each stage whether to stop or continue according to whether the optimal rule among those truncated k stages ahead stops or continues. Thus at stage n, if the optimal rule among those truncated at $n + k$ continues, the k-stage look-ahead rules continue; otherwise, the k-stage look-ahead rules stop. The stopping time N_k is defined as

$$N_k = \min\{n \geq 0 : y_n(x_1,\ldots,x_n)$$
$$\geq E(V_{n+1}^{(n+k)}(x_1,\ldots,X_{n+1},\ldots,X_{n+k})|X_1=x_1,\ldots,X_n=x_n)\}. \tag{13.10}$$

When $k = K - n$, it is optimal. This is the tradeoff between the degree of optimality and computational cost.

We use numerical results to show the performance of approximations. Figure 13.4 shows the approximation results with different setups. From Fig. 13.4a, we can see the difference between 1-stage look-ahead and k-stage look-ahead ($k > 1$) is small. According to the numerical results in Fig. 13.4b, 1-stage or 2-stage look-ahead is almost optimal. As a comparison, if a fixed number of channels is to be sensed, the results are much worse than the optimal and approximation ones, as shown in Fig. 13.4c. Optimal stopping and its approximation results are always better than the fixed one because their decision is based on the previous observation in each individual sensing process. In this paper, we approximate the optimal result using 1-stage or 2-stage look-ahead approach.

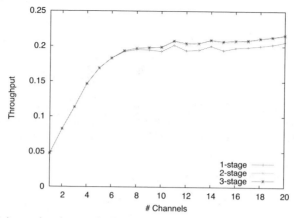

(a) Approximation results for 1-stage, 2-stage and 3-stage look-ahead

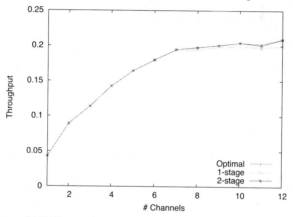

(b) Difference between 1-stage (2-stage) and optimal

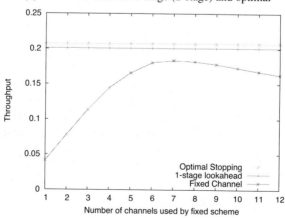

(c) Results for optimal stopping, 1-stage look-ahead, fixed channel scheme

Fig. 13.4. Numerical results for approximation rules ($c = 10, p = 0.5, W = 6, F = 2$).

13.5 HC-MAC: Hardware-Constrained Multi-Channel Cognitive MAC

In this section, we present the design for our proposed hardware-constrained multi-channel cognitive MAC protocol, HC-MAC, which will take the following challenges into consideration:

1. Spectrum sensing for existence of primary users before data transmission. Since the channel conditions are not known in advance, to protect primary users, the shared spectrum band must be sensed first. The best sensing decision for expected data rate is made according to the optimal stopping rule previously described.
2. Synchronization between transmitter and receiver. The sensing results at the transmitter and the receiver need to be exchanged because of the spectrum heterogeneity seen by them. The overhead for these information exchanges is also included in the calculation of the optimal stopping rule. Final sensing stopping and transmission decisions are after the last message exchange.
3. Multi-channel hidden terminal problem. In multi-channel systems, especially those consisting of single-radio equipped devices, new hidden terminal problems arise. This is because a single-radio device may listen to different channels, which makes it difficult to use virtual carrier sensing to avoid the hidden terminal problem.

Before we present the detailed HC-MAC, some necessary assumptions are summarized as follows:

1. There are totally N frequency channels of interest, $\{ch_i\}_N$. Here the term channel refers to the physical channel which is a spectrum band with a certain amount of bandwidth. We do not consider the logical channels such as the spreading codes in CDMA systems. For simplicity, we assume each channel has the same bandwidth B. These channels may not be continuous.
2. A common channel ch_0 is available for secondary users at any time. This can be in the unlicensed band in practice. This common channel is used as the control channel where secondary users make competition and collaboration as described later.
3. We consider a general case in which primary users are randomly distributed in an area, using N channels for their data transmissions. The state of N channels at time t is given by $\{X_1(t), X_2(t), \ldots, X_N(t)\}$, where $X_i(t) \in \{0 \text{ (occupied)}, 1 \text{ (idle)}\}$. If traffic of primary users follows Poisson traffic model, the probability of the states $\{X_i(t)\}$ can be determined.
4. Each secondary node is equipped with a single cognitive radio. The radio can either transmit or listen (sense), but cannot do both simultaneously. Based on the hardware costs, there may be limitations on the maximum number of idle channels and the maximum number of spectrum fragments a cognitive radio can use for transmission; a simple case is for a cognitive radio to utilize any idle channels for transmission. The time for primary signal detection depends on

different spectrum sensing mechanisms and also the primary signal type. We use t_s to denote the time to detect primary signal in a single channel and it cannot be neglected. The sensing results are assumed to be accurate.

5. There exists a certain degree of interference from secondary users' activity which is tolerable for primary users. Since our focus is on the overlay perspective of spectrum sharing, we use maximum tolerable interference time T as a hard protection criteria [7]. Let each primary activity in a channel lasts for a relative long time compared with T. Therefore, as long as a secondary user's data transmission ruled by the designed cognitive MAC protocol does not exceed the time limit T, it is considered safe for the primary users. In this chapter, the same T applies to all primary users.

With these assumptions, we present the challenges to design a cognitive MAC which explores the opportunity of transmission within multiple available channels.

13.5.1 Protocol Overview

We first give an overview of the protocol design. The time frame in HC-MAC is a unit of secondary operations depicted in Fig. 13.5. The whole time frame can be separated into three parts: contention, sensing, transmission. Three types of packets are introduced to facilitate these operations:

1. C-RTS/C-CTS: contention and spectrum reservation in contention part.
2. S-RTS/S-CTS: exchange channel availability information between sender and receiver in each sensing slot.
3. T-RTS/T-CTS: notify the neighboring nodes the completion of the transmission.

Figure 13.6 shows the state diagram for our HC-MAC. If one node wants to transmit, it first sends a C-RTS on ch_0 after random backoff. The intended receiver replies C-CTS on $ch0$. Any secondary node hearing either the C-RTS or C-CTS

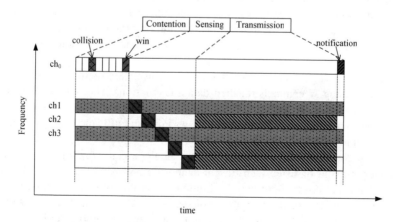

Fig. 13.5. HC-MAC operation phases.

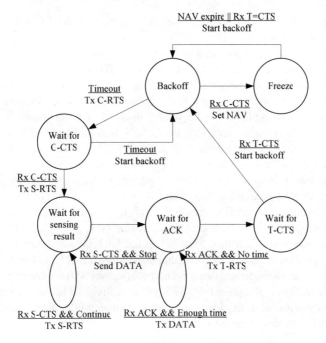

Fig. 13.6. State transition diagram of HC-MAC.

message will defer their operation and wait for the notification message on *ch0*. After reserving the sensing period, the transmission pair conducts sensing in each channel and exchange another S-RTS and S-CTS if that channel is available for both sides. A failure of such S-RTS/S-CTS in a channel indicates that the channel is not available. The optimal stopping rule described in the previous section is used to decide the time to stop probing. When an agreement is made between a sender and a receiver, data transmission is conducted in the selected channels. When the transmission finishes, they will switch back to the control channel and exchange T-RTS/T-CTS. This T-RTS/T-CTS exchange ends other neighbors' deferment and the neighboring node participates in another round of contention.

13.5.2 Protocol Design

13.5.2.1 Contention

HC-MAC does not require global synchronization. Any node entering the network first listens to control channel ch_0 for a time interval $t_d = t_p K + T$. This allows the new node to observe the current spectrum activities. Since any of the neighboring nodes cannot sense more than time $t_p K$ and transmit more than time T, a new node will not miss any control packet in its neighborhood. During the period, if a C-RTS (C-CTS) is received, it will defer and wait for the T-RTS (T-CTS). If T-RTS (T-

CTS) is received or time t_d is expired before receiving a T-RTS (T-CTS), new node participates in the contention process if it wants to transmit.

During the contention period, a multiple access scheme similar to IEEE 802.11 DCF model is used. A node reserves time for the following sensing and transmission operations within the neighborhood through the control channel by exchanging RTS/CTS messages with the target node. When a node wants to send packets to another node, it first sends a C-RTS packet to the destination through the control channel. The receiver, upon receiving the C-RTS, will reply a C-CTS packet. Other nodes overhear these packets defer their sensing and transmission, and wait for the notification from the transmitter/receiver pair or a timeout.

When a transmission is finished by a pair of nodes, other neighboring nodes contend the control channel with random backoff. Each of them chooses a back-off counter within a contention window. Each node maintains a variable cw, the contention windows size, which is reset to a value CW_{min} initially. The counter is deducted by one after each time slot. When the backoff counter reaches zero, the node will try to reserve the control channel by sending a C-RTS to the destination. If the C-RTS packets from neighboring nodes collide, they will double their contention window which lowers the probability of another collision. The node with the smallest contention window wins, and starts the next stage while other nodes freeze the counter until next contention period.

13.5.2.2 Sensing

A transmission pair which wins the contention will reverse the channels and starts to sense the spectrum. The sensing phase has one or several sensing slots, each of which includes the actual spectrum sensing and negotiation between sender and receiver. Since the transmitter and the receiver are now synchronized, they sense each channel with the same amount of time interval t_s. After getting the results, if the spectrum at the transmitter is available, it will send S-RTS to the receiver. If the spectrum is also available at the receiver side, the receiver will reply with S-CTS packet. Upon a successful exchange made between them, the spectrum availability for this channel is observed. When there is a channel occupied at any side of the transmission pair, either explicit message exchange with S-RTS (S-CTS) or timeout mechanism can be used. The negotiation message is quite short, so the interference for the primary user can be neglected. Since we use the timeout mechanism, i.e., no successful exchange before a timeout implies the occupation, another overhead comes from the exchange of another S-RTS and S-CTS if that channel is available for both sides, which is denoted by t_e. The total cost to obtain the status information of a channel is $t = t_s + t_e$.

A sensing stopping or continuing decision is made at the end of each spectrum sensing slot. The decision follows the optimal stopping rule described previously. The unit spectrum sensing time t, the maximum transmission time T and the hardware constraints (we assume they are identical for all nodes) are used to achieve the stopping decision. The decision is made by the sender and the receiver simultaneously and does not need any further negotiation.

For the probabilities of channel availability, they are assumed to be known for the secondary nodes. In the case that the probability for channel availability is not known in advance, the probability can be estimated with the information collected at each sensing of the channels. If channel conditions are similar for all the channels, the aggregated information for all channels is used to estimate the common availability probability; otherwise, separated probability is estimated for each of the channels. An estimation window with the size EW is used to approximate the probability with the information collected within the past EW sensing slots. The previous estimation between the transmitter and receiver must be synchronized, otherwise different decisions will occur. This is achieved by piggybacking RTS/CTS exchanges in contention and sensing stages. Each RTS/CTS exchanges the estimation, while the final decision uses the average of these two.

13.5.2.3 Transmission

After the transmission pair make the stopping decision, they begin to use a set of available channels to transmit packets. The transmission can include multiple data packets and corresponding ACK packets, when there is much data to transmit. The maximum transmission time is equal to T. After finishing the transmission, the transmitter will send a T-RTS to announce the completion of transmission; upon receiving the T-RTS, the receiver replies T-CTS. This information exchange ends the deferring of the neighboring node and starts the next round of contention.

One simple example is shown in Fig. 13.7, where pairs A–B and C–D contend for transmission. In the figure, node B is about to send packets to node A, while node C is targeting at node D. After pair A–B obtains the control channel ch_0 (indicated by the number in the parenthesis) via C-RTS/C-CTS control message exchange, pair A–B starts to sense while pair C–D freezes its state and backs off. When finishing the sensing of two channels (ch_1, ch_2) and exchanging the S-RTS/S-CTS messages, pair A–B makes a decision to stop sensing and enters to transmission stage. It uses the two available channels simultaneously to transmit two DATA packets and the associated ACK packets. Then the maximum transmission time T is almost used, it stops transmission and switches back to the control channel. To notify the completion of this round of spectrum usage, T-RTS/T-CTS messages are sent. Upon receiving of this last message exchange, pair C–D resumes its counting down of backoff timer and completes with pair A–B for the next round of spectrum access.

13.6 Performance Evaluation

In this section, we present the simulation results for the performance evaluation of the protocol. The simulations are conducted by ns-2 with version 2.29 [20]. We first consider a fully connected topology consisting of two transmission pair covered by a single primary user. The network throughput for HC-MAC and that for a fixed number of sensed channels are compared. The influence of different primary traffic usage, different transmission parameter setup on the secondary user's performance

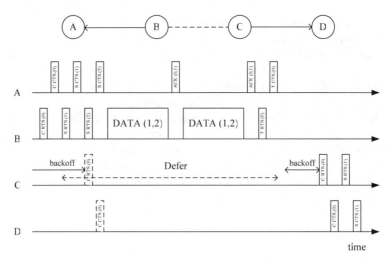

Fig. 13.7. An example of two competing flows.

is evaluated. The adaptation feature of HC-MAC is also demonstrated. After that, spectrum heterogeneity with fully connected topology is investigated with two primary users covering different sets of secondary users. Random topology is simulated to manifest the influence of primary user and secondary user density.

In all the following simulation setups, the bandwidth of each channel $B = 1$ MHz, and the secondary users have the same hardware constraints, maximum spread bandwidth $W = 6$ channels, and maximum fragments $F = 2$ fragments. Saturated CBR traffic flows are used by secondary users. In many of the simulations below, we compare our HC-MAC which makes intelligent sensing decision with the intuitive scheme which fixes the number of channels sensed.

13.6.1 Fully Connected, Spectrum-Homogeneous Topology

Figure 13.8 shows the first considered topology, where one primary user is covering two secondary transmission pairs. These two pairs are fully connected, thus the performance difference due to the topology is avoided. In addition, the spectrum opportunities exposed to two pairs are identical. The performance comparison for our MAC protocol with optimal stopping approximation (1-stage look-ahead) and with fixed number of sensed channels is given in Fig. 13.9. The approximation scheme is better than the fixed scheme which is consistent with our previous numerical results. We also examined the performance under different parameter settings. As shown in Fig. 13.10, with increasing number of total channels, the throughput of secondary users increases. This is because more bandwidth can be used simultaneously. In Fig. 13.11, when the probability of channel availability increases, the secondary throughput is also increased. Similar observation is shown in Fig. 13.12 for maximum transmission time interval.

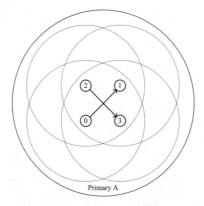

Fig. 13.8. one primary user and two secondary pairs.

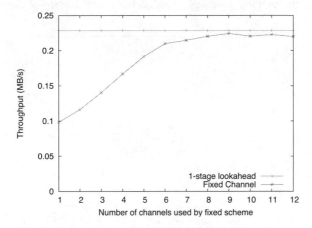

Fig. 13.9. Variation in throughput with different fixed numbers of sensing channels.

We further present the performance of HC-MAC under the situation when primary user's spectrum usage was alternating. The result is compared with the fixed scheme with a certain value for the number of sensed channels in Fig. 13.13. Since our scheme is adaptive in that the exploration of spectrum opportunity is according to the actual primary spectrum utilization, the throughput changes with the spectrum availability and is better than the fixed scheme.

13.6.2 Fully Connected, Spectrum-Heterogonous Topology

For the second considered topology, two primary users are covering two secondary transmission pairs shown in Fig. 13.14. The spectrum heterogeneity is examined with different spectrum availability for the two flows ($p = 0.4$ for flow 1–2, $p = 0.8$ for flow 3–4) while other parameters are the same as before (10 total channels, 0.01 s max transmission time). In Fig. 13.15, the performance is compared with fixed

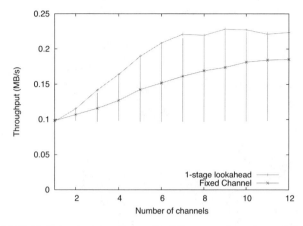

Fig. 13.10. Variation in throughput with different total numbers of channels.

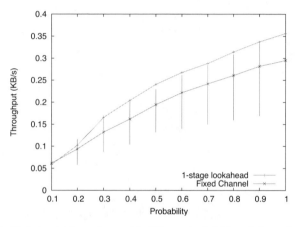

Fig. 13.11. Variation in throughput with different probability of channel availability.

scheme with five sensed channels. The adaptive decision makes our scheme outperform the fixed one. The overall throughput of flow 3–4 is greater than flow 1–2, since there exists more spectrum opportunity for flow 3–4. The fluctuation of the curves is due to the contention between these two flows.

13.6.3 Random Topology

We consider the random topology with the size of 1500×1500. Four non-overlaying primary users are located in the topology with same parameters for simplicity (spectrum availability probability p is 0.5). Secondary users are uniformly distributed within the area. We give the results of network throughput for secondary users with different numbers of secondary single hop flows shown in Fig. 13.16. Our scheme performs better than the fixed scheme with four sensed channels. The performance in

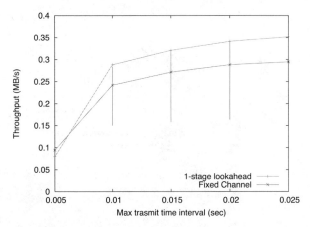

Fig. 13.12. Variation in throughput with different max transmission time interval.

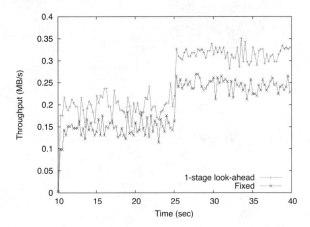

Fig. 13.13. Throughput comparison with time-varying primary spectrum usage ($p = 0.4$ during 10–25 s and $p = 0.8$ during 25–40 s).

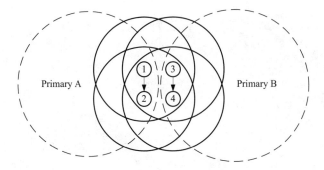

Fig. 13.14. Two primary users and two secondary pairs.

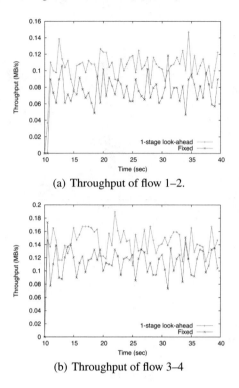

(a) Throughput of flow 1–2.

(b) Throughput of flow 3–4

Fig. 13.15. Individual throughput for two secondary pairs in Fig. 13.14.

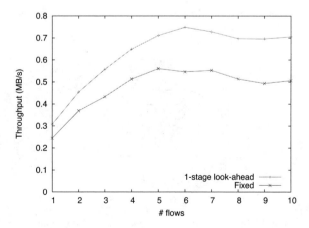

Fig. 13.16. Four primary users and random secondary single hop flows.

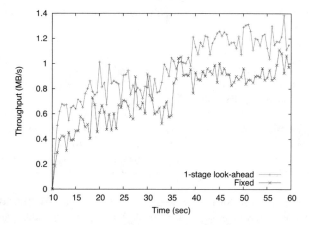

Fig. 13.17. A random topology with 15 secondary flows (p = 0.4 during 10–35 s, p = 0.8 during 35–60 s).

a single random topology with 15 secondary flows and time-varying primary traffic is shown in Fig. 13.17. Spectrum availability probability p is 0.4 during the first half-timeand 0.8 for the secondary one. Again our scheme is more efficient in capturing the spectrum opportunity than the fixed scheme.

Conclusion

In this chapter, we have proposed a MAC protocol that utilizes multiple spectrum opportunities (channels) to improve the cognitive network throughput and overall spectrum utilization. We took practical considerations of hardware constraints of cognitive radio used by secondary users: sensing constraints and transmission constraints. In our proposed system, the primary users had certain specifications of their maximum tolerable interference from the secondary users. We then identified the problem for each secondary user on how to maximize their throughput by optimizing the sensing decision in a sequence of sensing processes. This problem can be mapped to a well-defined optimal stopping problem. Both optimal solution and approximation rule were obtained. Based on this sensing decision, we designed a HC-MAC for hardware-constrained cognitive networks. Nearby transmission pairs were regulated so that the interference among the secondary users was mitigated. Simulation results showed the achievable throughput of secondary users for various system configurations.

References

1. FCC, ET Docket No 03-222. Notice of proposed rule making and order, 2003.

2. M. McHenry, "Spectrum white space measurements," in *New America Foundation Broad-Band Forum*, 2003.
3. DARPA XG WG, The XG Architectural Framework V1.0, 2003.
4. DARPA XG WG, The XG Vision RFC V1.0, 2003.
5. M. Buddhikot, P. Kolodzy, S. Miller, K. Ryan, and J. Evans, "DIMSUMnet: New directions in wireless networking using coordinated dynamic spectrum access," in *Proc. IEEE WoWMoM 2005*, pp. 78–85, 2005.
6. L. Xu, R. Tonjes, T. Paila, W. Hansmann, M. Frank, and M. Albrecht, "DRiVE-ing to the Internet: Dynamic radio for IP services in vehicular environments," in *Proc. 25th Annual IEEE Conference on Local Computer Networks (LCN 2000)*, pp. 281–289, 2000.
7. C. Cordeiro, K. Challapali, D. Birru, and S. Shankar, "IEEE 802.22: The first worldwide wireless standard based on cognitive radios," in *Proc. IEEE DySPAN 2005*, pp. 328–337, 2005.
8. L. Ma, X. Han, and C. C. Shen, *Proc. IEEE DySPAN 2005*, pp. 203–213, 2005.
9. P. Papadimitratos, S. Sankaranarayanan, and A. Mishra, "A bandwidth sharing approach to improve licensed spectrum utilization," *Proc. IEEE DySPAN 2005*, pp. 279–288, 2005.
10. A. Mishra, "A multi-channel MAC for opportunistic spectrum sharing in cognitive networks," in *Proc. IEEE MILCOM 2006*, pp. 1–6, 2006.
11. Q. Zhao, L. Tong, and A. Swami, "Decentralized cognitive MAC for dynamic spectrum access," in *Proc. IEEE DySPAN 2005*, pp. 224–232, 2005.
12. A. Shukla, B. Willamson, J. Burns, E. Burbidge, A. Taylor, and D. Robinson, "A study for the provision of aggregation of frequency to provide wider bandwidth services." Available at http://www.ofcom.org.uk/research/technology/overview/speclib/specagg.
13. D. Cabric, S. M. Mishra, and R. W. Brodersen, "Implementation issues in spectrum sensing for cognitive radios," in *Proc. Asilomar Conf. Signals, Systems, and Computers*, pp. 772–776, 2004.
14. J. So and N. H. Vaidya, "Multi-channel MAC for ad hoc networks handling multi-channel hidden terminals using a single transceiver," in *Proc. ACM MobiHoc 2004*, pp. 222–233, 2004.
15. S. L. Wu, C. Y. Lin, Y. C. Tseng, and J. P. Sheu, "A new multichannel MAC protocol with on-demand channel assignment for multi-hop mobile ad hoc networks," in *Proc. Intl Symposium on Parallel Architectures, Algorithms and Networks (I-SPAN 2000)*, pp. 232–237, 2000.
16. A. Sabharwal, V. Kanodia, and E. Knightly, "Opportunistic spectral usage: Bounds and a multi-band CSMA/CA protocol," in *IEEE/ACM Trans. Netw.*, 2006.
17. T. S. Rappaport, *Wireless communications: Principles and practice*. Englewood Cliffs, NJ: Prentice-Hall, 1996.
18. Y. S. Chow, H. Robbins, and D. Siegmund, *Great expectations: The theory of optimal stopping*. Houghton Mifflin Company, 1971.
19. T. Ferguson, "Optimal stopping time and applications," Available at http://www.math.ucla.edu/ tom/Stopping/Contents.html.

Microeconomic Models for Dynamic Spectrum Management in Cognitive Radio Networks

Dusit Niyato and Ekram Hossain

Department of Electrical and Computer Engineering
University of Manitoba, R3T 5V6 Canada, Winnipeg
{tao,ekram}@ee.umanitoba.ca

14.1 Introduction

Software-defined radio technique [1] was invented to improve adaptability and flexibility of wireless transmission so that the network performance can be improved. Developed based on software-defined radio, "cognitive radio" has been identified as a new paradigm for designing next generation wireless networks. A cognitive radio transceiver has the ability to observe, learn, optimize, and adapt the transmission parameters (e.g., transmission power, modulation level) according to the ambient environment [2]. Also, with this agility of the cognitive radio transceiver, frequency spectrum can be shared among licensed and unlicensed services, i.e., the primary and secondary services, respectively, to improve the spectrum utilization. The basic components/processes to achieve adaptability of wireless transmission in cognitive radio are described below.

- *Observation process:* The observation process typically consists of measurement and noise reduction mechanisms. The radio transceiver can silently listen to the environment, or the special messages and signals are transmitted and measured to obtain information about the surrounding environment. Estimation techniques play an important role in the observation process [3].
- *Learning process:* This refers to the process of extracting useful information from collected data. A learning process utilizes data obtained from observation process, previous decisions and actions.
- *Planning and decision making process:* This refers to the process of using the knowledge obtained from learning to schedule and prepare for transmission in the future. If multiple choices of actions are available, a transceiver must decide to choose the best strategy to achieve the objective. This planning and decision making process will change the current state of the transceiver, and subsequently, the surrounding environment which is observed by all the users.
- *Action:* This refers to the process of responding to the environment. The action of a transceiver is controlled by the planning and decision making process.

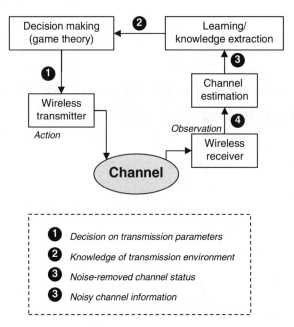

Fig. 14.1. Basic components and their interactions to achieve adaptivity in cognitive radio.

Figure 14.1 shows the interactions among these components.

Dynamic spectrum sharing is a challenging problem in cognitive radio network due to the requirement of "peaceful" co-existence of both licensed (primary) and unlicensed (secondary) users as well as the optimal utilization of radio spectrum. If the primary user cannot fully utilize the allocated spectrum, it results in spectrum hole which can be used by the secondary user(s) to improve the spectrum efficiency. In a scenario where the primary and the secondary services are provided by different operators, the secondary user(s) will require to pay the primary user(s) or service provider(s) for sharing the spectrum. Pricing is an important issue which affects dynamic spectrum sharing in cognitive radio networks. Channel allocation and spectrum sharing can be performed through the coordination of the service providers so that the spectrum owners can achieve their objectives. A negotiation protocol [4] is required for information exchange among the spectrum owners. The amount of shared bandwidth and the pricing should be determined such that the profit/utility of the service provider(s) is maximized while the quality of service (QoS) requirements of the the user(s) are satisfied. The dynamics of bandwidth sharing and pricing in a cognitive radio environment would depend on factors such as the number of primary users (or service providers), primary users' QoS requirements, and bandwidth demand of the secondary users.

14.2 Motivation, Contribution, and Organization of this Chapter

Game theory, widely used in microeconomics, can be effectively applied to address the problem of dynamic spectrum sharing, and in general, to the planning and decision-making process in a cognitive radio system. This is due to the fact that, in an environment where multiple agents interact with a view to achieve their own interests, in many cases, these objectives conflict with each other. Game theory can provide the basis to resolve this conflict so that all the agents are satisfied.

In this chapter, we investigate the problem of spectrum sharing and pricing in a cognitive radio environment using the game-theoretic *oligopoly model* from microeconomic theory. In microeconomics, oligopoly is defined as a situation where a small number of producers (i.e., oligopolists) dominate a particular market. In this market structure, producers compete with each other independently to achieve their objectives (i.e., maximize profit) by controlling the quantity or the price of the supplied commodity. The supply quantity and/or the price offered by one producer will affect the profit of other producers. For example, if one producer increases its supply to the market, the price for the entire market will decrease. As a result, profit of other producers tends to decrease.

Oligopoly is the more general case of a duopoly market (i.e., in duopoly market, the number of players is two), game theory can be used to analyze and predict the behavior of the producers. Each producer makes his decision independently, but the decision of one producer impacts the decision (i.e., profit) of other producers. The classical oligopoly models analyzed by game theory are *Cournot*, *Stackelberg*, and *Bertrand* game models. These models differ in market structure and competition. In particular, in Cournot model, producers compete in terms of quantity of supply to the market. All the producers make their decisions at the same time. On the other hand, in Stackelberg model, there are some producers (referred to as leaders) who are able to make decisions on the amount of supplied quantity before other producers (i.e., followers). Then, these followers make decision on the best amount of supplied quantity by taking into account the decision of the leader. Finally, in Bertrand model, all producers make decision simultaneously in terms of price. These different market structures result in different game formulations and also affect the behavior of the producers to achieve the best decision.

We demonstrate the applications of Cournot, Stackelberg, and Bertrand models of competition for spectrum/bandwidth sharing and pricing in cognitive wireless networks. Specifically, these three different models for oligopoly are applied to obtain the optimal size of spectrum/bandwidth sharing and the charging price. The oligopoly market models were well studied in economics. Also, they are computationally simple, and therefore, suitable for implementation in resource-limited software-defined radio transceiver.

The Cournot game model is used for the case where multiple secondary users share the spectrum/bandwidth with a primary user and the objective is to maximize the profit of the secondary users. Here, all secondary users can completely observe the strategies and the payoffs of other secondary users.

In the Bertrand model, several service providers (or primary users) compete with each other in terms of price to gain the highest profit under QoS constraints for the primary users. Here, the bandwidth demand of the secondary users is established based on a utility function which depends on the quality of transmission (i.e., channel quality) in the available spectrum. In addition, we consider *spectrum substitutability* which represents the ability of a secondary user to switch among the frequency spectra offered by different primary users.

Lastly, the Stackelberg leader-follower competition is used to model the problem of optimal sharing and pricing under elastic bandwidth demand from the secondary users. Numerical performance evaluation results are presented for these oligopoly competition models to show their efficacy in allocating radio resource in cognitive radio environments.

The rest of the chapter is organized as follows. Section 14.3 reviews the related works in the literature. The general characteristics of the three oligopoly competition models considered in this chapter are presented in Sect. 14.4. Section 14.5 presents the Cournot game model and its performance for dynamic spectrum sharing among multiple secondary users. The Bertrand game model for spectrum pricing under competition is presented in Sect. 14.6. Section 14.7 presents the Stackelberg game model for optimal pricing and sharing under elastic bandwidth demand. Then, the chapter is concluded.

14.3 Related Work

A partially observable Markov decision process (POMDP) was used for dynamic spectrum access in an ad hoc network [5]. An opportunistic spectrum access method was developed to allow secondary users to use the radio spectrum by using a decentralized cognitive medium access control (MAC) protocol. In the problem formulation, the state of the system was defined in terms of the availability of each channel and the action was defined as sensing and accessing the channel if available. The *reward* was defined as the amount of transmitted data. A heuristic algorithm was used to obtain the solution which was observed to be as good as the optimal algorithm but with much lower computational complexity.

In [6], a cognitive radio-based MAC layer scheduling algorithm was proposed for multihop wireless networks. An integer linear programming (ILP) model was formulated to solve the scheduling problem for time slot and channel allocation among the wireless nodes in the network. Also, to reduce the computational complexity, a distributed heuristic algorithm was devised to obtain the near optimal solution.

In [7], a pricing scheme for spectrum usage was presented where the price was described as a function of allocated spectrum, traffic intensity, and spectral efficiency of transmission. The pricing for spectral occupation under power constraints was obtained through an optimization formulation.

In [8], a game-theoretic adaptive channel allocation scheme was proposed to capture the selfish and the cooperative behaviors of the wireless nodes in the network.

The strategies of these players were defined in terms of channel selection. Two pay-off calculation schemes were used both of which depend on the level of interference. Also, a no-regret learning algorithm was used to learn the historical actions of other players. It was shown that the solution of this game formulation converges to the deterministic Nash equilibrium strategy.

In [9], the convergence dynamics of different types of games in cognitive radio was studied (i.e., coordinated behavior, best-response, and better response for dis-counted repeated games, S-modular games, and potential games, respectively). Also, a game theory framework was proposed for distributed power control to achieve agility in spectrum usage in a cognitive radio network. The problem of competitive channel allocation among multiradio devices was considered in [10]. Noncooperative game theory was used to analyze the dynamics of channel allocation where the strat-egy of a user was defined in terms of channel allocation and the payoff was obtained through a utility function of transmission rate. An algorithm was presented to achieve a channel allocation configuration which was shown to be both Pareto-optimal and system-optimal.

In [11], the problem of dynamic spectrum access in open spectrum wireless net-works was modeled by using a continuous-time Markov model. Also, a distributed algorithm modeled as a multiplayer game was proposed.

Oligopoly market model was used extensively to analyze the behavior of elec-tricity market [13–15]. In the electricity market, there are several producers who generate electricity to supply to the load (i.e., consumers). In general, the producers have to compete with each other by adjusting the price/supplied power to the load to achieve the maximum profit. A Cournot game model was used to analyze the power bidding in electricity market [13]. Since the transmission line from generator to the load is capacity limited, there is a constraint on the transmission network which was considered in the model [14].

An oligopoly model was used to analyze and develop network resource alloca-tion [16, 17]. In [16], the resource allocation problem in wired networks was for-mulated by using a Cournot model. In the considered system model, a user chooses the transmission rate and the links set the suitable price according to the marginal cost of the total rate allocation. In [17], a resource-trading mechanism for efficient distribution of large-volume contents in peer-to-peer networks was proposed. The objective of this mechanism was to maximize network capacity for higher revenue. The proposed mechanism was shown to be able to achieve Cournot equilibrium for resource-trading.

The problem of spectrum management and pricing can be formulated as an oligopoly market for which the product is the spectrum access opportunity (e.g., time, frequency, and code for time-division multiple access (TDMA), orthogonal frequency-division multiple access (OFDMA), and code-division multiple access (CDMA) networks, respectively). Game theory can be used to analyze the equilib-rium of sharing and pricing so that all the service providers are satisfied with the solution.

14.4 Oligopoly Market Models

The general description of the game formulation for an oligopoly competition presented in microeconomic literature is as follows [24]:

- Players: The players of an oligopoly competition are the producers (oligopolists).
- Strategies: The strategy for each producer corresponds to the supplied quantity (for the Cournot and the Stackelberg models) and the offered price (for Bertrand model).
- Payoffs: The payoff for the producer is the profit which can be determined based on the inverse demand function and the strategies adopted by all the producers in the market.

To illustrate the oligopoly market models (i.e., Cournot, Bertrand, and Stackelberg), we consider a market with only two producers (i.e., duopoly), so that the decisions (i.e., strategies) of the producers and their impacts can be presented by a two-dimensional graph. However, the same approach can be applied to the case of more than two producers (i.e., oligopoly). In order to study these oligopoly models, a demand function is required. In this case, we consider a linear inverse demand function in which the price of the product is determined from the total amount of supply to the market. This function can be defined as $P(Q) = A - Q$, where P is the price for unit amount of supplied quantity, Q is the total amount of supplied quantity, and $A > 0$ is the parameter of the inverse demand function. This demand function is shown in Fig. 14.2.

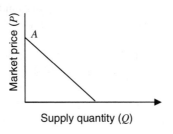

Fig. 14.2. Inverse demand function.

14.4.1 Cournot Competition

In Cournot competition, all producers who are the players of the game, make decisions (i.e., choose strategies) simultaneously on the amount of supplied quantity. Then, the total supplied quantity (i.e., aggregated supply) is used to determine the price which can be obtained from the given inverse demand function. The simplest case, to analyze this Cournot competition, assumes that all producers supply the same product, and therefore, there is no difference for the market to buy from a particular producer. Also, the cost of production for one unit of product is constant and is denoted by C and there is a fixed cost of production which is denoted by C_f. The objective of all of the producers is to maximize their profits by adjusting the supplied quantity to the market.

If Q_i and Q_j denote supply quantities from producer i and j respectively, the strategic form [12] of this game can be expressed as follows:

	0	$\cdots$	Q_j^m
0	$(\pi_1(0,0), \pi_2(0,0))$	$\cdots$	$(\pi_1(0, Q_j^m), \pi_2(0, Q_j^m))$
1	$(\pi_1(1,0), \pi_2(1,0))$	$\cdots$	$(\pi_1(1, Q_j^m), \pi_2(1, Q_j^m))$
$\vdots$	$\vdots$	$\cdots$	$\vdots$
Q_i^m	$(\pi_1(Q_i^m,0), \pi_2(Q_i^m,0))$	$\cdots$	$(\pi_1(Q_i^m, Q_j^m), \pi_2(Q_i^m, Q_j^m))$

$$(14.1)$$

where $\pi_1(Q_i, Q_j)$ and $\pi_2(Q_i, Q_j)$ denote the profit functions of producers i and j, respectively. This strategic form shows the market for which the supplied quantity is an integer, and producer i chooses a strategy in the rows and producer j chooses a strategy in the columns.

The Nash equilibrium, which is the solution of the game, can be used to determine the decision of the producer. This Nash equilibrium will provide the optimal strategy for each of the players where all the players are rational. This rationality indicates that all the players are willing to maximize their payoffs.

To obtain Nash equilibrium, best response or reaction function is typically used. This best response function is the optimal strategy of one producer given the strategies of other producers. If Q_j denotes the given strategy of producer j, profit of producer i can be expressed as follows:

$$\pi_i(Q_i, Q_j) = P(Q_i + Q_j)Q_i - CQ_i - C_f \tag{14.2}$$
$$= (A - Q_i - Q_j)Q_i - CQ_i - C_f. \tag{14.3}$$

The response function is the strategy that maximizes this profit. By differentiating the profit with respect to the available strategy Q_i, the best response function is obtained as follows:

$$\frac{\partial \pi_i(Q_i, Q_j)}{\partial Q_i} = A - 2Q_i - Q_j - C \tag{14.4}$$
$$0 = A - 2Q_i - Q_j - C \tag{14.5}$$
$$Q_i^*(Q_j) = \frac{A - Q_j - C}{2}. \tag{14.6}$$

Similarly, the best response function of producer j is obtained as $Q_j^*(Q_i) = \frac{A - Q_i - C}{2}$. As an example, Fig. 14.3 shows profit of producer i when for $A = 10$, $C = 0.5$, and $C_f = 1$. The best responses for which the maximum profit is achieved are also shown. We observe that if producer j increases its supply, profit of producer i decreases, and also the best response of producer i (in terms of supplied quantity)

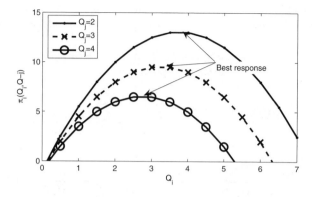

Fig. 14.3. Profit function.

decreases. This is due to the higher amount of total supply which results in lower market price. Therefore, the revenue becomes smaller while the cost remains the same.

The best response function of each of the producers given the other producer's strategy is shown in Fig. 14.4 for $C = 0.5$ and $C_{f=1}$. The best response of one producer decreases as the other producer increases its supply. When the value of the parameter A in the inverse demand function increases, the best response function shifts to the larger supplied quantity since the price is higher at the same aggregated supplied quantity. The same effect is observed when the cost per unit of product increases. However, this best response is not affected by the fixed cost.

The solution of the Cournot game model (i.e., the Nash equilibrium) gives the optimal supplied quantity that maximizes the profits of the firms. The *Nash equilibrium* of a game is a strategy profile (list of strategies, one for each player) with the property that no player can increase his payoff by choosing a different action, given the other players' actions [12]. In the context of Cournot model, this Nash equilibrium can be expressed as follows:

$$Q_i^*(Q_j^*) = Q_j^*(Q_i^*). \tag{14.7}$$

The Nash equilibrium is graphically shown in Fig. 14.4. This equilibrium is the point where the best responses intersect with each other, and it can be expressed as follows:

$$(Q_i^*, Q_j^*) = \left(\frac{A - C}{3}, \frac{A - C}{3}\right). \tag{14.8}$$

Note that, the profits of both the producers are the same, which is consistent with the assumption that both have the same information and they make decisions simultaneously.

At this Nash equilibrium, none of the producers can have better profit without adjustment in the supplied quantity of another producer. For example, if producer i

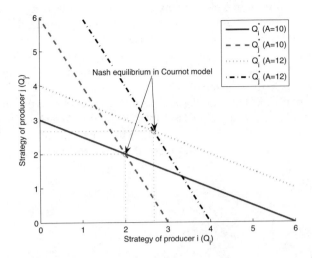

Fig. 14.4. Nash equilibrium in Cournot competition.

tries to increase its supplied quantity, the price will decrease. As a result, producer j must also increase its supplied quantity to gain higher profit. However, this will reduce the profit of both the producers. As a result, producer i is forced to reduce its supplied quantity. This process will repeat until the equilibrium is reached. This is referred to as the dynamic behavior of the Cournot model.

14.4.2 Bertrand Competition

Different from Cournot and Stackelberg competitions in which the producers compete in terms of supplied quantity, in Bertrand competition the producers compete by adjusting the price of the product. Before supplying the product to the market, all producers make decision on the price and announce to the market. Then, based on the demand function, a consumer decides the quantity to buy from each producer. The objective of this competition is again to maximize the profit of the producers.

However, in this Bertrand competition, the solution depends mainly on the substitutability of the products. If the products from the different producers are identical, then they are said to be totally substitutable. On the other hand, if the products are different, the products may be partly substitutable or may be completely unsubstitutable. The basic model of Bertrand competition considers the cases of identical and totally different products as described below.

In the case of identical or homogeneous products, the products from all the producers are totally substitutable, i.e., buying from one producer is not different from buying from others. Therefore, the consumer will alway choose to buy from the producer offering the lowest price. Furthermore, the entire market will buy from that producer, and other firms will have zero profit. Studies have shown that there is a unique Nash equilibrium in which the price charged by all producers are identical. In particular, at the Nash equilibrium, the price is equal to the production cost. When one producer decreases the price, that producer will capture the entire market. As a result, other producers will try to decrease their price to gain positive profit. Any price which is larger than the production cost is not equilibrium since one producer can gain higher profit by reducing the price.

In the case of differentiated products, the demand functions for the products from the different producers are different. Therefore, the market could buy different quantities of products from different producers where the prices are different. To describe the Bertrand competition in case of differentiated products, we consider the following demand functions:

$$Q_i(P_i, P_j) = A - P_i + BP_j \tag{14.9}$$
$$Q_j(P_i, P_j) = A - P_j + BP_i \tag{14.10}$$

where A and B are constants. Here, B also represents the substitutability of the products. Similar to the previous model, to obtain the Nash equilibrium of the game, the best response of producer i (i.e., which maximizes its profit) can be derived as follows:

$$\pi_i(P_i, P_j) = P_i Q_i - C Q_i - C_f \tag{14.11}$$
$$= (A - P_i + BP_j) P_i - C (A - P_i + BP_j) - C_f. \tag{14.12}$$

Differentiating $\pi_i(P_i, P_j)$ with respect to P_i we obtain the best response as follows:

$$\frac{\partial \pi_i(P_i, P_j)}{\partial P_i} = A + BP_j - 2P_i + C = 0 \tag{14.13}$$
$$P_i^* = \frac{A + C + BP_j}{2}. \tag{14.14}$$

Similarly, the best response function of producer j is

$$P_j^* = \frac{A + C + BP_i}{2}. \tag{14.15}$$

The Nash equilibrium for the charging price can be found to be

$$(P_i^*, P_j^*) = \left(\frac{A + C}{2 - B}, \frac{A + C}{2 - B} \right). \tag{14.16}$$

The Nash equilibrium of the above Bertrand competition is shown in Fig. 14.5 for $A = 5$, $C = 0.5$, and $C_f = 1$. Again, the Nash equilibrium is located at the point where the best responses of both the producers intersect with each other. Also, parameter B, which represents the substitutability of the products impacts the slope of the best response curves and hence the Nash equilibrium.

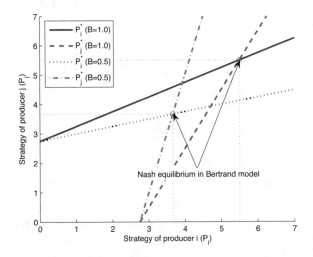

Fig. 14.5. Nash equilibrium for Bertrand competition.

14.4.3 Stackelberg Competition

In Stackelberg model, similar to Cournot model, the producers compete with each other in terms of supplied quantity. However, in the Stackelberg competition, there is at least one producer (referred to as the leader) who can commit the chosen strategy (i.e., supplied quantity) before other producers (referred to as the followers). An extensive form game [12] (as shown in Fig. 14.6) is used to present the Stackelberg competition model. This extensive form shows the sequence of decision making in which producer i is a leader and producer j is a follower. In this Stackelberg competition, since the leader will make the decision before the followers, the followers will choose their optimal strategy based on the observation from the leader. As a result, the solution of this game is a set of strategies where the profit of the leader is maximized for which the followers choose their best responses given the strategy of the leader.

In order to determine the equilibrium in a Stackelberg competition, *backward induction* is used. With backward induction, the best response of the follower is obtained at the last decision-making period. Again, the profit of the follower is computed from

$$\pi_j(Q_i, Q_j) = (A - Q_i - Q_j)Q_j - CQ_j - C_{\mathrm{f}}. \tag{14.17}$$

The best response of the follower is given as follows:

$$Q_j^*(Q_i) = \frac{A - Q_i - C}{2}. \tag{14.18}$$

Then, we backtrack to the decision of the leader. Here, the leader makes a decision based on the assumption that the follower will react with its optimal strategy

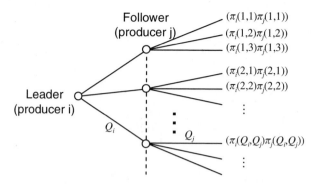

Fig. 14.6. Extensive form for the Stackelberg game.

(i.e., best response), and the objective of the leader is to maximize its profit. Therefore, we have

$$\pi_i(P_i, P_j) = (A - Q_i - Q_j)Q_i - CQ_i - C_f \tag{14.19}$$

$$= \left(A - Q_i - \frac{A - Q_i - C}{2} \right) Q_i - CQ_i - C_f. \tag{14.20}$$

Differentiating this profit function with respect to the strategy of the leader, which is Q_i, we obtain

$$\frac{\partial \pi_i(P_i, P_j)}{\partial P_i} = A - 2Q_i - \frac{A}{2} + Q_i + \frac{C}{2} - C \tag{14.21}$$

$$0 = \frac{A - C}{2} - Q_i \tag{14.22}$$

$$Q_i^* = \frac{A - C}{2}. \tag{14.23}$$

This is the subgame perfect Nash equilibrium or the optimal strategy for the leader if the leader can make a decision before the follower. Again, this optimal strategy for the leader will influence the decision of the follower. Based on the optimal strategy of the leader, the optimal strategy for the follower is

$$Q_j^* = \frac{A - C}{4}. \tag{14.24}$$

This Stackelberg equilibrium is graphically shown in Fig. 14.7, and it can be expressed mathematically as follows:

$$(Q_i^*, Q_j^*) = \left(\frac{A - C}{2}, \frac{A - C}{4} \right). \tag{14.25}$$

Note that, the optimal strategy of the leader is at the point where the leader predicts that the supplied quantity of a follower is zero. However, the follower will react with a non-zero supplied quantity.

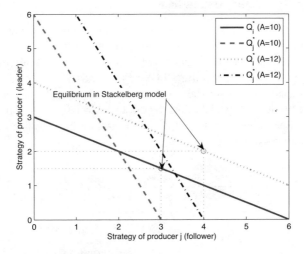

Fig. 14.7. Stackelberg equilibrium.

At the Stackelberg equilibrium, the leader will offer a larger amount of supplied quantity than that of a follower. Consequently, the profit of the leader is higher. This higher profit of the leader in a Stackelberg competition is also known as the *first-move advantage* in which the player of the game with the ability to make decision before other players will gain larger payoff.

In the following sections, we demonstrate the applications of the oligopoly market models to the spectrum/bandwidth sharing and pricing problem in cognitive radio networks. In particular, the three different oligopoly models described above are applied to obtain the optimal size of spectrum/bandwidth sharing and the charging price.

14.5 A Cournot Game Formulation for Dynamic Spectrum Sharing among Multiple Secondary Users

In this section, we formulate the problem of spectrum sharing among the primary user[1] and multiple secondary users as an oligopoly market competition. The objective of this spectrum sharing is to maximize the profit of secondary users by utilizing the concept of equilibrium. A Cournot game model is formulated for the case where a secondary user is assumed to have the knowledge on the strategies and the payoffs of other secondary users.

---

[1] We use "primary/secondary service" and "primary/secondary user" interchangeably.

14.5.1 System Model and Assumptions

We consider a wireless system with a primary user and multiple secondary users (i.e., total number of secondary users is denoted by N) who want to share the spectrum allocated to the primary user (Fig. 14.8) [18]. In this case, the primary user is willing to share some portion of the spectrum (Q_i) with secondary user i. The primary user charges the secondary user for the spectrum at a rate of $c(b)$ per unit bandwidth, where b is the amount of available bandwidth that can be shared. After allocation, the secondary users transmit in the allocated spectrum using adaptive modulation to enhance the transmission performance.

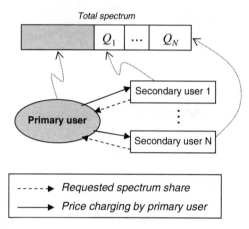

Fig. 14.8. System model for Cournot game spectrum sharing.

With adaptive modulation, the transmission rate can be dynamically adjusted based on the channel quality. For uncoded quadrature amplitude modulation (QAM) with square signal constellation (e.g., 4-QAM, 16-QAM) the bit-error-rate (BER) in single-input single-output Gaussian noise channel can be well approximated as follows [19]:

$$\text{BER} \approx 0.2 \exp\left(\frac{-1.5\gamma}{(2^k - 1)}\right) \tag{14.26}$$

where γ is the SNR at the receiver and k is the spectral efficiency of the modulation scheme used. Without loss of generality, we assume that the spectral efficiency is a non-negative real number (which can be obtained given any BER). To guarantee the quality of transmission, BER must be maintained at the target level (i.e., $\text{BER}_i^{\text{tar}}$). Therefore, spectral efficiency of transmission for secondary user i can be obtained from

$$k_i = \log_2(1 + K\gamma_i) \tag{14.27}$$

where

$$K = \frac{1.5}{\ln\left(0.2/\text{BER}_i^{\text{tar}}\right)}. \tag{14.28}$$

We assume that the received SNR information is available at the transmitter by channel estimation. In short, for secondary user i, given the received SNR γ_i, target BER$_i^{\text{tar}}$, and assigned spectrum Q_i, the transmission rate (in bits per second) can be obtained.

The revenue of secondary user i is denoted by r_i per unit of achievable transmission rate. It is assumed that a secondary user can communicate with the primary user but not with any other secondary users. Therefore, the adaptation for spectrum sharing is performed between each of the secondary users and the primary user only.

14.5.2 Cournot Game Formulation

Based on the above system model, a Cournot game can be formulated as follows:

- Players: The players in this game are the secondary users.
- Strategies: The strategy of each of the players is the spectrum size requested (denoted by Q_i for secondary user i) which is nonnegative.
- Payoffs: The payoff for each player is the profit (i.e., revenue minus cost) of secondary user i (denoted by π_i) in sharing the spectrum with the primary user and the other secondary users.

Note that, the commodity of this oligopoly market is the frequency spectrum.

For the primary user, we assume that the pricing function used to charge the secondary users is given by

$$P(\mathbb{Q}) = x + y \left(\sum_j Q_j \right)^\tau \tag{14.29}$$

where x, y, and τ are nonnegative constants, $\tau \geq 1$, and $\mathbb{Q}$ denotes the set of strategies of all secondary users (i.e., $\mathbb{Q} = \{Q_1, \ldots Q_N\}$). Let w denote the worth of the spectrum for the primary user. Then, the condition $P(\mathbb{Q}) > w \times \sum_j Q_j$ is necessary to ensure that the primary user is willing to share spectrum of size Q_j with the secondary users. Note that, the primary user charges all of the secondary users at the same price.

The revenue of secondary user i can be obtained from $r_i \times k_i \times Q_i$, while the cost of spectrum allocation is $Q_i P(\mathbb{Q})$. Therefore, the profit of the secondary user i can be obtained as follows:

$$\pi_i(\mathbb{Q}) = r_i k_i Q_i - Q_i P(\mathbb{Q}) \tag{14.30}$$

$$= r_i k_i Q_i - Q_i \left(x + y \left(\sum_j Q_j \right)^\tau \right). \tag{14.31}$$

The marginal profit function for secondary user i can be obtained from

$$\frac{\partial \pi_i(\mathbb{Q})}{\partial Q_i} = r_i k_i - x - y \left(\sum_j Q_j \right)^\tau - y Q_i \tau \left(\sum_j Q_j \right)^{\tau-1}. \tag{14.32}$$

Let $\mathbb{Q}_{-i}$ denote the set of strategies adopted by all except secondary user i (i.e., $\mathbb{Q}_{-i} = \{Q_j | j = 1, \ldots, N; j \neq i\}$ and $\mathbb{Q} = \mathbb{Q}_{-i} \cup \{Q_i\}$). In this case, the optimal allocated spectrum size to one secondary user depends on the strategies of other secondary users. Therefore, Nash equilibrium is considered as the solution of the game to ensure that all secondary users are satisfied with the solution.

In this case, we obtain the Nash equilibrium by using the best response function which is the best strategy of one player given others' strategies. The best response function of secondary user i given the allocated spectrum size to other secondary users Q_j, where $j \neq i$, is defined as follows:

$$\mathcal{BR}_i(\mathbb{Q}_{-i}) = \arg \max_{Q_i} \pi_i(\mathbb{Q}_{-i} \cup \{Q_i\}). \tag{14.33}$$

The set $\mathbb{Q}^* = \{Q_1^*, \ldots Q_N^*\}$ denotes the Nash equilibrium of this game if

$$Q_i^* = \mathcal{BR}_i(\mathbb{Q}_{-i}^*), \quad \forall i \tag{14.34}$$

where $\mathbb{Q}_{-i}^*$ denotes the set of best responses for secondary users j for $j \neq i$. Mathematically, to obtain the Nash equilibrium, we have to solve the following set of equations:

$$\frac{\partial \pi_1(\mathbb{Q})}{\partial Q_1} = 0 = r_1 k_1 - x - y \left(\sum_j Q_j \right)^\tau - y Q_1 \tau \left(\sum_j Q_j \right)^{\tau-1}$$

$$\vdots$$

$$\frac{\partial \pi_N(\mathbb{Q})}{\partial Q_N} = 0 = r_N k_N - x - y \left(\sum_j Q_j \right)^\tau - Q_N y \tau \left(\sum_j Q_j \right)^{\tau-1}.$$

14.5.3 Performance Evaluation

14.5.3.1 Parameter Setting

We consider a cognitive radio environment with a primary user and two secondary users sharing a frequency spectrum of size 15 MHz. The target BER for both the users is $\text{BER}_i^{\text{tar}} = 10^{-4}$. For the pricing function of primary user, we use $x = 0$ and $y = 1$, while τ is adjusted based on the evaluation scenario (e.g., $\tau = 1.0$), and the worth of spectrum for primary user is $w = 1$. The revenue of a secondary user per unit transmission rate is $r_i = 10$, $\forall i$.

14.5.3.2 Numerical Results

Figure 14.9 shows the best response of both secondary users in the Cournot game. The best response of each secondary user is a linear function of the other user's strategy. The Nash equilibrium is located at the point where the best responses of both

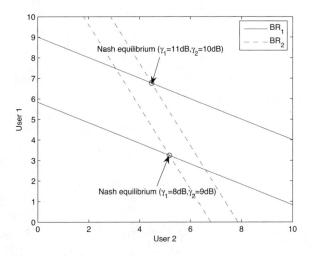

Fig. 14.9. Best responses and Nash equilibrium under different channel qualities.

the users intersect. We observe that the Nash equilibrium varies under different channel qualities. The adaptation of Nash equilibrium under different channel qualities is presented in Fig. 14.10. As expected, when the channel quality for a secondary user becomes better, the size of the spectrum allocated to that secondary users becomes larger. Also, we observe that the channel quality of one secondary user impacts the size of the allocated spectrum to other secondary user.

Figure 14.11 shows the impact of pricing function on the revenue of the primary user. When the value of the parameter τ increases, the primary user benefits from charging higher price to the secondary users. However, at a particular point (i.e., $\tau = 1.9$), the revenue gained by the primary user decreases since the price of the spectrum becomes too high and the secondary users request much smaller spectrum size. Therefore, the revenue from the secondary users increases at a rate smaller than the worth of spectrum to the primary users. Also, this result suggests that there is an optimal value for the pricing parameter τ which maximizes the revenue of the primary user.

14.6 Bertrand Game Model for Spectrum Pricing Under Competition

In this section, we consider a competitive situation for spectrum management where a few primary users offer spectrum access to the secondary users. For a primary user, the cost of sharing the frequency spectrum is modeled as a function of QoS degradation. The Nash equilibrium is considered as the optimal solution of this game.

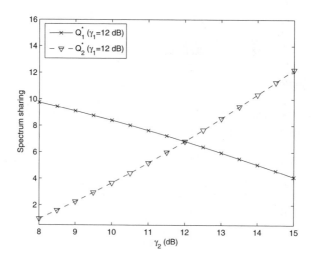

Fig. 14.10. Variation in Nash equilibrium of spectrum sharing under different channel qualities for user 2.

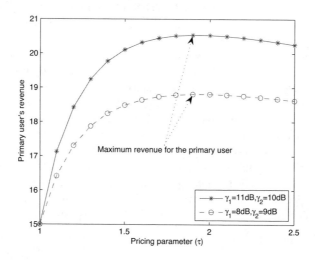

Fig. 14.11. Profit of primary user under different pricng parameters.

14.6.1 System Model and Assumptions

14.6.1.1 Primary and Secondary Users

We consider a wireless system with multiple primary users (total number of primary users is denoted by M) operating on the different frequency spectrum and a

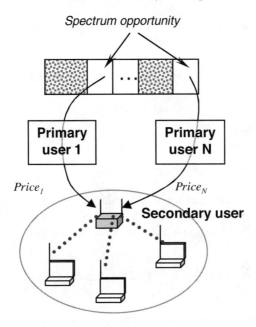

Fig. 14.12. System model of spectrum sharing and pricing for the Bertrand game model.

group of secondary users is willing to share these spectrum with the primary users (Fig. 14.12) [20]. In this case, primary user i wants to sell portions of the available bandwidth (e.g., time slots in a TDMA scheme) at price P_i (per unit bandwidth) to this group of secondary users. The spectrum can be shared by multiple terminals (i.e., secondary users) in which the base station (BS) or access point (AP) governs transmission in the allocated spectrum. The secondary users use adaptive modulation for transmissions in the allocated spectrum in a time-slotted manner. The spectral efficiency of the transmission for secondary user i is denoted by k_i. The spectrum demand of the secondary users depends on the transmission rate in the allocated frequency spectrum and the price charged by the primary user.

14.6.1.2 Cost Function of the Primary User

To develop a cost function, the QoS performance of the primary user needs to be considered. Degradation in the QoS performance of the primary user is expected if some portion of the frequency spectrum (i.e., in time domain or in frequency domain) is given to the secondary users. We consider the average delay as the QoS measure which is obtained for transmissions at the primary user based on an M/D/1 queueing model. Let λ_i denote the traffic arrival rate of the primary user, and $k_i^{(p)}(W_i - Q_i)$ denote the service rate, where $k_i^{(p)}$ and Q_i represent spectral efficiency of the wireless transmission by the primary user i, and the portion of frequency spectrum that is given to the secondary users. The average delay is defined as follows:

$$D_i(Q_i) = \frac{1}{2} \frac{\lambda_i}{\left(k_i^{(p)}(W_i - Q_i)\right)^2 - \lambda_i k_i^{(p)}(W_i - Q_i)} \tag{14.35}$$

where W_i is the total spectrum size of primary user i. The cost function can be simply defined as

$$C_i(Q_i) = dD_i(Q_i), \tag{14.36}$$

where d is a constant.

14.6.1.3 Utility of Secondary User

To quantify the spectrum demand, we consider the utility gained by the secondary users (e.g., if the spectrum creates high utility, the demand is high). We adopt a commonly used utility function in the economics defined as follows [21]:

$$U(\mathbb{Q}) = \sum_{i=1}^{M} Q_i k_i^{(s)} - \frac{1}{2} \left(\sum_{i=1}^{M} Q_i^2 + 2\Delta \sum_{i \neq j} Q_i Q_j \right) + J \tag{14.37}$$

where $\mathbb{Q}$ is the set consisting of the size of the spectrum available from all the primary users, i.e., $\mathbb{Q} = \{Q_1, \ldots, Q_i, \ldots, Q_M\}$, and

$$J = -\sum_{i=1}^{M} P_i Q_i \tag{14.38}$$

where P_i is the price offered by primary user i. Note that, $k_i^{(s)}$ denotes the spectral efficiency for transmission by the secondary user (e.g., BS/AP in Fig. 14.12) operating on the frequency spectrum offered by the primary user i. This utility function takes the spectrum substitutability into account through parameter Δ. That is, if the secondary users use multi-interface radio, they can switch among the frequency spectra freely depending on the offered price. This spectrum substitutability parameter (i.e., $\Delta \in [-1.0, 1.0]$) is defined as follows. When $\Delta = 1.0$, the secondary user cannot switch among the frequency spectrum, while for $\Delta = 0.0$ the secondary user can switch among the operating frequency spectra freely.

When $\Delta < 0$, the spectrum sharing by the secondary user is complementary. That is, when the secondary user wants to share one frequency spectrum, it will be required to buy one or more additional spectrum simultaneously (e.g., one spectrum for uplink transmission and another for downlink transmission) from the same or different primary users.

To derive the demand function of the secondary user who operates on the spectrum offered by primary user i,[2] we differentiate $U(\mathbb{Q})$ with respect to Q_i and then the optimal value of Q_i can be obtained.

---

[2] For brevity, we call it spectrum i.

The demand function is defined as the size of shared spectrum that maximizes the utility of the secondary user given the prices offered by the primary user [21], that is,

$$Q_i = \frac{k_i^{(s)} - P_i - \Delta(k_j^{(s)} - P_j)}{1 - \Delta^2}.$$ (14.39)

14.6.2 Bertrand Game Model

Based on this system model, a Bertrand game can be formulated as follows:

- Players: The players are the primary users.
- Strategies: The strategy of each of the players is the price per unit of spectrum (denoted by P_i) which is nonnegative.
- Payoffs: The payoff for each player is the profit (i.e., revenue minus cost) of primary user i (denoted by π_i) in selling spectrum to the secondary user.

Based on the demand function in (14.39) and the cost function in (14.36), the profit of each primary user/service provider can be expressed as follows:

$$\pi_i(\mathbb{P}) = Q_i P_i - C_i(Q_i)$$ (14.40)

where $\mathbb{P}$ denotes the set of prices offered by all players in the game (i.e., $\mathbb{P} = \{P_1, \ldots, P_i, \ldots, P_M\}$).

Again, the Nash equilibrium is considered as the solution of this game, and it is obtained by using the best response function. The best response function of primary user i given the prices of other primary users P_j, where $j \neq i$, is defined as follows:

$$\mathcal{BR}_i(\mathbb{P}_{-i}) = \arg \max_{P_i} \pi_i(\mathbb{P}_{-i} \cup \{P_i\})$$ (14.41)

where $\mathbb{P}_{-i}$ represents the set of prices offered by other players except player i (i.e., $\mathbb{P} = \mathbb{P}_{-i} \cup \{P_i\}$).

The set $\mathbb{P}^* = \{P_1^*, \ldots, P_M^*\}$ denotes the Nash equilibrium of this game if and only if

$$P_i^* = \mathcal{BR}_i(\mathbb{P}_{-i}^*), \quad \forall i$$ (14.42)

where $\mathbb{P}_{-i}^*$ denotes the set of best responses for player j for $j \neq i$. Mathematically, to obtain the Nash equilibrium, we have to solve the set of equations $\frac{\partial \pi_i(\mathbb{P})}{\partial P_i} = 0$ for all i where

$$\pi_i(\mathbb{P}) = P_i \frac{k_i^{(s)} - P_i - \Delta(k_j^{(s)} - P_j)}{1 - \Delta^2} - \frac{d\lambda_i}{2(W_i - Q_i)^2 - 2\lambda_i(W_i - Q_i)}.$$

We have to solve

$$0 = \frac{k_i^{(s)} - 2P_i - \Delta(k_j^{(s)} - P_j)}{1 - \Delta^2} + \frac{d\frac{\lambda_i}{1-\Delta^2}(4Q_i - \lambda_i)}{(2Q_i^2 - 2Q_i\lambda_i)^2}$$ (14.43)

where

$$Q_i = W_i - \frac{k_i^{(s)} - P_i - \Delta(k_j^{(s)} - P_j)}{1 - \Delta^2}.$$ (14.44)

14.6.3 Performance Evaluation

14.6.3.1 Parameter Setting

We consider a cognitive radio environment with two primary users and one secondary user/a set of secondary users (e.g., controlled by the the BS/AP in Fig. 14.12). The total frequency spectrum available to each primary user is 5 MHz. The target BER for the secondary user is $\text{BER}_i^{\text{tar}} = 10^{-4}$. Traffic arrival rate at a primary user is 1 Mbps, and we assume $d = 1$ for the cost function used by a primary user. The channel quality of the secondary user varies in the range of 10–20 dB.

14.6.3.2 Numerical Results

Figure 14.13 shows the demand function of the secondary user, and the revenue, cost, and profit of the first primary user under different pricing options. In this case, we set $\lambda_1 = 4$, $\gamma_1 = 15$ dB, $\gamma_2 = 18$ dB, $\Delta = 0.4$, $P_2 = 1$. As expected, when the first primary user increases the price, the secondary user demands a smaller spectrum size since the utility from the allocated spectrum decreases. Also, the cost for the primary user decreases since the secondary user demands smaller spectrum size. Therefore, the size of the remaining spectrum becomes bigger which results in smaller delay. However, the revenue and profit of the primary user first increase, and after a certain point it starts decreasing. Since at a small price the first primary user can sell a bigger spectrum size to the secondary user, the revenue and profit increase. In contrast, when the spectrum price becomes higher, a smaller amount of spectrum is sold to

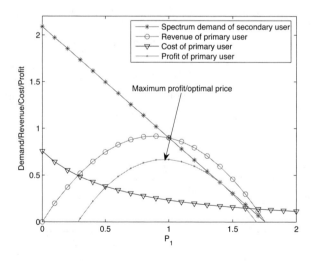

Fig. 14.13. Demand function of the secondary user, and revenue, cost, and profit of the first primary user.

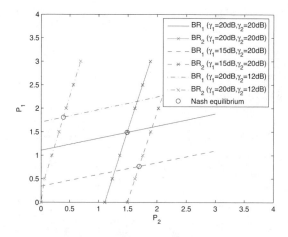

Fig. 14.14. Best response and Nash equilibrium under different channel quality (γ_1, γ_2).

the secondary user and this results in smaller revenue. We can observe that there is
an optimal price for which the profit is maximized and this price is referred to as the
best response of the corresponding primary user.

Then, the variations in the best response functions of both the primary users
are shown in Fig. 14.14 under different channel quality (γ_1, γ_2) for the secondary
user when $\Delta = 0.4$. As expected, when the channel quality becomes better, since
the secondary user can transmit at a higher rate due to the adaptive modulation, the
spectrum demand increases. As a result, the primary user can offer higher price. The
Nash equilibrium is located at the point where the best response functions of both
the primary users intersect.

Figure 14.15 shows the Nash equilibrium of the primary users under variations
in channel quality when $\Delta = 0.4$. As expected, the price at the Nash equilibrium is
higher for the spectrum with better channel quality. This is due to the larger demand
(which is a function of utility) generated by the secondary user. Also, we observe
that the channel quality of the spectrum offered by one player impacts the strategies
adopted by the other player. When the demand for spectrum offered by one player
changes, the other player must adapt the price to gain the highest profit.

Then, we investigate the impact of QoS requirements of the primary users on
the the Nash equilibrium. Figure 14.16 shows the Nash equilibrium of the secondary
user as functions of traffic arrival rate at the second primary user λ_2. In this case,
$\gamma_1 = 15$ dB and $\gamma_2 = 18$ dB. Since this arrival rate λ_2 affects the cost of the second
primary user in offering spectrum to the secondary user, at the Nash equilibrium
the price offered by the second primary user increases significantly. When the traffic
arrival rate increases, at the same spectrum size, traffic delay increases and the cost of
primary user increases accordingly. However, this traffic arrival rate has only small
impact on the price offered by the other player.

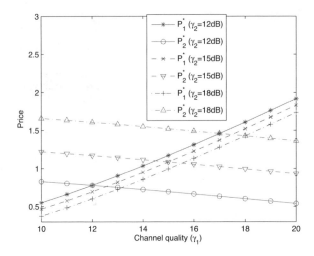

Fig. 14.15. Nash equilibrium under different channel qualities of the secondary user.

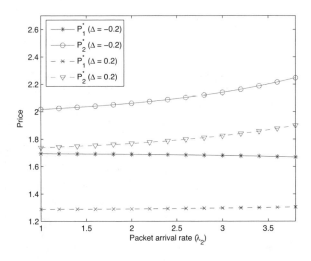

Fig. 14.16. Nash equilibrium under different traffic arrival rate of the second primary user.

14.7 Stackelberg Game Model for Optimal Pricing and Bandwidth Sharing Under Elastic Demand

In this section, we model the spectrum sharing problem between a primary user and multiple secondary users by a Stackelberg game model. The objective is to maximize the payoff of the service provider (leader) where the payoff considers price-elastic bandwidth demand of the secondary users. All the followers choose their best responses given the strategy of the leader.

14.7.1 System Model and Assumptions

The game model is described in the context of resource allocation/sharing in an integrated WiMAX/WiFi network where the WiMAX base stations (BSs) and the WiFi access points (APs)/routers are operated by different service providers. In the system model under consideration, the WiMAX BS charges the WiFi APs/routers for sharing the licensed WiMAX spectrum to provide mobile broadband Internet access to the WiFi clients. Each AP/router has a dual radio transceiver which can work by using both 802.11 and 802.16 interfaces. Traffic is transmitted from the BS using WiMAX radio interface and relayed through the WiFi AP/router using WiFi interface to the WiFi nodes.

The WiMAX subscriber stations (SSs) have fixed bandwidth demand, and therefore, subscribe at a flat rate to the WiMAX BS. On the other hand, the WiFi networks have elastic demand depending on the number of nodes and their preferences. Therefore, the WiMAX service provider charges the WiFi networks with adjustable pricing (i.e., P_1 and P_2 for WiFi router one and two, respectively, in Fig. 14.17). In this environment, the WiMAX and the WiFi service providers have to negotiate with each other to determine the optimal price such that their profits are maximized [22].

We formulate the pricing problem as a Stackelberg game in which the profit of the WiMAX BS is maximized and also the WiFi routers are satisfied with the bandwidth sharing and pricing. The WiMAX BS is the major player in this game – the decision on bandwidth allocation by the base station to the subscriber stations influences the decision of the WiFi APs/routers. Therefore, we consider this as a Stackelberg leader-follower game in which the WiMAX BS and the WiFi APs/routers are the leader and the followers, respectively. The solution of this game, i.e., the Stackelberg equilibrium, can be obtained easily if the information of all service providers and customers are available.

14.7.2 Stackelberg Game Model

14.7.2.1 Revenue and Elastic Demand

The revenue of the WiMAX BS from the service provided to the SSs is a function of the corresponding QoS performance. On the other hand, the WiMAX BS charges different prices to the different WiFi APs/routers depending on the bandwidth demand from the WiFi clients. This type of pricing model is particularly suitable for an environment in which the SSs serve real-time traffic (e.g., those for real-time polling service (rtPS)), while the WiFi networks serve best-effort traffic.

For the SSs, the queueing delay is the QoS metric and the revenue of the WiMAX BS is expressed as

$$r^{(\text{s})} = \sum_{i=1}^{N_{\text{ss}}} \left(a_i - e_i D(\lambda_i, Q_i^{(\text{s})}) \right) \tag{14.45}$$

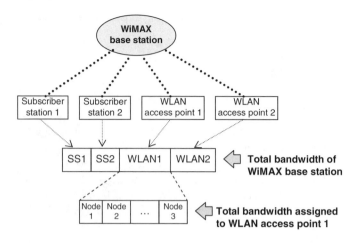

Fig. 14.17. An integrated WiMAX/WiFi network.

where a_i and e_i are constants (e.g., $a_i = 1$ and $e_i = 1$), $D(\lambda_i, Q_i^{(s)})$ is the queueing delay, λ_i is the traffic arrival rate at SS i (e.g., $\lambda_i = 0.2$ Mbps), $Q_i^{(s)}$ is the allocated bandwidth, and N_{ss} is the total number of SSs.

The bandwidth demand by a WiFi node depends on the price charged by the WiFi AP/router. We assume a linear demand function [23] which is expressed as follows:

$$\tilde{Q}_j = e_j - d_j P_k^{(wf)} \tag{14.46}$$

where $\tilde{Q}_j$ is the bandwidth demand of node j served by WiFi AP/router k, e_j and d_j are constants (e.g., $e_j = 2.0$ and $d_j = 0.4$), and $P_k^{(wf)}$ is the price charged at WiFi AP/router k. Therefore, the revenue of the WiFi network k is obtained from

$$r_k^{(wf)} = \sum_{j=1}^{N_k^{(wf)}} P_k^{(wf)} \tilde{Q}_j \tag{14.47}$$

and the cost is calculated from

$$C_k^{(wf)} = P_k^{(bs)} \sum_{j=1}^{N_k^{(wf)}} \tilde{Q}_j + F_k^{(wf)} \tag{14.48}$$

where $P_k^{(bs)}$ is the price charged by the WiMAX BS to the WiFi AP/router k, $N_k^{(wf)}$ is the number of WiFi nodes served by router k, and $F_k^{(wf)}$ denotes the fixed cost for WiFi router k. Note that, this demand function can be empirically obtained as in [23].

14.7.2.2 Stackelberg Game Formulation and the Equilibrium

We apply the Stackelberg game structure to obtain the equilibrium of bandwidth sharing and pricing between WiMAX and WiFi service providers. With the assumption that the WiMAX and the WiFi service providers are rational to maximize their profits, the game can be described as follows:

- The players: The WiMAX BS (i.e., leader) and the WiFi APs/routers (i.e., followers) are the players of this game.
- The strategies: For the WiMAX BS, the strategy is the price $P_k^{(bs)}$ charged to the WiFi APs and for a WiFi AP the strategy is the required bandwidth $Q_k^{(wf)} = \sum_{j=1}^{N_k^{(wf)}} \tilde{Q}_j$.
- The payoffs: For both the WiMAX BS and the WiFi APs/routers, the payoffs are the corresponding profits.

We first consider the payoff for a WiFi AP/router. Given the price charged by the WiMAX BS, $P_k^{(bs)}$, the profit of AP k is

$$\pi_k^{(wf)} = r_k^{(wf)} - C_k^{(wf)} \tag{14.49}$$

$$= \sum_{j=1}^{N_k^{(wf)}} P_k^{(wf)} \left(e_j - d_j P_k^{(wf)} \right) - P_k^{(bs)} \sum_{j=1}^{N_k^{(wf)}} \left(e_j - d_j P_k^{(wf)} \right)$$

$$- F_k^{(wf)}. \tag{14.50}$$

Therefore, the optimal price charged to a WiFi node (i.e., $P_k^{(wf)}$) can be obtained by differentiating the profit function and then setting it to zero. Then, given price $P_k^{(wf)}$, the bandwidth demand for all WiFi nodes in hotspot k can be obtained. Based on the best response of the WiFi AP/router, the WiMAX BS can adjust the price $P_k^{(bs)}$ charged to router k to achieve the highest payoff. The payoff (i.e., profit) of the WiMAX BS can be defined as follows:

$$\pi^{(bs)} = r^{(s)} + \sum_{k=1}^{N_r} r_k^{(wf)} \tag{14.51}$$

$$= \sum_{i=1}^{N_{ss}} \left(a_i - e_i D(\lambda_i, Q_i^{(s)}) \right) + \sum_{k=1}^{N_r} P_k^{(bs)} Q_k^{(wf)} \tag{14.52}$$

where N_r is the total number of WiFi APs/routers.

The Stackelberg equilibrium is defined as the strategy profile that maximizes the leader's payoff while the follower plays his/her best response [24]. We consider this equilibrium as the solution of the bandwidth sharing and pricing game to ensure that the profit of the WiMAX BS, which is the major player of this game, is maximized. In the case that all information on demand function are completely known, the equilibrium can be obtained easily by differentiating the profit function of the WiMAX BS and solving it for the price $P_k^{(bs)}$.

14.7.3 Performance Evaluation

14.7.3.1 Parameter Setting

We consider a single BS with multiple connections from SSs and WiFi APs/routers using the TDMA/TDD access mode based on single carrier modulation (e.g., WirelessMAN-SC). We consider downlink transmission from the WiMAX BS, and the frame size is assumed to be 5 ms. The total bandwidth of operation for the WiMAX BS is 20 MHz, and for transmission the BS uses QPSK modulation and a coding rate of 1/2.

14.7.3.2 Numerical Results

First, we show the bandwidth demand of the WiFi routers under different prices charged by the WiMAX BS (in Fig. 14.18). We consider the case of a homogeneous demand function for all WiFi nodes. This bandwidth demand represents the best response of the WiFi AP/router (i.e., follower) in the Stackelberg game formulation given the price charged by the WiMAX BS (i.e., leader). The best response for a WiFi AP/router can be obtained from the point at which the profit of the WiFi AP/router is maximized. The bandwidth demand decreases as the price increases since a WiFi router has to charge higher price to the WiFi nodes. As a result, the profit of the corresponding WiFi AP/router decreases. Also, as expected, when the number of WiFi nodes increases, the bandwidth demand of the WiFi AP/router increases.

Next, the profit of the WiMAX BS is shown in Fig. 14.19. Here, the WiFi routers serve 4 and 6 WiFi nodes, and the number of SSs is 10. The profit changes due to

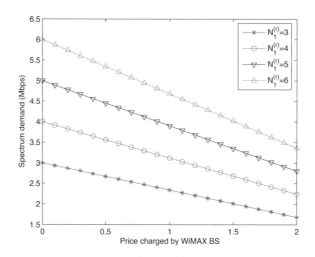

Fig. 14.18. Demand function which maximizes profit of the WiFi APs/routers.

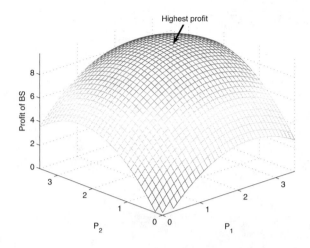

Fig. 14.19. Profit function of the WiMAX BS.

the different prices charged to the WiFi APs/routers, and also there is a point where the profit of the WiMAX BS is maximized. This point is the equilibrium of this Stackelberg game formulation. As is evident from Fig. 14.19, this profit, which is a function of price, is unimodel.

Figures 14.20a and b show optimal price for the WiMAX BS to charge the WiFi routers and the bandwidth to share with the WiFi routers. We consider the case when the first and the second WiFi AP/router serve 4 and 6 WiFi nodes, respectively. All the SSs have the same traffic arrival rate. Interestingly, even though the prices charged to the first and the second WiFi AP/router are formulated as different strategies in the game, at the equilibrium they are always equal. This implies that the WiMAX BS should charge the same price to the WiFi routers even though their bandwidth demands may be different. Also, as expected, when the traffic arrival rate increases, the WiMAX BS needs to increase the price charged to the WiFi routers to compensate the loss in revenue due to the degraded QoS performance (i.e., higher delay) for the SSs. Consequently, the bandwidth demand of both the WiFi APs/routers decreases. At the same price, bandwidth demand of the first WiFi router becomes smaller than that of the second router due to the smaller number of WiFi nodes.

Then, we vary the number of WiFi nodes served by router two and observe the price and the amount of bandwidth shared at the equilibrium (Fig. 14.21). We set the number of SSs to 16 and traffic arrival rate is assumed to be 0.5 Mbps. As expected, when the number of nodes increases, the bandwidth demand from the WiMAX BS increases. Consequently, the price charged to the WiFi APs/routers increases.

We observe that the bandwidth allocated to WiFi router two increases significantly while that to WiFi router one slightly decreases (which is due to the higher

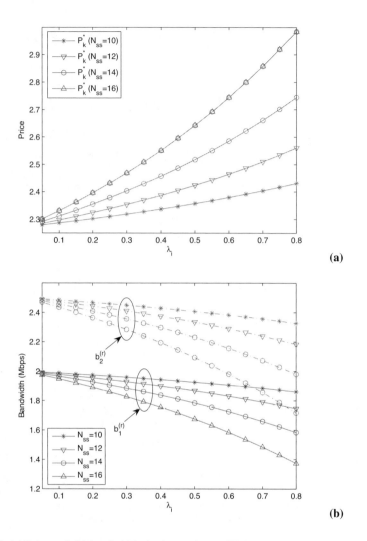

Fig. 14.20. (a) Price and (b) bandwidth sharing at the equilibrium under different traffic load at the subscriber stations.

price). We observe that with smaller number of SSs and lower traffic arrival rate (e.g., $\lambda_i = 0.1$ Mbps) the price does not change significantly. This is due to the fact that the WiMAX BS can take some bandwidth from the SSs (instead of taking bandwidth from other WiFi routers) with only slight degradation in their delay performances.

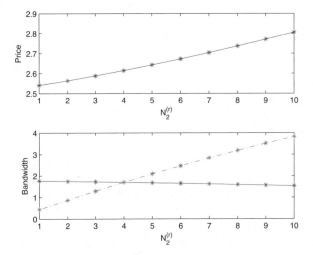

Fig. 14.21. Price and bandwidth sharing at the equilibrium under different number of WiFi nodes served by WiFi router two.

Conclusion

In this chapter, we have demonstrated the applications of oligopoly market models from microeconomics to solve the problem of spectrum/bandwidth sharing and pricing in cognitive radio environment. In microeconomics, oligopoly is used to describe a market situation which is composed of several producers, and the producers have their own interests to maximize their profits. We have considered three different oligopoly models, namely, *Cournot*, *Stackelberg*, and *Bertrand* models which have been analyzed by game theory techniques. In Cournot competition, producers compete in terms of supplied product quantity, and all of them make decisions simultaneously. In Stackelberg model, producers compete in terms of supplied quantity, but there are some producers who can make decision before the rest. In Bertrand competition, producers compete by varying the product price.

In the Cournot game model of bandwidth sharing, multiple secondary users (i.e., players) share the spectrum opportunity offered by a primary user. The Nash equilibrium of the game gives the bandwidth share for each secondary user such that the profits (or payoffs) for all the secondary users are maximized. Numerical performance evaluation results for this bandwidth sharing model in a cognitive wireless environment have been presented considering different channel qualities and prices (per unit bandwidth).

The Bertrand game model of bandwidth sharing and pricing considers multiple primary users (i.e., players) offering spectrum access to a secondary user and the objective is to maximize the profit of the primary users considering the degradation in QoS for the primary users due to spectrum sharing. The spectrum demand of the secondary user is obtained from its utility which is a function of spectrum price offered by the primary users and the corresponding channel quality such that the utility is maximized. The Nash equilibrium of the game is considered as the optimal solution of the game which gives the optimal price offered by the primary users. Numerical performance evaluation results have been presented under different channel qualities for the spectrum access by the secondary user, spectrum substitutability factor, and different traffic arrival rate at the primary users.

The Stackelberg game model of bandwidth sharing and pricing considers one leader (service provider) and multiple followers (secondary users) with time-varying demand function. The model has been used to obtain an optimal pricing scheme in an integrated WiMAX/WiFi network where the WiFi APs/routers share the WiMAX spectrum with the licensed WiMAX subscriber stations. The bandwidth demand of the WiFi APs/routers depends on the bandwidth demands by the WiFi nodes which depend on the price charged by the WiMAX BS. The Stackelberg equilibrium gives the optimal bandwidth share and the pricing for which the revenue of the service provider (i.e., leader) is maximized. The potential QoS degradation at the subscriber stations (i.e., licensed users) is considered while calculating the revenue for the service provider. Numerical performance evaluation results on price and bandwidth sharing at the equilibrium have been shown under different traffic load at the WiMAX subscriber stations and different number of WiFi nodes.

With oligopoly market model, the players in the game can adaptively adjust their strategies when only limited network information (e.g., the strategies adopted by other players, profit information) is available. Therefore, a learning algorithm would be required for the players to make their decisions effectively. Another possible research direction is the consideration of uncertainties in the model. In this case, the demand function can be random due to the time-varying traffic and QoS requirements of the users. Also, the payoff can be random due to the channel variation (e.g., fading). These issues should be addressed for competitive spectrum sharing in a practical cognitive radio environment.

References

1. J. Mitola, "Cognitive radio for flexible multimedia communications," in *Proc. MoMuC'99*, pp. 3–10, 1999.
2. S. Haykin, "Cognitive radio: Brain-empowered wireless communications," *IEEE J. Select. Areas Comm.*, vol. 23, no. 2, pp. 201–220, Feb. 2005.
3. L. N. Singh and G. R. Dattatreya, "Gaussian mixture parameter estimation for cognitive radio and network surveillance applications," in *Proc. IEEE VTC'05*, vol. 3, pp. 1993–1997, Sept. 2005.
4. S. Mangold, A. Jarosch, and C. Monney, "Operator assisted cognitive radio and dynamic spectrum assignment with dual beacons – detailed evaluation," in *Proc. IEEE Comsware'06*, pp. 1–6, Jan. 2006.
5. Q. Zhao, L. Tong, and A. Swami, "Decentralized cognitive MAC for dynamic spectrum access," in *Proc. IEEE DySPAN'05*, pp. 224–232, Nov. 2005.

6. M. Thoppian, S. Venkatesan, R. Prakash, and R. Chandrasekaran, "MAC-layer scheduling in cognitive radio based multi-hop wireless networks," in *Proc. IEEE WoWMoM'06*, pp. 191-202, June 2006.
7. N. S. Shankar, "Pricing for spectrum usage in cognitive radios," in *Proc. IEEE CCNR'06*, vol. 1, pp. 579–582, Jan. 2006.
8. N. Nie and C. Comaniciu, "Adaptive channel allocation spectrum etiquette for cognitive radio networks," in *Proc. IEEE DySPAN'05*, pp. 269–278, Nov. 2005.
9. J. O. Neel, J. H. Reed, and R. P. Gilles, "Convergence of cognitive radio networks," in *Proc. IEEE WCNC'04*, vol. 4, pp. 2250–2255, Mar. 2004.
10. M. Felegyhazi, M. Cagalj, and J.-P. Hubaux, "Multi-radio channel allocation in competitive wireless networks," in *Proc. IEEE ICDCS'06*, pp. 1545–0678, July 2006.
11. Y. Xing, R. Chandramouli, S. Mangold, and S. Shankar N, "Dynamic spectrum access in open spectrum wireless networks," *IEEE J. Select. Areas Commun.*, vol. 24, no. 3, pp. 626–637, Mar. 2006.
12. M. J. Osborne, *An Introduction to Game Theory*, Oxford University Press, 2003.
13. A. R. Kian, J. B. Cruz Jr., and R. J. Thomas, "Bidding strategies in oligopolistic dynamic electricity double-sided auctions," *IEEE Trans. Power Syst.*, vol. 20, no. 1, pp. 50–58, Feb. 2005.
14. Y. Liu and F. F. Wu, "Impacts of network constraints on electricity market equilibrium," in *Proc. IEEE Power Engineering Society General Meeting 2006*, June 2006.
15. Y. Chen and B. F.Hobbs, "An oligopolistic power market model with tradable NO/sub x/permits," *IEEE Trans. Power Syst.*, vol. 20, no. 1, pp. 119–129, Feb. 2005.
16. R. Johari and J. N. Tsitsiklis, "A scalable network resource allocation mechanism with bounded efficiency loss," *IEEE J. Select. Areas Commun.*, vol. 24, no. 5, pp. 992-999, May 2006.
17. S. G. M. Koo, C. S. G. Lee, and K. Kannan, "A resource-trading mechanism for efficient distribution of large-volume contents on peer-to-peer networks," in *Proc. IEEE ICCCN'05*, pp. 428–433, Oct. 2005.
18. D. Niyato and E. Hossain, "A game-theoretic approach to competitive spectrum sharing in cognitive radio networks," in *Proc. IEEE WCNC'07* (Hong Kong), 11–15 Mar., 2007.
19. A. J. Goldsmith and S.-G. Chua, "Variable rate variable power MQAM for fading channels," *IEEE Trans. Commun.*, vol. 45, no. 10, pp. 1218–1230, Oct. 1997.
20. D. Niyato and E. Hossain, "Optimal pricing under competition for dynamic spectrum sharing in cognitive radio: A game-theoretic approach," in *Proc. IEEE WCNC'07* (Hong Kong), 11–15 Mar., 2007.
21. N. Singh and X. Vives, "Price and quantity competition in a differentiated duopoly," *RAND J. Econ.*, vol. 15, no. 4, pp. 546–554, 1984.
22. D. Niyato and E. Hossain, "Integration of WiMAX and WiFi: Optimal pricing for bandwidth sharing," submitted to the *IEEE Commun. Mag.*
23. C. Thompson, "Supply and demand analysis in convergent networks," Master of Business Administration Thesis, Massachusetts Institute of Technology (MIT), June 2001.
24. G. Romp, *Game Theory: Introduction and Applications*, Oxford University Press, 1997.

15

Analysis of Cognitive Radio Dynamics*

Maria-Gabriella Di Benedetto[1], Maria Domenica Di Benedetto[2], and Guerino Giancola[1], Elena De Santis[2]

[1] University of Rome La Sapienza, Italy
{dibenedetto,giancola}@newyork.ing.uniroma1.it
[2] University of L'Aquila, Poggio di Roio, Italy
{dibenede,desantis}@ing.univaq.it

15.1 Introduction

The cognitive concept of a radio capable of adapting to the environment and of adjusting its operation as a function of both external and internal unpredictable events forms the conceptual basis for the design of future wireless communication systems. Designing and developing smart wireless devices able to sense the environment, and to modify accordingly spectral shape and other features of radiated signals is extremely appealing [1,2]. In particular, by defining and developing technologies that can enable a radio device to adapt its spectrum according to the operating environment, that is, to be aware of the scenario in which it operates, design innovation is taking its first step toward conceiving wireless networks that cooperatively coexist with other wireless networks and devices. The coexistence principle is intrinsic to innovative technologies such as ultra wide band (UWB) radio [3], although the concept has a rather broader acceptation.

Cognitive radio focuses on improving efficiency in the use of the wireless resource and applies basically to the behavior of one node. By introducing cognitive principles in the logic of the wireless network one extends the cognitive concept to rules of interaction between nodes. In order to optimize the design one must therefore model the set of wireless nodes as a social network forming one single entity.

Time scale is particularly important when conceiving adaptive mechanisms that should allow the update of system configuration. System operation is ruled by clocks, that must be tuned according to the granularity that is requested by a specific operation. In the case of spectrum sensing, for example, it might be desirable to force the system to being continuous, in order to incorporate the capability of detecting sudden and unpredicted changes in the environment. Conversely, it might be desirable for other operations to be ruled on different time scales.

* This work has been partially supported by the HYCON Network of Excellence, contract number FP6-IST-511368, Integrated Project PULSERS II, contract number FP6-IST-506897, and by Ministero dell'Istruzione, dell'Università e della Ricerca under Project SCEF (PRIN05).

The problem that must be analyzed is therefore related to asynchronicity of phenomena that would force a node to change its state of operation with respect to input–output dynamics of the node. Mobility is an example of such phenomena. Other types of perturbations, such as atmospheric changes, have the additional complex feature of being unpredictable. Continuous and discrete dynamics must be integrated in the mathematical model that describes the node. Hybrid systems (see, e.g., [4]) offer a challenging framework for formalizing such a complex system [5].

In this chapter, we propose a model that generalizes those proposed in [6] and [7] for UWB networks, to the case of a self-organizing network of nodes that operate under the coexisting principle. We formalize the model by using hybrid systems that offer in fact the analytical framework for modeling complex systems where continuous dynamics and discrete processes tightly interact.

The chapter is organized as follows. In Sect. 15.2, we define the problem and set the basis and main assumptions. Section 15.3 contains the rules for governing resource in the network, in terms of computing transmission power levels of nodes. Section 15.4 describes the application of hybrid system modeling to the system under consideration that incorporates the cognitive radio concept. In particular, we show how the proposed model represents the behavior of each node and of the population of nodes that form the network. In Sect. 15.4, we present some concluding remarks highlighting open problems that may be formally stated and analyzed using the proposed hybrid model.

15.2 Problem Statement and System Description

We consider the formation of a self-organizing network of nodes that adopt a multiple access scheme in which coexistence is foreseen, that is signals originating from different users share in principle a same resource in terms of time and frequency. Users separation is obtained by appropriate coding. Code division multiple access (CDMA) as well as time hopping multiple access (THMA) are possible access schemes. In general terms, the multiple access scheme may be based on any coding scheme that allows resource sharing while providing acceptable system performance at the receiver. In such a context, the receiver is supposed to operate in a correlation mode, that is to be capable of sensing the presence of a useful signal by appropriate synchronization in encoded time instances. The dominant noise is interference noise, with a dominant component formed by multi-user interference (MUI).

Call P_{TX} the average transmitted power. This power is upper-bounded by a maximum power P_{max} that can be determined from recommendations on emission levels as well as technological limitations. T_b is the bit repetition period. The impulse response of the pulse shaper is indicated by $p_w(t)$. Furthermore, we indicate by $v_{a_j}(t)$ the data-modulated multi-pulse signal made of the sum of N_S shifted, and eventually amplitude modified, versions of $p_w(t)$, where N_S is the number of chips forming one bit of the data sequence $\{a_j\}$. Chip duration is thus $T_S = T_b/N_S$. Note that $p_w(t)$ has a direct impact on the power spectral density of transmitted signals and that therefore by selecting a specific pulse shape one may adapt spectral features of radiated emissions as a function of the environment, i.e., for example specific interference patterns. An important hypothesis that is fundamental in our model is the possibility of selecting one pulse shaper among many [8]. We therefore assume W different pulse shapes $p_w(t)$, with $w = 1, ..., W$.

A general flat additive white Gaussian noise (AWGN) channel model is assumed. The impulse response for the channel between a reference transmitter TX and a reference receiver RX is indicated by $h(t) = \alpha\delta(t-\tau)$, and is characterized by a constant amplitude gain α and a constant delay τ. The signal at RX input writes:

$$
\begin{aligned}
r(t) &= \alpha\sqrt{P_{TX}T_S} \sum_j v_{a_j}(t - jT_b - \tau) + n(t) \\
&= \sqrt{P_{RX}T_S} \sum_j v_{a_j}(t - jT_b - \tau) + n(t)
\end{aligned}
\tag{15.1}
$$

where $P_{RX} = \alpha^2 P_{TX}$ is the average received power and $n(t)$ is the cumulative noise at the receiver input. It is well known that, under the above conditions, single-user reception is optimal when the receiver is composed of a coherent correlator followed by a maximum likelihood detector [3]. The output of the correlator within a bit period T_b is indicated by Z. It is on the basis of Z that the ML detector takes a decision, that is, we suppose decision is taken on a bit period, i.e., based on N_S pulses forming one multi-pulse (soft detection). According to this scheme, the received signal is thus cross-correlated with a correlation mask $m_w(t)$ that is matched with the train of pulses representing one bit. The correlator output Z is given by

$$
Z = \int_\tau^{\tau+T_b} r(t)\, m_w(t - \tau)\, \mathrm{d}t.
\tag{15.2}
$$

The decision variable Z in (15.2) is compared against a zero-valued threshold according to the following rule: when $Z > 0$, decision is "0," while when $Z < 0$, decision is "1," or vice versa. Then, for independent and equiprobable transmitted bits, given a transmitted bit $b_0 = 0$ the average bit error rate (BER) is:

$$
\mathrm{BER} = \mathrm{Prob}\left\{ Z < 0 | b_0 = 0 \right\}.
\tag{15.3}
$$

It is essential to set a hypothesis on the capability of the system to synchronize. By analyzing (15.2), we notice that a necessary condition for the receiver to properly function is that the correlator mask must be aligned with $s(t - \tau)$. In other terms, the receiver must be capable of estimating the delay introduced by propagation over the channel and synchronize with the received signal. In order to achieve this function one can suppose for example to send within each data packet a specific sequence to be used for synchronization purposes and that is known by both the transmitter and the receiver [9]. In any case, system performance depends on the accuracy that can be achieved in synchronization. This accuracy is in turn related to signal to noise ratio and in particular to the receiver ability of extracting one single pulse from noise. Detecting the first pulse of the synchronization trailer for example might be crucial for correct operation. One system specification is thus the level of signal to noise ratio that can be achieved on the single pulse. We call SNR_0 the minimum signal to noise ratio over a single pulse that is required by system specification. A condition for correct reception is therefore that signal to noise ratio on the pulse be $SNR_p > SNR_0$ [10].

The topology of the network is a star, that is, nodes communicate through the network controller. We suppose that the node that has the role of network controller also implements the cognitive radio paradigm. The role of controller can be played by any of the devices in the network but in our analysis we suppose that the controller is the node that starts the network by activating a beacon on a broadcast channel.

Multiple access is based on code division either in the time domain (time hopping) or amplitude-based (direct sequence). As a consequence, system performance is limited by multi-user interference (MUI). We suppose that the set-up of a link between a node and the coordinator occurs on a dedicated channel that is identified by a specific code.

The received signal of (15.1) that incorporates MUI can be expressed as follows:

$$r(t) = r_u(t) + n_e(t) + n_{mui}(t) \tag{15.4}$$

where $r_u(t)$ is the useful received signal, $n_{mui}(t)$ accounts for MUI, and $n_e(t)$ incorporates thermal noise and external interference introduced by wireless belonging to coexisting networks. In the present case, the decision variable at the output of the correlator is made of three terms: a useful contribution Z_u, external noise Z_e, and MUI Z_{mui} and writes $Z = Z_u + Z_e + Z_{mui}$. Signal to noise ratio SNR at the correlator output for one link is thus:

$$SNR = \frac{E_u}{\eta_e + \eta_{mui}} \tag{15.5}$$

where E_u is the received useful energy per bit for the reference link, and η_e and η_{mui} are the variance of Z_e and Z_{mui}. If all signals are received with same power,

$$E_u = (N_S)^2 P_{RX} T_S \tag{15.6}$$

and

$$\eta_e = N_S \, \eta_p(w) \tag{15.7}$$

$$\eta_{mui} = N_S \sigma_m^2(w)(N-1) P_{RX} \tag{15.8}$$

in which $\sigma_m^2(w)$ is a term that depends upon the mask shape of the correlator and $\eta_p(w)$ is noise variance on one pulse. Note that according to (15.7) and (15.8), both noise and interference depend on pulse shaping. If $R_b = 1/T_b$ is the bit rate for the link under examination, then the signal to noise ratio on the reference link can be obtained by combining (15.6), (15.7), (15.8), and (15.5) as follows:

$$\text{SNR} = \frac{N_s^2 T_S P_{RX}}{N_s \eta_p(w) + \sigma_m^2(w) N_s (N-1) P_{RX}} = \frac{1}{R_b} \frac{P_{RX}}{\eta_p(w) + \sigma_m^2(w)(N-1) P_{RX}}.$$

$$(15.9)$$

The BER of (15.3) can be expressed in a closed form as a function of SNR of (15.9) if $\eta_e(t)$ and $\eta_{mui}(t)$ have statistical properties that are known. If $\eta_e(t)$ and $\eta_{mui}(t)$ can be modeled as white Gaussian random processes, the relationship between BER and SNR becomes

$$\text{BER} = \frac{1}{2} \text{erfc} \left(\sqrt{\frac{\text{SNR}}{2}} \right) \tag{15.10}$$

where $\text{erfc}(x)$ is the complementary error function of x. Based on (15.9) the signal to noise ratio on the pulse SNR_p is

$$\text{SNR}_p = \frac{T_S P_{RX}}{\eta_p(w) + \sigma_m^2(w)(N-1) P_{RX}}. \tag{15.11}$$

Note that since we are considering an active connection, the following relationship holds:

$$\text{SNR}_p \geq \text{SNR}_0. \tag{15.12}$$

15.3 Rules by Which a Node Communicates with the Coordinator

We assume that data flows are grouped into packets and that each packet is segmented into MAC frames. These frames are transmitted over the radio interface. The MAC frames have standard format and each of the frames is composed of a header and a payload. The header typically contains the MAC address as well as the synchronization trailer.

The system supports the best-effort data sources which do not require the use of a minimum value for the transmission rate and are not bound by maximum delay specifications. In other terms, best-effort sources do not require any quality of service guarantee and are allowed to transmit, i.e., are admitted in the network, if their presence does not disturb the operation of other sources that cannot tolerate end-to-end delay D (s) greater than specific values, and that require at least a F percentage of packets to reach destination within D. These different sources, called quality of service sources are characterized by the characteristics of the traffic that they generate and by their required specifications D and F. We suppose that the traffic generated by the sources in the network is shaped by a standard dual leaky bucket (DLB) [11] that functions as an interface between the source and the system and that outputs traffic described by the four following parameters: the peak rate p (bits/s), the average rate r (bits/s), the token buffer dimension b (bits) and the maximum packet size M (bits). Rates p and r do not account for the overhead introduced at the MAC and physical layers, and can thus be lower than binary rate R_b of (15.9).

In this section we illustrate the principles of operation of a network of N nodes, and we define the basic principles of the admission control function. The admission control function for UWB nodes proposed in [6] can be generalized for the present analysis case. We summarize here the rules for power assignment for convenience of the reader. We refer to [6] for a full coverage of the topic and full description of the admission control function that includes the rules by which both best-effort and quality of service nodes compute their rate of transmission. As indicated above, a star topology is taken into consideration. As a consequence the analysis is focused on the uplink connections.

Different pulses $p_w(t)$, with $w = 1, ..., W$ can be used for shaping the spectra of radiated signals and adapt this shape to the features of the channel. It is the controller that has the capability of sensing the channel, and therefore, of appropriately selecting the "best" impulse response of the pulse shaper. The controller node computes $\eta_p(w)$ of (15.7) and $\sigma_m^2(w)$ of (15.8) for all possible pulse shapes that is $w = 1, ..., W$, and based on $\eta_p(w)$ and $\sigma_m^2(w)$, estimates $P_{min}(w)$ as follows:

$$P_{min}(w) = \frac{\eta_p(w)}{T_S} \left(\frac{1}{SNR_0} - \frac{\sigma_m^2(w)}{T_S}(N-1) \right)^{-1}. \qquad (15.13)$$

The above equations provide the minimum power that the controller must receive from each node in order to comply with (15.12). The "best" pulse shaper can be thus defined as the one that provides the lowest $P_{min}(w)$ value. As a consequence, each node j must use a transmission power P_j that can be computed as follows:

$$P_j = P_{min}(w^*) A_j \qquad j = 1, ..., N \qquad (15.14)$$

where A_j is the attenuation characterizing the link between node j and the controller.

15.4 Admission Control by Hybrid Modeling

Hybrid system formalism offers the framework for modeling the behavior of self-organizing networks. Thanks to this formalism, we can characterize self-organizing network dynamics as a discrete finite-state automaton where, for each state, state-specific rules of operation govern the evolution of the network itself. In this section, we first illustrate the fundamental principles of hybrid system modeling. We then describe the application of hybrid system modeling to the system under consideration that incorporates the cognitive radio concept.

15.4.1 Basic Principles of Hybrid Modeling

Hybrid systems are dynamical systems where continuous and discrete dynamics are embedded together to propositional logic. Continuous and discrete variables interact and determine the hybrid system evolution. The hybrid state of a hybrid system is made of two components: The discrete state belonging to a finite set Q and the continuous state belonging to a linear subspace of $\mathbb{R}^n$. The evolution of the discrete state is governed by an automaton, while the evolution of the continuous state is given by a dynamical system controlled by a continuous input and subject to continuous disturbances. Whenever a discrete transition occurs, the continuous state is instantly reset to a new value. Even if the intuitive notion of hybrid system is simple, the combination of discrete and continuous dynamics and the mechanisms that govern discrete transitions create serious difficulties in defining its operation precisely. Other complexity stems from the continuous state reset that occurs when the system undergoes a discrete transition. This is why we need formal definitions of the variables that characterize a hybrid system as well as of their evolution in time, as will be defined below:

1. The state variable of a hybrid system $\mathcal{H}$ is made of two components: the discrete state q and the continuous state x. The discrete state belongs to a finite set $\boldsymbol{Q} = \{q_i, i \in J\}$, $J = \{1, 2, ..., N\}$, $N \in \mathbb{N}$ and the continuous state takes value in $\mathbb{R}^n$. The set $\varXi = \boldsymbol{Q} \times \mathbb{R}$ is the hybrid state space of $\mathcal{H}$ and its elements $\xi = (q, x) \in \varXi$ are the hybrid states.
2. The control input variable of $\mathcal{H}$ is made of two components: the discrete control input σ_c and the continuous control input u. The discrete control input belongs to a finite set $\boldsymbol{\Sigma}^c$ and the continuous control input to the set $\mathbb{R}^m$, $m \in \mathbb{N}$. We assume that the input functions $\boldsymbol{u} : \mathbb{R} \to \mathbb{R}^m$ are piecewise continuous.
3. The disturbance variable of $\mathcal{H}$ is made of two components: the discrete disturbance σ_d and the continuous disturbance input d. The discrete disturbance takes value in a finite set $\boldsymbol{\Sigma}^d$ and the continuous disturbance in the set $\mathbb{R}^r$, $r \in \mathbb{N}$. We assume that the disturbance functions $\boldsymbol{d} : \mathbb{R} \to \mathbb{R}^r$ are piecewise continuous.
4. The output variable of $\mathcal{H}$ is made of two components: the discrete output p and the continuous output y. The discrete output is assumed to belong to a finite set $\boldsymbol{P}$ and the continuous output to the set $\mathbb{R}^s$, $s \in \mathbb{N}$. The continuous output functions $\boldsymbol{y} : \mathbb{R} \to \mathbb{R}^s$ are assumed to be piecewise continuous.

The evolution of the discrete state q of hybrid system $\mathcal{H}$ depends on the initial discrete state as well as on the discrete input σ_c, the discrete disturbance σ_d, and the continuous state x, and is driven by events forcing discrete states to jump. There are three types of discrete transitions:

1. Switching transition, forced by a discrete disturbance $\sigma_d \in \boldsymbol{\Sigma}^d$
2. Invariance transition, determined by the continuous state x reaching some regions of the continuous state space; events inducing invariance transitions are assumed to belong to the finite set $\boldsymbol{\Sigma}^i$ and are internally generated by the hybrid system
3. Controllable transition, determined by a discrete control input $\sigma_c \in \boldsymbol{\Sigma}^c$

We denote by $\boldsymbol{\Sigma}$ the set of all events causing discrete transitions of discrete states. A relation represents the collection of all discrete transitions $e = (q, \sigma, q\prime) \in E \subset \boldsymbol{Q} \times \boldsymbol{\Sigma} \times \boldsymbol{Q}$ taking the discrete state from q to $q\prime$ if the event $\sigma \in \boldsymbol{\Sigma}$ occurs. The evolution of the continuous state x depends on the initial continuous state and on the evolution in time of the continuous input u, the continuous disturbance d, and the discrete state q. The continuous state and output evolution between two consecutive discrete transitions is modeled by a dynamical system $S(q)$ that is assumed to be linear for simplicity and governed by the following equations:

$$\dot{x}(t) = Ax(t) + Bu(t) + Dd(t)$$
$$x(t) \in \mathbb{R}^n, u(t) \in \mathbb{R}^m, d(t) \in \mathbb{R}^r, t \geq 0. \tag{15.15}$$

During its evolution in time, the hybrid state $\xi = (q, x)$ has to satisfy the so-called invariance condition $x \in \text{Inv}(q)$, where $\text{Inv}(.)$ is called the invariance map. Whenever a discrete transition $e = (q, \sigma, q\prime) \in E$ occurs, the hybrid state $\xi = (q, x^-)$ has to satisfy the so-called guard condition $x^- \in G(e)$, where $G(.)$ is called the guard map and the continuous state instantly jumps from $x^- \in \mathbb{R}^n$ to a new value $x^+ \in R(e, x^+)$, where $R(., .)$ is called the reset map.

15.4.2 Hybrid Modeling of Self-Organizing Networks

We describe now how hybrid system formalism can be used for modeling the behavior of self-organizing networks. Different models are possible, depending on the role of the discrete state. In the model proposed in [6], each discrete state of the automaton corresponds to the presence in the network of N active nodes and one controller. In each discrete state, the system receives different inputs ranging from RF stimuli from the environment, that are processed by the controller, to indicators of the attenuation that is present over the N active links. These attenuation indicators are used by the active nodes for evaluating potential transmission parameters as well as their capability to comply with the above.

We generalize here the hybrid model proposed in [7] for UWB networks. Each discrete state of the automaton corresponds to an operation mode, described by the waveform used for pulse shaping. In addition, one particular state of the automaton corresponds to the admission control mode, where the controller evaluates the possibility of admitting a new node in the network. A transition to this state takes place when a new node is asking for admission in the network and, for simplicity, we assume that the control procedure requires a negligible time to be performed. The automaton is represented in Fig. 15.1 for the very simple case of two waveforms, namely w_1 and w_2.

A continuous variable $N(t)$, the current number of active nodes that are allowed to transmit data over the wireless channel, is associated to each discrete state of the automaton. This variable is reset to a new value whenever a transition occurs, as described in the next subsection.

In each state, the system receives different inputs ranging from radio frequencies (RF) stimuli from the environment, to indicators of the attenuation that is present over the active links. These attenuation indicators are used by the active nodes for evaluating at time t both potential transmission parameters as well as their capability to comply with the transmission constraints that are communicated by the controller through a time-dependent set of parameters named $K(t)$. This time-varying set of parameters $K(t)$, is formed as follows:

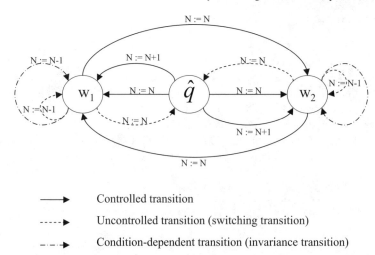

   ⟶   Controlled transition

----▶   Uncontrolled transition (switching transition)

–·–▶   Condition-dependent transition (invariance transition)

Fig. 15.1. Hybrid model.

1. The waveform w^* that must be used for pulse shaping. Different pulse shapes can be selected for transmitting data over the wireless channel; w^* is the one that better adapts with the environment, as well as with thermal noise and multi-user interference (MUI) patterns.
2. The power level $P_{\min}(w^*)$ that is required at the controller in order to comply with the requirement of a given signal to noise ratio threshold.
3. The noise level $\eta_p(w^*)$ that is currently measured at the coordinator.
4. The MUI weight $\sigma_m^2(w^*)$.
5. The number of active nodes N.

Within the above set, the first two parameters w^* and $P_{\min}(w^*)$ can be considered as constraints that are imposed to the nodes in the network. The noise level $\eta_p(w^*)$ can be interpreted as a continuous disturbance. The MUI weight $\sigma_m^2(w^*)$ and the number of active nodes N are information characterizing the current system state. The time-varying set of parameters $K(t)$ is evaluated at the coordinator. We suppose that the signal containing the above information is sent by the controller at a fixed power level that is pre-determined and known by all nodes.

Each active node j receives the signal conveying $K(t)$ and, on the basis of received power level, can estimate the attenuation A_j characterizing its path to the coordinator. Node j determines both power and rate to be used in its future transmissions (see Sect. 15.3). We assume for now that the possible variations in the environment that reflect in $K(t)$ are tolerable by all nodes.

In the next subsection, we formally describe the network dynamics using the hybrid systems formalism.

15.4.3 The Hybrid Model

A hybrid model $\mathcal{H}$ for the network can be defined by introducing the tuple $\mathcal{H} = (\Xi, \Sigma, S, E, R)$:

- $\Xi = Q \times \mathbb{R}$ is the hybrid state space; $Q = W \cup \{\widehat{q}\}$, W is the finite set of waveforms that can be used for pulse shaping; the state $\widehat{q}$ represents the fact that a candidate node is waiting for admission in the network.
- Σ is the set of discrete inputs; $\Sigma = \Sigma^c \cup \Sigma^d \cup \{\widehat{\sigma}\}$, where Σ^c is the set of discrete controls and Σ^d is the set of discrete disturbances and

$$\Sigma^c = \Sigma^c_W \cup \Sigma^c_a$$
$$\Sigma^c_W = \{\sigma_{ij}, i, j \in J\}$$
$$\Sigma^c_a = \{\text{OK}, \text{NO}\}$$
$$\Sigma^d = \{\sigma_a, \sigma_l\}.$$

The discrete controls σ_{ij} model the decision taken by the coordinator to commute from pulse shape w_i to pulse shape w_j. The discrete controls $\{\text{OK}, \text{NO}\}$ correspond, respectively, to the decision taken by the coordinator to accept or not to accept a candidate node in the network. The discrete disturbances σ_a and σ_l represent, respectively, the request by some candidate node to enter the network and the event that a node leaves the network. Finally, the discrete input $\widehat{\sigma}$ is an endogenous signal that is generated when changes in the environment and in radio propagation are no more compliant with node requirements.

- S is a map associating to every discrete state in Q, a dynamical system.
  - If the discrete state is $w_i \in W$, then the dynamical system $S(w_i)$ is described by the equations:

$$\dot{N}(t) = 0$$

$$\boldsymbol{K}_i(t) = \begin{pmatrix} w_i \\ p_i(t) \\ \eta_p(w_i) \\ \sigma^2_m(w_i) \\ N(t) \end{pmatrix}$$

where

$$p_i(t) = P_m(w_i, t)$$
$$P_m(w_i, t) = \frac{\eta_p(w_i)}{T_S} \left(\frac{1}{\text{SNR}_0} - \frac{\sigma^2_m(w_i)}{T_S}(N(t) - 1) \right).$$

$N(t)$ is the continuous state at time t, i.e., the number of current nodes in the network; $\boldsymbol{K}_i(t)$ is the output, where $p_i(t)$ is the power that the coordinator must receive from the $N(t)$ active nodes in order to comply with the requirements of a threshold SNR_0, when the pulse shape is w_i, and T_S is the pulse repetition period. $\eta_p(w_i)$ and $\sigma_m^2(w_i)$, represent the discrete state dependent disturbances. The initial value for the continuous state is $N(0) = 1$ (at the beginning, the coordinator is the only node of the network). The value of the state is reset whenever a transition occurs.

- If the discrete state is $\widehat{q}$, then $S(\widehat{q})$ is described by the equations

$$\dot{N}(t) = 0$$

$$\widehat{K}(t) = \begin{pmatrix} w^* = \arg\min_{w \in W} \widehat{P}_{\min}(w) \\ \widehat{p}(t) \\ \eta_p(w^*) \\ \sigma_m^2(w^*) \\ N(t) + 1 \end{pmatrix}$$

where

$$\widehat{p}(t) = \widehat{P}_m(w^*, t)$$

$$\widehat{P}_m(w^*, t) = \frac{\eta_p(w^*)}{T_S}\left(\frac{1}{\text{SNR}_0} - \frac{\sigma_m^2(w^*)}{T_S}N(t)\right).$$

$N(t)$ is the continuous state at time t, $\widehat{K}(t)$ is the output produced by the coordinator at time t. The state $\widehat{q}$ corresponds to the control admission mode and the role of the output $\widehat{K}(t)$ is therefore discussed in Sect. 15.4.4 describing the admission control algorithm.

- $E \subset Q \times \Sigma \times Q$ is a collection of transitions.

$$E = E^c \cup E^d \cup E^{\text{inv}}$$

where

$$E^c = E_W^c \cup E_a^c, \, E^d = E_a^d \cup E_l^d$$
$$E_W^c = \{(w_i, \sigma_{ij}, w_j), \sigma_{ij} \in \Sigma_W^c, w \in W\}$$
$$E_a^c = \{(\widehat{q}, \sigma, w), \sigma \in \Sigma_a^c, w \in W\}$$
$$E_a^d = \{(w, \sigma_a, \widehat{q}), w \in W\}$$
$$E_l^d = \{(w, \sigma_l, w), w \in W\}$$
$$E^{\text{inv}} = \{(w, \widehat{\sigma}, w), w \in W\}.$$

- The transitions in E^c are controlled (in Fig. 15.1 these transitions are represented by solid arrows).
- A transition (w_i, σ_{ij}, w_j) in E_W^c models the decision, taken at some time t by the coordinator, of commuting from pulse shape w_i to pulse shape w_j, for transmitting data over the wireless channel. This transition takes place when w_j is the pulse shape that, at time t, better adapts to the time-varying environment, as well as thermal noise and MUI patterns, or when $p_i(t) > P_m(w_j, t)$. In the latter case, w_j is such that $P_m(w_j, t) \leq P_m(w_h, t), \forall w_h \in W$.

- The transitions in E_a^c occur when the coordinator decides to accept or not to accept a candidate node in the network.
- The transitions $\{(w, \sigma_a, \widehat{q}), w \in W\}$ are not controlled (switching transitions) and represent the request of entering the network by some candidate node.
- The transitions $\{(w, \sigma_1, w), w \in W\}$ (dashed arrows in Fig. 15.1) are not controlled and represent the fact that a node could leave the network because its activity is terminated for reasons that range from no more data packets to transmit, to node failure, to power exhaustion.
- The transitions in E^{inv} (dot-line arrows in Fig. 15.1) occur because changes in the environment (as sensed by the coordinator) and in radio propagation (as perceived by the active nodes) are no more compliant with node's requirements. Then, the node leaves the network.

For simplicity, we assume that simultaneous transitions are not allowed.

- $R : \Xi \times E \to \Xi$ and

$R((q_i, x), e) = (q_h, x),$	$e = (q_i, \sigma, q_h) \in E_W^c$
$R((q_i, x), e) = (q_h, x + 1),$	$e = (q_i, \text{OK}, q_h) \in E_a^c$
$R((q_i, x), e) = (q_h, x),$	$e = (q_i, \text{NO}, q_h) \in E_a^c$
$R((q_i, x), e) = (\widehat{q}, x),$	$e = (q_i, \sigma, \widehat{q}) \in E_a^d$
$R((q_i, x), e) = (q_i, x - 1),$	$e = (q_i, \sigma, q_i) \in E_1^d$
$R((q_i, x), e) = (q_i, x - 1),$	$e = (q_i, \sigma, q_i) \in E^{inv}$

15.4.4 The Control Algorithm

The main control objective is to maximize the number of active nodes in the network, while preserving transmission requirements. In fact, when a node asks for admission, i.e., when the current discrete state at time t is $\widehat{q}$, the coordinator evaluates the possibility of admitting the new element in the network, by computing a hypothetical set of parameters $\widehat{K}(t)$. The use of this information is twofold. First, it serves to the current active nodes in order to check whether constraints for transition are compatible with their specifications and informs the coordinator. Willingness to transition of all nodes is a necessary condition for transition. Second, the information in $\widehat{K}(t)$ is used by the candidate node for evaluating its willingness to join the network. A candidate node that listens to $\widehat{K}(t)$ must agree in accepting those constraints for the transition to take place. The two conditions above correspond to guard conditions that must be satisfied in order for the transition $(\widehat{q}, \text{OK}, w^*) \in E_a^c$ to take place (where w^* is the first component of $\widehat{K}(t)$). If the above conditions are not fulfilled, then a transition $(\widehat{q}, \text{NO}, w_h)$ in the set E_a^c takes place, where $w_h = \arg\min_{w \in W} P_m(w)$. Therefore, a new node is admitted only if none of the current active nodes is forced to leave the network as a result of its admission.

At each time, the network is controlled so that the power level is minimum with respect to the number of active nodes, possible choices of pulse shaping, and environmental parameters. As a consequence, the described control strategy minimizes the energy consumption in the network, which is a beneficial effect in wireless communication.

Conclusion

Based on hybrid system formalism we described self-organizing network dynamics as a discrete finite-state automaton where, for each state, state-specific rules of operation govern the evolution of the network itself. By doing so, we modeled network of radio devices that must coexist with severely interfered environments, and therefore must control their behavior and adapt to ever-changing operating conditions in order to favor coexistence. In the proposed model, this is achieved by introducing cognitive mechanisms in the analysis process that is used by nodes for determining whether changes in the global network state are appropriate.

Several benefits are obtained by introducing the hybrid system model for the design of cognitive networks:

- The formal description resulting from the adoption of the hybrid system formalism allows a better understanding of some important properties of the system. As an example, it is possible to characterize the trade-off that exists between the complexity of a real-time and precise scanning of the external environment vs. the improvement in system efficiency that is achieved when the nodes can rapidly adapt themselves to the varying conditions of the operating scenario. Based on this trade-off, we could investigate the existence of suboptimal but computational-efficient strategies, where the capability of the nodes to adapt to the external environment is limited and depends upon the current state of the automaton.
- Using the hybrid system model, it is possible to optimize the distribution of functional specifications among the different components of the system. For example, we can analyze how system performance is affected whenever some of the functionalities that are related to active nodes are related to the coordinator and vice versa.
- The hybrid formalism may help to predict in which states the automaton will spend most of the time or the maximum number of nodes of the network. This information is of fundamental importance for network designers.
- The characterization of the wireless network as a hybrid system facilitates the analysis of the stability [8] of the overall system. This task is by no means trivial since we assume that the nodes dynamically adapt transmission parameters and rules of operation to external stimuli.

References

1. S. Haykin, "Cognitive radio: Brain-empowered wireless communications," *IEEE J. Select. Areas Commun.*, vol. 23, no. 2, pp. 201–220, Feb. 2005.

2. J. Mitola and G. Q. Maguire, "Cognitive radio: Making software radios more personal," *IEEE Pers. Commun.*, vol. 6, no. 4, pp. 13–18, Aug. 1999.
3. M.-G. Di Benedetto and G. Giancola, *Understanding ultra wide band radio fundamentals.* Prentice Hall, 2004.
4. *Proc. IEEE*, vol. 88, Special Issue on "Hybrid Systems," July 2000.
5. M. D. Di Benedetto, A. D'Innocenzo, G. Pola, C. Rinaldi, and F. Santucci, "A theoretical framework for control over wireless networks," Invited in Mini-Symposium on "Distributed decision-making over ad-hoc networks," in *Proc. of 17th International Symposium on Mathematical Theory of Networks and Systems* (Kyoto, Japan), July 24–28, 2006.
6. M.-G. Di Benedetto, G. Giancola, and M. D. Di Benedetto, "Introducing consciousness in UWB networks by hybrid modelling of admission control," *ACM/Springer J. Mobile Netw. Appl.*, Special Issue on "Ultra Wide Band for Sensor Networks," M.-G. Di Benedetto and F. Granelli, editors, vol. 11, no. 4, pp. 521–534, 2006.
7. E. De Santis, M. D. Di Benedetto, M.-G. Di Benedetto, and G. Giancola, "Application of hybrid models to the design of UWB self-organizing networks," in *Proc. of 17th International Symposium on Mathematical Theory of Networks and Systems* (Kyoto, Japan), pp. 1484–1487, July 24–28, 2006.
8. M.-G. Di Benedetto and L. De Nardis, "Tuning UWB signals by pulse shaping," *EURASIP J. Signal Process.*, Special Issue on "Signal Processing in UWB Communications," vol. 86, pp. 2172–2184, invited paper, Elsevier Publishers, 2006.
9. M.-G. Di Benedetto, L. De Nardis, M. Junk, and G. Giancola, "(UWB)2: Uncoordinated, wireless, baseborn, medium access control for UWB communication networks," *Mobile Netw. Appl. J.*, Special Issue on WLAN Optimization at the MAC and Network Levels, vol. 10, pp. 663–674, 2005.
10. G. Giancola, C. Martello, F. Cuomo, and M.-G. Di Benedetto, "Radio resource management in infrastructure-based and ad hoc UWB networks," *Wireless Commun. Mobile Comput.*, vol. 5, pp. 581–597, 2005.
11. P. P. White, "RSVP and integrated services in the Internet: A tutorial," *IEEE Commun. Mag.*, vol. 35, pp. 100–106, 1997.

Index

adaptive modulation, 243
auction, 250–253

bandwidth
  bandwidth-on-demand, 232

Cartel Maintenance, 248
coalition, 257
cognitive radio, 366
cognitive radios, 231
combination, 234, 258, 263
complexity, 231
concave, 261
constrained optimization, 260
convex function, 260, 261
convex hull, 239
convex optimization, 260, 261
correlated equilibrium, 236–241
Cournot game, 235
CSMA, 237
cutting planes, 263

data-transmission time, 146, 147, 149
delay
  delay requirement, 262
  propagation delay, 238
demodulator statistics, 151, 153, 154
deterministic, 234
dynamic channel, 140, 141, 157
dynamic programming, 263
dynamic spectrum access, 139, 140, 145,
    150

energy conservation, 141, 153

error-control coding, 141, 144, 151, 154
excess, 259

FCC, 231
frequency reuse, 140

game, 233
global optimum, 243, 260

half-duplex, 139, 140, 151
Hungarian Method, 255

interference level, 154
interference region, 148
interior-point method, 261

Jensen's inequality, 261

kernel, 259
KKT condition, 243
Knapsack, 263

Lagrangian
  Lagrangian method, 262
  Lagrangian multiplier, 262
least-resistance routing, 150
linear programming, 238, 240, 241, 260, 262

Markov decision process, 241
maximin, 239
min-max, 259
mixed strategy, 235
mobility, 232
MQAM, 243
mutual benefit, 233

Nash Bargaining Solution, 254

Nash equilibrium, 233–236, 238–240, 243
no-regret algorithm, 240
non-cooperative game, 234
normal form, 234
Nucleolus, 259

objective function, 262, 263
open spectrum, 365
open stopping, 374
opportunistic spectrum access, 236
optimality condition, 263
overhead, 263

packet radio, 139, 140
Pareto optimality, 233, 235
penalty, 241
polynomial time, 260
potential game, 235
pricing, 235
primary user, 236–238
protocol
    adaptive modulation and coding, 141,
        160, 161
    adaptive transmission, 140, 141, 151
    adaptive-rate coding, 141, 144, 147, 154,
        155, 158–160
    modulation-selection, 140

power-adjustment, 140, 147, 151, 153,
    154

quality-of-service (QoS), 140, 145, 150

receiver statistics, 151, 155, 156
repeated game, 247
resource consumption, 147–150
    normalized, 149

secondary user, 236–239
Shapley function, 258
share auction, 252
spectrum efficiency, 243, 245
spectrum etiquette, 140, 145, 147
stability, 258
static game, 234
strategic form, 234
strategy, 233

time-bandwidth product, 147
    normalized, 147, 149
time-varying propagation loss, 157–159, 161

utility function, 232–235, 237–240, 260

Vickery-Clarke-Groves (VCG) auction, 252
virtual referee, 242